Emily Lacroix

Managing Forest Ca
Climate

Mark S. Ashton • Mary L. Tyrrell
Deborah Spalding • Bradford Gentry
Editors

Managing Forest Carbon in a Changing Climate

Editors
Mark S. Ashton
School of Forestry
and Environmental Studies
Yale University
Prospect Street 360
New Haven, CT 06511
USA
mark.ashton@yale.edu

Deborah Spalding
Working Lands Investment Partners,
LLC, New Haven,
CT 06510,
USA
deborah.spalding@yale.edu

Mary L. Tyrrell
School of Forestry and Environmental
Studies
Yale University
Prospect Street 360
New Haven, CT 06511
USA
mary.tyrrell@yale.edu

Bradford Gentry
School of Forestry
and Environmental Studies
Yale University
Prospect Street 360
New Haven, CT 06511
USA
bradford.gentry@yale.edu

The Sinharaja Mana and the biosphere Reserve, Sri Lanka. A hill mixed dipterocarp forest with high endemicity, the source drinking water supply for millions of people and comprising some of the highest standing amounts of carbon in trees on the Indian subcontinent (Photo taken by Mark Ashton).

ISBN 978-94-007-2231-6 e-ISBN 978-94-007-2232-3
DOI 10.1007/978-94-007-2232-3
Springer Dordrecht Heidelberg London New York

Library of Congress Control Number: 2011941763

Printed on acid-free paper

Springer is part of Springer Science+Business Media (www.springer.com)

Preface

The goal of this book is to provide guidance for students, land managers and policymakers seeking to understand the complex science and policy of forest carbon as it relates to tangible problems of forest management and the more abstract problems of addressing drivers of deforestation and negotiating policy frameworks for reducing CO_2 emissions from forests. It is the culmination of three graduate seminars at the Yale School of Forestry & Environmental Studies focused on carbon sequestration in forest ecosystems and their role in addressing climate change. The seminars, part of the professional masters' degree curriculum, took place in 2008 and 2009. They were co-sponsored by the Global Institute of Sustainable Forestry, the Center for Industrial Ecology, and the Center for Business and Environment at Yale. The seminars were led by Professor Mark Ashton along with Professors Bradford Gentry, Thomas Graedel, Xuhui Lee, Reid Lifset, Deborah Spalding and Mary Tyrrell.

The purpose of the three seminars was to review and document what we know, what we do not know, and the implications for policy makers of: (i) the science of carbon sequestration in forests; (ii) the role of harvested wood products in the global carbon cycle; and (iii) the science, business, and policy aspects of managing forests to store carbon. An overarching goal was to develop an understanding of the complexity of forest carbon science and why forest carbon budgeting has been a particular challenge for policy makers.

The basis of each seminar was a thorough review of the current literature on the topics, followed by in-depth class discussion. Leaders in the field were invited to give seminal talks, followed by lengthy discussion and debate with the class, to help set the stage for the students' review and analysis.

The resulting review papers were written by the graduate students in the seminars with faculty from Yale and elsewhere. Students were carefully paired with appropriate Yale faculty and experts who are knowledgeable about their respective topics. The collection provides a unique synthesis of current knowledge about science and management, and current thinking about policy pertaining to the sequestration of carbon in forests globally. Overall, this book supplies what we feel is a much-needed scientific undergirding for discussions about carbon sequestration in forests. It contains recommendations for management and policy measures that reflect the scientific realities of how forests of many different types -- tropical, temperate, and boreal -- actually sequester carbon or do not, and under what circumstances.

We welcome comments and feedback – this is a work in progress amidst an evolving scientific understanding of a complex topic and an equally complex international dialogue on the role of forests in climate change mitigation.

Acknowledgements

The Yale School of Forestry & Environmental Studies (YF&ES) funded the three seminars and the initial costs for the publication of this book. We are very grateful to former Dean Gus Speth, who enthusiastically supported the idea.

The Global Institute of Sustainable Forestry (YF&ES) has been a strong partner throughout, co-sponsoring the first seminar, *Forest Carbon Science*, and taking the lead in producing this volume. The Center for Industrial Ecology (YF&ES) co-sponsored the second seminar, *The Role of Forest Products in the Global Carbon Cycle*, and the Center for Business and the Environment at Yale co-sponsored the third seminar, *Managing Forests for Carbon Sequestration: Science, Business, and Policy.*

We wish to acknowledge our colleagues on the F&ES faculty who co-led the seminars, hosted guest speakers, and reviewed many drafts of the papers. We appreciate their dedication to the success of the seminars and enjoyed the collaboration immensely.

Xuhui Lee, Professor of Meteorology

Thomas Graedel, Clifton R. Musser Professor of Industrial Ecology, Professor of Chemical Engineering, Professor of Geology and Geophysics, and Director of the Center for Industrial Ecology

Reid Lifset, Associate Research Scholar, Resident Fellow in Industrial Ecology, Associate Director of the Industrial Environmental Management Program, and Editor-in-Chief of the Journal of Industrial Ecology

F&ES faculty Graeme Berlyn, Ann Camp, Benjamin Cashore, Michael Dove, Timothy Gregoire, Florencia Montagnini, Chadwick Oliver, and Peter Raymond, also reviewed several of the science and policy papers, which benefited greatly from their suggestions.

We are grateful for the assistance and suggestions of F&ES students Lauren Goers, who helped with editing and compiled the glossary; and Laura Bozzi who was the teaching assistant for the spring '09 seminar.

We benefited tremendously from our guest speakers, not only from their presentations, but especially from the discussions that followed. They enthusiastically engaged with the students, probing the depth of the issues. We thank those who took time out of their busy schedules to travel to New Haven and all who shared their knowledge with the students.

Ralph Alig, USDA Forest Service
Jennifer Brady, US Environmental Protection Agency
Sandra Brown, Winrock International
Mark Ducey, University of New Hampshire
Stith (Tom) Gower, University of Wisconsin, Madison
Seiji Hashimoto, National Institute for Environmental Studies (Japan)
Linda Heath, USDA Forest Service, Northern Research Station
Paul Hanson, Oak Ridge National Laboratory
Richard Houghton, Woods Hole Research Center
Pekka Kauppi, University of Helsinki
Gregg Marland, Oak Ridge National Labs
Frank Merry, Woods Hole Research Center
Reid Miner, National Council for Air and Stream Improvement, Inc.
Brian Murray, Duke University
Catherine Potvin, McGill University
Daniel D. Richter, Duke University
Greg Norris, Sylvatica
Laurie Wayburn, Pacific Forest Trust
Steven Wofsy, Harvard University

Contents

Introduction 1

Lauren Goers, Mark S. Ashton, and Mary L. Tyrrell

This book provides guidance to students, land managers and policymakers seeking to understand the complex science and policy of forest carbon as it relates to both the technical issues of forest management and reforestation, and the more social and economic problems of addressing drivers of deforestation and forest degradation. It is an attempt at a comprehensive state-of-the-art review, encompassing the science of carbon sequestration in forests, management of forests for carbon and other values, and the socio-economic and policy implications and challenges of managing forests for carbon.

Forests are critical to mitigating the effects of global climate change because they are large storehouses of carbon and have the ability to continually absorb carbon dioxide from the atmosphere. But today, emissions from land use, and land use change mostly due to deforestation in the tropics, are estimated at 17% of total annual global CO_2 emissions, a figure larger than the transportation sector (IPCC 2007).

While the basic principles of the carbon cycle are well known (see Box 1), there are significant uncertainties about the actual behavior of many of its sinks and sources. This is a particular challenge in forested ecosystems due to the role played by biogeochemistry, climate, disturbance, and land use, as well as the spatial and temporal heterogeneity of carbon sequestration across regions and forest types. The subject of forest carbon is complex, encompassing the science of carbon in forests, the measurement of stored carbon and carbon flux, the economic drivers of deforestation, and the social and political contexts in which forests exist, making it a challenge to create comprehensive policies aimed at reducing CO_2 emissions from forests. Much work has been done on the science of forest carbon, deforestation, and various climate policy responses, including books, reports, symposia, and special journal issues (see for example, Streck et al. 2008; Griffiths and Jarvis 2005; Angelsen et al. 2008; IPCC 2000; Parker et al. 2008, among many others); however, what is lacking is a comprehensive review of all aspects of the challenge.

This book provides such a review by taking a holistic perspective on the subject. By creating a publication that outlines the research that has been done on forest carbon, pointing out what we know and what we don't know, and the implications for policy decisions, the hope is that land managers and policymakers alike will have a stronger foundation for making choices. The nature of the writing is meant to be accessible to a general audience and technical language has been simplified to the extent possible; nonetheless, this is a complicated topic with many "insider" terms. A glossary of scientific and technical terms is included at the end of the volume for quick reference.

L. Goers
World Resources Institute, Washington, DC, USA

M.S. Ashton (✉) • M.L. Tyrrell
Yale School of Forestry and Environmental Studies, Prospect Street 360, 06511 New Haven, CT, USA
e-mail: mark.ashton@yale.edu; mary.tyrrell@yale.edu

M.S. Ashton et al. (eds.), *Managing Forest Carbon in a Changing Climate*,
DOI 10.1007/978-94-007-2232-3_1,

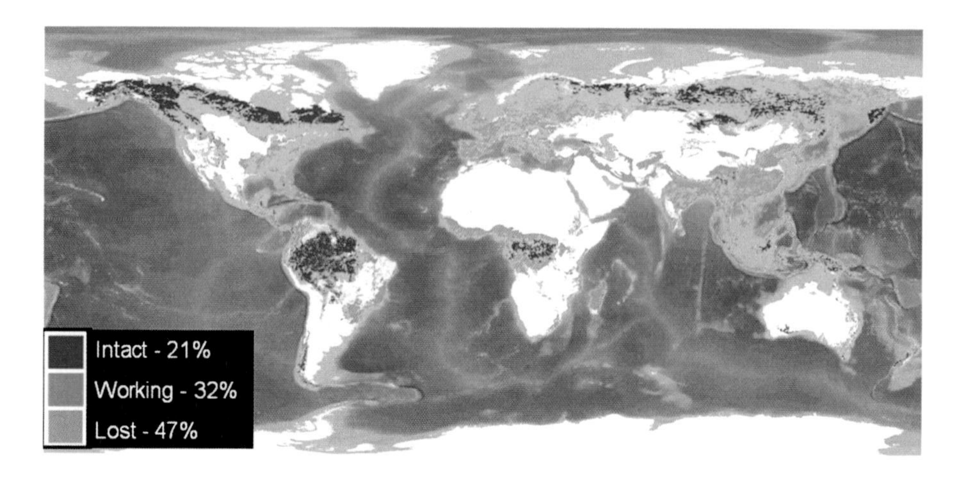

Fig. 1.1 Historical forest loss and current extent of intact and working forests (*Source*: World Resources Institute 2009. Forest Landscapes Initiative. Reprinted with permission)

1 Background

1.1 The Problem: Forest Degradation and Loss

Forests are enormously important to maintaining global carbon sinks because they contain 77% of all terrestrial above ground carbon (IPCC 2000; Houghton 2007). At the United Nations Framework Convention on Climate Change's Conference of the Parties in December 2005, the governments of Papua New Guinea and Costa Rica introduced an agenda item on "reducing emissions from deforestation in developing countries and approaches to stimulate action" (UNFCCC 2005). Since that introduction, the idea of addressing global increases in greenhouse gas emissions by reducing or avoiding tropical deforestation has been a topic that has sparked much debate in the international climate discussion. The need for a comprehensive strategy to reduce emissions from deforestation and degradation was subsequently reflected in the Bali Action Plan in December 2007 (UNFCCC 2007). Since that time, discussion of how to implement a mechanism to "Reduce Emissions from Deforestation and Degradation" (REDD) has centered around questions regarding both the science of forest carbon and the design of sound policy informed by that science to achieve verifiable and lasting reductions in greenhouse gas emissions from forests.

Currently forests occupy just under four billion hectares of the Earth's land area, or roughly 30% of its land base. However, worldwide forest cover today is only a fraction of its historical extent, with some research estimating that 47% of original forest cover has been lost (Fig. 1.1) (WRI 2009). The extent of current net annual tropical deforestation is estimated at 7.3 million hectares each year (FAO 2005). It is therefore imperative that forests be included in a global agreement to undertake actions for climate change mitigation and adaptation. Including forests as part of the global climate change mitigation strategy not only has climate benefits, but can help generate significant co-benefits, since keeping forests intact could also maintain biodiversity, preserve ecosystem services that many humans rely on, and help improve livelihoods of forest dwellers.

2 Contents of the Book

The book is organized in four parts: the science of carbon sequestration in forests; the science of measuring carbon in forests; management of forests and forest products for carbon storage; and the socio-economic, business, and policy aspects of managing forests for carbon.

2.1 Science

Part I focuses on forest carbon science. It examines carbon fluxes at varying spatial scales, from microsites to the global forest carbon budget, with particular attention on the impacts of such factors as climate, seasonality, disturbance patterns, and stand dynamics. It places this analysis within the context of broad forest types (boreal/temperate/tropical). It opens with Chap. 2, which analyzes research on carbon stocks and flows in forest soils, an important consideration for developing a forest carbon policy since two-thirds of the carbon in forests is in the soil (IPCC 2000). Chapter 3 explores the underlying drivers of forest development and the ways these drivers are affected by changes in atmospheric carbon dioxide concentrations, temperature, precipitation, and nutrient levels. Chapters 4, 5, and 6 focus on carbon stocks and flows in boreal, temperate, and tropical forests, respectively, by reviewing both the literature on experimental research on carbon storage and flux in each biome, and models of predicted changes in regional climate, disturbance drivers, and effect on forest regeneration and dynamics in each forest type.

2.2 Measurement

Part II reviews the different ways of measuring and estimating carbon in forests and summarizes the best known estimates of of storage and loss. Chapter 7 reviews methodologies for estimating carbon in above ground pools, a key topic for many nations in international policy discussions because of the need to develop standardized methods of carbon accounting with an emphasis on verifiable results. This part closes with Chap. 8, analyzing the relationship between forests and the global carbon budget and describing current estimates and trends in the different stocks and fluxes of forest carbon by region.

2.3 Management

Part III concerns the science and technology of managing forests for carbon sequestration and storage, including the life cycle of harvested wood products within managed forests and the advantages and disadvantages of accounting for wood carbon stored or lost outside the forest. Chapters 9 and 10 describe the management and stand dynamics of forests for temperate and boreal, and tropical regions, respectively. Both of these chapters focus on assessing the impacts of silvicultural and management practices on carbon stocks and flows in various forest types. Chapter 11 focuses on the science of managing plantations and addresses key factors of implementing afforestation/reforestation projects for carbon sequestration such as site and species selection.

Chapter 12 assesses the role of harvested wood products and the forest products industry within the context of global carbon stocks and flows, including life cycle analysis of forest products from harvest to end-of-life, and the implications for carbon storage.

2.4 Society, -Economics and Policy

Part IV concerns the socio-economic, business, and policy aspects of managing forests for carbon sequestration and storage. The first three chapters analyze threats to intact relatively undisturbed forests (Chap. 13), the economic drivers of deforestation, focusing on deforestation for agriculture in the tropics (Chap. 14), and development pressures on forests in the United States (Chap. 15).

The final two chapters provide an overview of existing mechanisms and proposals for forest carbon policy at the global and U.S. federal levels. They describe the scale, reference levels, and financing for carbon projects in an attempt to broaden the understanding of current proposals and highlight key concerns for designing policy on forest carbon. Chapter 16 reviews both voluntary market mechanisms and forest carbon legislation in the United States and analyzes the scope, reference level, and proposed financing mechanisms for carbon offset projects. Chapter 17 looks at the forest carbon regimes proposed at the international level for inclusion in the climate treaty that is intended to replace the Kyoto Protocol in 2012.

3 Concluding Remarks

At the end of each chapter and in the closing synthesis ideas, the authors have provided a summary of the most important conclusions from this review and their implications for forest carbon management or policy. These key points are designed to provide a guideline for developing strategies for managing forest carbon and developing a mechanism for reducing emissions from deforestation. The aim is to provide an accessible overview for resource professionals, such as land managers, to acquaint themselves with the established science and management practices that facilitate sequestration and allow for the storage of carbon in forests. The book has value for policymakers to better understand: (i) carbon science and management of forests and wood products; (ii) the underlying social mechanisms of deforestation; and (iii) the policy options in order to formulate a cohesive strategy for implementing forest carbon projects and ultimately reducing emissions from the forest and forestry sector.

References

Angelsen A, Brown S, Loisel C, Peskett L, Streck C, Zarin D (2008) REDD: An Options Assessment Report. Meridian Institute, Washington, DC

FAO (2005) Global Forest Resources Assessment 2005: progress towards sustainable forest management. United Nations Food and Agriculture Organization. ftp://ftp.fao.org/docrep/fao/008/A0400E/A0400E00.pdf. Accessed Aug 2009

Griffiths H, Jarvis PG (eds) (2005) The carbon balance of forest biomes. Taylor & Francis, New York, 356 p

Houghton RA (2007) Balancing the global carbon budget. Annu Rev Earth Planet Sci 35:313–347

IPCC (2000) In: Watson RT, Noble IR, Bolin B, Ravindranath NH, Verardo DJ, Dokken DJ (eds) Intergovernmental Panel on Climate Change Special Report: land use, land use change, and forestry. Cambridge University Press, Cambridge, 375 pp

IPCC (2007) Climate change 2007: synthesis report. Intergovernmental Panel on Climate Change

Larcher W (1983) Physiological plant ecology. Springer, New York, 303 pp

Parker C, Mitchell A, Trivedi M, Marda N (2008) The little REDD book: a guide to governmental and non-governmental proposals for reducing emissions from deforestation and degradation. Global Canopy Programme, Oxford

Streck C, O'Sullivan R, Janson-Smith T, Tarasofsky RG (eds) (2008) Climate change and forests: emerging policy and market opportunities. Chatham House/Brookings Institution Press, London/Washington, DC

UNFCCC (2005) United Nations Framework Convention on Climate Change Conference of the Parties 11. Montreal, Canada.

UNFCCC (2007) Decision -/CP.13 Bali Action Plan. United Nations Framework Convention on Climate Change

World Resources Institute (2009) Forest landscapes initiative. http://www.wri.org/project/global-forest-watch. Accessed Aug 2009

Part I

The Science of Forest Carbon

Section Summary

The following five papers build upon each other to provide a comprehensive synthesis and review of the science of carbon in forests. The papers highlight areas of research that are well known and areas that are lacking. The first two papers cover soils and above-ground physiology and growth. They highlight the fact that most studies have been done in the temperate forests of developed nations.

The next three papers review the regional differences in carbon among tropical, temperate and boreal forest biomes. Studies show that tropical forests comprise nearly half of the total terrestrial gross primary productivity and that in recent decades Amazonian and Central African old growth forests continue to increase in biomass, which may be a response to increased atmospheric CO_2. Temperate forests are mostly second growth and studies suggest that they are, on average, strong sinks for carbon, but a small change in temperature, rainfall or growing season length could change them from sink to source. The soil carbon pool plays a disproportionately large role in boreal forests, but increased fire frequency could greatly increase carbon release, with an even greater rate of heterotrophic respiration observed after fire.

Contributors toward organizing and editing this section were: *Mark S. Ashton, Mary L. Tyrrell, Deborah Spalding, and Xuhui Lee*

Characterizing Organic Carbon Stocks and Flows in Forest Soils

2

Samuel P. Price, Mark A. Bradford, and Mark S. Ashton

Executive Summary

Forests are expected to store additional carbon as part of the global initiative to offset the buildup of anthropogenic carbon dioxide (CO_2) in the atmosphere (IPCC 2007). Soil organic carbon (SOC) stored and cycled under forests is a significant portion of the global total carbon stock, but remains poorly understood due to its complexity in mechanisms of storage and inaccessibility at depth. This chapter first reviews our understanding of soil carbon inputs, losses from biotic respiration and the different soil carbon storage pools and mechanisms. Secondly, the paper evaluates methods of measurement and modeling of soil carbon. Thirdly, it summarizes the effects of diverse management histories and disturbance regimes that compound the difficulties in quantifying forest soil carbon pools and fluxes. Alterations of soil carbon cycling by land use change or disturbance may persist for decades or centuries, confounding results of short-term field studies. Such differences must be characterized and sequestration mechanisms elucidated to inform realistic climate change policy directed at carbon management in existing native forests, plantations, and agroforestry systems, as well as reforestation and afforestation. Such knowledge gains will also provide a theoretical basis for sound, stable investment in sequestration capacity. Lastly, the chapter provides recommendations for further research on those areas of soil carbon where knowledge is either scant or absent. Key findings of this review comprise what we do and do not know about soil organic carbon – inorganic carbon is also an important reservoir, especially in arid soils, but is not considered here.

What We Do Know About Soil Carbon
Substantial work has been done that provides knowledge on many processes of soil carbon dynamics, such as:

- our understanding of how dissolved organic matter (DOM) additions from litter infiltrate the mineral soil.
- fine roots are the main source of carbon additions to soils, whether through root turnover or via exudates to associated mycorrhizal fungi and the rhizosphere.
- the dynamic between nitrogen deposition and carbon storage in forest soils is different on low-quality, high-lignin litter than on high-quality, low-lignin litter, which provides an explanation for many contradictory studies on the effects of nitrogen deposition.
- roots and mycorrhizal fungi produce about half of total respired CO_2, with the balance from heterotrophic breakdown of organic matter.

S.P. Price • M.A. Bradford • M.S. Ashton (✉)
Yale University, School of Forestry and Environmental Studies, Prospect Street 360, 06511 New Haven, CT, USA
e-mail: mark.ashton@yale.edu

M.S. Ashton et al. (eds.), *Managing Forest Carbon in a Changing Climate*,
DOI 10.1007/978-94-007-2232-3_2, © Springer Science+Business Media B.V. 2012

- bacterial and fungal, as well as overall faunal community composition, are hypothesized to have significant affects (±) on soil carbon dynamics.
- fossil fuel burning, particulate deposition from forest fires, and wind erosion of agricultural soils is expected to affect microbial breakdown of organic matter and alter forest nutrient cycling.
- organic matter can be stabilized through microbial action, and from these actions by biochemical resistance or by physical protection within soil aggregates or microsites. Stabilization also occurs from poor drainage (water logging), fire and deep charcoal burial, and is dependent on texture and mineralogy of soil.

What We Do Not Know About Soil Carbon

More research is needed to understand how the processes of soil carbon dynamics, that are now becoming understood, vary across different forest regions and soil depths. New research is needed that characterizes:

- controls on the depth of the organic layer by leaching of dissolved organic carbon (DOC) into the mineral soil.
- rates of fine root turnover among species and biomes.
- patterns of bacterial, fungal and plant respiration and responses to physical and biotic factors and stresses (such as drought, increased temperature).
- dynamics of functionally-distinct soil carbon pools, rather than the most-easily measured and fractionated pools.
- the most accurate methods for quantifying forest soil carbon stocks and fluxes.

The global nature of the carbon cycle requires a globally-distributed and coordinated research program, but thus far research has been largely limited to the developed world, the top 30 cm of the soil profile, temperate biomes, and agricultural soils. Forest soils in tropical moist regions are represented by only a handful of studies, as are examinations of sequestration of carbon at depth but, perhaps most importantly, the dominant reservoir of soil carbon is at high latitudes and the response of this store to climate change is highly uncertain.

1 Introduction

Organic carbon enters the terrestrial biosphere primarily through photosynthesis, and is shunted to the soil system by leaf- and debris-fall, the turnover (cycle of death and new growth) of roots, and by the allocation of plant photosynthate to mycorrhizal fungi and saprotrophic microbes in soil immediately surrounding roots. Plant residues are broken down by bacteria and saprophytic fungi, resulting in a cascade of complex organic carbon compounds that leach deeper into the soil. Carbon that leaves the forest soil system exits via CO_2 respired by plants, bacteria and fungi (Fig. 2.1), and through leaching of dissolved organic matter (DOM) to groundwater and rivers (not shown in Fig. 2.1).

The divergent respiration pathways differ in rate, substrate preference (e.g. type of litter, root or woody debris), and response to environmental change, complicating our capacity to characterize them. Carbon that remains in soil does so because it is stabilized by its own intrinsic chemical properties, through production of secondary microbial compounds, by physical separation from microbial breakdown, by molecular interactions with metals or other bio-molecules, or by freezing, inundation from flooding or carbonization.

This is an introductory summary of the portion of the carbon cycle that is closely linked and affected by soil forming processes in terrestrial ecosystems, and most importantly, in forests. What is clear is that soil carbon, as a component of the ecosystem, varies enormously across different forest biomes (Fig. 2.2), and across different soil orders (Fig. 2.3).

In general, soil carbon is strongly associated with rainfall distribution and therefore there is more carbon stock in forests than in other terrestrial ecosystems (Fig. 2.4). The nature and condition of forests, by implication, can therefore play a critical role in soil carbon sequestration and storage processes. In this paper the carbon in soils is described in the form of inputs, losses, and as that portion of carbon that remains stable within soil. The paper proceeds with a review of methods of quantifying soil carbon processes and

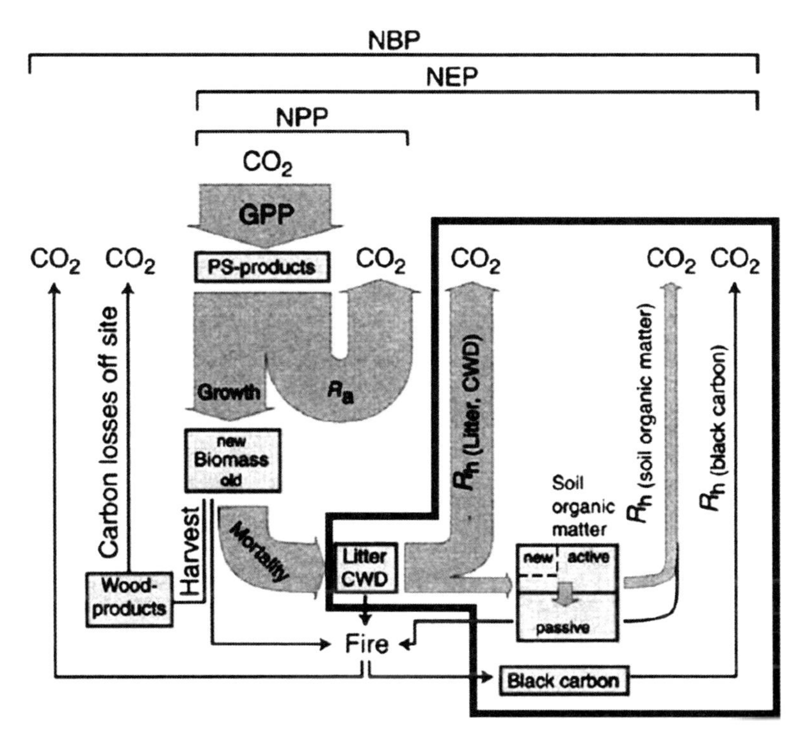

Fig. 2.1 Forest carbon flux. A conceptual diagram illustrating the limits of the belowground carbon cycle. Arrows represent fluxes and boxes indicate pools; the size of each indicates the relative rate of flux or size of pool. Litter and coarse woody debris on the forest floor are included in the belowground portion of the forest carbon cycle. *NBP* net biome productivity, *NEP* net ecosystem productivity, *NPP* net primary productivity, *GPP* gross primary productivity, *PS* photosynthesis, R_h heterotrophic (bacterial) respiration, R_a autotrophic (plant and associated mycorrhizal fungal) respiration, *CWD* coarse woody debris (*Source*: Schulze et al. 2000. Reprinted with permission from AAAS)

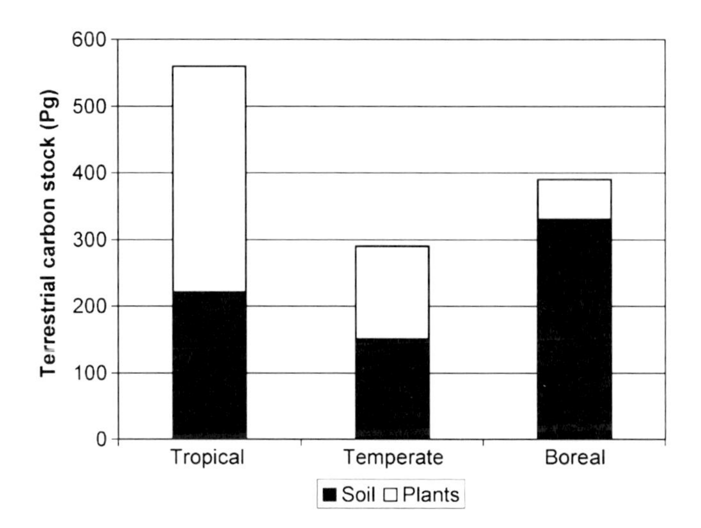

Fig. 2.2 Distribution of world forest carbon stocks by biome. Tropical forests worldwide contain approximately as much carbon in living plants (340 Pg) as boreal forests contain underground (338 Pg), indicating broad differences in carbon dynamics between biomes (*Source*: Data compiled from Vogt et al. 1998; Eswaran et al. 1995; Goodale et al. 2002; Guo and Gifford 2002)

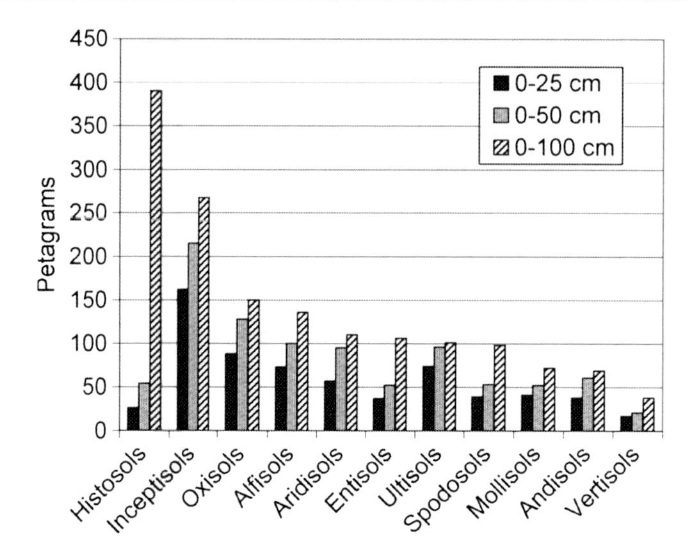

Fig. 2.3 Soil organic carbon (SOC) stocks worldwide, by soil order. Histosols store the majority of the world's SOC due to seasonal or continuous inundation, and do so at depths between 50 and 100 cm (*Source*: Adapted from Eswaran et al. 1995)

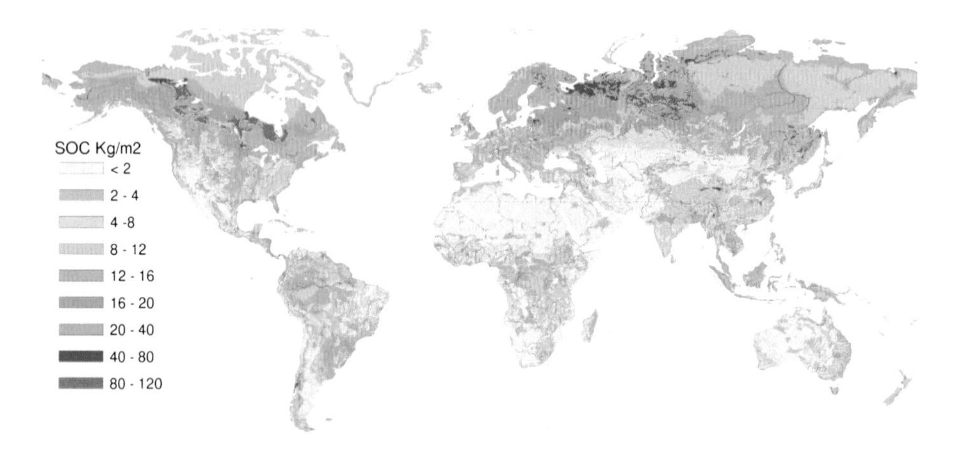

Fig. 2.4 Density of soil carbon stocks worldwide. Note the swaths of highest density across the boreal regions of North America, Europe and Asia. Across the boreal forest SOC stocks are spatially variable (*Source*: US Department of Agriculture, Natural Resources Conservation Service, Soil Survey Division. Washington D.C. http://soils.usda.gov/use/worldsoils/mapindex/soc.html. Reprinted with permission)

pools directly with measurements and through modeling. It concludes with a discussion of effects of management on the carbon in forest soils and finally makes recommendations on what further research and knowledge is needed and where.

2 Carbon Inputs to Forest Soils

Plants absorb CO_2 and produce sugars under photosynthesis. Photosynthetic products are used to drive cellular respiration and root exudation, or

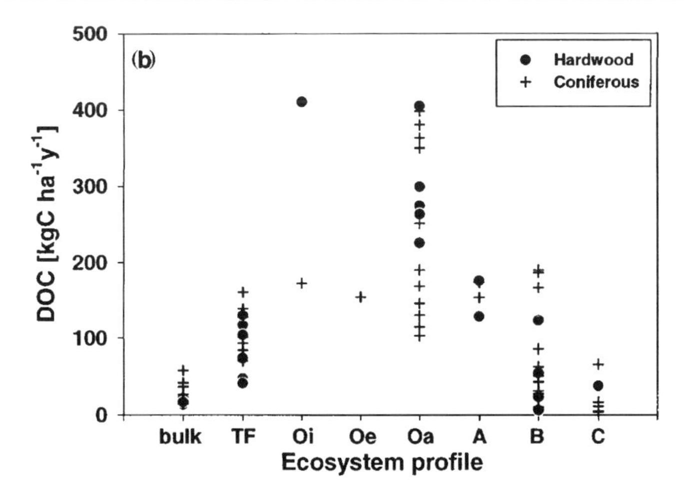

Fig. 2.5 Synthesis of 42 studies of DOC from the temperate forest biome showing annual fluxes of DOC through the organic and mineral soil profile. The greatest annual fluxes and greatest variability are for the lowest humified organic layer (Oa – soil organic layer). The figure depicts a lack of studies of DOC flux from Oi and Oe layers. DOC flux decreases with depth in the mineral soil. Note the significant contribution of DOC from throughfall (*TF*), a result of microbial breakdown of organic matter in the canopy. There was no significant difference between DOC fluxes under coniferous versus deciduous forest. *Bulk* bulk precipitation, *TF* throughfall precipitation, *Oi* litter layer, *Oe* fermented layer, *Oa* humic layer, *A*, *B* and *C* successively deeper mineral soil horizons (*Source*: From Michalzik et al. 2001. Reprinted with permission)

are stored for consumption, reproduction, and/or allocation to root, shoot and wood growth. When leaves, branches or roots outlast their useful life and cease to provide a net contribution to plant growth, they senesce (i.e. cease to live). Plants thus control the input of carbon to the soil system via above- and below-ground carbon inputs into forest soils from plant litter, coarse woody debris, fine root turnover, and root exudates.

2.1 Aboveground Carbon Inputs: Litter and Coarse Woody Debris

Carbon from aboveground sources enters the soil system when it falls to the forest floor in the form of dead leaves, bark, wood and/or herbivore inputs such as greenfall, carcasses and frass. Carbon is lost from surface organic matter as CO_2 by microbial respiration, by mixing and incorporation of surface organic matter into mineral soil horizons by soil fauna, and by leaching of dissolved organic matter (DOM) of which dissolved organic carbon (DOC) is an important constituent.

In a synthesis of 42 studies from temperate forests, Michalzik et al. (2001) reported that precipitation was strongly positively correlated with the flux rate of DOC from the forest floor into the mineral soil. The concentration of DOC in leachate from the forest floor to the mineral soil was positively correlated with pH, suggesting that more basic conditions favor microbial decomposition and thus DOC production. They also found that the greatest annual fluxes and greatest variability were in the lowest humified organic layer (Oa). There were very few studies of DOC flux from the upper organic layers. DOC flux decreases with depth in the mineral soil. There was a significant contribution of DOC from throughfall (TF), a result of microbial breakdown of organic matter in the canopy and also potentially from sap-sucking herbivores. There was no significant difference between DOC fluxes under coniferous versus deciduous forest (Fig. 2.5). More recent ^{14}C labeling studies from both Sweden and Tennessee, USA, corroborated these results (Froberg et al. 2007a, b), indicating that most litter-derived DOC is either respired before it reaches the mineral soil or immobilized in the Oe and Oa surface layers of the soil (Fig. 2.5).

In a litter manipulation study at a hardwood forest in Bavaria, Germany the net loss of DOC from organic horizons was related to depth of those horizons rather than microbial respiration. DOM is continually leaching through the soil profile, such that leachate at any depth will be a combination of new litter-derived DOM and older DOM released from humic or lower layers (Park and Matzner 2003). DOC from older litter showed a higher contribution of carbon from lignin and lower biodegradability relative to fresh litter (Don and Kalbitz 2005; Kalbitz et al. 2006). Conflicting results from laboratory and field studies have been hard to reconcile because of lack of experimental control across studies for hydrology, as well as nitrogen and phosphorus availability (Kalbitz et al. 2000).

Different physical properties of litter affect microbial colonization rates and thus litter breakdown rates (Hyvonen and Agren 2001). Litter, coarse woody debris, and roots of trees show differences in chemistry, rates of mass loss of litter due to decomposition, and nitrogen dynamics by species. It was suggested that increased atmospheric CO_2 might lead to altered degradability of organic matter due to chemical changes in leaf or root chemistry (Hyvonen and Agren 2001). But it appears from free-air CO_2 enrichment (FACE) studies that species-specific differences in organic chemistry (e.g. pine versus birch) outweigh changes due to CO_2 enrichment. For temperate forests at least, changes in species competitive growth advantages due to heightened CO_2 will be the real driver of change in litter decomposition dynamics (King et al. 2001, 2005; Finzi and Schlesinger 2002; Hagedorn and Machwitz 2007). Barring limiting nutrients or water, litterfall (leaf productivity and turnover) for any one species is expected to increase under heightened atmospheric CO_2 without a concomitant change in litter chemistry (Allen et al. 2000).

Recent work suggests an interesting dynamic between nitrogen deposition and carbon storage in forest soils. Specifically, under nitrogen deposition on low-quality, high-lignin litter, decomposition of the organic layer slows, while nitrogen deposition on high-quality, low-lignin litter tends to accelerate decomposition (Knorr et al. 2005a). This dynamic provides an explanation for many contradictory studies on the effects of nitrogen deposition. A long-term study in Michigan, USA, demonstrated that chronic nitrogen additions increase soil carbon storage through reduced mineralization of surface and soil organic matter (Pregitzer et al. 2008), although a contrasting study indicated increased litter mass loss under high-nitrogen inputs using experimental microcosms (Manning et al. 2008). Furthermore, the progressive nitrogen limitation hypothesis suggests that CO_2 fertilization effects on forest productivity will not be realized if nitrogen is bound to aggrading organic matter pools. Recent work suggests that this hypothesis may be falsified if plants compensate by shunting more carbon belowground, to fuel the breakdown of organic matter by soil microbes (see Phillips et al. 2011).

2.2 Belowground Carbon Inputs: Fine Root Turnover and Exudates

Fine roots are the main source of carbon additions to soils, whether through root turnover or via exudates to associated mycorrhizal fungi and the rhizosphere (the soil immediately surrounding the roots). Quantifying fine root turnover *in situ* is important but difficult because of their variable turnover rates. Previous studies had indicated an extremely rapid turnover of fine roots, on the order of months to just a few years (Vogt et al. 1998). More recent studies using radiocarbon dating, however, indicated that fine roots were turning over on a 5–10 year cycle (Trumbore 2006). These opposing observations can be reconciled if the distribution of root ages is assumed to be positively skewed, with many small and ephemeral roots turning over in a matter of weeks, with a long tail of older roots surviving upwards of two decades (Trumbore and Gaudinski 2003). Results underline the need to conceptualize and model root turnover with multiple root pools rather than a single pool with a universally-applied turnover time (see below for a discussion of problems encountered in determining the rate of fine root turnover).

While DOM additions from leaf litter have been extensively researched and reviewed, the fate of DOM additions from root litter has been investigated rarely. Uselman et al. (2007) found that root litter at the soil surface lost most carbon, with decreasing percentage loss with depth of litter addition, suggesting an important role for deep roots in adding stable carbon to the soil system. A large scale tree girdling experiment in a Scots pine forest in Sweden resulted in a 40% drop in DOC, suggesting that current photosynthate contributes significantly to soil DOC through ectomycorrhizal fungi growing in association with roots (Giesler et al. 2007). This finding contrasts with the popular paradigm that DOC is primarily the product of root decomposition, since DOC should have increased following girdling had decomposition been the primary avenue for DOC production (Hogberg and Hogberg 2002; Giesler et al. 2007). A recent Free Air Carbon Dioxide Enrichment (FACE) experiment documented 62% of carbon entering the SOM pool through mycorrhizal turnover (Godbold et al. 2006), which may explain the close link between recent photosynthesis and DOC additions to soil. In addition to this, there is increasing evidence that roots directly supply low-molecular weight carbon compounds to rhizosphere soils, and that this flux may fuel from 30 to sometimes 100% of heterotrophic soil respiration (Van Hees et al. 2005). In contrast to anticipated increases in this rhizosphere flux, results are inconclusive as to the impact of elevated CO_2 on fine root production and turnover, with some studies indicating modest positive increases in root productivity (Luo et al. 2001b; Wan et al. 2004), while others show little or no increase (Pritchard et al. 2001; King et al. 2005; Pritchard et al. 2008). A recent study also showed that elevated atmospheric CO_2 does not cause changes in fine root chemistry; the concern was that elevated CO_2 would increase recalcitrant compounds that might slow decomposition (King et al. 2005).

3 Carbon Loss Through Root, Fungal, and Bacterial Respiration

Our increasing ability to measure accurately respiration of microorganisms in field circumstances has focused attention on understanding processes of carbon loss from soils. This section summarizes the more recent work done on root exudates, decomposition, and fungal and bacterial activities contributing to carbon loss from forest soils.

3.1 Root, Fungal, and Bacterial Respiration

Roots and mycorrhizal fungi produce about half of total respired CO_2, with the balance from heterotrophic breakdown of organic matter (Ryan and Law 2005). Soil respiration is commonly partitioned between autotrophic (plant) and heterotrophic (decomposition) respiration. These lumped categories simplify complex relationships in the soil system. For example, ectomycorrhizal fungi are clearly not primary producers, yet respiration products from ectomycorrhizal fungi are lumped with autotrophic respiration due to their close coupling with root processes and dependence on recent photosynthate (Hogberg and Read 2006). Respiration is more accurately viewed as a continuum from fully autotrophic photosynthesizers to fully heterotrophic predators and decomposers (Ryan and Law 2005). Conceptual models and new techniques for partitioning soil respiration among sources are needed (Table 2.1).

Lumping of soil respiration under heterotrophic and autotrophic respiration also neglects daily and seasonal differences in CO_2 flux as a result of physiological differences among bacteria, fungi and plants (e.g. Bradford et al. 2008; Allison et al. 2010). Radiocarbon dating is proving useful (Cisneros-Dozal et al. 2006; Hahn et al. 2006; Schuur and Trumbore 2006), but there are big differences between results from radiocarbon dating, ^{13}C labeling and CO_2 efflux studies (Hogberg et al. 2005). New research suggests tight coupling of current photosynthesis with soil respiration, possibly via the supply of labile carbon at the root-soil interface (Bond-Lamberty et al. 2004; Sampson et al. 2007; Stoy et al. 2007).

Ectomycorrhizal fungi make up a large proportion of soil biomass and contribute significantly to soil respiration but respond differently to environmental change compared to either

Table 2.1 Experimental methods employed to date for partitioning soil respiration among autotrophic and heterotrophic sources

Category	Technique	
Root exclusion	Trenching	All roots crossing the perimeter of the treatment plot are severed; membrane installed to prevent regrowth
	Girdling	Girdled trees near or within treatment plots cannot allocate photosynthate to roots
	Gap	Compare soil CO_2 efflux in clearcut stand to control stand
Physical separation of components	Components	Separate litter, roots and mineral soil rom a soil core; incubate separately; measure CO_2 efflux from each component
	Root excising	Remove roots from a fresh soil core; measure CO_2 efflux immediately
	Live root respiration	Excavate roots while still attached to tree; isolate and measure CO_2 efflux *in situ*
Isotopic techniques	Isotopic labelling	^{13}C labelling in FACE or chambers; switch C3 with C4 plants
	Radiocarbon	Radioactive decay of ^{14}C permits dating of photosynthetic event
Indirect techniques	Modelling	Bottom-up simulation of response of soil components to biotic and abiotic factors
	Mass balance	Assume soil C is at steady state; measure rates of C addition to soil from above- and belowground sources; subtract soil CO_2 efflux
	Subtraction	Soil CO_2 efflux minus other flux components from ecosystem NPP models and published values
	Root mass regression	Regress CO_2 efflux at multiple sites against root biomass; y-intercept is heterotrophic respiration

Source: Derived from Subke et al. (2006) and Hanson et al. (2000)

roots or bacteria, suggesting a need to separately model bacterial, mycorrhizal, and root respiration (Hogberg and Hogberg 2002; Langley and Hungate 2003; Fahey et al. 2005; Groenigen et al. 2007; Hogberg et al. 2005; Heinemeyer et al. 2007; Blackwood et al. 2007). Bacterial and fungal dominance, as well as overall faunal community composition, affects soil carbon dynamics (Jones and Bradford 2001; Bradford et al. 2002b, 2007). However, their differing responses may cancel each other out (Bradford et al. 2002a) and the evidence that bacterial-fungal ratios directly affect soil carbon stocks is only correlative (see Strickland and Rousk 2010). Further studies must clarify understanding of underlying mechanisms and environmental factors that characterize differing microbe responses (e.g. fungi, bacteria) (Chung et al. 2006; Monson et al. 2006; Blackwood et al. 2007; Fierer et al. 2007; Hogberg et al. 2007).

In addition, earthworm effects on carbon and nitrogen cycling are significant (Li et al. 2002; Marhan and Scheu 2006). Fresh inputs of carbon (i.e. priming) from organisms such as earthworms may allow soil microbes to mine old carbon deeper in the profile (Dijkstra and Cheng 2007; Fontaine et al. 2007). This suggests that increased input from leaf productivity may boost soil heterotrophic respiration and CO_2 flux from soils. Priming can lead to rapid shifts in community composition (Cleveland et al. 2007; Montano et al. 2007). Low molecular weight compounds, including organic acids, amino acids and sugars, are products of microbial breakdown and root exudation, and represent a small fraction of the total mass of carbon cycling through soil. However, breakdown of low molecular weight carbon compounds may contribute up to 30% of total soil CO_2 efflux because of extremely rapid turnover, with residence times estimated at 1–10 h (Van Hees et al. 2005).

3.2 Respiration Responses to Environmental Change

Under global change scenarios, nitrate deposition from fossil fuel burning, particulate deposition from forest fires, and wind erosion of agricultural soils, are expected to alter forest nutrient cycling. The addition of nitrogen has been shown to affect microbial breakdown of litter and SOM, the results varying with litter type and microbial community composition (Sinsabaugh et al. 2004, 2005; Waldrop et al. 2004). Litter quality is important, at least for temperate forests, where litter in high-lignin systems shows unchanged or decreased rates of decomposition under nitrogen deposition, while low-lignin, low-tannin systems tend to increase decomposition rates (Magill and Aber 2000; Gallo et al. 2005). In turn, it was shown in northern temperate forests that the composition of the microbial communities changed in response to nitrogen deposition (Waldrop et al. 2004). Based on these observations, elucidating carbon dynamics under elevated nitrogen scenarios for other biomes and across canopy tree associations should be a priority.

Fertilization by increased atmospheric CO_2 and the deleterious effects of ozone (O_3), both resulting from burning of fossil fuels, are also expected to alter forest soil carbon cycling. In a 4-year study in experimental temperate forest stands, Loya et al. (2003) found that a simultaneous 50% increase in CO_2 and O_3 resulted in significantly lower soil carbon formation, possibly due to reduced plant detritus inputs and/or increased consumption of recent carbon by soil microbes. In another study of temperate forest soils, soil faunal communities changed composition under exposure to CO_2 or O_3 singly but, when combined, there was no main effect (Loranger et al. 2004). Fungal community composition was significantly altered as a response to elevated O_3 in a FACE study in Wisconsin, USA (Chung et al. 2006). The response of fungal respiration to elevated CO_2 is so far equivocal. One study indicated a rise in fungal activity (Phillips et al. 2002) while another recorded a decrease (Groenigen et al. 2007). As in the divergent responses under nitrogen deposition, the result may depend heavily on litter chemical properties. Soil respiration is expected to increase under increased CO_2 and O_3 (Andrews and Schlesinger 2001; King et al. 2004; Luo et al. 2001b; Pregitzer et al. 2006), but some studies show conflicting results (Suwa et al. 2004; Lichter et al. 2005) or a decrease when combined with fertilization (Butnor et al. 2003). Overall, FACE studies indicate a net increase in carbon storage, mostly in litter and fine root mass, despite soil respiration increases (Allen et al. 2000; Hamilton et al. 2002), although cycling through litter is especially rapid and sequestration in litter is likely limited (Schlesinger and Lichter 2001).

Respiration response to temperature changes, especially pertaining to a still-hypothetical positive feedback of warming to carbon mineralization, is highly uncertain (Denman et al. 2007). The original uncertainty centered on whether the soil carbon pool should be lumped or split by rate of turnover (Davidson et al. 2000; Giardina and Ryan 2000), since (often small) portions of the soil carbon pool cycle very quickly, especially low molecular weight organic acids, and therefore may be more responsive to temperature than larger, older or adsorbed compounds. Later, evidence mounted for an acclimation of soils to heightened temperatures over time, although it now seems clear that depletion of the fast-cycling labile carbon pool under increased initial mineralization rates is partly responsible for the apparent downshifting in respiration over time (Luo et al. 2001a; Melillo et al. 2002; Eliasson et al. 2005). Further experiments are needed to test the acclimation hypothesis (see Bradford et al. 2008). Reworking the data of Giardina and Ryan (2000), others found that the response of the fast pool over experimental scales obscured the slower but ultimately more important response of the large pool of stable carbon (Knorr et al. 2005b). A consensus is still in the making concerning the impact of warming on soil respiration, although it now seems clear that the complex nature of SOC, and confounding factors, including soil water content, complicate a simple determination of the temperature effect (Davidson and Janssens 2006). Indeed, inclusion of microbial physiology in soil models can negate the projected positive feedback between warming and loss of soil carbon (Allison et al. 2010). Complicating the interpretation of field data, soil respiration is closely coupled to photosynthesis of the canopy, explaining some of the apparent causal correlation between temperature and respiration *in situ* (Sampson et al. 2007).

The effects of soil freezing, compounded by a decreased or absent snow pack predicted for some temperate and boreal regions, may decrease winter soil respiration (Monson et al. 2006). More attention has been given to drying and wetting cycles recently, which appear to substantially increase annual decomposition (Fierer and Schimel 2002; Borken et al. 2003; Jarvis et al. 2007).

4 Stabilization of Carbon in Forest Soils

Plant-derived organic molecules are stabilized from microbial action by biochemical resistance or by physical protection within soil aggregates or microsites (Table 2.2 and Fig. 2.6). Microbial decomposition can also, itself, facilitate stabilization through the production of secondary microbial products and the majority of carbon that is most resistant to decay has microbial – and not plant – signatures (Grandy and Neff 2008). When assessing environmental correlates, in order of importance for the measured stabilization of organic matter in soils across Ohio, USA, drainage class was the only significant determinant of SOM content in the upper 30 cm in forest soils, whereas the significance of individual site variables on SOM content in non-forested soils was firstly soil taxon, then drainage class, and lastly texture. The low significance of these other factors on forest soils suggests different drivers of SOM dynamics in forests (Tan et al. 2004).

The importance of anoxic conditions for preservation of organic matter in boreal peatlands is without doubt (Fierer and Schimel 2002; Borken et al. 2003; Jarvis et al. 2007). But in aerobic soils, dissolved organic carbon (DOC) that leaches from decomposing material is vulnerable to mineralization and respiration as CO_2 by bacteria or saprophytic fungi. The portion of DOM that escapes mineralization by microbes generally does so by sorption to soil minerals where it is stabilized as SOC. Carbon is also stabilized when fire produces black charcoal from organic matter. Finally, some DOC may be flushed from the soil system during periods of high soil water flow.

4.1 Sorption and Complexation of Dissolved Organic Matter

In aerobic soils, texture is considered to be the most important driver of DOC stabilization in soil, with mineralogy an important factor that is dependent on texture (Bird et al. 2002). Across a 1,000 km latitudinal transect in Siberia, SOC stocks on fine-textured soils were approximately double the stocks on coarse-textured soils (Bird et al. 2002). The layered structure of clay results in an extremely high surface area to volume ratio, and clay interlayers host a multitude of cations which provide binding sites for DOC. Clay content exerts a powerful control on the size of the older soil carbon pool in Amazonian soils (Telles et al. 2003). However, Giardina et al. (2001)

Table 2.2 Mechanisms of carbon stabilization in forest soils (some mechanisms are specific to soil order or biome, while others are active in all soils)

Selective preservation	Inherent stability due to e.g. alkyl-C chains in lipids, aromatic structures, phenolics
Spatial segregation	Occlusion inside soil aggregates
	Sequestration within soil micropores
	Coating with hydrophobic aliphatic compounds
	Intercalation within phyllosilicates (clay)
Molecular interaction	Complexation with metal ions
	Interaction with other organic molecules through ligand exchange, polyvalent cation bridges or weak interactions
Inundation	Anoxic conditions prevent abiotic oxidation and aerobic microbial respiration
Freezing	Sub-freezing temperature stifles microbial respiration
Carbonization	Relatively inert carbon is broken down only at millennial timescales

Source: After Lorenz et al. (2007) and Lutzow et al. (2006)

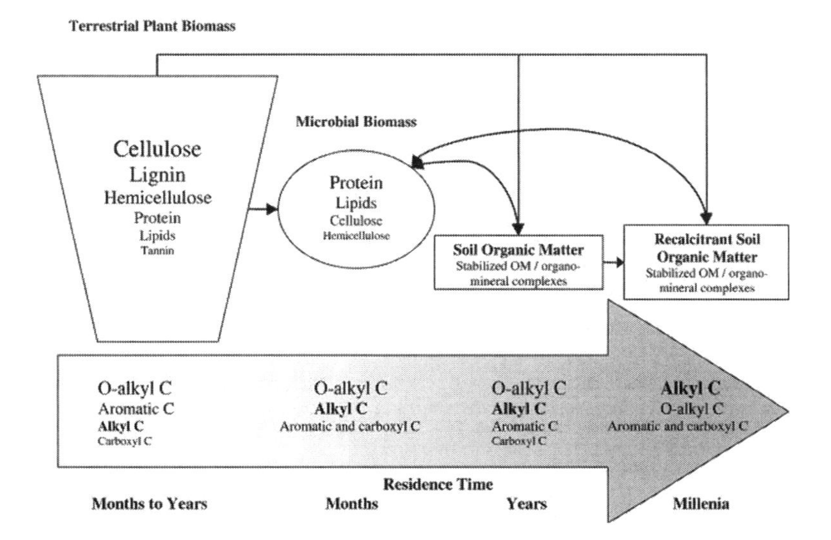

Fig. 2.6 Carbon flux through terrestrial organic matter pools (*top*) and relative enrichment of recalcitrant alkyl carbon during breakdown of more-labile compounds (*bottom*). The relative composition of organic matter by carbon structure is represented by size of type (*Source*: Reprinted from Lorenz et al. 2007, with permission from Elsevier)

found no relation between carbon mineralization rates and clay content in laboratory-incubated upland forest soils. Clay content of soil is well-correlated with SOC generally, although other factors dominate DOC stabilization in cold or wet climates. Sorption to the mineral matrix has been shown to strongly preserve DOM (Kaiser and Guggenberger 2000). Aluminum (Al) and iron (Fe) cations are the most important interlayer mineral binders for DOM (Zinn et al. 2007). Besides binding to clay particles, colloidal and soluble organic matter can form insoluble complexes with Al and Fe cations, which precipitate (Schwesig et al. 2003; Rasmussen et al. 2006; Scheel et al. 2007). These results suggest that whole-ecosystem carbon cycle models should account for both soil texture and soil mineralogy when modeling carbon fluxes (Table 2.3). Labile DOM high in carbohydrate has a large increase in stability due to sorption, but for DOM with a greater proportion in complex aromatic organic compounds stability due to sorption is relatively small because such compounds are already relatively stable. However, irrespective of proportional increases, gross sorption of recalcitrant compounds was much larger than sorption of labile compounds, in fact as much as four times larger (Kalbitz et al. 2005). Stabilization of OM by sorption therefore depends on particulars of the organic compounds sorbed, strong chemical bonds to the mineral soil, and/or a physical inaccessibility of OM to microorganisms (Kalbitz et al. 2005).

4.2 Fire as a Sequestration Mechanism

The many effects of fire on forest soils have been reviewed by Certini (2005). In areas with frequent fires, 35–40% of SOC was fire-derived black carbon. Fire can sterilize soils to depths of 10 cm or more, and effects of sterilization may last a decade, resulting in decreased microbial respiration. When fire does not remove carbon from the soil system through combustion, it tends to increase the stability of the carbon remaining through carbonization, reduction in water solubility, and relative enrichment in aromatic groups (Certini 2005). Czimczik et al. (2003) found that in a boreal Siberian Scots pine forest, black carbon contributed a small percentage of the SOC pool while the fire reduced the mass of the forest floor by 60%. A wildfire

Table 2.3 Characteristics of six process-based forest soil carbon models

	Yasso	ROMUL	SOILN	RothC	Forest-DNDC	CENTURY
Time-step	Year	Month	Day	Month	Day	Month
Simulation depth	Organic layer +1 m mineral soil	Organic layer +1 m mineral soil	Any depth	Adjustable from 0 to 1 m	Adjustable from 1 to 1.5 m	20 cm
Organic matter pools						
Stand	–	–	Roots, stems, leaves, grains	–	Canopy, understory, groundstory	Leaves, fine roots, fine branches, coarse wood, coarse roots
Litter	Fine and coarse woody litter	Aboveground and belowground pools divided by N and ash contents	1–2 per soil layer, 10–15 soil layers	Resistant and decomposable pools	Very labile, labile, and resistant pools for each soil layer	Aboveground and belowground pools divided into metabolic and structural, senescent litter pooled based on lignin:N ratio
Soil	Extractives, celluloses, lignin-like compounds, 2 humus pools	Six or more	One humus, one microbe pool per soil layer, 10–15 soil layers	Living, humic and insoluble OM pools	Two humads and humus per layer	Active, slow and passive SOM pools
Different pools for organic and mineral soil?	No	Yes	Yes	No	Yes	No
Nutrient input	–	N deposition	N deposition, fertilization, N content of plant parts	–	N deposition and fertilization	N deposition and fertilization, organic N inputs, P, S
Soil texture input	–	Clay content	Hydraulic properties	Clay content	Clay content	Sand, silt and clay content
Limitations	Upland forest soils only	Well or excessively drained mineral soils only	Substantial input information required, not for peatlands	Upland forest soils only	Substantial input information required	Very shallow, not for peatlands, does not separate humified litter from mineral soil
Measurability of pools	Only extractives, celluloses and the sum of the other pools are measurable	Yes, all pools measurable	No, pools are conceptual and cannot be directly measured	Yes, all pools measurable	No, pools are conceptual and cannot be directly measured	SOC, litter pools measurable, sub-pools conceptual only

Source: Adapted from Peltoniemi et al. (2007)

in boreal Alaska burned polysaccharide-derived compounds preferentially, resulting in a relative enrichment of lipid- and lignin-derived compounds (Neff et al. 2005). There appears to be an inverse relationship between fire frequency and complete combustion: infrequent fire return intervals and high intensity may result in less carbonization and more complete combustion than in regions with shorter fire return intervals that experience lower-intensity fires, increased carbonization, and therefore increased storage (Czimczik et al. 2005).

5 Quantifying the Carbon Under Forests

In the northern hemisphere, the carbon in soils remains the highest uncertainty in global budgeting (Goodale et al. 2002) and partitioning soil respiration among sources to identify carbon leakage/loss has proved to be one of the most difficult tasks (Ryan and Law 2005). Failure to close the soil carbon budget stems from discrepancies between measured bulk CO_2 fluxes and the predictions of process models of autotrophic and heterotrophic respiration (Trumbore 2006). Different methods used to accommodate study objectives and resources make comparison difficult (Wayson et al. 2006) and, as a result, many budgets leave out soil carbon and litter carbon accumulations completely (Liski et al. 2003).

5.1 Quantifying Carbon Additions

The turnover of organic matter in surface soil layers can be quantified by direct measurement of mass loss through litterbag studies or by ^{14}C enrichment, litter sampling and mass spectrometer analysis. However, there are known problems with both litterbag studies and ^{14}C enrichment as methodologies for measuring carbon addition to soils. Litterbags limit breakdown of litter by soil macrofauna. The ^{14}C signature measures the mean residence time of carbon in the surface layer, but not the lifetime of various recognizable litter components (e.g. from fine roots, leaves, bark) (Hanson et al. 2005). And it has been found that ^{14}C-labeled carbon residence time in fine roots, estimated at >4 years, is much longer than the <1 year root longevity estimated by using the minirhizotron, a small camera lowered through a clear plastic tube to monitor the growth of roots over time (Strand et al. 2008).

It was previously thought that radiocarbon dating and the turnover time for roots – estimated by laborious sorting and weighing of root production year-to-year and then dividing total root biomass by annual production – could be reconciled if the age distribution of roots was positively skewed (Tierney and Fahey 2002; Trumbore and Gaudinski 2003). Current thinking on important sources of discrepancy in estimating fine root turnover are outlined by Strand et al. (2008) and include: (1) the presence of different root pools cycling at different rates; (2) the confounding effect of stored carbohydrates, which throw off radiocarbon estimates of age; (3) the skewed nature of root age distribution as pointed out in Trumbore and Gaudinski (2003); (4) lingering effects of minirhizotron installation on root growth; and (5) the use of median root longevity as an inaccurate substitute for mean longevity in minirhizotron studies. These sources of error cause radiocarbon methods to underestimate the importance of fine root turnover to soil carbon cycling and the minirhizotron method to overestimate this importance (Strand et al. 2008). Work is underway to address these shortcomings and to deepen understanding of root turnover, e.g. by partitioning root pools by branching order (Guo et al. 2008).

5.2 Partitioning Soil Respiration

Hanson et al. (2000) discuss in detail and outline the major classes of soil partitioning techniques (summarized in Table 2.1). A comprehensive review of research needs in measuring and modeling soil respiration has been done by Ryan and Law (2005), while Subke et al. (2006) provide an exhaustive list of soil CO_2 efflux partitioning studies through 2006 across all terrestrial biomes. They show that many of the techniques for partitioning have inherent methodological biases (Subke et al. 2006). For example, detection of changes, especially depletion, of the large, slow-cycling pool of recalcitrant soil carbon represents a significant challenge and is almost always underrepresented, but it is essential to quantifying the carbon exchange between soil and the atmosphere. Additionally, multi-factor experiments must be of sufficient length to allow adjustment to treatment conditions (Ryan and Law 2005). It is therefore better to resample the same points than to randomly select new ones in long-term sampling studies and inventories. Due to the spatial variability of SOC processes, 15–20% changes in soil carbon stocks may be overlooked (Yanai et al. 2003). Site variability therefore confounds broad-scale application of flux data (Hibbard et al. 2005). There is also a trade-off between spatial and temporal resolution when using manual vs. automated CO_2 flux measurements. Used in combination, the two systems provide combined resolution in both dimensions, but manual measurements are sufficient for measuring integrated seasonal fluxes (Savage and Davidson 2003).

5.3 Modeling Soil Carbon Dynamics

Many SOC models have been created for agricultural systems, and may be modified to simulate forested systems, in order to accommodate important differences in management, disturbance regime, vegetation, and biota. To date there has been only one published model comparison for forest SOC dynamics. Peltoniemi et al. (2007) review and compare six process-based, multiple-SOC-pool models of forest SOC dynamics. The review includes an extensive comparison of model inputs and modeled processes. More work is needed to assess the accuracy of forest soil carbon models, and to adapt them or develop new ones for diverse biomes. Critically, very few countries and regions have published long-term soil carbon datasets with the ancillary data needed to verify model accuracy, and there is a total lack of such inventories for tropical regions. Most soil carbon process models do not deal explicitly with peatlands, severely limiting their applicability in some boreal regions. Wetlands versions of the *RothC* and *ROMUL* models are expected in the near future, and *Forest-DNDC* includes a wetland component (Peltoniemi et al. 2007). The move toward process-based models of SOC dynamics is hindered by poor understanding of the different mechanisms of sequestration of diverse classes of organic biomolecules in soils. Model soil carbon pools must be derived from functional classes of compounds (which must first be characterized) with similar sequestration mechanisms, rather than from the most easily differentiated classes of SOC based on *in situ* measurement techniques or fractionation (Lutzow et al. 2006) (Table 2.3).

Empirical models must be careful to parameterize at the same time step as the output (Janssens and Pilegaard 2003). The concentration of labile carbon, its rapid turnover, and the resultant large CO_2 efflux can obscure the sensitivity of heterotrophic respiration to soil temperature change. Care should be taken to control for labile carbon concentrations when extrapolating field measurements of bulk soil respiration to global change scenarios (Gu et al. 2004).

The temperature dependence of soil is often described by the Q_{10} value, which is defined by the difference in respiration rates over a 10°C interval. Q_{10} has been found to be extremely variable, with a range from 1 (no effect of temperature on respiration) to 5 (five times higher respiration rate with a 10° rise in temperature) under different combinations of soil moisture and soil temperature (Reichstein et al. 2003). Kinetic properties of the many organic compounds in soils, plus environmental constraints such as limiting soil moisture or nutrients, complicate efforts to fully explain the temperature sensitivity of microbial respiration (Davidson and Janssens 2006). In an analysis of sources of uncertainty in the soil carbon model *SWIM*, Post et al. (2008) identified the carbon mineralization rate, carbon use efficiency, Q_{10}, soil bulk density, and initial carbon content as the most critically sensitive parameters. Better models will have to differentiate the direct effects of drying, wetting, and carbon substrate supply to soil microbes from the indirect effects of soil water content and temperature on diffusion of carbon substrates to the microbial population (Davidson et al. 2006). Work in this area indicates that models incorporating realistic spatial relationships, hourly time steps, and mechanistic workings give the most accurate results (Hanson et al. 2004). Not all applications will be suited to process models, however, due to the extensive inputs required (Liski et al. 2005). Yet, leaving out process may yield projections for soil carbon stocks that differ to when biological response is explicitly modeled. For example, Allison et al. (2010) showed that by making carbon use efficiency temperature sensitive, microbial biomass decreases with warming. This decrease then negates the loss of soil carbon with temperature increase that is projected with conventional models, because of the reduction in microbes that actually carry out decomposition. Indeed, the IPPC identify such responses as significant areas of uncertainty, with all eleven models coupling climate with carbon cycling omitting the soil microbiology.

5.4 The Superficial Nature of Soil Carbon Research

Studies of soil organic matter under conventional and no-till soil management in agriculture have been largely limited to the top 30 cm of soil. Now, some are suggesting the need to consider SOM deeper in the profile (Baker et al. 2007). The same argument ought to be made for forest soil research: rooting depths are far greater for many tree species than field crops. Soil depth confounds warming studies by insulating deeper soil layers (Pavelka et al. 2007) and delaying CO_2 efflux (Jassal et al. 2004; Drewitt et al. 2005). Also, a significant portion of below-ground carbon is deeper than 1 m (Jobbagy and Jackson 2000) and recent research indicates that roots exert powerful influences on redox activity in their vicinity, with important implications for carbon cycling deep in the soil profile (Fimmen et al. 2008).

5.5 Quantifying Carbon Stocks After Land Use Change

Long-term soil experiments and inventories can elucidate SOC dynamics in ways that shorter ones cannot. Peltoniemi et al. (2007) point out the importance of repeated soil surveys for SOC model verification and validation. Given the importance of chronosequence studies for area-based carbon budgeting under land use change (Woodbury et al. 2007), the first unified global network of long-term soil experiments (LTSEs) has been formed (http://ltse.env.duke.edu/). This network will ideally address the lack of uniformity of measurements of soil carbon that so complicate comparison and synthesis. There are other problems: only 20% of soil studies measuring SOC are in forested biomes; therefore, boreal, tropical and warm-temperate forests are under-represented; soil studies measuring SOC are heavily concentrated in developed countries; and long-term SOC studies on alfisols and mollisols dominate, while long-term changes on oxisols, histosols and gelisols are still poorly understood (Fig. 2.7). Chronosequences, or space-for-time substitutions, though useful for characterizing soil change over centuries or millennia, may confuse the effects of land use with weathering. Land use history can be difficult to properly control for (Richter et al. 2007) (Table 2.4). Yet perhaps of most concern, if we can standardize methods for quantifying soil carbon stocks, it is in selecting a method that can robustly quantify both absolute carbon stocks and their change with time. The currently accepted practice of measuring soil carbon concentrations in the surface 30 cm of soil suffers from the fact that soils are not static entities. For example, soil compaction following forest removal – or soil expansion following afforestation – both change the mass of soil carbon in the surface 30 cm even when the absolute mass of soil carbon at a location does not change. This means that depth-independent (i.e. mass dependent) sampling is required to resolve change in carbon stocks with time, which involves measurement of soil carbon in at least two soil cores (a surficial core and one immediately below) at a location (Gifford and Roderick 2003). In addition to this, it must be recognized that soil carbon stocks are spatially variable at the scale of only a few meters. This means that multiple samples are required to gain a spatial estimate for a site. Also, this variability causes lack of statistical power to

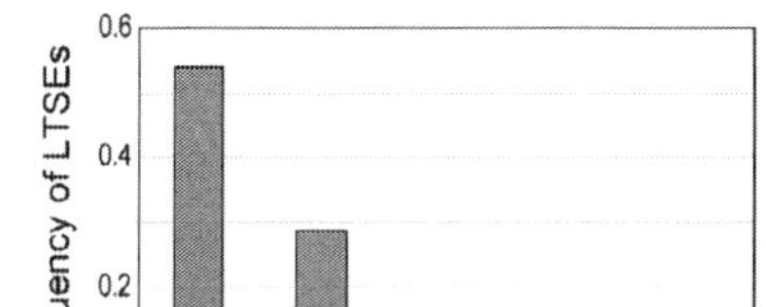
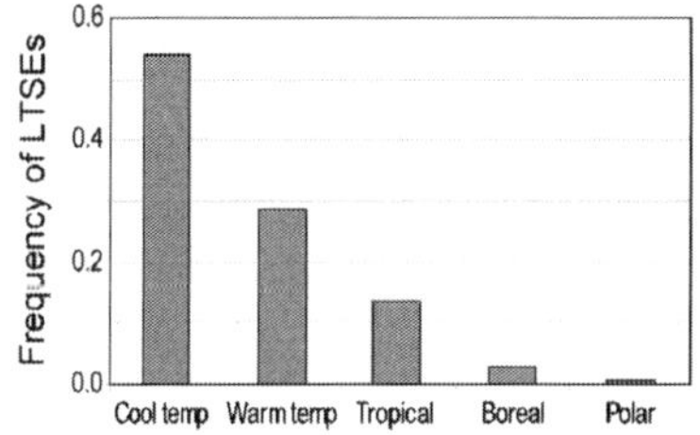

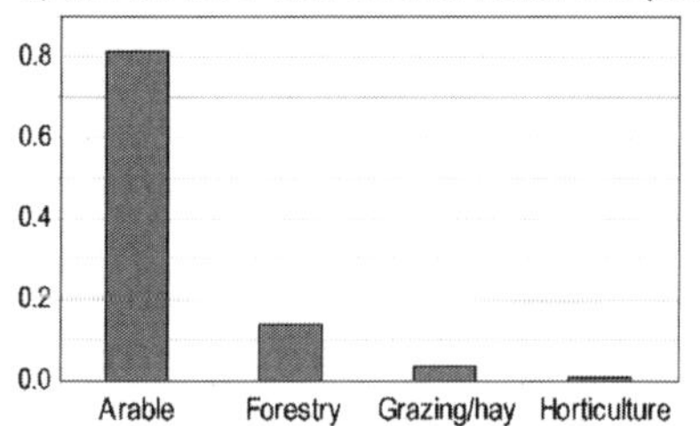

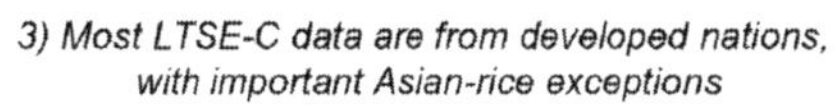
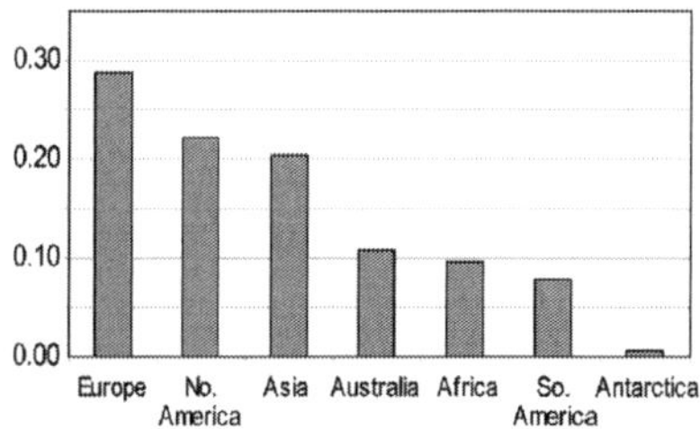

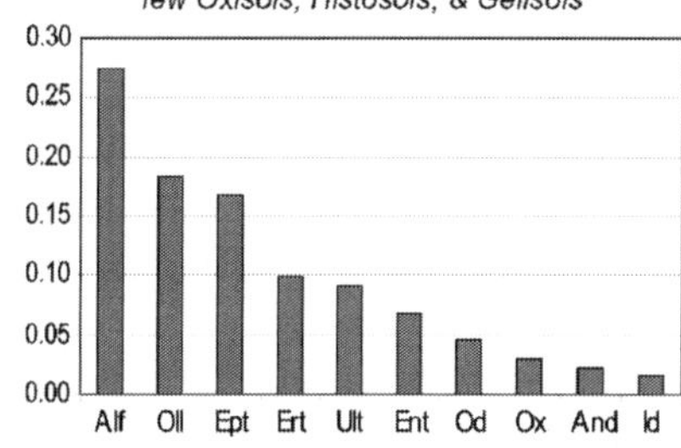

Fig. 2.7 Distribution of long term soil experiments (LTSE) measuring SOC across climate zones, land uses, continents and soil orders. Note the lack of forest LTSEs despite the importance of land use change, specifically deforestation, to national and global carbon budgets (*Source*: Graphs produced by Richter et al. from data at http://ltse.env.duke.edu. Used with permission)

Table 2.4 Types of soil experiments that may be used to elucidate carbon dynamics and changes in carbon stocks under land-use change

Approach	Time scale (year)	Uses and strengths	Challenges and limitations
Short-term soil experiments	<1–10	Field or lab based, experimental control, versatile, short-term processes	Extrapolation to larger scales of space and time, reductionist
Long-term soil experiments	>10	Field based, direct soil observation, experimental control, sample archive	Duration before useful data, vulnerable to loss or neglect, extrapolation to larger scales
Repeated soil surveys	>10	Field based, direct soil observation, regional perspective, sample archive	Planning and operational details, very few yet conducted, monitoring
Space-for-time-substitution	>10 to>>1,000	Field based, highly time efficient	Space and time confounded
Computer models	<1 to>>1,000	Versatile, heuristic and predictive, positively interact with all approaches	Dependent on observational data

Source: Richter et al. (2007). Reprinted with permission

detect change in soil carbon; so statistically non-significant change in soil carbon does not mean that there has not been a biologically significant change (Throop and Archer 2008; Strickland et al. 2010). Lastly, soil carbon must be sampled at a landscape scale – loss of soil carbon through erosion from one area of a forest patch might simply be redistributed within that patch in another area that is aggrading carbon.

6 Effects of Management Regime on Soil Carbon Cycling

The Fourth Assessment Report by the IPCC Working Group III projects that, initially, reduction in deforestation will lead to the greatest positive increase in global carbon sequestration, due to the current rapid rate of deforestation and the

Table 2.5 The generalized impact of forest management actions on carbon stocks Summary of the effects of specific forest management actions on ecosystem C stocks ('+'...increases C stock, '–'...decreases C stock; '±' neutral with respect to C stock)

Afforestation
+ Accumulation of aboveground biomass formation of a C-rich litter layer and slow build-up of the C pool in the mineral soil
± Stand stability depends on the mixture of tree species
– Monotone landscape, in the case of even-aged mono-species plantations
Tree species
+ Affects stand stability and resilience against disturbances; effect applies for entire rotation period; positive side-effect on landscape diversity, when mixed species stands are established
– Effect on C storage in stable soil pools controversial and so far insufficiently proven
Stand management
+ Long rotation period ensures less disturbance due to harvesting, many forest operations aim at increased stand stability, every measure that increases ecosystem stability against disturbance
± Different conclusions on the effect of harvesting, depending if harvest residues are counted as a C loss or a C input to the soil
– Forests are already C-rich ecosystems – small increase in C possible; thinning increases stand stability at the expense of the C pool size; harvesting invariably exports C
Disturbance
+ Effects such as pest infestation and fire can be controlled to a certain extent
± Low intensity fires limit the risk of catastrophic events
– Catastrophic (singular) events cannot be controlled; probability of disturbance can rise under changed climatic conditions, when stands are poorly adapted
Site improvement
+ N fertilization affects aboveground biomass; effect on soil C depends on interaction of litter production by trees and carbon use efficiency of soil microbes
± Drainage of peatland enables the establishment of forests (increased C storage in the biomass) and decreases CH_4 emissions from soil, but is linked to the increased release of CO_2 and N_2O from the soil
– Liming and site preparation always stimulate soil microbial activity. The intended effect of activating the nutrient cycle is adverse to C sequestration; N fertilization leads to emission of potent greenhouse gases from soils

Source: Reprinted from Jandl et al. (2007), with permission from Elsevier

large associated CO_2 loss to the atmosphere. Over the long term, sustainable forest management that increases forest growing stock while also providing timber, fiber and energy will provide the greatest mitigation benefit at the lowest cost to society (IPCC 2007). But the link between different forest management activities, deforestation, reforestation and afforestation and the net carbon flux between soils and the atmosphere is not well characterized (Table 2.5).

Productivity of the forest increases litter fall and potentially sequestration; less disturbance of soil tends to preserve soil carbon stocks; and mixed species forests are more resilient and therefore better systems for securing carbon in forest soils. On the other hand, planting on agricultural soils increases carbon accumulation by soils for both conifers and broadleaf trees (Morris et al. 2007). Although the rate of carbon accumulation and sequestration within the soil profile differs by tree species, no species effect on SOM stability has yet been reported (Jandl et al. 2007). Differences in plant anatomy lead to changes in the vertical distribution of minerals and soil carbon when there is land use or land cover change (Jackson et al. 2000). For example, in Fujian, China, conversion of natural forests to plantations has been linked to carbon loss (Yang et al. 2007). However, combined CO_2 sequestration and timber production can be economically maximized (Thornley and Cannell 2000). In addition, during reforestation, soils are a slower but more persistent sink than aboveground carbon, and are more stable pools than aboveground pools for actively harvested forests (Thuille et al. 2000).

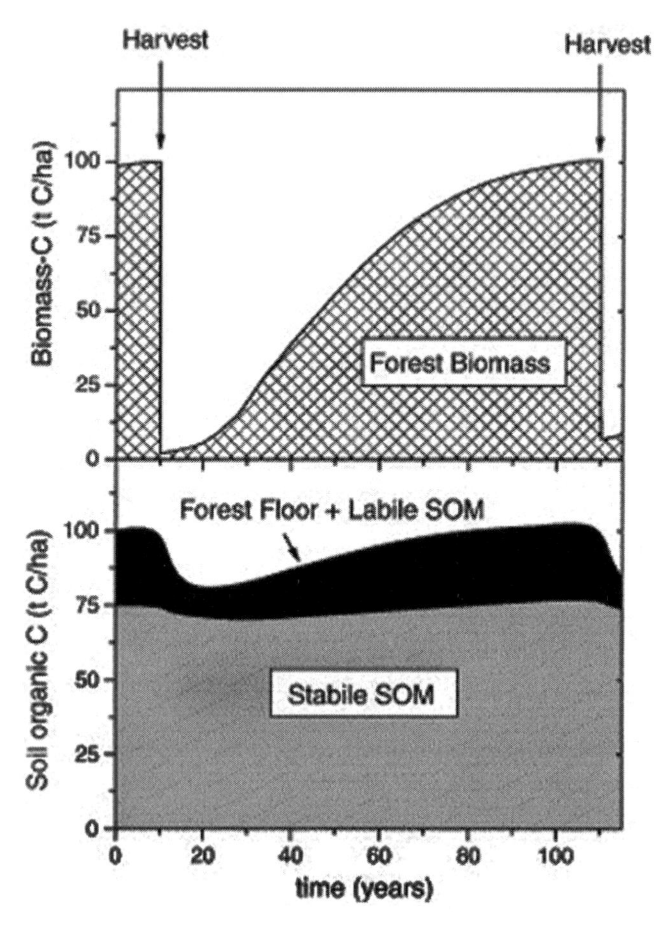

Fig. 2.8 A simulation of carbon stocks above- and belowground before and after forest harvesting, for a typical Central European Norway spruce forest. Assumptions include a 100 year rotation for a typical Norway spruce stand with 25% labile SOM (*Source*: Reprinted from Jandl et al. 2007, with permission from Elsevier)

Studies in boreal forests have demonstrated that tree harvesting generally has little long-term effect on stabile soil carbon stocks (Martin et al. 2005), although evidence from temperate forests of the southeastern USA indicates that whole tree harvesting can be detrimental by removing nitrogen (Johnson et al. 2002). There are no obvious long-term effects from clearcutting that can be detected on in-stream DOC chemistry (Johnson et al. 2002), although clearcutting released a pulse of labile DOC at Hubbard Brook in New Hampshire, USA (Dai et al. 2001), probably from humic substances at the forest floor (Ussiri and Johnson 2007). Shortened rotations from 90 to 60 years in Finland increased soil carbon by increasing input of litter but did not maximize system-wide carbon sequestration because of increased frequency of harvest operations (Liski et al. 2001); although others have found that fresh carbon additions due to harvesting operations can stimulate microbial populations to mineralize ancient deep soil carbon (Fontaine et al. 2007; Jandl et al. 2007) (Fig. 2.8).

Plant diversity and composition effects on net primary productivity (NPP) are becoming apparent and must be accounted for (Catovsky et al. 2002). Oak forests turn SOM over faster compared to pine, which locks up more litter for longer in the surface layers (Quideau et al. 2001). Broadleaf tree plantations replacing natural forest or pasture tend not to change soil carbon stocks, while pine plantations reduce soil carbon stocks 12–15% (Guo and Gifford 2002). Conversion of forest to pasture results in a slow but marked increase in soil carbon stocks, but this is the reverse for tilled agriculture (Cerri et al. 2003, 2004). Pasture systems are very productive and thus larger carbon fluxes from them indicate greater allocation of carbon belowground (Johnson and Curtis 2001; Johnson et al. 2002; Paul et al. 2002; Salimon et al. 2004; Thuille and Schulze 2006).

7 Conclusion and Summary Recommendations

This review outlined the most critical issues and impediments to characterizing belowground carbon cycling in forested biomes. To further our understanding of belowground carbon dynamics in forests, more work is needed to characterize the following:

- controls on the depth of the forest floor organic layer by leaching of dissolved organic carbon (DOC) to the mineral soil.
- responses of root carbon inputs to environmental change, such as nitrogen deposition.
- rates of fine root turnover across species and biomes.
- patterns of bacterial, fungal and plant respiration and responses to physical and biotic forcing.
- dynamics of functionally-distinct soil carbon pools, rather than the most easily measured and fractionated pools.
- the most accurate methods for quantifying forest soil carbon stocks and fluxes.

The global nature of the carbon cycle requires a globally-distributed and coordinated research program, but has thus far been largely limited to:

- the developed world
- the top 30 cm of the soil profile
- temperate biomes
- agricultural soils

Political and financial resources are being mobilized to increase the stock of carbon in forest soils despite minimal research to date about the long-term effects of land use on SOC stocks. Key research needs are to reduce the uncertainty in environmental response of the mechanisms that stabilize soil carbon inputs, and to develop and implement appropriate methods to estimate stocks and their change with time. Such efforts will inform management strategies, ensuring effectiveness in their intended goal of sequestering carbon in forest soils.

References

Allen AS, Andrews JA, Finzi AC, Matamala R, Richter DD, Schlesinger WH (2000) Effects of free-air CO_2 enrichment (FACE) on belowground processes in a *Pinus taeda* forest. Ecol Appl 10:437–448

Allison SD, Wallenstein MD, Bradford MA (2010) Soil-carbon response to warming dependent on microbial physiology. Nat Geosci 3:336–340

Andrews JA, Schlesinger WH (2001) Soil CO_2 dynamics, acidification, and chemical weathering in a temperate forest with experimental CO_2 enrichment. Glob Biogeochem Cycles 15:149–162

Baker JM, Ochsner TE, Venterea RT, Griffis TJ (2007) Tillage and soil carbon sequestration-what do we really know? Agric Ecosyst Environ 118:1–5

Bird MI, Santruckova H, Arneth A, Grigoriev S, Gleixner G, Kalaschnikov YN, Lloyd J, Schulze ED (2002) Soil carbon inventories and carbon-13 on a latitude transect in Siberia. Tellus B 54:631–641

Blackwood CB, Waldrop MP, Zak DR, Sinsabaugh RL (2007) Molecular analysis of fungal communities and laccase genes in decomposing litter reveals differences among forest types but no impact of nitrogen deposition. Environ Microbiol 9:1306–1316

Bond-Lamberty B, Wang C, Gower ST (2004) A global relationship between the heterotrophic and autotrophic components of soil respiration? Glob Change Biol 10:1756–1766

Borken W, Davidson EA, Savage K, Gaudinski J, Trumbore SE (2003) Drying and wetting effects on carbon dioxide release from organic horizons. Soil Sci Soc Am J 67:1888–1896

Bradford MA, Jones TH, Bardgett RD, Black HIJ, Boag B, Bonkowski M, Cook R, Eggers T, Gange AC, Grayston SJ, Kandeler E, McCaig AE, Newington JE, Prosser JI, Setala H, Staddon PL, Tordoff GM, Tscherko D, Lawton JH (2002a) Impacts of soil faunal community composition on model grassland ecosystems. Science 298:615–618

Bradford MA, Tordoff GM, Eggers T, Jones TH, Newington JE (2002b) Microbiota, fauna, and mesh size interactions in litter decomposition. Oikos 99:317–323

Bradford MA, Tordoff GM, Black HIJ, Cook R, Eggers T, Garnett MH, Grayston SJ, Hutcheson KA, Ineson P, Newington JE, Ostle N, Sleep D, Stott A, Jones TH (2007) Carbon dynamics in a model grassland with functionally different soil communities. Funct Ecol 21:690–697

Bradford MA, Davies CA, Frey SD et al (2008) Thermal adaptation of soil microbial respiration to elevated temperature. Ecol Lett 11:1316–1327

Butnor JR, Johnsen KH, Oren R, Katul GG (2003) Reduction of forest floor respiration by fertilization on both carbon dioxide-enriched and reference 17-year-old loblolly pine stands. Glob Change Biol 9:849–861

Catovsky S, Bradford MA, Hector A (2002) Biodiversity and ecosystem productivity: implications for carbon storage. Oikos 97:443–448

Cerri CEP, Coleman K, Jenkinson DS, Bernoux M, Victoria R, Cerri CC (2003) Modeling soil carbon from forest and pasture ecosystems of Amazon, Brazil. Soil Sci Soc Am J 67:1879–1887

Cerri CEP, Paustian K, Bernoux M, Victoria RL, Melillo JM, Cerri CC (2004) Modeling changes in soil organic matter in Amazon forest to pasture conversion with the century model. Glob Change Biol 10:815–832

Certini G (2005) Effects of fire on properties of forest soils: a review. Oecologia 143:1–10

Chung H, Zak DR, Lilleskov EA (2006) Fungal community composition and metabolism under elevated CO_2 and O_3. Oecologia 147:143–154

Cisneros-Dozal LM, Trumbore S, Hanson PJ (2006) Partitioning sources of soil-respired CO_2 and their seasonal variation using a unique radiocarbon tracer. Glob Change Biol 12:194–204

Cleveland CC, Nemergut DR, Schmidt SK, Townsend AR (2007) Increases in soil respiration following labile carbon additions linked to rapid shifts in soil microbial community composition. Biogeochemistry 82:229–240

Czimczik CI, Preston CM, Schmidt MWI, Schulze ED (2003) How surface fire in Siberian Scots pine forests affects soil organic carbon in the forest floor: stocks, molecular structure, and conversion to black carbon (charcoal). Glob Biogeochem Cycles 17:20–21

Czimczik CI, Schmidt MWI, Schulze ED (2005) Effects of increasing fire frequency on black carbon and organic matter in podzols of Siberian Scots pine forests. Eur J Soil Sci 56:417–428

Dai KH, Johnson CE, Driscoll CT (2001) Organic matter chemistry and dynamics in clear-cut and unmanaged hardwood forest ecosystems. Biogeochemistry 54:51–83

Davidson EA, Janssens IA (2006) Temperature sensitivity of soil carbon decomposition and feedbacks to climate change. Nature 440:165–173

Davidson EA, Trumbore SE, Amundson R (2000) Soil warming and organic carbon content. Nature 408: 789–790

Davidson EA, Janssens IA, Lou Y (2006) On the variability of respiration in terrestrial ecosystems: moving beyond Q10. Glob Change Biol 12:154–164

Denman KL, Brasseur G, Chidthaisong A et al (2007) Couplings between changes in the climate system and biogeochemistry. In: Solomon S, Qin D, Manning M et al (eds) Climate change 2007: the physical science basis. Contribution of working group I to the fourth assessment report of the Intergovernmental Panel on Climate Change. Cambridge University Press, Cambridge

Dijkstra FA, Cheng WX (2007) Interactions between soil and tree roots accelerate long-term soil carbon decomposition. Ecol Lett 10:1046–1053

Don A, Kalbitz K (2005) Amounts and degradability of dissolved organic carbon from foliar litter at different decomposition stages. Soil Biol Biochem 37: 2171–2179

Drewitt GB, Black TA, Jassal RS (2005) Using measurements of soil CO_2 efflux and concentrations to infer the depth distribution of CO_2 production in a forest soil. Can J Soil Sci 85:213–221

Eliasson PE, McMurtrie RE, Pepper DA, Stromgren M, Linder S, Agren GI (2005) The response of heterotrophic CO_2 flux to soil warming. Glob Change Biol 11:167–181

Eswaran H, Van den Berg E, Reich P, Kimble J (1995) Global soil carbon resources. In: Lal R, Kimble JM, Levine E (eds) Soils and global change. CRC Press, Boca Raton, pp 27–43

Fahey TJ, Tierney GL, Fitzhugh RD, Wilson GF, Siccama TG (2005) Soil respiration and soil carbon balance in a northern hardwood forest ecosystem. Can J Forest Res 35:244–253

Fierer N, Schimel JP (2002) Effects of drying-rewetting frequency on soil carbon and nitrogen transformations. Soil Biol Biochem 34:777–787

Fierer N, Bradford MA, Jackson RB (2007) Toward an ecological classification of soil bacteria. Ecology 88:1354–1364

Fimmen RL, Richter DD Jr, Vasudevan D, Williams MA, West LT (2008) Rhizogenic Fe-C redox cycling: a hypothetical biogeochemical mechanism that drives crustal weathering in upland soils. Biogeochemistry 87:127–141

Finzi AC, Schlesinger WH (2002) Species control variation in litter decomposition in a pine forest exposed to elevated CO_2. Glob Change Biol 8:1217–1229

Fontaine S, Barot S, Barre P, Bdioui N, Mary B, Rumpel C (2007) Stability of organic carbon in deep soil layers controlled by fresh carbon supply. Nature 450:277–280

Froberg M, Jardine PM, Hanson PJ, Swanston CW, Todd DE, Tarver JR, Garten CT Jr (2007a) Low dissolved organic carbon input from fresh litter to deep mineral soils. Soil Sci Soc Am J 71:347–354

Froberg M, Kleja DB, Hagedorn F (2007b) The contribution of fresh litter to dissolved organic carbon leached from a coniferous forest floor. Eur J Soil Sci 58:108–114

Gallo ME, Lauber CL, Cabaniss SE, Waldrop MP, Sinsabaugh RL, Zak DR (2005) Soil organic matter and litter chemistry response to experimental N deposition in northern temperate deciduous forest ecosystems. Glob Change Biol 11:1514–1521

Giardina CP, Ryan MG (2000) Evidence that decomposition rates of organic carbon in mineral soil do not vary with temperature. Nature 404:858–861

Giardina CP, Ryan MG, Hubbard RM, Binkley D (2001) Tree species and soil textural controls on carbon and nitrogen mineralization rates. Soil Sci Soc Am J 65: 1272–1279

Giesler R, Hogberg MN, Strobel BW, Richter A, Nordgren A, Hogberg P (2007) Production of dissolved organic carbon and low-molecular weight organic acids in soil solution driven by recent tree photosynthate. Biogeochemistry 84:1–12

Gifford RM, Roderick ML (2003) Soil carbon stocks and bulk density: spatial or cumulative mass coordinates as a basis of expression? Glob Change Biol 9:1507–1514
Godbold DL, Hoosbeek MR, Lukac M, Cotrufo MF, Janssens IA, Ceulemans R, Polle A, Velthorst EJ, Scarascia-Mugnozza G, De Angelis P, Miglietta F, Peressotti A (2006) Mycorrhizal hyphal turnover as a dominant process for carbon input into soil organic matter. Plant Soil 281:15–24
Goodale CL, Apps MJ, Birdsey RA, Field CB, Heath LS, Houghton RA, Jenkins JC, Kohlmaier GH, Kurz W, Liu S, Nabuurs GJ, Nilsson S, Shvidenko AZ (2002) Forest carbon sinks in the Northern Hemisphere. Ecol Appl 12:891–899
Grandy AS, Neff JC (2008) Molecular C dynamics downstream: the biochemical decomposition sequence and its impact on soil organic matter structure and function. Sci Total Environ 404:297–307
Groenigen KJ, Six J, Harris D, Kessel CV (2007) Elevated CO_2 does not favor a fungal decomposition pathway. Soil Biol Biochem 39:2168–2172
Gu L, Post WM, King AW (2004) Fast labile carbon turnover obscures sensitivity of heterotrophic respiration from soil to temperature: a model analysis. Glob Biogeochem Cycles 18:GB1022
Guo LB, Gifford RM (2002) Soil carbon stocks and land use change: a meta analysis. Glob Change Biol 8:345–360
Guo D, Li H, Mitchell RJ, Han W, Hendricks JJ, Fahey TJ, Hendrick RL (2008) Fine root heterogeneity by branch order: exploring the discrepancy in root turnover estimates between minirhizotron and carbon isotopic methods. New Phytol 177:443–456
Hagedorn F, Machwitz M (2007) Controls on dissolved organic matter leaching from forest litter grown under elevated atmospheric CO_2. Soil Biol Biochem 39: 1759–1769
Hahn V, Hogberg P, Buchmann N (2006) ^{14}C - A tool for separation of autotrophic and heterotrophic soil respiration. Glob Change Biol 12:972–982
Hamilton JG, DeLucia EH, George K, Naidu SL, Finzi AC, Schlesinger WH (2002) Forest carbon balance under elevated CO_2. Oecologia 131:250–260
Hanson PJ, Edwards NT, Garten CT, Andrews JA (2000) Separating root and soil microbial contributions to soil respiration: a review of methods and observations. Biogeochemistry 48:115–146
Hanson PJ, Amthor JS, Wullschleger SD, Wilson KB, Grant RF, Hartley A, Hui D, Hunt ER Jr, Johnson DW, Kimball JS, King AW, Luo Y, McNulty SG, Sun G, Thornton PE, Wang S, Williams M, Baldocchi DD, Cushman RM (2004) Oak forest carbon and water simulations: model intercomparisons and evaluations against independent data. Ecol Monogr 74:443–489
Hanson PJ, Swanston CW, Garten CT Jr, Todd DE, Trumbore SE (2005) Reconciling change in Oi-horizon carbon-14 with mass loss for an oak forest. Soil Sci Soc Am J 69:1492–1502
Heinemeyer A, Hartley IP, Evans SP, De la Fuente JAC, Ineson P (2007) Forest soil CO_2 flux: uncovering the contribution and environmental responses of ectomycorrhizas. Glob Change Biol 13:1786–1797
Hibbard KA, Law BE, Reichstein M, Sulzman J (2005) An analysis of soil respiration across northern hemisphere temperate ecosystems. Biogeochemistry 73:29–70
Hogberg MN, Hogberg P (2002) Extramatrical ectomycorrhizal mycelium contributes one-third of microbial biomass and produces, together with associated roots, half the dissolved organic carbon in a forest soil. New Phytol 154:791–795
Högberg P, Read DJ (2006) Towards a more plant physiological perspective on soil ecology. Trends Ecol Evol 21:548–554
Hogberg P, Nordgren A, Hogberg MN, Ottosson-Lofvenius M, Bhupinderpal S, Olsson P, Linder S (2005) Fractional contributions by autotrophic and heterotrophic respiration to soil-surface CO_2 efflux in boreal forests. SEB Exp Biol Ser 2005:251–267
Hogberg MN, Hogberg P, Myrold DD (2007) Is microbial community composition in boreal forest soils determined by pH, C-to-N ratio, the trees, or all three? Oecologia 150:590–601
Hyvonen R, Agren GI (2001) Decomposer invasion rate, decomposer growth rate, and substrate chemical quality: how they influence soil organic matter turnover. Can J Forest Res 31:1594–1601
IPCC (2007) Forestry. In: Climate change 2007: mitigation. Contribution of Working Group III to the Fourth Assessment Report of the Intergovernmental Panel on Climate Change
Jackson RB, Schenk HJ, Jobbagy EG, Canadell J, Colello GD, Dickinson RE, Field CB, Friedlingstein P, Heimann M, Hibbard K, Kicklighter DW, Kleidon A, Neilson RP, Parton WJ, Sala OE, Sykes MT (2000) Belowground consequences of vegetation change and their treatment in models. Ecol Appl 10:470–483
Jandl R, Lindner M, Vesterdal L, Bauwens B, Baritz R, Hagedorn F, Johnson DW, Minkkinen K, Byrne KA (2007) How strongly can forest management influence soil carbon sequestration? Geoderma 137:253–268
Janssens IA, Pilegaard K (2003) Large seasonal changes in Q10 of soil respiration in a beech forest. Glob Change Biol 9:911–918
Jarvis P, Rey A, Petsikos C, Wingate L, Rayment M, Pereira J, Banza J, David J, Miglietta F, Borghetti M, Manca G, Valentini R (2007) Drying and wetting of Mediterranean soils stimulates decomposition and carbon dioxide emission: the "Birch effect". Tree Physiol 27:929–940
Jassal RS, Black TA, Drewitt GB, Novak MD, Gaumont-Guay D, Nesic Z (2004) A model of the production and transport of CO_2 in soil: predicting soil CO_2 concentrations and CO_2 efflux from a forest floor. Agric Forest Meteorol 124:219–236
Jobbagy EG, Jackson RB (2000) The vertical distribution of soil organic carbon and its relation to climate and vegetation. Ecol Appl 10:423–436
Johnson DW, Curtis PS (2001) Effects of forest management on soil C and N storage: meta analysis. Forest Ecol Manag 140:227–238
Johnson DW, Knoepp JD, Swank WT, Shan J, Morris LA, Van Lear DH, Kapeluck PR (2002) Effects of forest management on soil carbon: results of some long-term resampling studies. Environ Pollut 116:S201–S208

Jones TH, Bradford MA (2001) Assessing the functional implications of soil biodiversity in ecosystems. Ecol Res 16:845–858

Kaiser K, Guggenberger G (2000) The role of DOM sorption to mineral surfaces in the preservation of organic matter in soils. Org Geochem 31:711–725

Kalbitz K, Solinger S, Park JH, Michalzik B, Matzner E (2000) Controls on the dynamics of dissolved organic matter in soils: a review. Soil Sci 165:277–304

Kalbitz K, Schwesig D, Rethemeyer J, Matzner E (2005) Stabilization of dissolved organic matter by sorption to the mineral soil. Soil Biol Biochem 37:1319–1331

Kalbitz K, Kaiser K, Bargholz J, Dardenne P (2006) Lignin degradation controls the production of dissolved organic matter in decomposing foliar litter. Eur J Soil Sci 57:504–516

King JS, Pregitzer KS, Zak DR, Kubiske ME, Ashby JA, Holmes WE (2001) Chemistry and decomposition of litter from *Populus tremuloides* michaux grown at elevated atmospheric CO_2 and varying N availability. Glob Change Biol 7:65–74

King JS, Hanson PJ, Bernhardt E, Deangelis P, Norby RJ, Pregitzer KS (2004) A multiyear synthesis of soil respiration responses to elevated atmospheric CO_2 from four forest FACE experiments. Glob Change Biol 10:1027–1042

King JS, Pregitzer KS, Zak DR, Holmes WE, Schmidt K (2005) Fine root chemistry and decomposition in model communities of north-temperate tree species show little response to elevated atmospheric CO_2 and varying soil resource availability. Oecologia 146:318–328

Knorr M, Frey SD, Curtis PS (2005a) Nitrogen additions and litter decomposition: a meta-analysis. Ecology 86:3252–3257

Knorr W, Prentice IC, House JI, Holland EA (2005b) Long-term sensitivity of soil carbon turnover to warming. Nature 433:298–301

Langley JA, Hungate BA (2003) Mycorrhizal controls on belowground litter quality. Ecology 84:2302–2312

Li X, Fisk MC, Fahey TJ, Bohlen PJ (2002) Influence of earthworm invasion on soil microbial biomass and activity in a northern hardwood forest. Soil Biol Biochem 34:1929–1937

Lichter J, Barron SH, Bevacqua CE, Finzi AC, Irving KF, Stemmler EA, Schlesinger WH (2005) Soil carbon sequestration and turnover in a pine forest after six years of atmospheric CO_2 enrichment. Ecology 86:1835–1847

Liski J, Pussinen A, Pingoud K, Makipaa R, Karjalainen T (2001) Which rotation length is favourable to carbon sequestration? Can J Forest Res 31:2004–2013

Liski J, Korotkov AV, Prins CFL, Karjalainen T, Victor DG, Kauppi PE (2003) Increased carbon sink in temperate and boreal forests. Climatic Change 61:89–99

Liski J, Palosuo T, Peltoniemi M, Sievanen R (2005) Carbon and decomposition model yasso for forest soils. Ecol Model 189:168–182

Loranger GI, Pregitzer KS, King JS (2004) Elevated CO_2 and O3 concentrations differentially affect selected groups of the fauna in temperate forest soils. Soil Biol Biochem 36:1521–1524

Lorenz K, Lal R, Preston CM, Nierop KGJ (2007) Strengthening the soil organic carbon pool by increasing contributions from recalcitrant aliphatic bio(macro) molecules. Geoderma 142:1–10

Loya WM, Pregitzer KS, Karberg NJ, King JS, Giardina CP (2003) Reduction of soil carbon formation by tropospheric ozone under increased carbon dioxide levels. Nature 425:705–707

Luo Y, Wan S, Hui D, Wallace LL (2001a) Acclimatization of soil respiration to warming in a tall grass prairie. Nature 413:622–625

Luo Y, Wu L, Andrews JA, White L, Matamala R, Schafer K, Schlesinger WH (2001b) Elevated CO_2 differentiates ecosystem carbon processes: deconvolution analysis of Duke Forest FACE data. Ecol Monogr 71: 357–376

Lutzow MV, Kogel-Knabner I, Ekschmitt K, Matzner E, Guggenberger G, Marschner B, Flessa H (2006) Stabilization of organic matter in temperate soils: mechanisms and their relevance under different soil conditions - a review. Eur J Soil Sci 57:426–445

Magill AH, Aber JD (2000) Dissolved organic carbon and nitrogen relationships in forest litter as affected by nitrogen deposition. Soil Biol Biochem 32:603–613

Manning P, Saunders M, Bardgett RD, Bonkowski M, Bradford MA, Ellis RJ, Kandeler E, Marhan S, Tscherko D (2008) Direct and indirect effects of nitrogen deposition on litter decomposition. Soil Biol Biochem 40:688–698

Marhan S, Scheu S (2006) Mixing of different mineral soil layers by endogeic earthworms affects carbon and nitrogen mineralization. Biol Fert Soils 42:308–314

Martin JL, Gower ST, Plaut J, Holmes B (2005) Carbon pools in a boreal mixedwood logging chronosequence. Glob Change Biol 11:1883–1894

Melillo JM, Steudler PA, Aber JD, Newkirk K, Lux H, Bowles FP, Catricala C, Magill A, Ahrens T, Morrisseau S (2002) Soil warming and carbon-cycle feedbacks to the climate system. Science 298:2173–2176

Michalzik B, Kalbitz K, Park JH, Solinger S, Matzner E (2001) Fluxes and concentrations of dissolved organic carbon and nitrogen - a synthesis for temperate forests. Biogeochemistry 52:173–205

Monson RK, Lipson DL, Burns SP, Turnipseed AA, Delany AC, Williams MW, Schmidt SK (2006) Winter forest soil respiration controlled by climate and microbial community composition. Nature 439:711–714

Montano NM, Garcia-Oliva F, Jaramillo VJ (2007) Dissolved organic carbon affects soil microbial activity and nitrogen dynamics in a Mexican tropical deciduous forest. Plant Soil 295:265–277

Morris SJ, Bohm S, Haile-Mariam S, Paul EA (2007) Evaluation of carbon accrual in afforested agricultural soils. Glob Change Biol 13:1145–1156

Neff JC, Harden JW, Gleixner G (2005) Fire effects on soil organic matter content, composition, and nutrients in boreal interior Alaska. Can J Forest Res 35: 2178–2187

Park JH, Matzner E (2003) Controls on the release of dissolved organic carbon and nitrogen from a deciduous forest floor investigated by manipulations of aboveground litter inputs and water flux. Biogeochemistry 66:265–286

Paul KI, Polglase PJ, Nyakuengama JG, Khanna PK (2002) Change in soil carbon following afforestation. Forest Ecol Manag 168:241–257

Pavelka M, Acosta M, Marek MV, Kutsch W, Janous D (2007) Dependence of the Q10 values on the depth of the soil temperature measuring point. Plant Soil 292: 171–179

Peltoniemi M, Thurig E, Ogle S, Palosuo T, Schrumpf M, Wutzler T, Butterbach-Bahl K, Chertov O, Komarov A, Mikhailov A, Gardenas A, Perry C, Liski J, Smith P, Makipaa R (2007) Models in country scale carbon accounting of forest soils. Silva Fenn 41:575–602

Phillips RL, Zak DR, Holmes WE, White DC (2002) Microbial community composition and function beneath temperate trees exposed to elevated atmospheric carbon dioxide and ozone. Oecologia 131: 236–244

Phillips RL, Finzi AC, Bernhardt ES (2011) Enhanced root exudation induces microbial feedbacks to N cycling in a pine forest under long-term CO_2 fumigation. Ecol Lett 14:187–194

Post J, Hattermann FF, Krysanova V, Suckow F (2008) Parameter and input data uncertainty estimation for the assessment of long-term soil organic carbon dynamics. Environ Modell Softw 23:125–138

Pregitzer K, Loya W, Kubiske M, Zak D (2006) Soil respiration in northern forests exposed to elevated atmospheric carbon dioxide and ozone. Oecologia 148: 503–516

Pregitzer KS, Burton AJ, Zak DR, Talhelm AF (2008) Simulated chronic nitrogen deposition increases carbon storage in Northern Temperate forests. Glob Change Biol 14:142–153

Pritchard SG, Rogers HH, Davis MA, Van Santen E, Prior SA, Schlesinger WH (2001) The influence of elevated atmospheric CO_2 on fine root dynamics in an intact temperate forest. Glob Change Biol 7:829–837

Pritchard SG, Strand AE, McCormack ML, Davis MA, Finzi AC, Jackson RB, Matamala R, Rogers HH, Oren R (2008) Fine root dynamics in a loblolly pine forest are influenced by free-air- CO_2-enrichment: a six-year-minirhizotron study. Glob Change Biol 14:588–602

Quideau SA, Chadwick OA, Trumbore SE, Johnson-Maynard JL, Graham RC, Anderson MA (2001) Vegetation control on soil organic matter dynamics. Org Geochem 32:247–252

Rasmussen C, Southard RJ, Horwath WR (2006) Mineral control of organic carbon mineralization in a range of temperate conifer forest soils. Glob Change Biol 12: 834–847

Reichstein M, Rey A, Freibauer A, Tenhunen J, Valentini R, Banza J, Casals P, Cheng Y, Grunzweig JM, Irvine J, Joffre R, Law BE, Loustau D, Miglietta F, Oechel W, Ourcival JM, Pereira JS, Peressotti A, Ponti F, Qi Y, Rambal S, Rayment M, Romanya J, Rossi F, Tedeschi V, Tirone G, Xu M, Yakir D (2003) Modeling temporal and large-scale spatial variability of soil respiration from soil water availability, temperature and vegetation productivity indices. Glob Biogeochem Cycles 17:15, 11

Richter DD, Hofmockel M, Callaham MA Jr, Powlson DS, Smith P (2007) Long-term soil experiments: keys to managing Earth's rapidly chancing ecosystems. Soil Sci Soc Am J 71:266–279

Ryan MG, Law BE (2005) Interpreting, measuring, and modeling soil respiration. Biogeochemistry 73:3–27

Salimon CI, Davidson EA, Victoria RL, Melo AWF (2004) CO_2 flux from soil in pastures and forests in southwestern Amazonia. Glob Change Biol 10: 833–843

Sampson DA, Janssens IA, Yuste JC, Ceulemans R (2007) Basal rates of soil respiration are correlated with photosynthesis in a mixed temperate forest. Glob Change Biol 13:2008–2017

Savage KE, Davidson EA (2003) A comparison of manual and automated systems for soil CO_2 flux measurements: trade-offs between spatial and temporal resolution. J Exp Bot 54:891–899

Scheel T, Dorfler C, Kalbitz K (2007) Precipitation of dissolved organic matter by aluminum stabilizes carbon in acidic forest soils. Soil Sci Soc Am J 71:64–74

Schlesinger WH, Lichter J (2001) Limited carbon storage in soil and litter of experimental forest plots under increased atmospheric CO_2. Nature 411:466–469

Schulze ED, Wirth C, Heimann M (2000) Managing forests after Kyoto. Science 289:2058–2059

Schuur EAG, Trumbore SE (2006) Partitioning sources of soil respiration in boreal black spruce forest using radiocarbon. Glob Change Biol 12:165–176

Schwesig D, Kalbitz K, Matzner E (2003) Effects of aluminium on the mineralization of dissolved organic carbon derived from forest floors. Eur J Soil Sci 54: 311–322

Sinsabaugh RL, Zak DR, Gallo M, Lauber C, Amonette R (2004) Nitrogen deposition and dissolved organic carbon production in northern temperate forests. Soil Biol Biochem 36:1509–1515

Sinsabaugh RL, Gallo ME, Lauber C, Waldrop MP, Zak DR (2005) Extracellular enzyme activities and soil organic matter dynamics for northern hardwood forests receiving simulated nitrogen deposition. Biogeochemistry 75:201–215

Stoy PC, Palmroth S, Oishi AC, Siqueira MB, Juang JY, Novick KA, Ward EJ, Katul GG, Oren R (2007) Are ecosystem carbon inputs and outputs coupled at short time scales? a case study from adjacent pine and hardwood forests using impulse-response analysis. Plant Cell Environ 30:700–710

Strand AE, Pritchard SG, McCormack ML, Davis MA, Oren R (2008) Irreconcilable differences: fine-root life spans and soil carbon persistence. Science 319: 456–458

Strickland MS, Rousk J (2010) Considering fungal: bacterial dominance in soils – methods, controls, and ecosystem implications. Soil Biol Biochem 42:1385–1395

Strickland MS, Devore JL, Maerz JC, Bradford MA (2010) Grass invasion of a hardwood forest is associated with declines in belowground carbon pools. Glob Change Biol 16:1338–1350

Subke JA, Inglima I, Cotrufo MF (2006) Trends and methodological impacts in soil CO_2 efflux partitioning: a metaanalytical review. Glob Change Biol 12: 921–943

Suwa M, Katul GG, Oren R, Andrews J, Pippen J, Mace A, Schlesinger WH (2004) Impact of elevated atmospheric CO_2 on forest floor respiration in a temperate pine forest. Glob Biogeochem Cycles 18:GB2013

Tan ZX, Lal R, Smeck NE, Calhoun FG (2004) Relationships between surface soil organic carbon pool and site variables. Geoderma 121:187–195

Telles ECC, de Camargo PB, Martinelli LA, Trumbore SE, da Costa ES, Santos J, Higuchi N, Oliveira RC Jr (2003) Influence of soil texture on carbon dynamics and storage potential in tropical forest soils of Amazonia. Glob Biogeochem Cycles 17:9, 1

Thornley JHM, Cannell MGR (2000) Managing forests for wood yield and carbon storage: a theoretical study. Tree Physiol 20:477–484

Throop HL, Archer SR (2008) Shrub (*Prosopis velutina*) encroachment in a semidesert grassland: spatial-temporal changes in soil organic carbon and nitrogen pools. Glob Change Biol 14:2420–2431

Thuille A, Schulze ED (2006) Carbon dynamics in successional and afforested spruce stands in Thuringia and the Alps. Glob Change Biol 12:325–342

Thuille A, Buchmann N, Schulze ED (2000) Carbon stocks and soil respiration rates during deforestation, grassland use and subsequent Norway spruce afforestation in the Southern Alps, Italy. Tree Physiol 20:849–857

Tierney GL, Fahey TJ (2002) Fine root turnover in a northern hardwood forest: a direct comparison of the radiocarbon and minirhizotron methods. Can J Forest Res 32:1692–1697

Trumbore S (2006) Carbon respired by terrestrial ecosystems - recent progress and challenges. Glob Change Biol 12:141–153

Trumbore SE, Gaudinski JB (2003) The secret lives of roots. Science 302:1344–1345

Uselman SM, Qualls RG, Lilienfein J (2007) Contribution of root vs. leaf litter to dissolved organic carbon leaching through soil. Soil Sci Soc Am J 71:1555–1563

Ussiri DA, Johnson CE (2007) Organic matter composition and dynamics in a northern hardwood forest ecosystem 15 years after clear-cutting. Forest Ecol Manag 240:131–142

Van Hees PAW, Jones DL, Finlay R, Godbold DL, Lundstrom US (2005) The carbon we do not see - the impact of low molecular weight compounds on carbon dynamics and respiration in forest soils: a review. Soil Biol Biochem 37:1–13

Vogt KA, Vogt DJ, Bloomfield J (1998) Analysis of some direct and indirect methods for estimating root biomass and production of forests at an ecosystem level. Plant Soil 200:71–89

Waldrop MP, Zak DR, Sinsabaugh RL (2004) Microbial community response to nitrogen deposition in northern forest ecosystems. Soil Biol Biochem 36:1443–1451

Wan S, Norby RJ, Pregitzer KS, Ledford J, O'Neill EG (2004) CO_2 enrichment and warming of the atmosphere enhance both productivity and mortality of maple tree fine roots. New Phytol 162:437–446

Wayson CA, Randolph JC, Hanson PJ, Grimmond CSB, Schmid HP (2006) Comparison of soil respiration methods in a mid-latitude deciduous forest. Biogeochemistry 80:173–189

Woodbury PB, Heath LS, Smith JE (2007) Effects of land use change on soil carbon cycling in the conterminous United States from 1900 to 2050. Glob Biogeochem Cycles 21:GB3006

Yanai RD, Stehman SV, Arthur MA, Prescott CE, Friedland AJ, Siccama TG, Binkley D (2003) Detecting change in forest floor carbon. Soil Sci Soc Am J 67: 1583–1593

Yang Y, Chen G, Guo J, Xie J, Wang X (2007) Soil respiration and carbon balance in a subtropical native forest and two managed plantations. Plant Ecol 193:71–84

Zinn YL, Lal R, Bigham JM, Resck DVS (2007) Edaphic controls on soil organic carbon retention in the Brazilian Cerrado: texture and mineralogy. Soil Sci Soc Am J 71:1204–1214

The Physiological Ecology of Carbon Science in Forest Stands

3

Kristofer R. Covey, Joseph Orefice, and Xuhui Lee

Executive Summary

In order to better understand the ways in which future forests will change and be changed by shifting climates, it is necessary to understand the underlying drivers of forest development and the ways these drivers are affected by changes in atmospheric carbon dioxide concentrations, temperature, precipitation, and nutrient levels. Successional forces lead to somewhat predictable changes in forest stands throughout the world. These changes can lead to corresponding shifts in the dynamics of carbon uptake, storage, and release.

Many studies have attempted to elucidate the effects of changing climate conditions on forest ecosystem dynamics; however, the complexity of forest systems, long time horizons, and high costs associated with large-scale research have limited the ability of scientists to make reliable predictions about future changes in forest carbon flux at the global scale. Free Air Carbon Exchange (FACE) experiments are suggesting that forest net primary productivity, and thus carbon uptake, usually increases when atmospheric carbon dioxide levels increase, likely due to factors such as increased nitrogen use efficiency and competitive advantages of shade tolerant species. Experiments dealing with drought and temperature change are providing evidence that water availability, may be the most important factor driving forest carbon dynamics. Forest ecosystem experiments, such as FACE programs, have not been operating long enough to predict long term responses of forest ecosystems to increases in carbon dioxide. The expense and time constraints of field experiments force scientists to rely on multifactor models (the majority of which account for five or fewer variables) leading to results based on large assumptions.

If predictions are made regarding stand level carbon within forest ecosystems, it is important to have an understanding of what scientific research has or has not established. Key findings of this review summarize what we do and do not know about stand dynamics and carbon.

What We Do Know About Stand Dynamics and Carbon Assimilation

- Forests have relatively predictable stages of development that have been termed initiation, stem exclusion, understory initiation, and old growth.
- The nature of type, scale, and frequency of disturbances and their effects on forests are well documented and their effects on the nature of the origin of new or released regeneration well understood.

K.R. Covey (✉) • X. Lee
Yale School of Forestry and Environmental Studies, Prospect Street 360, 06511 New Haven, CT, USA

J. Orefice
Paul Smiths College, Paul Smiths, NY, USA

M.S. Ashton et al. (eds.), *Managing Forest Carbon in a Changing Climate*, DOI 10.1007/978-94-007-2232-3_3, © Springer Science+Business Media B.V. 2012

- Most studies support the notion that the stem exclusion stage is a period of high carbon assimilation, water uptake, and nutrient acquisition.
- Recent studies are showing that old growth forests are not just storing carbon,but are also sequestering significant amounts – particularly in large tropical basins such as the Congo and the Amazon.
- Free Air Carbon Exchange Experiments (FACE) have provided insights into our understanding of the physiological and stand level responses (± feedbacks) to elevated carbon dioxide over short periods of time (15 years). Stands in the stem exclusion stage are expected to increase sequestration, with increase in water use and nutrient use efficiencies, and a potential to favor shade tolerant species.
- Stand level rainfall exclusion and addition experiments have provided insight into carbon reallocation, carbon respiration and storage processes, drought aversion and avoidance adaptations, and ± feedbacks with other soil resources (e.g. soil fertility). Results suggest that timing of drought (growing versus non-growing season) and species composition change are two factors to consider.

What We Do Not Know About Stand Dynamics and Carbon Assimilation

- Although we understand the stages of stand development, there is considerable unpredictability in the actual nature of species composition, stocking, and rates of development because of numerous positive and negative feedbacks that make precise understanding of future stand development difficult.
- Carbon stocks and fluxes across and within different forest biomes – particularly in the tropics – have not been well documented.
- While informative, FACE experiments are limited to temperate and boreal stands that are mostly in the stem exclusion stage – only some of this information can be applied to tropical regions and other developmental stages.
- More studies are needed that investigate the multiple interactions of limiting and non-limiting resources of soil nutrients, soil water availability, and temperature fluctuations in elevated carbon dioxide environments.

1 Introduction

Understanding how future forests will affect and be affected by changing climates requires an understanding of the principles governing the development of forests over time. In an effort to provide a comprehensive understanding of stand level changes in forest carbon with relation to climatic conditions, we present a synthesis of the literature.

Although there are many forest types composed of seemingly infinite combinations of species, similarities in stand development produce analogous stand structures in most of the world's forests (Oliver 1992). Successional processes alter both forest structure and accompanying ecological processes in predictable ways (Cowles 1911; Odum 1969, 1971; Shugart and West 1980; Bormann and Likens 1979; Oliver 1981; Hibbs 1983; Glenn-Lewin et al. 1992; Oliver and Larson 1996; Barnes et al. 1998) and regulate changes in forest biomass (Odum 1969) in systems as seemingly disparate as the tropical rainforests of the Amazon and the boreal forests of the Canadian Shield (Oliver 1992). The amount of carbon within a forest stand is a factor of both forest structure and competition between individuals.

In this paper we first describe the concept of stand dynamics, the stages of stand development, and their relevance to our understanding of carbon assimilation and storage in forests. We then describe the physiological processes of photosynthesis and carbon dioxide assimilation, and the effects of other limiting resources on this process (soil water availability, soil nutrients). We then describe the experimental approaches used to manipulate resources (soil, water, air) and monitor such effects on stand developmental and physiological processes – especially carbon assimilation and storage. In this section we describe the Free Air Carbon Exchange Experiments (FACE) with a review of the results so far and their limitations. We also describe several other stand-scale experiments that have manipulated precipitation – another key climate effect on forests. We then conclude with summary recommendations on further work that is needed.

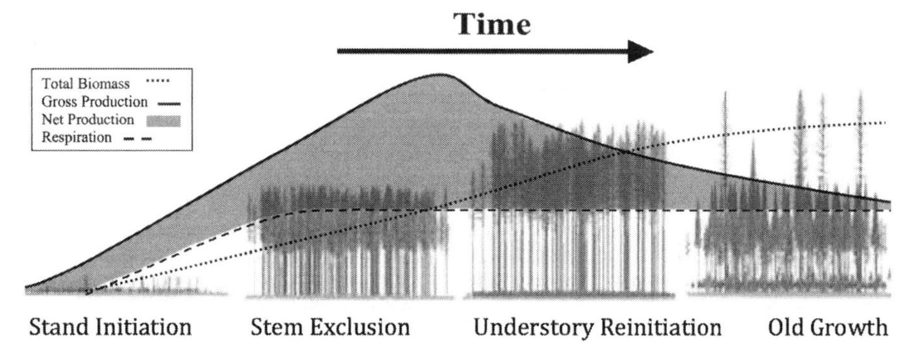

Fig. 3.1 Stand dynamics, respiration, production, and total biomass. As stands age they move through predictable stages of development, with predictable consequences for production. Stand level carbon stocks in the form of biomass and coarse woody material increase as a stand progresses through successional stages. The rate of increase in biomass is also not constant over the life of a stand; early stages of stand development have low rates of biomass accumulation due to trees re-establishing themselves on the site. During stand initiation, net production steadily increases and peaks during the stem exclusion stage. Different stands will move through these stages at different rates influenced by species composition, climate, disturbance and other site factors. Stands with the same species composition growing on favorable sites will not only accumulate carbon at a higher maximum rate but they will also reach this maximum rate sooner than stands on poor sites (Adapted from Oliver and Larson 1996)

2 The Concept of Stand Dynamics

Relatively predictable changes in forest stand structures over time occur in continuous sequential stages (Bormann and Likens 1979; Oliver 1981; Peet and Christensen 1987; Oliver and Larson 1996; Franklin et al. 2002) which various authors have described using differing terminology. However, they all outline a progressive shift in community dynamics from colonization to competition to peak growth and then slow decay (Fig. 3.1).

The four stages of stand dynamics as described by Oliver and Larson (1996) are:

1. **Stand initiation** takes place following disturbance and is usually characterized by large numbers of young trees growing from seed or sprouts to rapidly occupy newly available growing space. This period of invasion is critical in determining the trajectory of a developing stand. During this stage the environment in the stand transforms relatively quickly as the influence of re-vegetation, site parameters, disturbance type, and the return to biogeochemical balance all shape the rapidly developing stand (Bormann and Likens 1979; Canham and Marks 1985).
2. **Stem exclusion** is the period of intense competition for resources (e.g. light, soil water, nutrients) and for physical space, characterized by high rates of mortality and rapid assimilation of nutrients and carbon. Maximum assimilation rates of carbon and biomass occur during this stage.
3. **Understory initiation** begins as the survivors of stem exclusion grow older and weaken in resource acquisition. The remaining trees are not able to fully utilize the released growing space and new cohorts establish in the understory.
4. **Old growth** follows as the overstory trees of the initiating cohort die, breaking the uniformity of the canopy, allowing for their slow replacement by the new cohorts established during and after understory initiation. This process leads to the characteristic presence of multiple cohorts. This stage has foliage distributed throughout the vertical layers of the canopy with "horizontal heterogeneity, often evident as canopy gaps and dense reproduction patches" (Franklin and Van Pelt 2004).

Forests move through these successional stages at varying rates and along a multitude of possible trajectories depending on stem density (competition), species composition and available resources (site factors), climate, disturbance patterns and human activity.

Growth and uptake of carbon are quantified in different ways. When comparing growth activity in different ecosystems and stand structures it is important to use the same measurement and methodological approach. Measures of net primary production (NPP) and net ecosystem production (NEP) are used to quantify ecosystem uptake of carbon. NPP is the overall net uptake of carbon by primary producers (organisms that photosynthesize) in an ecosystem per unit of time. NEP is the overall net uptake or release of carbon by an ecosystem per unit of time. Ecosystems are often stratified and NPP and NEP can refer to all or just part of an ecosystem. Biomass is another way of monitoring change in forest stands; it is the mass of organic matter in an ecosystem. Biomass can be stratified into many groups including, but not limited to: living biomass, woody biomass, and above and below ground biomass. The importance of understanding what measure is being used to quantify carbon, and for what part of an ecosystem, cannot be stressed enough, because confusing these will lead to false conclusions.

3 The Physiology of Trees and Forest Stands

Trees and other vegetation can uptake and sequester atmospheric carbon and draw up moisture from the soil, transpiring it to the atmosphere, profoundly influencing climate (Chapin et al. 2002) (Fig. 3.2).

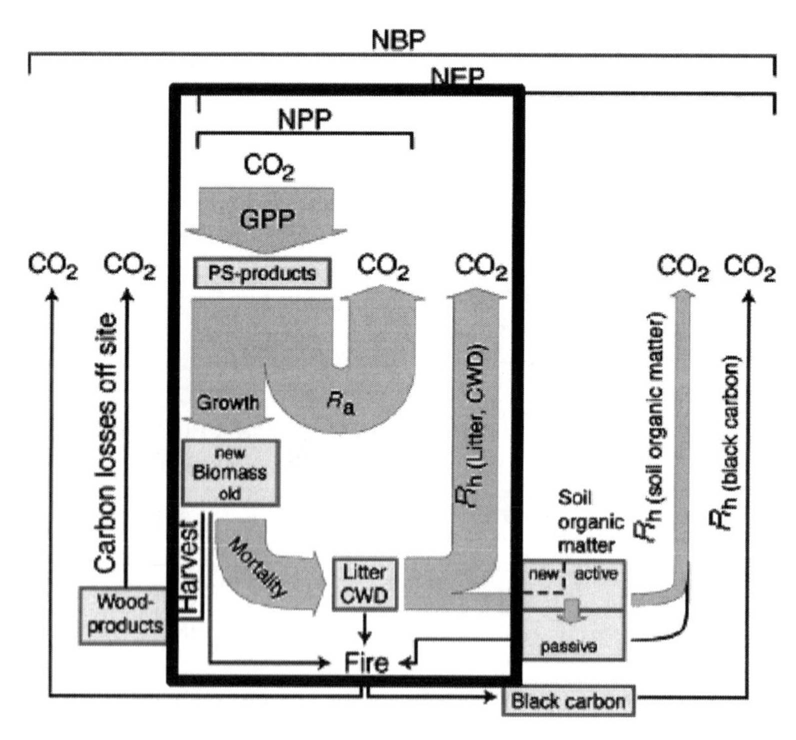

Fig. 3.2 Forest carbon flux. A conceptual diagram illustrating the limits of the above-ground carbon cycle. Arrows represent fluxes and boxes indicate pools; the size of each indicates the relative rate of flux or size of pool (see Chap. 2 for complete diagram). Half of all carbon dioxide absorbed by forests is used for respiration maintenance; the remainder is stored as biomass. Branch, leaf and root turnover, eventual tree death, and inevitable decomposition transfer carbon back to the atmosphere or into the soil carbon pool (Adapted from Schulze et al. 2000)

Understanding the basic physiological processes of photosynthesis and transpiration is essential if reliable assumptions are to be made about the effects of elevated CO_2 on future forests (Long 1998). Individual trees share similarities at both the micro and macro level in physiology, morphology, requirements for survival and patterns of photosynthate allocation. Although they are more complex in both structure and function than other plants, they are physiologically similar (Oliver 1992; Oliver and Larson 1996).

All carbon allocation in plants – and subsequently trees – can be divided into three categories:

- Respiration, both for the ongoing maintenance of tissues and for the synthesis of compounds used in the growth of new tissues.
- Vegetative growth of roots, stems, and leaves.
- Reproductive growth used to produce flowers, cones, fruit, and seeds.

The relative carbon allocation priorities vary from species to species, with age, by stage of stand development, and with biotic, edaphic, climatic, and physiographic site factors (Grime 1977; Keyes and Grier 1981; Tritton and Hornbeck 1982; Dickson 1986; Ericsson et al. 1996; Lacointe 2000; Gower et al. 2001; Larcher 2003; Lockhart et al. 2008). The complexities inherent in this shifting priority have been known to plant physiologists for some time and have been well demonstrated in trees. Factors such as the availability of light, water and nutrients, atmospheric CO_2 concentrations, or variations in temperature can lead to significant changes in both the proportional allocation and the total rate of fixation of carbon in trees (Aber et al. 1985; Chapin et al. 1987; Steeves and Sussex 1989). The effect of any one or any combination of these factors on carbon uptake is predictable (Farrar 1999), but the magnitude of the effect varies greatly both between and within species (Raghavendra 1991). For example, in a temperate forest in North Carolina, winged elm (*Ulmus alata*) regeneration had a 21% relative increase in growth under elevated carbon dioxide while black locust (*Robinia pseudoacacia*) had a 230% relative increase under the same amount of carbon dioxide elevation (Mohan et al. 2007). While predicting how environmental changes may affect a single tree is challenging on its own, estimating the effects of similar changes at the stand or landscape scale is extremely difficult (Lavigne 1992; Schulze 2000).

Plants take in the CO_2 necessary for photosynthesis by opening leaf stomata, allowing access to the gas exchange sites located in the mesophyll (Larcher 2003). In doing so, they also transpire moisture into the atmosphere. Plants, therefore, face a delicate balance between the loss of water – forcing the expenditure of energy to replace it – and the need for the CO_2 necessary to fuel photosynthesis. The demand for water can be extreme, in some cases as much 400 units of water loss for every unit of CO_2 gained (Chapin et al. 2002). An increased amount of atmospheric CO_2 allows for more efficient uptake of CO_2 and thus lower rates of stomatal water conductance at the leaf and individual organism levels (Curtis 1996; Farnsworth et al. 1996; Urban 2003; Herrick et al. 2004). However, just as it is both difficult and unreliable to extrapolate changes in carbon uptake from the single tree to the forest, it is also difficult and unreliable to predict changes in transpiration at larger scales (Long 1998).

4 Forest Carbon and Stand Dynamics

4.1 Stand Development and Carbon

Forest stands are dynamic components of the ecosystem in which carbon flux changes with size, age and species composition of trees. Although different species will influence stand development and carbon flux, general patterns exist for forest stands throughout the world (Oliver 1992). Determining the developmental stage of a forest stand provides insight into the amount and nature of carbon flux, as different structural and age conditions influence photosynthetic rates and decomposition activity.

An important rule of thumb is that carbon allocation changes as tree size and stand structure increases. For example, seedlings allocate much of their carbon to shoot and root growth. One study found that paper birch (*Betula papyrifera*) and Douglas fir (*Pseudotsuga menziesii*) seedlings allocated 49% and 41% of absorbed isotopic carbon to their roots, respectively, and over 55% of this carbon was allocated to fine roots (Simard et al. 1997). Once seedlings have produced a sufficient root and foliar system, they are able to allocate carbon to stem height and then to diameter increment. At Hubbard Brook Experimental Forest in New Hampshire, the sapling stage of trees was found to contain the highest percent of dry weight biomass in stem or bole, and in another study, five-year-old loblolly pine (*Pinus taeda*) saplings were found to have allocated the majority of their carbon to stem growth (Whittaker et al. 1974; Retzlaff et al. 2001). As a tree matures, the percentage of total biomass held in stem and bole wood diminishes as the relative amount of biomass in woody branches increases (Whittaker et al. 1974). The amount of time it takes for a seedling to begin rapid height growth is dependent on stand species composition, temperature, light, soil and moisture conditions. For example, eastern hemlock (*Tsuga canadensis*) advance regeneration is able to survive without significant growth in the understory for decades until overstory conditions are right for it to continue height and diameter growth, in contrast to species such as eastern white pine (*Pinus strobus*) which will have high mortality at low light intensities (Burns and Honkala 1990).

As saplings develop into poles and then mature trees, increasingly large quantities of carbon are stored in the stem. This process has been demonstrated by a study in which entire eastern white pine trees in Ontario, Canada were destructively sampled; researchers found that mature 65 year old trees contained 69% of their total biomass in their stem while only 25% of total tree biomass was in the stems of 2 year old trees (Peichl and Arain 2007). Mature trees will eventually sequester less and less carbon as they become larger due to physical growth limitations such as water stress (Whittaker et al. 1974). Carbon is constantly lost over the life of a tree due to respiration and leaf, root, branch, and bark senescence; it may, however, be partially retained in the stand as coarse woody debris, leaf and branch litter, and soil organic matter.

4.2 Stand Level Carbon Stocks

Stand level carbon stocks in the form of biomass and coarse woody material increase as a stand progresses through succession stages (Odum 1969; Whittaker et al. 1974; Acker et al. 2000; Taylor et al. 2007). The rate of increase in biomass is not constant over the life of a stand (Song and Woodcock 2003; Taylor et al. 2007); early stages of stand development have low rates of biomass accumulation due to trees re-establishing themselves on the site. A study of Siberian Scots pine (*Pinus sylvestris*) stands found that stand age had the largest influence on above ground net primary production (Wirth et al. 2002a). During stand initiation, net production (i.e., biomass accumulation) steadily increases and peaks during the stem exclusion stage (Odum 1969; Whittaker et al. 1974; Acker et al. 2000). A study on a Douglas fir- and western hemlock (*Tsuga heterophylla*) – dominated stand in the Pacific Northwest of the United States found net primary productivity to be greatest during stem exclusion at 30–40 years (Song and Woodcock 2003). Estimates of ponderosa pine (*Pinus ponderosa*) carbon uptake in newly developed stands was shown to increase exponentially as trees increased in size and recruitment of trees into the stand continued, with rates of increase ranging from 0.09 to 0.7 Mg of carbon per hectare per year depending on stand slope, aspect and soil conditions (Hicke et al. 2004). Stands with the same species composition growing on favorable sites will not only accumulate carbon at a higher maximum rate, but they will also reach this maximum rate sooner than stands on poor sites (Chen et al. 2002).

Mature stands continue to accumulate carbon but at slower rates than stands going through the early stages of succession (Odum 1969; Acker et al. 2000). Carbon storage in the living and dead biomass of a red spruce (*Picea rubens*) stand in Nova Scotia, Canada was found to follow a sigmoidal pattern across stand development, with 94 Mg of carbon per hectare in the youngest age class and 247 Mg of carbon per hectare in the 81–100 year age class, with lower amounts in the oldest age classes (Taylor et al. 2007). In the study on a Douglas fir- and western hemlock-dominated forest in the Pacific Northwest of the United States, a stand development model projected a gradual decrease in net primary production from 40 years until 300 years, when net primary production levels off (Song and Woodcock 2003). The decreased rate of uptake is correlated with decreased woody biomass growth as stands age (Chen et al. 2002). Historically these old growth stands were considered neither sources nor sinks for atmospheric carbon. Although their rates of sequestration are lower, some may sequester far more carbon than previously thought; Carey et al. (2001) suggest that old growth forests in the Pacific Northwest are sequestering 145 Tg more carbon than terrestrial carbon models have predicted in the past for these forests. Similar results come from a model of a 200 year old eastern hemlock stand in central Massachusetts, which predicts that this forest has the ability to annually sequester more carbon in the living biomass with future climate change, because of higher atmospheric concentrations of CO_2, than younger coniferous and deciduous stands had done in historical climates (Hadley and Schedlbauer 2002).

While the magnitude of the carbon flux associated with mature stands is still being debated, it is important to consider that mature forest stands store far more carbon than early successional stands (Thuille and Schulze 2006). For example, a mature eastern white pine stand in southern Ontario, Canada held nearly double the carbon, both above (100 tons per hectare) and below ground (56 tons per hectare), than a similar stand going through stem exclusion, which held 40 tons per hectare above ground and 39 tons per hectare below ground (Peichl and Arain 2006).

4.3 Stand Disturbance Effects

Natural and anthropogenic disturbances such as fire, disease, insect outbreaks, logging, and windthrow can alter the rate and/or direction of successional change and subsequently affect carbon flux in forest systems (Table 3.1).

The severity and frequency of naturally occurring disturbances vary greatly within and between different forest types. The return interval for a forest is the approximate number of years between two disturbances. For major forest types, fire return intervals range from 2 to 14,000 years; insect outbreaks occur from 6 to 117 years; and wind throw is a perturbation that has a broad return interval ranging from 5 to 1,300 years (Table 3.1). The enormous variation in the size and type of disturbance and intervals between them within and across different forests, and the climates that drive them, is critical to understanding and managing stand dynamics and by implication carbon sequestration and storage.

Whenever forests are disturbed, they become sources of carbon as woody tissue dies, decomposes and releases stored carbon. The length of time it takes after a disturbance for a stand to become a carbon sink depends on the growth rate of newly established vegetation and the decomposition rate of downed woody material. When a forest is disturbed, some portion of available

Table 3.1 Disturbance return intervals (in years) among different forest types

	Forest type					
Disturbance	Boreal	Temperate hardwoods	Western N.A. conifers	Tropical rain forests	Mediterranean climate	Tropical savanna
Fire	20–500	14–14,000	8–600	400–900	2–125	2–100
Insect outbreaks	10–50	6–34	25–117			
Severe wind throw	50–75	150–1,300	5–15	9–20		

Source: (Fitzgerald 1988; Huff 1995; Lassig and Mocalov 1998; Newbery 1998; Walker 1999; McKenzie et al. 2000; Thonicke et al. 2001; Lorimer and White 2002; Ne'eman et al. 2002; Sinton and Jones 2002; Burton et al. 2003; Ryerson et al. 2003; Felderhof and Gillieson 2006; Fry and Stephens 2006; Spetich and He 2006; Shang et al. 2007; Bouchard et al. 2008)

growing space is left unoccupied for a period of time while new and surviving vegetation grows to fill the site (Campbell et al. 2004; Humphreys et al. 2006). This lag time can range from months to centuries depending on disturbance type, climate, and site conditions. For example, 400 square meter logging gaps in a forest of the Bolivian Amazon were visually indistinguishable in aerial imagery from undisturbed forest after just 3 months (Broadbent et al. 2006) while boreal Scots pine stands may never reach previous stocking levels after low intensity ground fires (Schulze et al. 1999; Wirth et al. 2002b).

Estimates of carbon flux within forests must therefore take into account disturbances in their various forms and frequencies (Cook et al. 2008). But, because each disturbance is unique, and species may respond to the same disturbance in different ways, determining the effects of a particular perturbation on stand level carbon budgets can be both difficult and imprecise. For example, in a study of a boreal forest fire in Canada by Randerson et al. (2006), analysis showed that when all the integrating effects of the fire (e.g. greenhouse gases, aerosols, carbon deposition on snow and sea ice, and post-fire changes in surface albedo) are accounted for, a decrease in radiative energy is expected when the fire cycle is over 80 years because surface albedo had a proportionately greater effect than fire-emitted greenhouse gases that only dramatically spiked radiation during the years immediately after the fire. This suggests that increases in boreal fires may not contribute to climate warming.

5 Response of Forests to Increased Carbon Dioxide

5.1 Free Air Carbon Dioxide Enrichment (FACE) Experiments

Carbon dioxide enrichment studies provide insight into what the future may hold for the world's forests. Experiments are being conducted on a wide variety of terrestrial ecosystems in response to a predicted, continual increase in atmospheric carbon dioxide (IPCC et al. 2007). A review by McLeod and Long (1999) cited 145 references related to carbon dioxide enrichment experiments in multiple terrestrial ecosystems. These studies examined the response of ecosystem processes, including tree growth, to elevated levels of carbon dioxide in the ecosystem's local atmosphere by elevating ambient carbon dioxide to levels predicted for a specific year in the future. It is commonly believed that carbon dioxide enrichment will lead to an increase in vegetative growth in forest systems similar to that observed in carbon dioxide fertilized greenhouses. Such an increase would indicate that the growth of the stands being studied is currently limited by the concentration of atmospheric CO_2 (Millard et al. 2007).

There are two principal types of carbon dioxide enrichment experiments – free air carbon dioxide enrichment (FACE) and chamber carbon dioxide enrichment. FACE experiments elevate ambient levels by either releasing carbon dioxide gas into the air surrounding the study site or by releasing carbon enriched air into the study area (McLeod and Long 1999). Other carbon dioxide enrichment experiments work with either fully enclosed chambers or open topped chambers which hold carbon enriched air on the site. FACE experiments are generally preferred over chambers when modeling ecosystem processes because they do not alter as many other environmental variables (Gielen and Ceulemans 2001).

FACE experiments began in the 1980s with much of the research being done in agricultural systems (McLeod and Long 1999). Forest ecosystems are still not well represented, primarily due to the difficulties and costs of creating and running carbon dioxide enrichment towers in a forest. The annual cost of just the carbon dioxide necessary to operate a forest FACE experiment in the United States is over $650,000, and represents one third of the annual budget of a site (DOE 2002). Of the FACE experiments in forested ecosystems, only three are being conducted on sites larger than 5 ha (Table 3.2). Only the Web FACE in Switzerland is being conducted in a forest stand that originated before 1980. The next oldest is the Duke Forest FACE which is in a 25 year old loblolly pine plantation (Asshoff et al. 2006; Keel et al. 2006; Oren 2008). Globally, eight FACE

Table 3.2 Global Forest FACE areas 1

Name	Location	Year CO_2 enrichment began	Area (hectares)	Ecosystem	Stand initiation year	CO_2 elevation	Reference
Duke Forest FACE	Chapel Hill, NC, USA	1996	90	Planted *Pinus toedo*, Hardwood understory	1983	Ambient +200 ppm	Oren (2008)
Oak Ridge FACE	Tennessee, USA	1995	1.7	Planted *Liquidambar styraciflua*	1988	Ambient +200 ppm	ORNL (2003)
Aspen FACE	Rhinelander, WI, USA	1998	32	Planted *Populus tremuloides*	1997	Ambient +200 ppm	
EuroFACE	Viterbo Providence, Italy	1999	9	Planted *Populus*	1999 (coppiced in 2001)	550 ppm	Pikkarainen and Karnosky (2008)
OZFACE	Yabulu, OLD, Australia	2001	0.1	Planted tropical savanna	2001	460 and 550 ppm	CSIRO (2005)
Web-FACE	Hofstetten, Switzerland	2000	0.28	Temperate deciduous forest	Circa 1900	600 ppm	Asshoff et al. (2006)
Hokkaido FACE	Sapporo Japan	2003	0.014	Planted temperate deciduous forest	2003	500 ppm	Eguchi et al. (2005)
Bangor FACE	Bangor, UK	2004	2.36	Planted *Betula pendula, Alnus glutinosa, fagus sylvatica*	2004	Ambient +200 ppm	Lukac (2007)

Source: (ORNL 2003; CSIRO 2005; Eguchi et al. 2005; Asshoff et al. 2006; Lukac 2007; Oren 2008; Pikkarainen and Karnosky 2008)
ppm parts per million, additional reference information from: http://public.ornl.gov/face/global_face.shtml

experiments have been conducted in forested ecosystems (Table 3.2). There are three forest FACE experiments in the USA, three in Europe, one in Australia, and one in Japan. Of these, the Duke Forest FACE was the earliest; carbon dioxide enrichment began there in 1996 (Oren 2008).

While informative, FACE studies are limited. There are large parts of the world and whole forest types in which no FACE studies have been conducted, notably Africa, mainland Asia, and South America. Of the eight studies, only one is in a tropical ecosystem (OZFACE), and none are in tropical moist forests or boreal forests (Table 3.2). In addition to spatial gaps, FACE studies also lack structural diversity. Only the Web FACE is operating in a naturally regenerated forest (Asshoff et al. 2006; Keel et al. 2006), all others are being conducted in forest plantations, five sixths of which are younger than 20 years old (Table 3.2). Many of these forests are not only young but also small, with some studies occupying less than 1 ha. Although FACE studies provide us the best insight we currently have into ecosystem responses to elevated carbon dioxide, each of these shortcomings limits the scale and certainty of using results to predict ecosystem responses to carbon enrichment.

5.2 Results of FACE Experiments

There is more room to explore the dynamics of forest carbon in relation to elevated atmospheric carbon dioxide, but what has been found so far is intriguing. Elevated carbon dioxide experiments have provided evidence that forest net primary productivity (NPP), and thus carbon uptake, increases when atmospheric carbon dioxide levels are increased. A study that analyzed the results from the Duke Forest FACE, Aspen Experiment, Oak Ridge, and EuroFACE experiments found that when atmospheric carbon dioxide levels were increased to a level predicted for the middle part of the twenty first century, NPP increased by an average of 23(± 2)% (Norby et al. 2005). A follow-up study determined that nitrogen use increased on the three nitrogen limited sites (Duke Forest, Aspen Experiment, Oak Ridge) and nitrogen use efficiency increased on the EuroFACE site (Finzi et al. 2007). It is reasonable to attribute at least some of this increased nitrogen uptake to a reallocation of growth priority to root development (Chapin et al. 1987; Norby and Iversen 2006; Brunner and Godbold 2007). These results raise questions about the long-term sustainability of NPP increases. As stands increase NPP, nitrogen may become progressively limiting and restrain future growth response (Finzi et al. 2006). The Duke Forest FACE – the longest running forest FACE program – has not yet shown such limitation, although the ecosystem level carbon-nitrogen ratio has increased (Finzi et al. 2006).

While FACE studies conducted in plantation forests in the stand initiation and stem exclusion phases showed significant increase in NPP, 4 year results in a mature temperate forest at the Web FACE showed that while the shoot length of some trees exposed to elevated carbon dioxide did increase, elevated carbon dioxide had no significant effect on stem growth (Asshoff et al. 2006). These results are difficult to extrapolate to larger scales and other mature stands as only 11 trees were exposed to elevated carbon dioxide, with 32 control trees.

A regeneration study at the Duke Forest FACE looked at the effect of elevated carbon dioxide on planted seedlings under low light conditions. Fourteen species of seedlings were planted, with a diverse light tolerance range between species. Only shade tolerant species were found to have better growth under elevated carbon dioxide and certain shade tolerant species were found to have higher survivorship (Mohan et al. 2007), a result that indicates that only those species not already limited by light were able to respond to carbon dioxide fertilization. Indeed, several studies have concluded that while it is possible that trees will increase use efficiency to overcome nutrient limitation driven by nutrient paucity (Ceulemans et al. 1999; Suter et al. 2002; Norby et al. 2005; Luo et al. 2006; Norby and Iversen 2006; Springer and Thomas 2007), light limitation appears to be insuperable (Teskey and Shrestha 1985; Kerstiens 2001; Urban 2003). This will likely result in competitive advantages for shade tolerant species under elevated CO_2 (Hattenschwiler and Korner 2000; Kerstiens 2001; Mohan et al. 2007). These conclusions provide insight into potential future stand development patterns in forest systems; higher survivorship of shade tolerant regeneration may mean that total biomass for individual stands will increase and/or the understory reinitiation stage of stand development could occur sooner.

FACE studies have helped elucidate the interactions between carbon and stand dynamics in forested ecosystems. Some interactions are far too complex to understand in just a few years, such as how carbon dioxide fertilization will interact with nitrogen limitation in future stands (Finzi et al. 2006; Millard et al. 2007; Iversen and Norby 2008). It will take decades of studies to determine the true effects of increased atmospheric carbon dioxide on forest stand dynamics.

6 Precipitation and Temperature as Other Climate Effects

6.1 Temperature Change Experiments in Forest Ecosystems

Over the next century, global temperatures are predicted to change at rates faster than at any time in historical records (IPCC et al. 2007). These rapid changes in temperature will alter future forest stand development and carbon cycling (Walther 2004). Researchers have begun field experiments that simulate forests under predicted temperature changes in order to provide insight into the effects of climate change on these ecosystems (Ayres and Lombardero 2000; Hanson et al. 2005; Danby and Hik 2007; Hyvonen et al. 2007; Bronson et al. 2008; Lellei-Kovacs et al. 2008; Yin et al. 2008). Due to their ability to make large-scale predictions and the expense of on-the-ground experiments, there has been a heavy reliance on mathematically-based computer models (Plochl and Cramer 1995; Sykes and Prentice 1996; Iverson and Prasad 1998; Beerling 1999; Keller et al. 2000; Kirilenko et al. 2000; Bachelet et al. 2001; Schwartz et al. 2001; Dullinger et al. 2004; Iverson et al. 2004; Gibbard et al. 2005; Goldblum and Rigg 2005; Hanson et al. 2005; He et al. 2005; Matala et al. 2006; Notaro et al. 2007; Xu et al. 2007; Delire et al. 2008; Leng et al. 2008).

The threat of warmer climates causes concern that higher temperatures will negatively affect species that have adapted to historical climate patterns, leading to shifts in species composition. While there is relative certainty that shifts in species' existing ranges will occur (Saxe et al. 2001; Walther 2004; Wilmking et al. 2004; Hyvonen et al. 2007; Yin et al. 2008), predictions of which species will be affected and to what degree remain unreliable (Thuiller 2004). One area where temperature-driven change will likely be dramatic is in boreal forests, where temperature is often a limiting factor, and species tolerant of low temperatures dominate the landscape (Hyvonen et al. 2007; Xu et al. 2007). Increased temperatures are likely to increase respiration in many species due to a longer growing season, and drought stress will occur in stands lacking the soil moisture needed to support the increased respiration (Saxe et al. 2001). Any drought stress may be moderated by plant reductions in stomatal conductance experienced at elevated CO_2 levels (Curtis 1996; Heath 1998; Herrick et al. 2004; Ainsworth and Long 2005); however, the degree of response is highly species specific (Urban 2003). Experiments in the boreal forest have shown reductions in growth induced by warmer temperatures and less relative moisture (Barber et al. 2000; D'Arrigo et al. 2004; Wilmking et al. 2004). A study of North American black spruce (*Picea mariana*), however, found no change in net ecosystem uptake of carbon dioxide despite a longer growing season, possibly due to increased respiration of the forest as a whole (Dunn et al. 2007). Where water is not a limiting resource, increased temperatures will likely allow for increased carbon uptake by trees; however, the total ecosystem response will be species and ecosystem specific (Boisvenue and Running 2006; Matala et al. 2006).

6.2 Precipitation Fluctuation Experiments in Forest Ecosystems

Water availability may be the most important factor driving NPP and consequently forest carbon dynamics (Tian et al. 1998; Del Grosso et al. 2008). As the Earth's climate continues to change, water availability in forest stands will change with temperature and precipitation. Precipitation regimes are predicted to change around the globe in relation to many factors including El Nino/Southern Oscillation (ENSO) events (Trenberth and Hoar 1997). How these precipitation changes will affect carbon cycling and forest stand dynamics is unknown, but current drought studies provide some insight.

Seasonal changes in precipitation are likely to occur due to ENSO and other climatic events, thus creating seasonal droughts in some areas, such as the tropical forests in Borneo (Potts 2003). Plant physiology tells us that if droughts occur in water-limited forests during the growing season, then carbon uptake by forests will decrease; if more rainfall occurs during the growing season in areas where water is limiting, we might expect more uptake of carbon by forest stands (Larcher 2003). This effect has been observed in a Scots pine forest in the Rhine plain, where a relatively cool/moist growing season led to a near doubling in the carbon sink as compared to a relatively warm/dry year (Holst et al. 2008). Care should be taken when making broad generalizations on the degree to which growth and subsequently carbon uptake of forest trees is affected by changes in precipitation because these responses are highly species- and site-specific and wider trends remain unclear. What is clear, however, is that soil moisture plays an important role in controlling carbon storage in forests (Tian et al. 1998).

A precipitation study done in a temperate forest ecosystem in the Appalachian mountains found that forest growth in wet years was as much as three times greater than growth in dry years (Hanson et al. 2001). This same study found that spring droughts reduce growth to a greater degree than droughts later in the growing season, with greater mortality in saplings as compared to mature trees. This study suggests that the timing of droughts will play a major role in controlling future stand development. If droughts occur during dormant seasons, the effects of precipitation regime changes may be minimal or only expressed in long-term soil drying. Also noteworthy from the Hanson et al. (2001) study was the fact that mature trees were less affected by drought than were understory saplings, demonstrating the resilience of current forest stands to climatic changes. This is consistent with established ideas about the relative sensitivity of regenerating stands (Finegan 1984). Mature Scots pine stands in Siberia were found to have a positive correlation between above ground net primary productivity and growing season precipitation (Wirth et al. 2002a). These results suggest that current stands will endure climatic changes in part through changes in species composition, which is partially as a result of changes in moisture conditions (Hanson et al. 2001; Thuiller 2004; Frey et al. 2007).

6.3 Combined Effects of Climate Change on Forest Ecosystems

Experiments investigating the combined effects of climate change – increased carbon dioxide, temperature changes, and precipitation changes – are more realistic than those exploring any single factor (Hyvonen et al. 2007). Despite this, single factor studies are far more common than those exploring multiple climate variables. The true dynamics of these systems are unknown, and models predicting carbon cycling using multiple variables are often very sensitive to changes in site, species, and productivity of forests (Hanson et al. 2005). Results from a model representing dynamics in an upland oak forest in the eastern United States demonstrate the sensitivity of model results to external factors affecting forest stand dynamics, such as nutrient availability, temperature and water availability (Hanson et al. 2005). The model combined the effects of increased atmospheric carbon dioxide, temperature, precipitation and ground level ozone. Results showed that without any other changes the forest would reduce its net exchange of carbon dioxide by 29%, therefore increasing sequestration. However, when physiological adjustments (such as longer growing seasons) were incorporated into the model, results showed net exchange of carbon dioxide increasing by 20%, and therefore *releasing* of carbon. While helpful, the results of stand development models are based on a series of assumptions, and will vary widely as those assumptions are revised. An example might be the variability in different possible combinations of temperature flux across seasons. Increases or decreases in temperature (summer temperatures may increase, winter ones may stay the same or vice-versa) will have repercussions on phenology, herbivory, snow melt, and many other interacting biological and physical factors that make predictions so hard to make regarding the ultimate effects on carbon flux and storage in ecosystems.

The complexity of forest stand dynamics makes controlling variables in manipulation experiments extremely challenging and expensive and for that reason scientists rely heavily on multifactor models (Luo et al. 2008); however, the majority of climate change models for forests account for just 1–5 variables (Curtis et al. 1995; BassiriRad et al. 2003; Hanson et al. 2005; Bandeff et al. 2006). Long-term studies are needed to investigate the interactions of changes in carbon dioxide, temperature, and precipitation, as these will provide the best data for making predictions about carbon cycling in future forest stands (Karnosky 2003).

7 Conclusion and Summary Recommendations

Carbon cycling in forests is a complex process with many variables. General patterns of stand carbon cycling are universal, but the temporal dynamics of these patterns are very site specific. We suggest that the following findings are important to consider:

- Stands accumulate carbon as they progress through succession. Most studies show that the greatest rate of carbon uptake occurs during the stem exclusion stage, but even mature stands sequester and store significant quantities of carbon. Recent studies have shown that this can be significant – even for old growth – particularly when such old forests represent significant portions of large areas such as the Amazon and Congo basins (Lewis et al. 2009).
- Disturbances to forest ecosystems cause a release of carbon as woody vegetation and soil organic matter decompose. Future climatic conditions will play a major role in carbon cycling in forest stands, and conversely, future stand conditions will influence climate.
- FACE experiments provide evidence that with increased atmospheric carbon dioxide some species will have increased growth; however, further research is clearly needed.
- Future precipitation patterns and moisture regimes will shape forest structure, species composition and productivity, but those changes will vary greatly with both site and timing.
- The combined effects of climate change are being investigated, but often there are too few variables being considered, making the global application of results from these studies somewhat questionable.

Areas of uncertainty in forest carbon science at the stand level provide numerous opportunities for future research. A major area of uncertainty in current research is the long-term effect of changing climates on forest ecosystems. The majority of FACE studies (Table 3.2) and drought studies are less than 20 years old. Further investigation of below ground carbon dynamics in forest systems is also needed (Ceulemans et al. 1999; Curtis et al. 2002). There are entire forest types with little research related to carbon cycling at the stand level, such as the tropics (Clark 2007; Stork et al. 2007), and the scale of the existing studies also leaves much to be investigated. The largest FACE study currently operating is the Duke Forest FACE on 90 ha; the second largest is Aspen FACE at 32 ha (Table 3.2). With only two large-scale (greater than 30 ha) FACE studies in forest ecosystems, extrapolating the effects of increasing concentrations of atmospheric carbon dioxide to global scales is untenable. What we know about forest stand carbon cycling provides a quality base for future experiments, but leaves much to be desired in predicting forest carbon budgets (Karnosky 2003).

References

Aber JD, Melillo JM, Nadelhoffer KJ, McClaugherty CA, Pastor J (1985) Fine root turnover in forest ecosystems in relation to quantity and form of nitrogen availability: a comparison of two methods. Oecologia 66:317–321

Acker SA, Harcombe PA, Harmon ME, Greene SE (2000) Biomass accumulation over the first 150 years in coastal Oregon *Picea-tsuga* forest. J Veg Sci 11: 725–738

Ainsworth EA, Long SP (2005) What have we learned from 15 years of free-air CO_2 enrichment (FACE)? A meta-analytic review of the responses of photosynthesis, canopy properties and plant production to rising CO_2. New Phytol 165:351–371

Asshoff R, Zotz G, Korner C (2006) Growth and phenology of mature temperate forest trees in elevated CO_2. Glob Change Biol 12:848–861

Ayres MP, Lombardero MJ (2000) Assessing the consequences of global change for forest disturbance from herbivores and pathogens. Sci Total Environ 262:263–286

Bachelet D, Neilson RP, Lenihan JM, Drapek RJ (2001) Climate change effects on vegetation distribution and carbon budget in the United States. Ecosystems 4:164–185

Bandeff JM, Pregitzer KS, Loya WM, Holmes WE, Zak DR (2006) Overstory community composition and elevated atmospheric CO_2 and O_3 modify understory biomass production and nitrogen acquisition. Plant Soil 282:251–259

Barber VA, Juday GP, Finney BP (2000) Reduced growth of Alaskan white spruce in the twentieth century from temperature-induced drought stress. Nature 405:668–673

Barnes BV, Zak DR, Denton SR, Spurr SH (1998) Forest ecology. Wiley, New York

BassiriRad H, Constable JVH, Lussenhop J, Kimball BA, Norby RJ, Oechel WC, Reich PB, Schlesinger WH, Zitzer S, Sehtiya HL, Silim S (2003) Widespread foliage delta n-15 depletion under elevated CO_2: inferences for the nitrogen cycle. Glob Change Biol 9:1582–1590

Beerling DJ (1999) Long-term responses of boreal vegetation to global change: an experimental and modelling investigation. Glob Change Biol 5:55–74

Boisvenue C, Running SW (2006) Impacts of climate change on natural forest productivity - evidence since the middle of the 20th century. Glob Change Biol 12:862–882

Bormann FH, Likens GE (1979) Pattern and process in a forested ecosystem. Springer, New York

Bouchard M, Pothier D, Gauthier S (2008) Fire return intervals and tree species succession in the north shore region of eastern Quebec. Can J Forest Res 38:1621–1633

Broadbent EN, Zarin DJ, Asner GP, Pena-Claros M, Cooper A, Littell R (2006) Recovery of forest structure and spectral properties after selective logging in lowland Bolivia. Ecol Appl 16:1148–1163

Bronson DR, Gower ST, Tanner M, Linder S, Van Herk I (2008) Response of soil surface CO_2 flux in a boreal forest to ecosystem warming. Glob Change Biol 14:856–867

Brunner I, Godbold DL (2007) Tree roots in a changing world. J Forest Res 12:78–82

Burns RM, Honkala BH (1990) Silvics of North America. Agriculture handbook 654, vol 2. U.S. Department of Agriculture, Forest Service, Washington, DC, 877 p

Burton PJ, Messier C, Smith DW, Adamowicz WL (2003) Towards sustainable management of the boreal forest. NRC Research Press, Ottawa

Campbell JL, Sun OJ, Law BE (2004) Disturbance and net ecosystem production across three climatically distinct forest landscapes. Glob Biogeochem Cycles 18, 12 p

Canham CD, Marks PL (1985) The response of woody plants to disturbance: patterns of establishment and growth. In: Pickett STA, White PS (eds) The ecology of natural disturbance and patch dynamics. Academic, New York, pp 197–216

Carey EV, Sala A, Keane R, Callaway RM (2001) Are old forests underestimated as global carbon sinks? Glob Change Biol 7:339–344

Ceulemans R, Janssens IA, Jach ME (1999) Effects of CO_2 enrichment on trees and forests: lessons to be learned in view of future ecosystem studies. Ann Bot 84:577–590

Chapin FS III, Bloom AJ, Field CB, Waring RH (1987) Plant responses to multiple environmental factors. Bioscience 37:49–57

Chapin FS III, Matson PA, Mooney HA (2002) Principles of terrestrial ecosystem ecology. Springer Science, New York

Chen WJ, Chen JM, Price DT, Cihlar J (2002) Effects of stand age on net primary productivity of boreal black spruce forests in Ontario, Canada. Can J Forest Res 32:833–842

Clark DA (2007) Detecting tropical forests' responses to global climatic and atmospheric change: current challenges and a way forward. Biotropica 39:4–19

Cook BD, Bolstad PV, Martin JG, Heinsch FA, Davis KJ, Wang WG, Desai AR, Teclaw RM (2008) Using light-use and production efficiency models to predict photosynthesis and net carbon exchange during forest canopy disturbance. Ecosystems 11:26–44

Cowles HC (1911) The causes of vegetational cycles. Ann Assoc Am Geogr 1:3–20

CSIRO (2005) Ozface. CSIRO Sustainable ecosystems, Crace, Australia. Http://www.Cse.Csiro.Au/research/ras/ozface/index.Htm. Accessed May 2008

Curtis PS (1996) A meta-analysis of leaf gas exchange and nitrogen in trees grown under elevated carbon dioxide. Plant Cell Environ 19:127–137

Curtis PS, Vogel CS, Pregitzer KS, Zak DR, Teeri JA (1995) Interacting effects of soil fertility and atmospheric CO_2 on leaf-area growth and carbon gain physiology in *Populus x euramericana* (dode) guinier. New Phytol 129:253–263

Curtis PS, Hanson PJ, Bolstad P, Barford C, Randolph JC, Schmid HP, Wilson KB (2002) Biometric and eddy-covariance based estimates of annual carbon storage in five eastern North American deciduous forests. Agric Forest Meteorol 113:3–19

Danby RK, Hik DS (2007) Responses of white spruce (*Picea glauca*) to experimental warming at a subarctic alpine treeline. Glob Change Biol 13:437–451

D'Arrigo RD, Kaufmann RK, Davi N, Jacoby GC, Laskowski C, Myneni RB, Cherubini P (2004) Thresholds for warming-induced growth decline at elevational tree line in the Yukon Territory, Canada. Glob Biogeochem Cycles 18:1–7

Del Grosso S, Parton W, Stohlgren T, Zheng DL, Bachelet D, Prince S, Hibbard K, Olson R (2008) Global potential net primary production predicted from vegetation class, precipitation, and temperature. Ecology 89: 2117–2126

Delire C, Ngomanda A, Jolly D (2008) Possible impacts of 21st century climate on vegetation in central and west Africa. Glob Planet Change 64:3–15

Dickson RE (1986) Forest tree physiology. Ann Forest Sci 46(suppl. 16):1–11

DOE (2002) An evaluation of the Department of Energy's Free-air Carbon Dioxide Enrichment (FACE) experiments as scientific user facilities.US Department Of Energy Biological and Environmental Research Advisory Committee, 41 p

Dullinger S, Dirnbock T, Grabherr G (2004) Modelling climate change-driven treeline shifts: relative effects of temperature increase, dispersal and invasibility. J Ecol 92:241–252

Dunn AL, Barford CC, Wofsy SC, Goulden ML, Daube BC (2007) A long-term record of carbon exchange in a boreal black spruce forest: means, responses to interannual variability, and decadal trends. Glob Change Biol 13:577–590

Eguchi N, Koike T, Ueda T (2005) Free air CO_2 enrichment experiment in northern Japan. In: Vaisala News

Ericsson T, Rytter L, Vapaavuori E (1996) Physiology of carbon allocation in trees. Biomass Bioenergy 11: 115–127

Farnsworth EJ, Ellison AM, Gong WK (1996) Elevated CO_2 alters anatomy, physiology, growth, and reproduction of red mangrove (*Rhizophora mangle l*). Oecologia 108:599–609

Farrar JF (1999) Aquisition, partitioning and loss of carbon. In: Press MC, Scholes JD, Barker MG (eds) Physiolocal plant ecology. Blackwell Science, Cornwall, pp 25–43

Felderhof L, Gillieson D (2006) Comparison of fire patterns and fire frequency in two tropical savanna bioregions. Austral Ecol 31:736–746

Finegan B (1984) Forest succession. Nature 312: 109–114

Finzi AC, Moore DJP, DeLucia EH, Lichter J, Hofmockel KS, Jackson RB, Kim HS, Matamala R, McCarthy HR, Oren R, Pippen JS, Schlesinger WH (2006) Progressive nitrogen limitation of ecosystem processes under elevated CO_2 in a warm-temperate forest. Ecology 87:15–25

Finzi AC, Norby RJ, Calfapietra C, Gallet-Budynek A, Gielen B, Holmes WE, Hoosbeek MR, Iversen CM, Jackson RB, Kubiske ME, Ledford J, Liberloo M, Oren R, Polle A, Pritchard S, Zak DR, Schlesinger WH, Ceulemans R (2007) Increases in nitrogen uptake rather than nitrogen-use efficiency support higher rates of temperate forest productivity under elevated CO2. Proc Natl Acad Sci USA 104:14014–14019

Fitzgerald, T.D., 1988. Social caterpillars web page. State University of New York, Cortland. http://facultyweb.cortland.edu/fitzgerald/Foresttentcaterpillaroutbreaks.html. Accessed May 2008

Franklin JF, Van Pelt R (2004) Spatial aspects of structural complexity in old-growth forests. J For 102: 22–28

Franklin JF, Spies TA, Pelt RV, Carey AB, Thornburgh DA, Berg DR, Lindenmayer DB, Harmon ME, Keeton WS, Shaw DC, Bible K, Chen J (2002) Disturbances and structural development of natural forest ecosystems with silvicultural implications, using douglas-fir forests as an example. Forest Ecol Manag 155:399–423

Frey BR, Ashton MS, McKenna JJ, Ellum D, Finkral A (2007) Topographic and temporal patterns in tree seedling establishment, growth, and survival among masting species of southern New England mixed-deciduous forests. Forest Ecol Manag 245:54–63

Fry DL, Stephens SL (2006) Influence of humans and climate on the fire history of a ponderosa pine-mixed conifer forest in the southeastern Klamath mountains, California. Forest Ecol Manag 223: 428–438

Gibbard S, Caldeira K, Bala G, Phillips TJ, Wickett M (2005) Climate effects of global land cover change. Geophys Res Lett 32:L23705

Gielen B, Ceulemans R (2001) The likely impact of rising atmospheric CO2 on natural and managed *Populus*: a literature review. Environ Pollut 115:335–358

Glenn-Lewin DC, Peet RK, Veblen TT (1992) Plant succession. Springer, Berlin

Goldblum D, Rigg LS (2005) Tree growth response to climate change at the deciduous-boreal forest ecotone, Ontario, Canada. Can J Forest Res 35:2709–2718

Gower ST, Krankina O, Olson RJ, Apps M, Linder S, Wang C (2001) Net primary production and carbon allocation patterns of boreal forest ecosystems. Ecol Appl 11:1395–1411

Grime JP (1977) Evidence for the existence of three primary strategies in plants and its relevance to ecological and evolutionary theory. Am Nat 111:1169–1194

Hadley JL, Schedlbauer JL (2002) Carbon exchange of an old-growth eastern hemlock (*Tsuga canadensis*) forest in central New England. Tree Physiol 22:1079–1092

Hanson PJ, Todd JDE, Amthor JS (2001) A six-year study of sapling and large-tree growth and mortality responses to natural and induced variability in precipitation and throughfall. Tree Physiol 21:345–358

Hanson PJ, Wullschleger SD, Norby RJ, Tschaplinski TJ, Gunderson CA (2005) Importance of changing CO_2, temperature, precipitation, and ozone on carbon and water cycles of an upland-oak forest: incorporating experimental results into model simulations. Glob Change Biol 11:1402–1423

Hattenschwiler S, Korner C (2000) Tree seedling responses to in situ CO_2-enrichment differ among species and depend on understorey light availability. Glob Change Biol 6:213–226

He HS, Hao ZQ, Mladenoff DJ, Shao GF, Hu YM, Chang Y (2005) Simulating forest ecosystem response to climate warming incorporating spatial effects in north-eastern China. J Biogeogr 32:2043–2056

Heath J (1998) Stomata of trees growing in CO_2-enriched air show reduced sensitivity to vapour pressure deficit and drought. Plant Cell Environ 21:1077–1088

Herrick JD, Maherali H, Thomas RB (2004) Reduced stomatal conductance in sweetgum (*Liquidambar styraciflua*) sustained over long-term CO_2 enrichment. New Phytol 162:387–396

Hibbs DE (1983) Forty years of forest succession in central New England. Ecology 64:1394–1401

Hicke JA, Sherriff RL, Veblen TT, Asner GP (2004) Carbon accumulation in Colorado ponderosa pine stands. Can J Forest Res 34:1283–1295

Holst J, Barnard R, Brandes E, Buchmann N, Gessler A, Jaeger L (2008) Impacts of summer water limitation on the carbon balance of a scots pine forest in the southern upper Rhine plain. Agric Forest Meteorol 148:1815–1826

Huff MH (1995) Forest age structure and development following wildfires in the western Olympic mountains, Washington. Ecol Appl 5:471–483

Humphreys ER, Black TA, Morgenstern K, Cai TB, Drewitt GB, Nesic Z, Trofymow JA (2006) Carbon dioxide fluxes in coastal douglas-fir stands at different stages of development after clearcut harvesting. Agric Forest Meteorol 140:6–22

Hyvonen R, Agren GI, Linder S, Persson T, Cotrufo MF, Ekblad A, Freeman M, Grelle A, Janssens IA, Jarvis PG, Kellomaki S, Lindroth A, Loustau D, Lundmark T, Norby RJ, Oren R, Pilegaard K, Ryan MG, Sigurdsson BD, Stromgren M, van Oijen M, Wallin G (2007) The likely impact of elevated [CO_2], nitrogen deposition, increased temperature and management on carbon sequestration in temperate and boreal forest ecosystems: a literature review. New Phytol 173: 463–480

IPCC, Nabuurs GJ, Masera O, Andrask K, Benitez-Ronce P, Borer R, Dutscheke M, Dlsiddig E, J F-R, Frumhoff P, Karjalainen T, Krankina O, Kurz WO, Matsumonto M, Oyhantcabal W, Ravindranath NH, Sanz Sanchesz MJ, Zhang X (2007) Forestry. In: Metz B, Davidson OR, Bosch PR, Dave R, Meyer LA (eds) Climate change 2007: Mitigation. Contribution of working group iii to the fourth assessment report of the intergovernmental panel on climate change. Cambridge University Press, United Kingdom and New York, pp 543–578

Iversen CM, Norby RJ (2008) Nitrogen limitation in a sweetgum plantation: implications for carbon allocation and storage. Can J Forest Res 38:1021–1032

Iverson LR, Prasad AM (1998) Predicting abundance of 80 tree species following climate change in the eastern United States. Ecol Monogr 68:465–485

Iverson LR, Schwartz MW, Prasad AM (2004) How fast and far might tree species migrate in the eastern United States due to climate change? Glob Ecol Biogeogr 13:209–219

Karnosky DF (2003) Impacts of elevated atmospheric CO_2 on forest trees and forest ecosystems: knowledge gaps. Environ Int 29:161–169

Keel SG, Siegwolf RTW, Korner C (2006) Canopy CO_2 enrichment permits tracing the fate of recently assimilated carbon in a mature deciduous forest. New Phytol 172:319–329

Keller T, Edouard JL, Guibal F, Guiot L, Tessier L, Vila B (2000) Impact of a climatic warming scenario on tree growth. CR Acad Sci III-VIE 323:913–924

Kerstiens G (2001) Meta-analysis of the interaction between shade-tolerance, light environment and growth response of woody species to elevated CO_2. Acta Oecol 22:61–69

Keyes MR, Grier CC (1981) Above- and below-ground net production in 40-year-old douglas-fir stands on low and high productivity sites. Can J Forest Res 11:599–605

Kirilenko AP, Belotelov NV, Bogatyrev BG (2000) Global model of vegetation migration: incorporation of climatic variability. Ecol Model 132:125–133

Lacointe A (2000) Carbon allocation among tree organs: a review of basic processes and representation in functional-structural tree models. Ann Forest Sci 57: 521–533

Larcher W (2003) Physiological plant ecology. Springer, Berlin

Lassig R, Mocalov SA (1998) Frequency and characteristics of severe storms in the Urals and their influence on the development, structure and management of the boreal forests. In: Conference on Wind and Other Abiotic Risks to Forests. Elsevier Science Bv, Jownsuu, pp 179–194

Lavigne MB (1992) Exploring the possibilities of developing a physiological model of mixed stands. In: Kelty MJ (ed) The ecology and silviculture of mixed-species forests; a festschrift for David M. Smith. Kluwer Academic Publishers, Dordrecht, pp 143–161

Lellei-Kovacs E, Kovacs-Lang E, Kalapos T, Botta-Dukat Z, Barabas S, Beier C (2008) Experimental warming does not enhance soil respiration in a semiarid temperate forest-steppe ecosystem. Community Ecol 9:29–37

Leng WF, He HS, Bu RC, Dai LM, Hu YM, Wang XG (2008) Predicting the distributions of suitable habitat for three larch species under climate warming in northeastern China. Forest Ecol Manag 254:420–428

Lewis SL, Lopez-Gonzalez G, Sonké B, Affum-Baffoe K, Baker TR, Ojo LO, Phillips OL, Reitsma JM, White L, Comiskey JA, Djuikouo KMN, Ewango CEN, Feldpausch TR, Hamilton AC, Gloor M, Hart T, Hladik A, Lloyd J, Lovett JC, Makana J-R, Malhi Y, Mbago FM, Ndangalasi HJ, Peacock J, Peh KS-H, Sheil D, Sunderland T, Swaine MD, Taplin J, Taylor D, Thomas SC, Votere R, Wöll H (2009) Increasing carbon storage in intact African tropical forests. Nature 457:1003–1006

Lockhart BR, Gardiner ES, Hodges JD, Ezell AW (2008) Carbon allocation and morphology of cherrybark oak seedlings and sprouts under three light regimes. Ann Forest Sci 65:1–8

Long SP (1998) Understanding the impacts of rising CO_2: the contribution of environmental physiology. In: Press MC, Scholes JD, Barker MG (eds) Physiological plant ecology. Blackwell Science, London, pp 263–283

Lorimer CG, White AS (2002) Scale and frequency of natural disturbances in the northeastern US: Implications for early successional forest habitats and regional age distributions. In: Conference on Shrublands and Early-Successional Forests. Elsevier Science Bv, Durham, pp 41–64

Lukac M (2007) Bangorface. School of Agricultural and Forest Sciences, University of Wales. http://www.bangor.ac.uk/~afsa0e/. Accessed May 2008

Luo YQ, Hui DF, Zhang DQ (2006) Elevated CO_2 stimulates net accumulations of carbon and nitrogen in land ecosystems: a meta-analysis. Ecology 87:53–63

Luo YQ, Gerten D, Le Maire G, Parton WJ, Weng ES, Zhou XH, Keough C, Beier C, Ciais P, Cramer W, Dukes JS, Emmett B, Hanson PJ, Knapp A, Linder S, Nepstad D, Rustad L (2008) Modeled interactive effects of precipitation, temperature, and [CO_2] on ecosystem carbon and water dynamics in different climatic zones. Glob Change Biol 14:1986–1999

Matala J, Ojansuu R, Peltola H, Raitio H, Kellomaki S (2006) Modelling the response of tree growth to temperature and CO_2 elevation as related to the fertility and current temperature sum of a site. Ecol Model 199:39–52

McKenzie D, Peterson DL, Agee JK (2000) Fire frequency in the interior Columbia river basin: building regional models from fire history data. Ecol Appl 10:1497–1516

McLeod AR, Long SP (1999) Free-air carbon dioxide enrichment (FACE) in global change research: a review. Adv Ecol Res 28:1–56

Millard P, Sommerkorn M, Grelet GA (2007) Environmental change and carbon limitation in trees: a biochemical, ecophysiological and ecosystem appraisal. New Phytol 175:11–28

Mohan JE, Clark JS, Schlesinger WH (2007) Long-term CO_2 enrichment of a forest ecosystem: implications for forest regeneration and succession. Ecol Appl 17:1198–1212

Ne'eman G, Goubitz S, Nathan R (2002) Reproductive traits of *pinus halepensis* in the light of fire - a critical review. In: 2nd International Conference on Mediterranean Pines. Kluwer Academic Publishers, Chania, pp 69–79

Newbery DM (1998) Dynamics of tropical communities. Cambridge University Press

Norby RJ, Iversen CM (2006) Nitrogen uptake, distribution, turnover, and efficiency of use in a CO_2-enriched sweetgum forest. Ecology 87:5–14

Norby RJ, DeLucia EH, Gielen B, Calfapietra C, Giardina CP, King JS, Ledford J, McCarthy HR, Moore DJP, Ceulemans R, De Angelis P, Finzi AC, Karnosky DF, Kubiske ME, Lukac M, Pregitzer KS, Scarascia-Mugnozza GE, Schlesinger WH, Oren R (2005) Forest response to elevated CO_2 is conserved across a broad range of productivity. Proc Natl Acad Sci USA 102:18052–18056

Notaro M, Vavrus S, Liu ZY (2007) Global vegetation and climate change due to future increases in CO_2 as projected by a fully coupled model with dynamic vegetation. J Climate 20:70–90

Odum EP (1969) Strategy of ecosystem development. Science 164:262–270

Odum EP (1971) Fundamentals of ecology. W.B. Saunders Company, Philadelphia

Oliver CD (1981) Forest development in North America following major disturbances. Forest Ecol Manag 3:153–168

Oliver CD (1992) Similarities of stand structure patterns based on uniformities of stand development processes throughout the world–some evidence and the application to silviculture through adaptive management. In: Kelty MJ, Larson BC, Oliver CD (eds) The ecology and silviculture of mixed-species forests; a Festschrift for David M. Smith. Kluwer Academic Publishers, Boston, pp 11–26

Oliver CD, Larson BC (1996) Forest stand dynamics. Wiley, New York

Oren R (2008) Duke University FACE Experiment. Duke University. http://www.nicholas.duke.edu/other/AMERIFLUX/amerflux.html. Accessed May 2008

ORNL (2003) Oak Ridge experiment on CO_2 enrichment of sweetgum. Oak Ridge National Laboratory

Peet RK, Christensen NL (1987) Competition and tree death. Bioscience 37:586–595

Peichl M, Arain AA (2006) Above- and belowground ecosystem biomass and carbon pools in an age-sequence of temperate pine plantation forests. Agric Forest Meteorol 140:51–63

Peichl M, Arain MA (2007) Allometry and partitioning of above- and belowground tree biomass in an age-sequence of white pine forests. Forest Ecol Manag 253:68–80

Pikkarainen J, Karnosky DB (2008) Facts ii: the aspen FACE experiment. Michigan Technological University

Plochl M, Cramer W (1995) Possible impacts of global warming on tundra and boreal forest ecosystems:

comparison of some biogeochemical models. J Biogeogr 22:775–783
Potts MD (2003) Drought in a Bornean everwet rain forest. J Ecol 91:467–474
Raghavendra AS (1991) Physiology of Trees. Wiley, New York
Randerson JT, Liu H, Flanner MG, Chambers SD, Jin Y, Hess PG, Pfister G, Mack MC, Treseder KK, Welp LR, Chapin FS, Harden JW, Goulden ML, Lyons E, Neff JC, Schuur EAG, Zender CS (2006) The impact of boreal forest fire on climate warming. Science 314: 1130–1132
Retzlaff WA, Handest JA, O'Malley DM, McKeand SE, Topa MA (2001) Whole-tree biomass and carbon allocation of juvenile trees of loblolly pine (*Pinus taeda*): influence of genetics and fertilization. Can J Forest Res 31:960–970
Ryerson DE, Swetnam TW, Lynch AM (2003) A tree-ring reconstruction of western spruce budworm outbreaks in the San Juan mountains, Colorado, USA. Can J Forest Res 33:1010–1028
Saxe H, Cannell MGR, Johnsen B, Ryan MG, Vourlitis G (2001) Tree and forest functioning in response to global warming. New Phytol 149:369–399
Schulze R (2000) Transcending scales of space and time in impact studies of climate and climate change on agrohydrological responses. Agric Ecosyst Environ 82:185–212
Schulze ED, Lloyd J, Kelliher FM, Wirth C, Rebmann C, Luhker B, Mund M, Knohl A, Milyukova IM, Schulze W, Ziegler W, Varlagin AB, Sogachev AF, Valentini R, Dore S, Grigoriev S, Kolle O, Panfyorov MI, Tchebakova N, Vygodskaya NN (1999) Productivity of forests in the Eurosiberian boreal region and their potential to act as a carbon sink - a synthesis. Glob Change Biol 5:703–722
Schulze ED, Wirth C, Heimann M (2000) Managing forests after Kyoto. Science 289:2058–2059
Schwartz MW, Iverson LR, Prasad AM (2001) Predicting the potential future distribution of four tree species in Ohio using current habitat availability and climatic forcing. Ecosystems 4:568–581
Shang ZB, He HS, Lytle DE, Shifley SR, Crow TR (2007) Modeling the long-term effects of fire suppression on central hardwood forests in Missouri Ozarks, using LANDIS. Forest Ecol Manag 242:776–790
Shugart HH Jr, West DC (1980) Forest succession models. Bioscience 30:308–313
Simard SW, Durall DM, Jones MD (1997) Carbon allocation and carbon transfer between *Betula papyrifera* and *Pseudotsuga menziesii* seedlings using a C-13 pulse-labeling method. Plant Soil 191:41–55
Sinton DS, Jones JA (2002) Extreme winds and windthrow in the western Columbia river gorge. Northwest Sci 76:173–182
Song CH, Woodcock CE (2003) A regional forest ecosystem carbon budget model: impacts of forest age structure and landuse history. Ecol Model 164:33–47
Spetich MA, He HS (2006) Oak decline in the Boston Mountains, Arkansas, USA: spatial and temporal patterns under two fire regimes. In: International Workshop on Forest Landscape Modeling. Elsevier Science Bv, Beijing, pp 454–462
Springer CJ, Thomas RB (2007) Photosynthetic responses of forest understory tree species to long-term exposure to elevated carbon dioxide concentration at the Duke forest FACE experiment. Tree Physiol 27: 25–32
Steeves TA, Sussex IM (1989) Patterns in plant development. Cambridge University Press, New York
Stork NE, Balston J, Farquhar GD, Franks PJ, Holtum JAM, Liddell MJ (2007) Tropical rainforest canopies and climate change. Austral Ecol 32:105–112
Suter D, Frehner M, Fischer BU, Nosberger J, Luscher A (2002) Elevated CO_2 increases carbon allocation to the roots of *Lolium perenne* under free-air CO_2 enrichment but not in a controlled environment. New Phytol 154:65–75
Sykes MT, Prentice IC (1996) Climate change, tree species distributions and forest dynamics: a case study in the mixed conifer/northern hardwoods zone of northern Europe. Climatic Change 34:161–177
Taylor AR, Wang JR, Chen HYH (2007) Carbon storage in a chronosequence of red spruce (*Picea rubens*) forests in central Nova Scotia, Canada. Can J Forest Res 37:2260–2269
Teskey RO, Shrestha RB (1985) A relationship between carbon dioxide, photosynthetic efficiency and shade tolerance. Physiol Plant 63:126–132
Thonicke K, Venevsky S, Sitch S, Cramer W (2001) The role of fire disturbance for global vegetation dynamics: coupling fire into a dynamic global vegetation model. Glob Ecol Biogeogr 10:661–677
Thuille A, Schulze ED (2006) Carbon dynamics in successional and afforested spruce stands in Thuringia and the Alps. Glob Change Biol 12:325–342
Thuiller W (2004) Patterns and uncertainties of species' range shifts under climate change. Glob Change Biol 10:2020–2027
Tian HQ, Melillo JM, Kicklighter DW, McGuire AD, Helfrich JVK, Moore B, Vorosmarty CJ (1998) Effect of interannual climate variability on carbon storage in Amazonian ecosystems. Nature 396:664–667
Trenberth KE, Hoar TJ (1997) El nino and climate change. Geophys Res Lett 24:3057–3060
Tritton LM, Hornbeck JW (1982) Biomass equations for major tree species of the northeast. Gen. Tech. Rep. NE-69. U.S. Department of Agriculture, Forest Service, Northeastern Forest Experimental Station, Broomall, 46 p
Urban O (2003) Physiological impacts of elevated CO_2 concentration ranging from molecular to whole plant responses. Photosynthetica 41:9–20
Walker LR (1999) Ecosystems of disturbed ground. Elsevier, New York
Walther GR (2004) Plants in a warmer world. Perspect Plant Ecol Evol Syst 6:169–185
Whittaker RH, Bormann FH, Likens GE, Siccama TG (1974) The hubbard brook ecosystem study: forest biomass and production. Ecol Monogr 44:233–254

Wilmking M, Juday GP, Barber VA, Zald HSJ (2004) Recent climate warming forces contrasting growth responses of white spruce at treeline in Alaska through temperature thresholds. Glob Change Biol 10:1724–1736

Wirth C, Schulze ED, Kusznetova V, Milyukova I, Hardes G, Siry M, Schulze B, Vygodskaya NN (2002a) Comparing the influence of site quality, stand age, fire and climate on aboveground tree production in Siberian scots pine forests. Tree Physiol 22:537–552

Wirth C, Schulze ED, Luhker B, Grigoriev S, Siry M, Hardes G, Ziegler W, Backor M, Bauer G, Vygodskaya NN (2002b) Fire and site type effects on the long-term carbon and nitrogen balance in pristine Siberian scots pine forests. Plant Soil 242:41–63

Xu CG, Gertner GZ, Scheller RM (2007) Potential effects of interaction between CO_2 and temperature on forest landscape response to global warming. Glob Change Biol 13:1469–1483

Yin HJ, Liu Q, Lai T (2008) Warming effects on growth and physiology in the seedlings of the two conifers *Picea asperata* and *Abies faxoniana* under two contrasting light conditions. Ecol Res 23:459–469

Carbon Dynamics of Tropical Forests

4

Kyle Meister, Mark S. Ashton, Dylan Craven, and Heather Griscom

Executive Summary

Tropical forests, a critical resource affecting world climate, are very diverse, largely because of variations in regional climate and soil. For purposes of this analysis they have been divided in four broad forest types – ever-wet, semi-evergreen, dry deciduous, and montane. Existing literature on climate and tropical forests suggests that, compared to temperate and boreal forest biomes, tropical forests play a disproportionate role in contributing to emissions that both affect and mitigate climate. This chapter describes the geographical extent of tropical forests and their role in terrestrial carbon storage, uptake (through processes of photosynthesis), and loss (through plant respiration and microbial decomposition of dead biomass). A review is provided of current knowledge about the role of disturbance (natural and human caused) in affecting the carbon balance of tropical forests. The chapter concludes with an analysis of the threats to tropical forests and how they may influence climate change and elevated CO_2. Findings of this review are summarized in the section below under "what we know" and "what we don't know" about the carbon dynamics of tropical forests.

What We Do Know About Carbon Storage and Flux in Tropical Forests

- Tropical forests contribute nearly half of the total terrestrial gross primary productivity. About 8% of the total atmospheric carbon dioxide cycles through these forests annually.
- Tropical forests contain about 40% of the stored carbon in the terrestrial biosphere (estimated at 428 Gt of carbon), with vegetation accounting for 58% and soil 41%. This ratio of vegetation carbon to soil carbon varies greatly by tropical forest type.
- If tropical ever-wet forest soils become drier, the few studies that have been done suggest that litter decomposition and release of CO_2 from soil may slow. However, some studies show that release of methane, which has a higher global warming potential than CO_2, increases as soils dry. The cause of the methane increase is suspected to be related to increased termite activity.
- Tropical ever-wet and semi-evergreen forests in the Amazon and southeastern Asia typically suffer from droughts during ENSO events (El Niño – La Niña). In the short-term, tropical forests may be resilient to drought. However,

K. Meister • M.S. Ashton (✉)
Yale School of Forestry and Environmental Studies, Prospect Street 360, 06511 New Haven, CT, USA
e-mail: mark.ashton@yale.edu

D. Craven
Yale School of Forestry and Environmental Studies, Greeley Memorial Laboratory, 06511, New Haven, CT, USA

H. Griscom
James Madison University, Harrisonburg, VA, USA

M.S. Ashton et al. (eds.), *Managing Forest Carbon in a Changing Climate*,
DOI 10.1007/978-94-007-2232-3_4,

this may be offset by increased vulnerability to fire after both short- and long-term droughts. These droughts are more severe during strong El Niño years. In tropical ever-wet forests, where droughts are rare, mortality may increase during strong El Niño years due to severe drought, while seasonal semi-evergreen forests may experience relatively little change.

- Expanding crop and pasture lands have a profound effect on the global carbon cycle as tropical forests typically store 20–100 times more carbon per unit area than the agriculture that replaces them. The use of fire to clear forested lands may exacerbate changes to carbon cycling since fire fills the atmosphere with aerosols, thereby reducing transpiration.
- There are proportionately higher amounts of fine root biomass (as compared to other vegetative parts – e.g. leaves, stem) in infertile soils as compared to fertile soils. Infertile soils (e.g. oxisols) make up a greater proportion of the African and South American upland ever-wet and semi-evergreen forest than any other soil type.
- CO_2 production in tropical soils is positively correlated with both temperature and soil moisture, suggesting that tropical rain forest oxisols are very sensitive to carbon loss with land use change.

What We Do Not Know About Carbon Storage and Flux in Tropical Forests

- Uncertainties in both the estimates of biomass and rates of deforestation contribute to a wide range of estimates of carbon emissions in the tropics. More studies are needed.
- Some studies show old growth ever-wet and semi-evergreen forests of Amazonia and Africa are increasing in biomass in response to elevated CO_2. Other studies (from Asia and Central America) suggest that this likely reflects a natural growth response to previous disturbance events. More long-term plot research is needed.
- In response to elevated CO_2, many models predict increased forest productivity, but recent studies suggest that stem growth rate actually decreased in the last 20 years largely due to increased nighttime temperature, decreased total precipitation, and increased cloudiness.
- Direct measurement of below-ground carbon stored in roots is often very difficult even with the most thorough root collection. Current estimates of root soil carbon in tropical forests could be underestimated by as much as 60%. Contrary to past assumptions, a significant portion of stored carbon exists below ground in tropical forests.
- Since many climate models predict further soil drying and increased litter fall in tropical forests, understanding the role of soil microbial communities in processes within the litter layer, belowground biomass, and soil carbon is key.
- Only three studies have analyzed land surface-atmosphere interactions in tropical forest ecosystems. It is essential to understand how carbon is taken up by plants and the pathways of carbon release, and how increasing temperatures could affect these processes and the balance between them.
- Better estimates are needed on the amount of mature, secondary, and disturbed forests in the tropics in order to better predict changes in carbon storage trends and the threat of release of this terrestrial sink.
- The effects of elevated atmospheric CO_2 and global climate change on herbivory and other plant/animal interactions in tropical forests are not well understood. Little research has been done in this area.
- Tropical dry deciduous and montane forests are almost a complete unknown because so little research has been done on these forest types. While the majority of dry deciduous forests in the Americas and Asia have been cleared, there is still a significant amount remaining in Africa and Mexico.

What We Think Are the Influences on Carbon Storage and Flux in Tropical Forests

- First and foremost, the primary risk to the carbon stored in tropical forests is deforestation, particularly converting forests to agriculture. Current estimates of carbon emissions from tropical deforestation vary greatly and are difficult to compare due to differences in data sources, assumptions, and methods. Developing and incorporating multiple variables into new and existing ecosystem models for tropical forests is essential to determining carbon fluxes and future effects of deforestation and climate change.

- Combined climate-carbon cycle models predict that tropical forests are vulnerable to both short- and long-term droughts. The effects of drought will vary, depending on the forest type, whether or not the forest is water-limited, and the counter-effects of increased sunlight. At least in the short-term, tropical forests may be resilient to drought. However, this may be offset by increased vulnerability to fire after both short- and long-term droughts
- Changes in soil moisture affect not only the response of plant species and communities, but also the population dynamics of animals, fungi and microbes, which in turn will affect herbivory and decomposition. Elevated CO_2 reduces nitrogen-based defenses (e.g., alkaloids) and causes an increase in carbon-based defenses (e.g., tannins). As leaves exhibit lower nutritional value, herbivory may increase substantially to compensate.
- All large-scale wind and rain events are episodic and occur at relatively long time intervals that are difficult to predict. However, they drive the successional dynamics of forests, and therefore by implication, the above- and below-ground carbon stocks. Little to no work has been done on assessing and including this dynamism in the development of regional carbon models predicting future change. The assumption is that small-scale disturbances in old-growth forests will remain the dominant phase of growth.

How Might the Carbon Status of Tropical Forests Change with Changing Climate?

- The difference between the annual stand level growth (uptake: 2%) and mortality (release: 1.6%) of Amazonia is currently estimated to be 0.4%, which is just about enough carbon sequestered to compensate for the carbon emissions of deforestation in the region. This means that either a small decrease in growth or a small increase in mortality in mature forests could convert Amazonia from a sink to a source of carbon.
- It is difficult to model carbon flux and productivity in tropical forests due to their structural and age complexity and species composition. As a result, few ecosystem process models have been developed, parameterized, and applied within tropical forest systems. Nevertheless, it is a reasonable assumption that rising temperatures will increase the rate of most if not all biochemical processes in tropical plants and soils.
- In response to elevated CO_2, many models predict increased productivity, both in semi-evergreen forests of the Amazon and central Africa. However, on ever-wet sites in Panama and Malaysia, stem growth rate actually decreased from 1981 to 2005 largely due to increased nighttime temperature, decreased total precipitation, and increased cloudiness.
- Old-growth tropical forests are experiencing accelerated stand dynamics and increasing biomass. Most climate models and forest carbon balance models do not take forest composition into account. Forests with accelerated or "faster" dynamic have less biomass due in part to ecophysiological differences in plant growth.
- A warmer climate could drive low elevation forests to higher elevations or extend the range of tropical seasonal forests. However, if there is more deforestation in these seasonal and dry areas, there may be fewer species available that can migrate and adapt to warmer climates with drier soils.
- Many future climate scenarios predict soil drying in Amazonia and a general reduced capacity of the ecosystem to take up carbon. Understanding how tropical forests respond to water stress could be important because canopy-to-air vapor deficits and stomatal feedback effects could determine how tropical forest photosynthesis responds to future climate change.
- As tropical forest soils become drier, litter decomposition and its release of CO_2 from soil may slow in response to less water avail-

ability. However, there is also some evidence that methane release may increase as soils dry out.

- If drought becomes more common in tropical ever-wet and semi-evergreen forests, as some climate models predict, the likelihood of human-induced fires escaping and impacting large portions of the landscape increases.

1 Introduction

This chapter reviews current literature about carbon cycling in tropical forests. First the different types of tropical forests are described. This includes where they are found, their current and past extent, and their role in terrestrial carbon storage. Secondly we review how and where carbon is allocated in tropical forests, how carbon cycles, and how climate change may affect this cycling. Thirdly, we discuss how changes in carbon storage may occur through uptake, via photosynthesis, and through loss, via respiration and decomposition. The role of disturbance and its potential effects on stored carbon is also considered. Finally, the chapter concludes with a review of some of the threats to tropical forests and how they may influence climate change and elevated CO_2.

The level of interest in tropical forests has increased in recent decades due to global issues of climate change, biodiversity loss, and land use change (predominantly conversion of forest to agriculture). Globally, the tropical rain forest regions of Southeast Asia, South America, and Central Africa are some of the most rapidly developing areas of the world in terms of population growth, land conversion, and urbanization (Houghton 1991a; Soepadmo 1993; Nightingale et al. 2004). Tropical deforestation is one of the main contributing factors to the increase of CO_2 in the atmosphere (Houghton 1991a, b). Despite their importance and impact on the global carbon cycle, there is a lack of systematic assessment, and therefore knowledge, about the carbon pools and fluxes in tropical forests (Dixon et al. 1994; Lal and Kimble 2000; Nightingale et al. 2004). Although some generalizations can be made about tropical forest biomes across the globe, such highly diverse, complex systems warrant closer attention in order to make better estimates and predictions of global carbon budgets. Moreover, there is a tendency in carbon-related policy making to overlook the carbon cycle's interconnectedness with other biogeochemical cycles, such as water and nitrogen. None of these cycles occur in isolation; it is important to remember that carbon is related to biodiversity, water storage and filtration, and other ecosystem values.

Tropical forests occupy a broad range between the Tropic of Cancer and the Tropic of Capricorn, where moist air rising from the equatorial region loses this moisture in the form of precipitation as it descends over the tropics and subtropics (Heinsohn and Kabel 1999). These forests cover approximately 12% of the land surface and account for 50% of global forest area (Fig. 4.1). Approximately 8% of total atmospheric carbon dioxide cycles through these regions annually (Malhi et al. 1998). Tropical forests are responsible for nearly half of the total terrestrial gross primary productivity (Malhi et al. 1998). Consequently, they play a major, yet poorly understood, role in the cycling of carbon (Frangi and Lugo 1985; Soepadmo 1993; Foody et al. 1996; Malhi et al. 1998).

2 Tropical Forest Systems

Tropical forests can be divided into four broad types: (i) ever-wet (often called rainforest); (ii) semi-evergreen; (iii) dry deciduous; and (iv) montane (Fig. 4.1). Forests types have been categorized in relation to both the amount of precipitation and degree of seasonality as the main driver of productivity and decomposition, and hence carbon sequestration and loss.

2.1 Forest Type Descriptions

2.1.1 Ever-Wet Forests

Tropical ever-wet forests receive at least 100 mm of precipitation each month and at least 2,000 mm per year (Ricklefs 2001). Vegetation

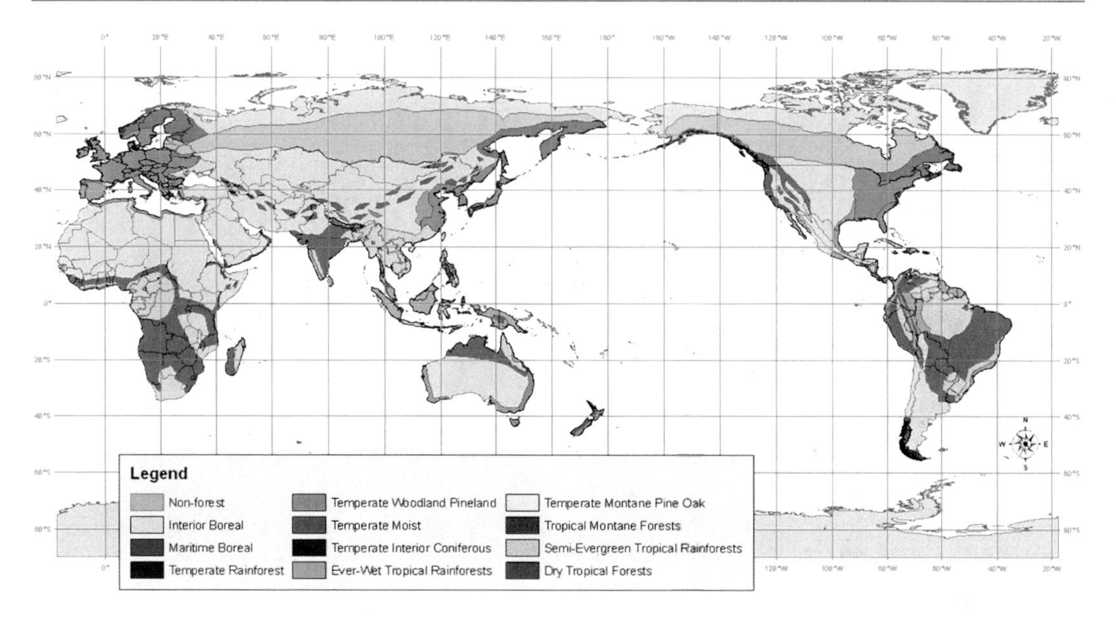

Fig. 4.1 Original extent of boreal, temperate, and tropical forest types of the world prior to land clearing

tends to be dense and of several strata (e.g., canopy emergents, canopy, lianas, epiphytes, treelets, shrubs, and herbs). The highly productive vegetation has adapted to this climate with the ability to immediately incorporate nutrients. As a result, many of the nutrients of tropical rain forest ecosystems are retained by the vegetation. Poorly planned and intensive logging or land clearance and burning can result in the loss of nutrients to the atmosphere and as run-off, thereby reducing the potential productivity of the landscape (Ricklefs 2001; Vandermeer and Perfecto 2005).

The majority of soils in ever-wet forests tend to be well-weathered ultisols, which are acidic, vary in fertility depending upon underlying geology, have relatively high cation exchange capacity, and are very susceptible to erosion. However, this is by no means consistent across the biome. Inceptisols predominate on young foothills, and andisols dominate on volcanic substrates. Both are characteristic of Central America and volcanic islands such as Sumatra, and both are fertile but strongly erodable (Fig. 4.2).

In West Africa, the ever-wet forest occurs along a thin strip of coast from Liberia to Ghana. It starts again in southeastern Nigeria, expanding across Cameroon and around the Gulf of Guinea. The wettest area of the region is the Cameroon Highlands, where rain fall at the base of Mt. Cameroon can reach over 12,000 mm per year. However, most of the area would be classified as marginally ever-wet, with rainfall in most of the range barely over 2,000 mm. Because of its ease of access for human populations, most of the coastal forest that historically spanned Cote d'Ivoire, Ghana, Nigeria, and Cameroon has been lost during the periods of French and British colonization with the commercialization of plantation crops such as coffee and cocoa (1930–1960), and now oil palm. Forests in these countries are now largely reduced to small, degraded patches.

The ever-wet rainforests were once expansive, covering all of eastern Central America (Atlantic Coast) from northern Costa Rica south through Panama, and along the Pacific coastal mountains of Columbia and northern Ecuador (Fig. 4.1). The other wet evergreen forest of the Americas covers the eastern foothills of the Andes and forms the upper basin of the Amazon. The wettest forests in Latin America are those straddling the Andes in the region known as the Chocó on the Pacific coast range of Colombia, and the

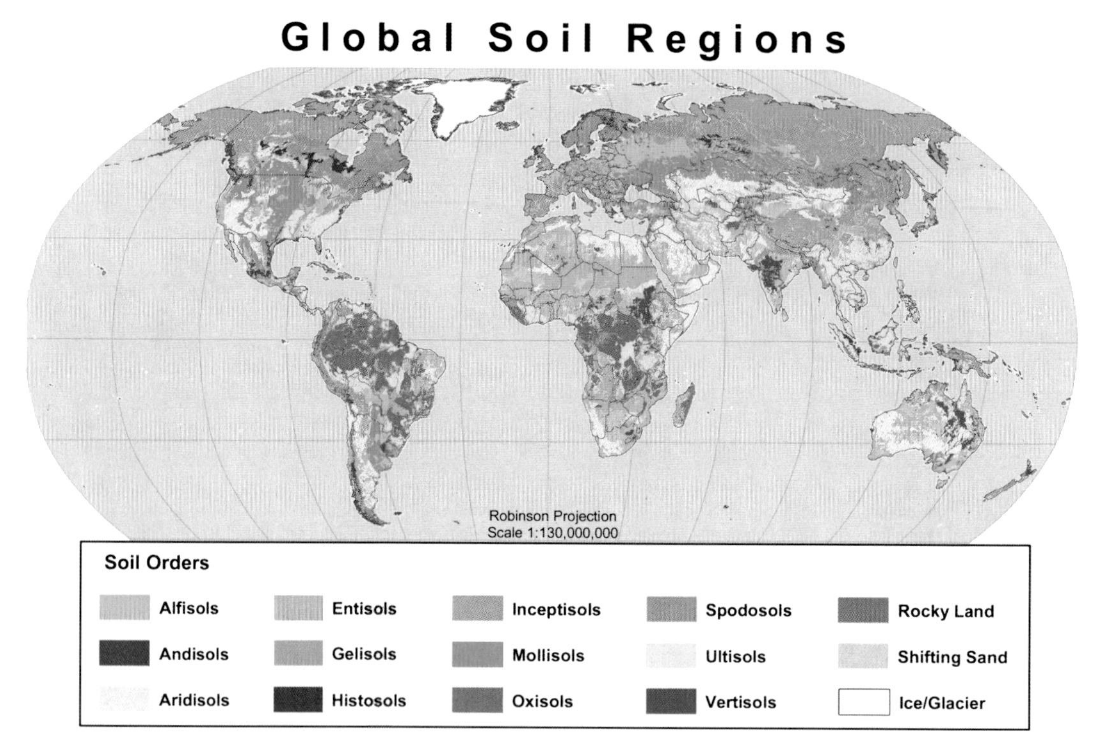

Fig. 4.2 A map depicting the major soil orders of the world (*Source*: From USDA 2005. Reprinted with permission)

upper Amazon of Ecuador. The Atlantic region of Central America has been difficult for people to access and still remains extensively forested, particularly in Panama and Nicaragua, as well as the upper Amazon regions bordering Colombia, Ecuador, and Peru.

The core Asian ever-wet forest can be considered the most moderated in seasonality largely because the land-sea margin and north–south mountain ranges serve as important sources of convectional and orographic precipitation during inter-monsoonal wet seasons. The heart of the ever-wet rain forest is in Borneo, Sumatra, New Guinea, and the Malay Peninsula, an area that makes up the largest extent of ever-wet rainforest in the world. Small areas also exist in southwest Sri Lanka, parts of the Western Ghats of India, and Mindanao in the Philippines.

Asia has had the longest legacy of rainforest commercialization (dating back 2,000 years), largely through maritime trade between Indian, Arab, and Chinese traders and the regional peoples. India's and Sri Lanka's forests are now largely restricted to the mountains and uplands of the countries, where historical land use for intensive rice cultivation, private tree garden systems, and plantation agriculture (tea, rubber, coconut – 1850–1950) has been happening much longer than elsewhere in Asia. Most of the ever-wet forest in the Philippines and Thailand is now confined to degraded patches, first logged over, and then subsequently and incrementally converted to village agricultural projects (1940–1985), many of which subsequently failed and are now wastelands. The Malay Peninsula had most of its lowland forest converted to plantation crops (rubber and oil palm) starting with the British (1900) but accelerating post independence (1948) such that most of the lowlands had been converted by 1980.

Substantial forest remains in the highlands but it is heavily cut over. A similar story exists for Sarawak and Sabah, the two east Malay states on Borneo. However, for these states, land conversion of the lowland forests occurred very rapidly and recently (1970–2000). Indonesia embarked on rapid logging and land conversion of its wet evergreen forests in Kalimantan (Indonesian Borneo) and Sumatra initially for colonization schemes (1970–1980), then more substantively as logging concessions. Subsequently, much of the logged over forest has been converted to oil palm plantations. In Borneo and Sumatra, both countries (Malaysia and Indonesia) have now embarked on clearing the remaining logged over forest for *Acacia mangium* pulp plantations or for oil palm (1995-ongoing). The remaining forested areas are restricted to the most unproductive soils and upland regions that are difficult to access. New Guinea (Papua and Irian Jaya, Indonesia) can be considered the last frontier of remaining large intact forest within the region, although much of it has been allocated for logging concessions (1990-ongoing).

2.1.2 Semi-Evergreen Forests

Tropical seasonally moist forests, also known as tropical semi-evergreen, like ever-wet forests, receive greater than 2,000 mm per year of rain. However, the forest type is more strongly seasonal (in Asia – monsoonal) with extended dry periods and then high periods of rain. Wet periods are generally longer than dry periods.

Soils are usually oxisols (or sometimes spodosols). They are both infertile soils, and acidic. Oxisols are highly weathered, with high clay content, and low cation exchange capacity (Clark et al. 1999; Vitousek and Sanford 1986). Oxisols dominate the uplands of the core Amazon and Congo basins. Alfisols, which are relatively more fertile, are usually found in seasonally drier climates that are not so strongly monsoonal. They predominate particularly in Indochina (India, Burma, Thailand, Cambodia, and Vietnam) (USDA 2002; Vitousek and Sanford 1986) (Fig. 4.2).

The greatest extent of semi-evergreen forest was that of the central and lower Amazon basin and the upper Orinoco of southern Venezuela. Much of the forest in the heart of the Amazon remains largely intact, but has been logged over through the use of the extensive river systems. Coastal and floodplain forests of the major rivers that flow into the Amazon have largely been converted to agriculture. The outer periphery of the basin (particularly on the southern side) and the coastal Atlantic forest of Brazil has retreated considerably because of colonization schemes and large land conversion to commercial soybean and ranching (1970-ongoing).

In West Africa, semi-evergreen forest dominates behind the band of coastal ever-wet rainforest, and can be considered a transition zone to dry deciduous forest further inland. Semi-evergreen forests also predominate in the central Congo River basin. The forests are generally more seasonal than those of the Amazon, with greater levels of deciduousness exhibited by some canopy species. Because of the difficulty of access, the inner core region of the Congo (Central African Republic, Republic of Congo, Democratic Republic of Congo, Gabon) largely remains whole, though current timber extraction is high (1990-ongoing). Both Amazon and Congo semi-evergreen forest can be considered by far the most important and largest tracts of tropical forest left in relation to forest carbon and climate change. However, Indonesia should be recognized for its significant peat forests that amount to the same amount of carbon emissions as the Amazon.

Indochina is the third region with semi-evergreen forest. The forest is found in parts of the Philippines, southern Thailand, northeast India/Burma, southeast Cambodia, southern Vietnam, northeast Sri Lanka, and the Western Ghats of India. The forest type is highly fragmented because of the physical geography and climate. This is an area of high soil fertility so most of the forest has been cut down and converted to agricultural use. The remaining forest patches are mostly degraded.

2.1.3 Tropical Dry Forests

Tropical dry deciduous forests can be defined as forests that shed their leaves during a dry season due to low water availability (Ricklefs 2001). They are located in the tropics and subtropics, mainly in Latin America, Africa, India, Australia and parts of Southeast Asia (Bullock et al. 1995). They can be located in rain shadows of mountainous regions and near mid-latitudes of convergence. Longer and more severe dry seasons support tropical dry seasonal forests and savannah ecosystems (Ricklefs 2001). Soils tend to be alfisols, entisols, and inceptisols (USDA 2002).

This gradient in moisture regimes across the biome has led to much debate over the extent of dry deciduous forests vs. savannahs in the drier tropics (Bullock et al. 1995). Dry deciduous forests are found on more fertile sites than savannas, although they can occur in the same climate zone. In many places, human intervention and fire govern the line between forest and savannah (Bullock et al. 1995). Tropical dry forests receive far less attention than tropical ever-wet and semi-evergreen forests, even though conservation concerns are high due to increased land use conversion, habitat fragmentation, and high levels of biodiversity and structural diversity (Bullock et al. 1995).

In Central America dry deciduous forest used to dominate the Pacific side of Nicaragua, El Salvador, Costa Rica, and Panama. Most of this forest has been cleared for ranching, but some is now coming back as secondary forest because ranching cannot be sustained, or along the coast because of land gentrification. Dry deciduous forest still dominates much of the Yucatan (Mexico, Guatemala) (Griscom and Ashton 2011). In South America, dry deciduous forests were extensive across the coast range and interior Pacific sides of the Andes in Colombia and Ecuador and in the Caribbean coastal mountains and interiors of the lower Orinoco. Most of this forest has now been converted to pasture, although in places second growth is coming back (Griscom and Ashton 2011). In the southern rim of the Amazon basin in Brazil, Bolivia, and northern Argentina, dry deciduous forests have been cleared for plantation agriculture and ranching. Little forest exists today except for some remnant patches in more remote areas.

Africa has the largest dry deciduous forest remaining, making up the miombo woodlands of Malawi, Zimbabwe, Tanzania, Angola, southern Democratic Republic of Congo, Mozambique and Botswana. It is an important resource for local people for firewood, timber, and grazing, and in some areas is heavily deforested. Nevertheless, the woodland in many areas remains relatively intact.

Dry deciduous forests also exist as small residual patches in what was extensive woodland in south India (east of the Ghats) across central India, Central Burma and Thailand, and interior Cambodia. Most forest is now converted to smallholder farms and degraded forest patches. Australia has considerable dry *Eucalyptus* woodland remaining across West and South Australia and in the north (Queensland and Northern Territories). However, a still greater portion has been cleared for raising sheep and for commercial agriculture.

2.1.4 Montane Tropical Forests

Montane tropical forest is the smallest in area (current and historical) compared to the other tropical forest types. Montane forest occurs above 3,000 m above sea level and is characterized by high precipitation (>2,000 mm per year) and lower amounts of radiation because of cloud cover. Epiphytes, particularly bromeliads, often characterize the groundstory and canopy. The greatest amount of forest of this type is in Latin America down the Cordierra of Central America and along the northern Andes from Venezuela to Peru. Asia has montane forests that are numerous but small, being largely confined to the tops of the Western Ghats (India), the central range of Sri Lanka, the highlands of Thailand, Cambodia, and Vietnam and the Ginteng Highlands of Peninsula Malaysia. Larger extents of montane forest exist as the backbone of the islands of Borneo and Sumatra, and the volcanoes of the Philippines. The greatest extent is on the plateaus and the jagged mountains of Papua New Guinea. Africa has only small amounts of montane forest on the slopes of the inland mountain systems of Central (Rwanda, Burundi) and East Africa (Kenya, Uganda and Tanzania).

The soils of montane forests are often some of the most productive and are mostly classified as inceptisols, which are high in soil organic matter (soil carbon), but are erosion-prone because of steep slopes. Many of the mountain regions adjacent to cities (Kuala Lumpur, Malaysia; Colombo, Sri Lanka; David, Panama; Quito, Ecuador; Bogota, Colombia; Nairobi, Kenya) have had their forests cleared for vegetable production, tea and coffee cultivation, and dairy. Much of the organic matter is lost through decomposition, and once depleted; such areas often revert to fire-prone invasive grass and fern lands.

3 Pools of Carbon in Tropical Forests

Tropical forests contain about 40% of the carbon in the terrestrial biosphere, an estimated 428 Gt of carbon, with vegetation accounting for 58%, soil 41%, and litter 1% (Soepadmo 1993; Watson et al. 2000). The carbon budget across tropical forest types can be further broken down into interrelated components: aboveground biomass, belowground biomass, litter, and soil carbon (Table 4.1). Aboveground biomass consists of live stems and large branches and often includes coarse woody debris (Malhi et al. 2004). Belowground biomass includes all root mass (Robinson 2007). Litter usually includes twigs, leaves, reproductive parts and other small biotic debris with short residence times (Malhi et al. 2004). What is included in soil carbon measurements, and how it is allocated within these categories, can vary from study to study. For example, some studies include the litter layer with the soil carbon analysis (e.g., Schwendenmann and Veldkamp 2005). Other researchers separate roots, large organic debris, and rocks from soil for analysis (e.g., Cleveland et al. 2007). No one method is superior. Each method comes with its own advantages and disadvantages depending on the research question. While the use of categories helps to facilitate measurement and analysis, it is also necessary to understand the level of flux between the various carbon pools. This is important not only to correctly measure each component of the carbon cycle, but also to determine the strengths of the links between pools and other biogeochemical cycles.

3.1 Aboveground Biomass

Aboveground biomass is generally derived from field inventory and forest cover data, extrapolated to forest biomass. Uncertainties in the estimates of both biomass and forest cover contribute to a wide range of estimates of carbon stocks in the tropics (Houghton 2005). Many analysts use the FAO estimates of aboveground biomass. These estimates are derived from national data provided by each country. Since countries often use different inventory systems and methods, comparisons between countries can be difficult. For example, the increase in biomass estimates in tropical forests of Latin America and tropical Africa seen in FAO data from the 1980s to the 2000s is most likely attributed to more forests being inventoried (Houghton 2005).

Biomass estimates also vary widely because different tropical forests allocate biomass in different ways in response to environmental conditions, and forest composition and structure. Some of variability, however, derives from factors related to how the data is collected, particularly data that are used to extrapolate from ground measurements to forest biomass. Another source of variability are the models that are used to predict biomass (eg those that do and do not use wood density as a proxy for carbon content). For example, measurements taken at the buttresses of individual trees and then extrapolated to total tree biomass have tended to inflate estimates of biomass in some past studies. (Malhi et al. 2004). Table 4.2 highlights some of the historical variability in above ground biomass estimates.

In response to elevated CO_2, many models predict increased productivity (Laurance et al. 2004; Lewis et al. 2004), both in semi-evergreen forests of the Amazon and central Africa. Feeley et al. (2007) found, however, that on seasonally wet sites in Panama and Malaysia, stem growth rate actually decreased from 1981 to 2005 largely due to increased nighttime temperature, decreased total precipitation, and increased cloudiness.

Decreases in stem growth rate may not be indicative of overall productivity decline, however. Trees could be shifting their allocation of resources from stem growth to root growth, leaf production and/or reproduction (LaDeau and Clark 2001). Nevertheless, even if overall productivity is increasing, decreased stem growth could affect carbon sequestration if, for example, the residence time of carbon in fine roots, leaves,

Table 4.1 A summary of carbon studies in tropical forests

	Site characteristics			Carbon pools				
Source	Location	Forest type	Age	Aboveground biomass	Coarse/fine woody debris stock	Coarse woody debris annual input	Belowground biomass	Soil
Malhi et al. (2004)	Lowland Amazon	Ever-wet/ semi-evergreen	Mature/old growth	Increment of 1.5–5.5 Mg C ha^{-1} a^{-1}				
Baker et al. (2007)	Upper Amazon/Peru	Ever-wet	Mature/old growth		24.4 ± 5.5 Mg C ha^{-1}	3.8 ± 0.2 Mg C ha^{-1} y^{-1} with a 4.7 ± 2.6 y^{-1} tunover		
Clark et al. (2001)	Costa Rica	Ever-wet	Mature/old growth	Increment of 1.7–11.8 Mg C ha y^{-1} (lower bounds); 3.1–21.7 Mg C ha^{-1} y^{-1} (upper bounds)				
Clark et al. (2002)	Costa Rica	Ever-wet	Mature/old growth		Fallen – 22.3 Mg C ha^{-1}; standing – 3.1 Mg C ha^{-1}	2.4 Mg C Ha^{-1} with a 9 y turnover		
Nepstad et al. (1994)	Lowland Amazon	Semi-evergreen	Mature/old growth					Between 1 and 8 m soil depth more soil carbon than above ground biomass; 15% soil carbon turnover
Feeley et al. (2007)	Lowland Malaya; Panama	Ever-wet	Mature/old growth	Decreases in growth recorded in 24–71% trees in Panama; 58–95% trees in Malaya				
Espeleta and Clark (2007)	Costa Rica	Ever-wet	Mature/old growth				Ten fold variation over 7 year period; four fold change across edaphic gradient of soil water availability/fertility	

Houghton et al. (2001)	Amazon	Ever-wet/ Semi-evergreen	Mature/old growth	Mean total standing and below ground biomass 177 Mg C ha^{-1}				
Lewis et al. (2004)	Amazon	Ever-wet/ Semi-evergreen	Mature/old growth	Basal area has been increasing at 0.10 ± 0.04 M2 ha^{-1} y^{-1} between 1971 and 2002				
Lewis et al. (2009)	Central Africa	Ever-wet/ Semi-evergreen	Mature/old growth	Above-ground biomass has been increasing at 0.63 Mg C ha^{-1} y^{-1}				
Phillips et al. (2008)	Amazon	Ever-wet/ Semi-evergreen	Mature/old growth	Above-ground biomass has been increasing at 0.62 Mg C ha^{-1} y^{-1}				
Robinson (2007)	Tropical Forests						68% higher amounts of below-ground biomass than previously estimated	
Paoli and Curran (2007)	Borneo	Ever-wet	Mature/old growth	Above-ground biomass increment 5.8–23.6 Mg ha^{-1} y^{-1}	Annual fine litter input 5.1–11.0 Mg ha^{-1} y^{-1}			Total amounts of annual NPP related to soil fertility – phosphorus
Whigham et al. (1991)	Yucatan, Mexico	Semi-evergreen	Early secondary		A hurricane can increase dead and downed coarse woody debris by 50%			
Wilcke et al. (2004)	Ecuador	Montane	Mature/old growth		9.1 Mg biomass ha^{-1}			

Table 4.2 Estimates of total biomass from various studies in tropical forests (mg dry weight per ha)

Above ground living biomass (mg dry weight)	Total above ground biomass (mg dry weight)	Below ground biomass (mg dry weight)	Reference
413.4	425.2	104.0	Russell (1983)
406.3		67.0	Klinge and Rodrigues (1973)
358.0	396.2		Delaney et al. (1997)
347.7	371.2	56.5	Grimm and Fassbender (1981)
346.0	395.0		Delaney et al. (1997)
343.0	351.0		Overman et al. (1994)
314.0	353.8		Delaney et al. (1997)
306.2	348.0		Uhl et al. (1988)
296.0	308.0		Delaney et al. (1997)
285.0	325.0		Brown et al. (1995)
267.0	320.0	68.0	Salomao et al. (1996)
264.0		35.4	Nepstad (1989)
221.0	247.3	58.2	Saldarriaga et al. (1988)
242.2	264.6	46.0	Fearnside et al. (1993)
140.0	155.2		Delaney et al. (1997)

Source: Modified from Houghton et al. (2001)

flowers, or fruits is shorter than in coarse woody material (Pregitzer et al. 1995).

Studies have found large differences in productivity between Southeast Asian tropical forests and those in the neotropics. In a meta analysis of 39 diverse neotropical forests (dry to wet, lowland to montane, nutrient-rich to nutrient-poor soils), total net primary productivity (NPP – above and below-ground) ranged from 1.7 to 11.8 Mg C/ha/year (lower bounds) and from 3.1 to 21.7 Mg C/ha/year (upper bounds) (Clark et al. 2001). In a tropical Asian ever-wet forest in southwest Borneo, however, Paoli and Curran (2007) found that above ground NPP alone ranged from 11.1 to 32.3 Mg C/ha/year, which implies that total NPP is much higher than in neotropical forests. Paoli and Curran (2007) also found that the spatial pattern of productivity in the lowland Bornean forests was significantly related to soil nutrients, particularly phosphorus. It is important to note that almost all the work cited here is from semi-evergreen and ever-wet forests of the Amazon, Central America, and Malaysia/Borneo. Little work has been done in other regions on this topic, especially in dry deciduous or montane forest types.

3.2 Belowground Biomass

Measuring belowground biomass is very difficult because roots are embedded in the soil. Not only is uncertainty in inventory data problematic for belowground biomass estimates, but direct measurement is often very difficult even with the most thorough root collection (Robinson 2007). Attempts to remove entire trees and their root systems tend to underestimate root biomass because many of the fine roots remain in the soil. Current estimates of root masses could be understated by as much as 60% according to Robinson (2007), who provides adjusted values for biomes to reflect this discrepancy. These findings suggest that root mass for tropical forests worldwide could contain up to 49 more Pg of carbon than found in previous studies, with a subsequent increase in total carbon sink of 9% for tropical forests (Robinson 2007). More belowground biomass could account for some of the "missing" global carbon sink and has implications for soil carbon estimates as well.

Understanding how belowground carbon allocation varies with soil and topographic conditions and across different climates is crucial to linking the different carbon pools in forests. Belowground biomass allocation can differ significantly both spatially and temporally in tropical forests. Spatial variation in belowground fine root biomass for an ever-wet forest at La Selva research station in Costa Rica was similar to that of studies done in other tropical and temperate forests (Espeleta and Clark 2007). Higher fine root biomass in the soil profile was associated with less fertile oxisols, lower in phosphorous, and with less soil water availability across a landscape gradient, while lower fine root biomass was associated with greater fertility and soil water availability in the soil profile. Espeleta and Clark (2007) produced the first belowground dataset for tropical forests to sufficiently assess temporal variation of fine root stocks. They found that sites on slope crests had greater live and dead fine-root variation in turnover due to changes in soil water content and its effect on nutrient acquisition. Drier years led to increased litter fall, and tree and root mortality. This has implications for how belowground biomass allocation and nutrient cycling may be impacted in a changing climate. If tropical forest soils dry as predicted by many models (e.g., Cox et al. 2000; Friedlingstein et al. 2006; Notaro et al. 2007), then fine roots located in the driest portions of the soil profile should die. If water stress does not lead to mortality, then plants should respond by allocating more root biomass to wetter areas of the soil profile.

3.3 Epiphytes, Litter and Logs

There have been numerous studies on the role of coarse woody debris in temperate forests, particularly old growth (Harmon et al. 1986). However few such studies have been done within tropical forests. Dry deciduous and semi-evergreen forests might have larger proportional loads of coarse woody debris than ever-wet and montane forests because of proneness to hurricanes and fire and greater impacts from swidden/fallow cultivation systems. For example, Eaton and Lawrence (2006) found that in the northern Yucatan, the largest amounts of downed debris were recorded post land clearance (88% of above ground biomass). Studies by others have shown that hurricanes can create large amounts of coarse debris, not directly, because most vegetation survives and re-sprouts, but indirectly through susceptibility to fire (Whigham et al. 1991).

Studies of coarse woody debris in ever-wet forests are also rare. One study in Costa Rica found no difference in standing dead and downed wood (>10 cm in diameter) in relation to topography and soil, but that overall dead biomass contributed to 33% of the above-ground biomass, with a turnover of about 9 years (Clark et al. 2002). In a semi-evergreen forest in the Brazilian Amazon, downed coarse woody debris was recorded between 50 and 55 Mg biomass per ha (Keller et al. 2004). For ever-wet forests in Costa Rica (Clark et al. 2002) and the Peruvian upper Amazon (Baker et al. 2007) stocks were about the same (22 and 24 Mg C per ha respectively or 46 and 50 Mg biomass per ha). In an Ecuadorian montane forest, Wilcke et al. (2004) found much lower woody debris biomass stocks (9 Mg biomass per ha) but it was highly variable and represented only 4% of the total estimated carbon in the forest.

Litter production in tropical forests is likely to increase in an elevated CO_2 environment as it is linked to higher respiration rates (Sayer et al. 2007). Litter production in the tropics, and indeed aboveground productivity, is related to soil nutrients, especially phosphorous, in addition to carbon (Paoli and Curran 2007). CO_2 enrichment tends to have a positive effect on plant growth up to a certain point before plants begin to exhaust other resources and reach a limit of enhanced growth, at which point litter production levels off.

3.4 Soil Carbon

Most soil carbon in tropical forests is located in the uppermost layers where root density is generally the highest. In a soil respiration measurement experiment comparing sites in Paragominas, Brazil (semi-evergreen) and La Selva, Costa Rica (ever-wet), Schwendenmann and Veldcamp (2005) found that more than 75% of the CO_2 was produced in the upper 0.5 m (including the litter layer) while less than 7% came from soil below 1 m depth. CO_2 production was positively correlated with both temperature and soil moisture in the top 0.5 m (Schwendenmann and Veldcamp 2005). In the Paragominas site, beyond 2 m in soil depth CO_2 production increased greatly with increasing temperature (Schwendenmann and Veldkamp 2005). Nevertheless, this is still a much lower amount of flux than in the upper layers. The increases in CO_2 production observed by Schwendenmann and Veldcamp (2005) indicate a strong positive feedback between ecosystem warming and CO_2 flux from moist tropical forest soils, but further studies need to verify this.

This study also highlights how differences in local climate, soil, and forest type can affect soil carbon flux. Paragominas is a tropical deciduous forest with a long dry season. Its forests have deep roots to a depth of at least 18 m (Nepstad et al. 1994) that enable trees to extract water stored at greater depths. Active soil water extraction occurs with root respiration, which can explain the high CO_2 production observed in the deep soil at the site in Paragominas (Schwendenmann and Veldkamp 2005). In contrast, the forest at La Selva does not experience an intense seasonal drought and the water content below 0.75 m is always above field capacity. The La Selva Forest also has a low root biomass below 2 m (Veldkamp et al. 2003). The contribution of root respiration to CO_2 produced beyond 2 m in depth at La Selva is minimal. Deep soil CO_2 at La Selva is principally from decomposition of soil organic carbon and/or dissolved organic carbon by soil microbes (Schwendenmann and Veldkamp 2005). The sheer contrast in CO_2 production at different depths of different soil and forest types highlights the complexity of soils and the need to further examine microbial and plant biochemical processes in deeper soil layers over longer periods (see Chap. 2 for further details).

The dynamic changes in the composition of the soil microbial community in response to inputs of organic matter may increase soil respiration rates and drive soil carbon losses in the form of carbon dioxide to the atmosphere (Cleveland and Townsend 2006; Cleveland et al. 2007). Since many climate models predict further soil drying and increased litter fall in tropical forests (e.g., Cox et al. 2000; Friedlingstein et al. 2006; Notaro et al. 2007), understanding the role of the soil microbial community and its function within the litter layer, belowground biomass and soil carbon is key. Changes in climate, the concentration of CO_2 in the atmosphere, and the nutrient content of litter could all have an effect on soil biota and decomposition rates (Coley 1998).

4 Biotic Drivers of Uptake and Release

Since the early 1980s, only three studies have analyzed land surface-atmosphere interactions in tropical forest ecosystems: the Anglo-Brazilian Climate Observation Study (ABRACOS; 1990–95); the Large-scale Biosphere/Atmosphere Experiment in Amazonia (LBA; 1996–2003); and the GEWEX Asian Monsoon Experiment (GAME; since 1996) (Nightingale et al. 2004). All three studies were conducted in semi-evergreen forests. It is difficult to model carbon flux and productivity in tropical forests due to their structural and age complexity and species composition. As a result, few ecosystem process models have been developed, parameterized, and applied within tropical forest systems (Nightingale et al. 2004). Nevertheless, it is a reasonable assumption that rising temperatures will increase the rate of most if not all biochemical process in tropical plants and soils (Lloyd and Farquhar 1996). Therefore, it is essential to understand how carbon is taken up by plants and the pathways of carbon release, and how increasing temperatures could affect these processes and the balance between them.

4.1 Photosynthesis and Autotrophic Respiration

Photosynthesis is the process through which plants assimilate carbon in the form of carbon dioxide (CO_2). Specifically, photosynthesis requires CO_2, sunlight, and water as inputs to produce glucose (carbohydrates), oxygen, and water. If the carbon uptake of photosynthesis exceeds the carbon efflux of respiration, intact forests are thought to remain a carbon sink (Phillips et al. 2008). However, the increases in productivity observed in Amazonian and Central African semi-evergreen and tropical forests over the past few decades by Phillips et al. (2008) and Lewis et al. (2009) cannot continue indefinitely. Lewis et al. (2009) estimate that one fifth of the CO_2 currently produced globally by land conversion and industrial emissions is absorbed by the tropical forest regions through increased productivity. However, if the increased atmospheric level of CO_2 is the cause for this increased productivity, then trees will eventually reach a saturation point and become limited by some other resource (Phillips et al. 2008). Thus, it is critical to consider the role of other biogeochemical cycles in relation to carbon.

Many future climate scenarios predict soil drying in Amazonia and a general reduced capacity of the ecosystem to take up carbon due to lack of water (Friedlingstein et al. 2006; Notaro et al. 2007). Understanding how tropical forests respond to water stress could be important because canopy-to-air vapor deficits and stomatal feedback effects could determine how tropical forest photosynthesis responds to future climate change (Lloyd and Farquhar 1996).

4.2 Heterotrophic Respiration and Decomposition

Respiration requires oxygen, carbohydrates, and water to release energy, CO_2 and water. Autotrophic respiration occurs when plants release CO_2 during biochemical processes, such as growth and production of chemical defenses. Heterotrophs (e.g., animals) also contribute to CO_2 release in a similar process. Like photosynthesis, respiration is linked to temperature fluctuations and other environmental factors (Phillips et al. 2008).

Decomposition is a type of respiration in which dead organic matter, oxygen, and water are transformed into CO_2, and water. Barring poor access to moisture and oxygen, decomposition in the humid tropics tends to be rapid, which limits the accumulation of detritus on the forest floor (Ricklefs 2001; Vandermeer and Perfecto 2005). Where moisture stress or oxygen stress inhibit aerobic respiration, however, detritus can accumulate, such as in peat swamps and other poorly drained areas or certain areas of tropical dry forests. When oxygen stress limits aerobic respiration, microbes and fungi responsible for decomposition rely on anaerobic respiration – a less efficient method of respiration in which methane is often a byproduct.

As tropical forest soils become drier, litter decomposition and its release of CO_2 from soil may slow in response to less water availability. However, Cattânio et al. (2002) found that greenhouse gas release in the form of methane, which has a higher global warming potential than CO_2, increased as soils dried in experimental water exclusion plots. This is surprising, as methane production requires anaerobic microsites that are uncommon in dry soils. Dry plots had more litter and woody debris; there was also anecdotal evidence of increased termite activity, which may explain the release of methane (Cattânio et al. 2002). Changes in soil moisture not only affect the response of plant species and communities, but also the population dynamics of animals, fungi and microbes, which in turn will have impacts on herbivory and decomposition. Thus, it is important to remember that changes in ecosystems rely on the interaction of all of its components, not just a few.

5 Disturbance: Abiotic Drivers of Uptake and Release

Disturbance is a natural process of all ecosystems, to which most organisms have some form of adaptation. Tropical forests experience tree mortality from old age, earthquakes or storms, which open up the forest floor to light and allow younger trees to attain the canopy. When trees die, they decompose and release CO_2 and nutrients to the soil and atmosphere. Nutrients may be taken up quickly by other plants, stored in soil for a period of time, or leached from the system during rain events. In large scale disturbances, especially fires, landslides, land clearance, or logging, large amounts of nutrients are lost from the ecosystem. It may take hundreds of years to recover from this nutrient loss. At the same time, land-use conversion to non-forest uses, such as farming and urban development, releases carbon to the atmosphere, further altering the carbon budget of the landscape.

Many studies use old growth forests that have not experienced major disturbances for a long period of time (e.g., Malhi and Phillips 2004). This has led to unexpected results in carbon flux measurements. In one 3-year study of old growth forests in the Amazon, carbon was released in the wet season and taken up in the dry season, in opposition to the seasonal cycles of both tree growth and model predictions (Saleska et al. 2003). This disconnect was attributed to decomposition and soil moisture availability, and transient effects of recent disturbance. This has important implications for carbon budgeting in the Amazon. If studies tend to be concentrated in undisturbed, old growth forests versus recently or regularly disturbed sites, predictions of future carbon sequestration rates are likely to be overestimated (Saleska et al. 2003).

5.1 Drought and El Niño-Southern Oscillation (ENSO) Events

ENSO events and droughts are part of the planet's natural climate cycles. Although there has been much research into ENSO events and their effect on droughts in the tropics, droughts can be independent of ENSO events. Combined climate-carbon cycle models predict that the Amazon forests are vulnerable to both short- and long-term droughts (Samanta et al. 2010). When water is initially limited, vegetation responds by reducing transpiration and photosynthesis, which in turn reduces the amount of water recycled to the atmosphere.At least in the short-term, some tropical forests may be resilient to drought through deep roots. However, this may be offset by increased vulnerability to fire after both short- and long-term droughts (Saleska et al. 2007; Nepstad et al. 2007).

Tropical rainforests in the Amazon and southeastern Asia typically suffer from droughts during ENSO events. These droughts are more severe during strong El Niño years (Lyons 2004). Ever-wet forests and semi-evergreen forests response to drought varies. In one study of the ever-wet forests of Borneo, where droughts are rare, mortality increased during strong El Niño years due to severe drought, while semi evergreen forests experienced relatively little change (Potts 2003). In addition to forest type, position in the landscape, soil texture and rooting depth play a role in the vegetation's response to drought (Sotta et al. 2007). For example, temporarily flooded valleys and lowlands often receive drainage from upslope areas and are able to retain moisture longer than uplands (Ashton 1992; Ashton et al. 1995; Grogan et al. 2003; Ediriweera et al. 2008). Areas with finer soil textures retain more water for longer time periods than those with coarser textures. Texture can vary within the soil profile, which means that the texture of soil at lower depths could be an important indicator for a site's water retention capacity during droughts (Grogan et al. 2003; Sotta et al. 2007). In addition, the location of roots within the soil profile determines where a plant can take up water. During drought events, the surface tends to dry first, giving plants with deeper roots or the ability to quickly respond to drought by allocating root growth to deeper soils an advantage (Sotta et al. 2007). Increased water stress during drought is linked to higher tree and liana mortality, which suggests that more carbon will be released through decay and increased probability of fire (Nepstad et al. 2007). These differing responses are significant because ENSO events are expected to become more frequent in response to climate change (Tsonis et al. 2005).

5.2 Wind and Rain

Wind throw and snap-off of trees from winds can occur in a variety of forms from large landscape level effects (hurricanes and typhoons) to more landscape-specific convectional windstorms that affect multiple trees (stand scale) to individual wind throw and branch breakage (Whigham et al. 1999). Most winds come with rain, either before or after these events. Rain-soaked soils are less firm, and roots insecure, making trees more prone to windthrow.

Seasonality provides another axis for differentiation. Subtropical forests and regions more than 10° north or south of the equator experience greater variation in climate, and therefore stronger trade winds, monsoons (and hurricanes), particularly on the eastern sides of continents (e.g. Honduras, Belize, Yucatan-Mexico, Guatemala, Nicaragua, southeastern Africa, Madagascar, Vietnam, eastern coast of the Philippines, southeast China, the Caribbean islands, the southwest Pacific Islands, northeast Australia) (Whigham et al. 1999). These regions can be exposed to periodic large scale wind events which leads to rapid regrowth following such events, due to the large quantity of tree species that vigorously resprout (Whitmore 1989; Brokaw and Walker 1991; Whigham et al. 1991; Vandermeer 1996; Vandermeer et al. 1997, 1998; Eaton and Lawrence 2006). Most forests in these regions would be considered semi-evergreen or dry deciduous – meaning that periods of drying can promote fire (often human caused) for land clearance. In fact, many swidden systems are cleared during the dry season and then burned prior to the rains to take advantage of the pulse of nutrients for crop cultivation in the wet season.

In the more equatorial regions where ever-wet forest dominates, winds often occur with the onset of rains through vigorous frontal or convectional thunderstorms that can knock over large swaths of forest with strong down drafts (Whigham et al. 1999). On steeper and often younger more erosion-prone hills and mountains, large amounts of rain can cause landslides, riparian flooding and bank erosion (e.g. in the Andes, Central American Cordierra, central ranges of Sumatra, Borneo, and Malay Peninsula). All large scale wind and rain events are episodic and occur at relatively long time intervals that are difficult to predict. However, they drive the successional dynamic of forests, and therefore by implication, the above- and below-ground carbon stocks. Little to no work has been done on assessing and including this dynamism in the development of regional carbon models predicting future change superimposed upon which, is young second-growth forest, originating after agricultural cessation.

5.3 Fire

Fires in tropical forests are typically the result of drought and human land management practices (Bush et al. 2008). Indeed, fire is thought to be a more imminent threat to tropical forests than climate change (Barlow and Peres 2004; Nepstad et al. 2004; Bush et al. 2008). In contrast to other biomes, such as certain jack pine (*Pinus banksiana*) boreal forests where fire events tend to be naturally occurring, humans have been the primary ignition source of fires in tropical forests since pre-Columbian times. In fact, natural fire in the Amazon has been so rare since the mid-Holocene that the presence of charcoal in soil is an indicator of human activity (Bush et al. 2008). Under normal moisture conditions, the likelihood of fire decreases exponentially with distance from roads and clearings (Cochrane and Laurance 2002). This supports the view that fire is a direct result of human activity in tropical systems. If drought becomes more common in tropical forests as some climate models predict (e.g., Cox et al. 2000), the likelihood of human- induced fires escaping and impacting large portions of the landscape increases. This was seen during the ENSO-induced drought in tropical Indonesia and Amazonia in 1997–1998 where drought caused many human-ignited fires to escape and become wildfires (Bush et al. 2008).

In addition to climate, the impact of fire on tropical forests is also highly linked to forest structure. Nepstad et al. (2007) found that mortality of large trees and lianas following an experimental drought increased. Large trees not only store significant amounts of carbon, but also provide shade, which helps to keep litter moist. The absence of this shade dries out the litter layer and the dead lianas become ladder fuels, thus increasing the probability that an escaped fire will burn the litter layer and reach the canopy (Nepstad et al. 2007). This in turn is likely to impact what types of plants can regenerate and colonize after a fire.

5.4 Herbivory

The effects of elevated atmospheric CO_2 and global climate change on herbivory and other plant/animal interactions in tropical forests are not well understood. Little research has been done in this area. One seminal piece, Coley's "Possible effects of climate change on plant/herbivore interactions in moist tropical forests" (1998) addressed the interdependent roles of climate change and herbivory in tropical forests. More research into how climate change and elevated greenhouse gases will affect herbivore-plant and other predator–prey dynamics is needed. Indeed, although the Coley study is a core research paper on this topic, even this study is not adequately supported by direct experimentation.

6 Climate Change Impacts on Tropical Forest Dynamics

6.1 Increased Productivity Versus Increased Respiration

Many old-growth tropical forests are showing increases in biomass (Malhi and Phillips 2004; Lewis et al. 2009). Studies have shown that there has been a net increase in biomass in recent decades in several Amazonian and Central African semi-evergreen and ever-wet forests. Several studies have addressed methodological challenges in measuring biomass (Baker et al. 2004; Chave et al. 2004). According to new estimations by Malhi and Phillips (2004), the net carbon sink of intact old growth forests of Amazonia is 0.9 ± 0.2 Mg C per ha per year. Applying this rate to the area of moist forest in Amazonia, the Amazon rain forest is thought to sequester nearly 0.6 Pg C per year (Baker et al. 2004).

Like many ecological processes, biomass growth does not occur in isolation. Forest turnover rates in Amazonia have accelerated (Phillips et al. 2004; Laurance et al. 2004). In particular, the greatest increases in turnover rates have occurred on more fertile soils in western Amazonia. This increase in recruitment has been greater than the increase in mortality, which has actually lagged behind this acceleration in growth (Phillips et al. 2004). Similarly, Laurance et al. (2004) found that forests of central Amazonia have experienced changes in dynamics and composition that are not due to any detectable disturbance. In a network of 18 permanent study plots, not only have mortality, recruitment, and growth rates increased over time, but 27 of 115 relatively abundant tree genera have changed significantly in population density or basal area.

However, in a study based on large-scale plots in other tropical regions (mostly Asia, Central America) results do not support the hypothesis that fast-growing species are consistently increasing in dominance in tropical tree communities (Chave et al. 2004). Instead, results suggest that the forests are simultaneously recovering from past disturbances and affected by changes in resource availability. More long-termstudies are necessary to clarify the contribution of global change to the functioning of tropical forests (Chave et al. 2004). What is certain, however, is that these changes in dynamics and composition could have important impacts on the carbon storage and biota of Amazonian forests (Laurance et al. 2004).

In addition, as mentioned previously, most climate models and forest carbon balance models do not take forest composition into account (Phillips et al. 2008). Forests with accelerated or "faster" dynamic have less biomass due in part to ecophysiological differences in plant growth. For example, fast growing species have less dense wood, and therefore less stored carbon, compared to slow-growing species which have denser wood, with more carbon. Early successional forest therefore has less carbon stored than late sucessional forest (Phillips et al. 2008). A summary of how increasing CO_2 concentrations may or may not affect tropical forest growth is provided in Table 4.3 (Malhi and Phillips 2004).

Table 4.3 Arguments to expect, or not, substantial effects of increasing CO_2 concentrations on tropical forest growth and carbon balance (direct and indirect effects are considered, including climate change)

Zero or negative effect on growth and carbon storage	Positive effect on growth and carbon storage
Leaf level	
Increased respiration and photorespiration caused by rising temperatures	Direct fertilization of photosynthesis by high CO_2
Warming temperatures lead to increased evaporative demand, inducing stomatal closure and reducing photosynthesis	Reduced photorespiration caused by high CO_2
Increased emissions of volatile hydrocarbons at higher temperatures consume assimilated carbon	Improved water-use efficiency caused by high CO_2
	Photosynthetic rates increase with moderate warming optimum temperature for photosynthesis rises with rising CO_2
Plant level	
Plants are often saturated with respect to non-structural carbohydrates?	Excess carbon may be used preferentially above ground to acquire rate-limiting resource (light) by investing in wood
Plant growth limited by nutrients other than carbon (N, P, K, Ca)	Excess carbon may be used preferentially below ground to acquire rate-limiting resource (nutrients) through fine root development or supporting P-scavenging mycosymbionts
Rising soil temperatures increase soil acidification and mobilize aluminium, reducing soil nutrient supply	Rising temperatures increase soil mineralization rates and improve nutrient supply
Plant carbon balance is limited by respiration costs rather than by photosynthesis gains?	
Acclimation (downregulation of photosynthesis) limits any response to increasing CO_2	
Stand level	
Biomass ultimately limited by disturbance (e.g. windthrow risk) rather than by resources	Faster growth leads to some biomass gains, with mortality gains lagging
Forest canopies are close to physical limits of forest structure which cannot be increased (e.g. maximum tree height is limited by hydraulics or mechanics)	Rising CO_2 improves water-use efficiency and reduces tension in the water column, allowing an increase in maximum tree height for a given cross-sectional area
Faster growth and turnover may favour disturbance-adapted taxa, with less dense wood	Faster growth and turnover may prevent stand dominance by senescent 'over-mature' trees with high respiration costs, creating positive feedback on stand-level growth rates
Lianas may benefit from increased CO_2 and disturbance, limiting biomass gains by trees	Improved forest water balance leads to reduced drought mortality and fire incidence
Mortality rates increase because of climatic warming and/or drying, or increased climatic variability	
Climatic drying combined with forest fragmentation and degradation lead to increased fire frequency	

Source: From Malhi and Phillips (2004). Reprinted with permission

Tropical forests are resilient to many types of environmental change. However, given the human footprint in many of these forests, the expected resiliency may not materialize (Cowling and Shin 2006). Evergreen rain forests have dominated the Amazon Basin since the last glacial maximum (Beerling and Mayle 2006). Historically, climate change has driven biome shifts in transition or ecotonal zones, while CO_2 changes have led to increased carbon storage (Beerling and Mayle 2006). Many transition zones (e.g., montane forests) and tropical seasonal forests are areas that have experienced rampant deforestation and other types of land-use change in the past century. This may yield some insight into how tropical forests might change in both composition and range in response to climate change (Malhi and Phillips 2004). For example, a warmer climate could drive low elevation forests to higher elevations or extend the range of tropical seasonal forests. However, if there is more deforestation in these seasonal and dry areas, there may be fewer species available that can migrate and adapt to warmer climates with drier soils (Malhi and Phillips 2004).

Many models have predicted decreased forest cover and soil drying over Amazonia in response to the radiative effect of rising CO_2 concentrations in the atmosphere (Friedlingstein et al. 2006; Notaro et al. 2007). What happens to soil carbon pools and the dead biomass from this reduced forest cover is of great importance to researchers studying carbon fluxes under climate change. The fate of these carbon pools under the most extreme scenario modeled – wide-spread tree die-off – depends on drought conditions and elevated soil respiration under higher temperatures (Cox et al. 2000). As air temperature rises, respiration increases, while carbon uptake from photosynthesis continues until it reaches some threshold. In short, the current carbon sink that intact tropical forests provide cannot continue in the same manner indefinitely. How this carbon balance could change, apart from the immediate threats of land use change, habitat fragmentation and fire, is uncertain.

Phillips et al. (2008) provide three scenarios about the future of this carbon sink in the Amazon based on an extensive network of research sites: mature Amazonian forests will either (i) continue to be a carbon sink for decades (Cramer et al. 2001); (ii) quickly become neutral (i.e., uptake equals release) or a small carbon source (Cramer et al. 2001; Körner 2004; Laurance et al. 2004) or (iii) become a mega-carbon source (Cox et al. 2000; Lewis et al. 2006). The difference between the annual stand level growth (uptake: 2%) and mortality (release: 1.6%) of Amazonia is currently estimated to be 0.4%, which is just about enough carbon sequestered to compensate for the carbon emissions of deforestation in the region (Phillips et al. 2008). This means that either a small decrease in growth or a small increase in mortality in mature forests could convert Amazonia from a sink to a source of carbon (Phillips et al. 2008). Better estimates are needed of the amount of mature, secondary and disturbed forests in the Amazon in order to better predict changes in carbon storage trends and the threat of release of this terrestrial sink.

6.2 Changes in Precipitation Amounts and Patterns

Some models have predicted that reduced forest cover and soil drying over Amazonia, in response to the radiative effect of rising CO_2 concentrations in the atmosphere, will result in a reduction in the land's capacity to take up carbon (Friedlingstein et al. 2006; Notaro et al. 2007). In simulation modeling of ecosystem threshold responses to changes in temperature, precipitation, and CO_2, Cowling and Shin (2006) found the 'natural,' intact Amazonian rainforest to be resilient to environmental change, particularly to decreases in temperature and precipitation. However, they also warn that humans have changed these forests so quickly in the past several decades that the resiliency of the Amazonian rain forest is at risk (Cowling and Shin 2006). Asian ever-wet forests are thought to be considerably more sensitive to drying conditions (Paoli and Curran 2007). However, the interactions with other human disturbance factors such as land clearance, edge effects and fragmentation, and fire need to be considered and could have important negative feedback influences (Leighton and Wirawan 1986).

6.3 Land Use Change

Human intervention through deforestation and forest degradation has been the leading cause of perturbation to the carbon cycle in tropical forests (Houghton 1991a; Sampson et al. 1993). As a result, by the year 2050, the tropics could be a source of atmospheric CO_2 (Sampson et al. 1993). Land use change is perhaps the most imminent threat to the ecosystem services that tropical forests provide. It is believed that land use change could lead to the release of 40–80 Pg C per year over the next 50 years (Nightingale et al. 2004).

6.3.1 Deforestation

Deforestation affects the carbon balance of tropical forests and climate feedback cycles in two principal ways: carbon emissions from deforestation and the albedo effect of deforested lands (Bala et al. 2007; Ramankutty et al. 2007). Moreover, for every ton of carbon released to the atmosphere through deforestation, 0.6 additional tons of carbon are released through degradation of the remaining forest (Houghton 1991a). However, current estimates of carbon emissions from tropical deforestation vary greatly and are difficult to compare due to differences in data sources, assumptions, and methods (Ramankutty et al. 2007). Developing and incorporating multiple variables into new and existing ecosystem models for tropical forests is essential to determining carbon fluxes and future effects of deforestation and climate change (Nightingale et al. 2004). In order to fully quantify the carbon emissions from tropical deforestation, one must account for initial carbon stock of vegetation and soils, influence of historical land use, rates and dynamics of land-cover changes, methods of land clearing and the fate of the carbon from cleared vegetation, response of soils following land-cover change, and the representation of processes in ecosystem and climate models used to integrate all of these components (Ramankutty et al. 2007).

While it is a fact that deforestation releases CO_2 to the atmosphere, which in turn has a warming effect on climate, there is another important piece of the deforestation equation that some models neglect. Deforestation comes with biophysical effects on climate, such as changes in land surface albedo, evapotranspiration, and cloud cover. Simulations out to 2150 by Bala et al. (2007), using a three-dimensional model representing physical and biogeochemical interactions between land, atmosphere, and ocean, found that at a global level, deforestation has a net cooling effect on climate. This is because the net cooling influence of changes in albedo and evapotranspiration outweigh the warming effects associated with carbon release (Bala et al. 2007). It is noteworthy that the model predicted different effects associated with the deforestation of tropical vs. temperate and boreal forests. According to the model results, afforestation in the tropics would be beneficial because of the greater role of tropical forests in increasing evapotranspiration, CO_2 sequestration, and cloud cover and thus reducing the heating impacts of global warming. In contrast, deforestation of higher latitude boreal forests would greatly increase albedo relative to evapotranspiration and CO_2 sequestration having an overall positive effect on climate (Bala et al. 2007). It must be emphasized that this is a single study involving simulations so caution needs to be used in interpreting these results.

6.3.2 Agriculture

Expanding crop and pasture lands have a profound effect on the global carbon cycle as tropical forests typically store 20–100 times more carbon per unit area than the agriculture that replaces them (Houghton 1991a). In the Amazon, the growing profitability of large-scale industrial agriculture and cattle ranching has led to significant deforestation. This will only increase forest fragmentation and degradation and subsequent climate effects as it continues to expand (Nepstad et al. 2008). The use of fire to clear forested lands may exacerbate changes to carbon cycling since fire fills the atmosphere with aerosols, thereby reducing transpiration (IPCC 2007).

Within Southeast Asia, the conversion of tropical forests to oil palm plantations is accelerating. This land use change results in a significant net loss of carbon to the atmosphere since the aboveground biomass of oil palm plantations stores less carbon (<36–48 tons C/ha) than tropical primary forests (235 tons C/ha) (Reijinders and Huijbregts 2008). Including carbon releases for fire, which is the primary method for land clearing, the net carbon loss from the system may be as much as 187–199 tons C/ha. If such fires are of high intensity, there is even greater loss of soil carbon (Reijinders and Huijbregts 2008). A full life cycle analysis of forest conversion and carbon loss, and then cultivation and production of biofuels from oil palm, puts into question the assertion that oil palm reduces CO_2 emissions (Reijinders and Huijbregts 2008).

7 Conclusion and Summary Recommendations

Tropical forests account for almost half the gross primary productivity of the world's terrestrial ecosystems.

- Tropical ever-wet and semi-evergreen forests in the Amazon and southeastern Asia typically suffer from droughts during ENSO events. In the short term, tropical forests may be resilient to drought, but increased susceptibility to anthropogenic fire may negate this.
- In tropical ever-wet forests, where droughts are rare, mortality may dramatically increase. Seasonal semi-evergreen forests may show little change.

In tropical forest regions humans have caused most fires. If climate model predictions are correct, increased drought will promote the escape of human-caused fires that will impact large portions of the remaining forest.

- More work should be done to investigate the negative and positive feedbacks of drought, windstorms, insects/pathogens, fire and humans, and their interactions, on the forest dynamic – and in particular, their effects on carbon.

According to long-term permanent plot data, old growth ever-wet and semi-evergreen forests show increasing biomass in recent decades in Amazonian and Central African forests. In contrast to climate model predictions, this increase in forest biomass may reflect forest response to previous natural and anthropogenic disturbances (Chave et al. 2004).

Uncertainties in estimates of both biomass and deforestation contribute to a wide range of estimates of carbon emissions in the tropics.

- Better estimates are needed of the amount of mature, secondary, and disturbed forest in order to better predict changes in carbon storage trends.
- Dry deciduous and montane forests are almost a complete unknown because so little work has been done on these forest types; therefore, much more research needs to be carried out in these areas.
- Even though in the Americas and Asia many dry deciduous forests have been cleared, significant amounts remain in Africa.

References

Ashton PMS (1992) Some measurements of the microclimate within a Sri Lankan tropical rain forest. Agric Forest Meteorol 59:217–235

Ashton PMS, Gunatilleke CVS, Gunatilleke IAUN (1995) Seedling survival and growth of four *Shorea* species in a Sri Lankan rainforest. J Trop Ecol 11:263–279

Baker TR, Phillips OL, Malhi Y, Almeida S, Arroyo L, Di Fiore A, Erwin T, Higuchi N, Killeen TJ, Laurance SG, Laurance WF, Lewis SL, Monteagudo A, Neill DA, Vargas PN, Pitman NCA, Silva JNM, Martinez RV (2004) Increasing biomass in Amazonian forest plots. Philos Trans R Soc Lond B Biol Sci 359:353–365

Baker TR, Coronado ENH, Phillips OL, Martin J, van der Heijden GMF, Garcia M, Espejo JS (2007) Low stocks of coarse woody debris in a southwest Amazonian forest. Oecologia 152:495–504

Bala G, Caldeira K, Wickett M, Phillips TJ, Lobell DB, Delire C, Mirin A (2007) Combined climate and carbon-cycle effects of large-scale deforestation. Proc Natl Acad Sci USA 104:6550–6555

Barlow J, Peres C (2004) Ecological responses to El Nino-induced surface fires in central Brazilian Amazonia: management implications for flammable tropical forests. Philos Trans R Soc Lond B Biol Sci 359:367–380

Beerling DJ, Mayle FE (2006) Contrasting effects of climate and CO_2 on Amazonian ecosystems since the last glacial maximum. Glob Change Biol 12:1977–1984
Brokaw NVL, Walker LR (1991) Summary of the effects of Caribbean hurricanes on vegetation. Biotropica 23:442–447
Bullock SH, Mooney HA, Medina E (1995) Seasonally dry tropical forests. Cambridge University Press, Cambridge
Bush MB, Silman MR, McMichael C, Saatchi S (2008) Fire, climate change and biodiversity in Amazonia: a late-Holocene perspective. Philos Trans R Soc B Biol Sci 363:1795–1802
Cattanio JH, Davidson EA, Nepstad DC, Verchot LV, Ackerman IL (2002) Unexpected results of a pilot throughfall exclusion experiment on soil emissions of CO_2, CH_4, N_2O, and NO in eastern Amazonia. Biol Fert Soils 36:102–108
Chave J, Condit R, Aguilar S, Hernandez A, Lao S, Perez R (2004) Error propagation and scaling for tropical forest biomass estimates. Philos Trans R Soc Lond B Biol Sci 359:409–420
Clark DB, Palmer MW, Clark DA (1999) Edaphic factors and the landscape scale distributions of tropical rain forest trees. Ecology 80:2662–2675
Clark D, Brown S, Kicklighter D, Chambers J, Thomlinson I, Ni J, Holland E (2001) Net primary production in tropical forests: an evaluation and synthesis of existing field data. Ecol Appl 11:371–384
Clark DB, Clark DA, Brown S, Oberbaur SF, Veldekamp E (2002) Stocks and flows of coarse woody debris across a tropical rain forest nutrient and topography gradient. Forest Ecol Manag 164:237–248
Cleveland CC, Townsend AR (2006) Nutrient additions to a tropical rain forest drive substantial soil carbon dioxide losses to the atmosphere. Proc Natl Acad Sci USA 103:10316–10321
Cleveland CC, Nemergut DR, Schmidt SK, Townsend AR (2007) Increases in soil respiration following labile carbon additions linked to rapid shifts in soil microbial community composition. Biogeochemistry 82:229–240
Cochrane M, Laurance W (2002) Fire as a large-scale edge effect in Amazonian forests. J Trop Ecol 18:311–325
Coley PD (1998) Possible effects of climate change on plant/herbivore interactions in moist tropical forests. Climatic Change 39:455–472
Cowling SA, Shin Y (2006) Simulated ecosystem threshold responses to co-varying temperature, precipitation and atmospheric CO_2 within a region of Amazonia. Glob Ecol Biogeogr 15:553–566
Cox PM, Betts RA, Jones CD, Spall SA, Totterdell IJ (2000) Acceleration of global warming due to carbon-cycle feedbacks in a coupled climate model. Nature 408:184–187
Cramer W, Bondeau A, Woodward FI, Prentice IC, Betts RA, Brovkin V, Cox PM, Fisher V, Foley JA, Friend AD, Kucharik C, Lomas MR, Ramankutty N, Sitch S, Smith B, White A, Young-Molling C (2001) Global response of terrestrial ecosystem structure and function to CO_2 and climate change: results from six dynamic global vegetation models. Glob Change Biol 7:357–373
Dixon RK, Brown S, Houghton RA, Solomon AM, Trexler MC, Wisniewski J (1994) Carbon pools and flux of global forest ecosystems. Science 263:185–190
Eaton J, Lawrence D (2006) Woody stocks and fluxes during succession in a dry tropical forest. Forest Ecol Manag 232:46–55
Ediriweera S, Singhakumara BMP, Ashton MS (2008) Variation in canopy structure, light and soil nutrition across elevation of a Sri Lankan tropical rain forest. Forest Ecol Manag 256:1339–1349
Espeleta JF, Clark DA (2007) Multi-scale variation in fine-root biomass in a tropical rain forest: a seven-year study. Ecol Monogr 77:377–404
Feeley KJ, Wright SJ, Supardi MNN, Kassim AR, Davies SJ (2007) Decelerating growth in tropical forest trees. Ecol Lett 10:461–469
Foody G, Palubinskas G, Lucas R, Curran P, Honzak M (1996) Identifying terrestrial carbon sinks: classification of successional stages in regenerating tropical forest from Landsat TM data. Remote Sens Environ 55:205–216
Frangi J, Lugo A (1985) Ecosystem dynamics of a subtropical floodplain forest. Ecol Monogr 55:351–369
Friedlingstein P, Cox P, Betts R, Bopp L, Von Bloh W, Brovkin V, Cadule P, Doney S, Eby M, Fung I, Bala G, John J, Jones C, Joos F, Kato T, Kawamiya M, Knorr W, Lindsay K, Matthews HD, Raddatz T, Rayner P, Reick C, Roeckner E, Schnitzler KG, Schnur R, Strassmann K, Weaver AJ, Yoshikawa C, Zeng N (2006) Climate-carbon cycle feedback analysis: results from the (CMIP)-M-4 model intercomparison. J Climate 19: 3337–3353
Grogan J, Ashton MS, Galvao J (2003) Big-leaf mahogany (*Swietenia macrophylla*) seedling survival and growth across a topographic gradient in southeast Para, Brazil. For Ecol Manag 186:311–326
Griscom HP, Ashton MS (2011) Restoration of dry tropical forests in Central America: a review of pattern and process. Forest Ecol Manag 261:1564–1579
Harmon ME, Franklin JF, Swanson PJ, Sollins P, Gregory SV, Lattin JD, Anderson NH, Cine SP, Aumen NG, Sedell SR, Lienkaemper GW, Cromack K, Cummins KW (1986) Ecology of coarse woody debris in temperate ecosystems. Adv Ecol Res 15:133–156
Heinsohn RJ, Kabel RL (1999) Sources and control of air pollution. Prentice Hall, Upper Saddle River
Houghton R (1991a) Releases of carbon to the atmosphere from degradation of forests in tropical Asia. Can J Forest Res 21:132–142
Houghton R (1991b) Tropical deforestation and atmospheric carbon-dioxide. Climatic Change 19:99–118
Houghton RA (2005) Aboveground forest biomass and the global carbon balance. Glob Change Biol 11:945–958
Houghton RA, Lawrence KT, Hackler JL, Brown S (2001) The spatial distribution of forest biomass in the Brazilian Amazon: a comparison of estimates. Glob Change Biol 7:731–746

IPCC (2007) Climate change: synthesis report. In: Pachauri RK, Reisinger A (eds) Contribution of working groups I, II and III to the fourth assessment report of the intergovernmental panel on climate change. IPCC, Geneva, 104 p

Keller M, Palace M, Asner GP, Pereira R, Silva JN (2004) Coarse woody debris in undisturbed and logged forests in the eastern Brazilian Amazon. Glob Change Biol 10:784–795

Korner C (2004) Through enhanced tree dynamics carbon dioxide enrichment may cause tropical forests to lose carbon. Philos Trans R Soc Lond B Biol Sci 359:493–498

LaDeau S, Clark J (2001) Rising CO_2 levels and the fecundity of forest trees. Science 292:95–98

Lal R, Kimble JM (2000) What do we know and what needs to be known and implemented for C sequestration in tropical ecosystems. Glob Climate Change Trop Ecosyst 417–431

Laurance W, Oliveira A, Laurance S, Condit R, Nascimento H, Sanchez-Thorin A, Lovejoy T, Andrade A, D'Angelo S, Ribeiro J, Dick C (2004) Pervasive alteration of tree communities in undisturbed Amazonian forests. Nature 428:171–175

Leighton M, Wirawan N (1986) Catastrophic drought and fire in Borneo tropical rain forest associated with the 1982–1983 El Nino southern oscillation event. In: Prance GT (ed) Tropical rain forests and the world atmosphere. Westview Press, Boulder, pp 75–102

Lewis SL, Malhi Y, Phillips OL (2004) Fingerprinting the impacts of global change on tropical forests. Philos Trans R Soc Lond B Biol Sci 359:437–462

Lewis SL, Phillips OL, Baker TR (2006) Impacts of global atmospheric change on tropical forests. Trends Ecol Evol 21:173–174

Lewis SL, Lopez-Gonzalez G, Sonke B, Affum-Baffoe K, Baker TR, Ojo LO, Phillips OL, Reitsma JM, White L, Comiskey JA, Djuikouo MN, Ewango CEN, Feldpausch TR, Hamilton AC, Gloor M, Hart T, Hladik A, Lloyd J, Lovett JC, Makana JR, Malhi Y, Mbago FM, Ndangalasi HJ, Peacock J, Peh KSH, Sheil D, Sunderland T, Swaine MD, Taplin J, Taylor D, Thomas SC, Votere R, Woll H (2009) Increasing carbon storage in intact African tropical forests. Nature 457:1003–1006

Lloyd J, Farquhar GD (1996) The CO_2 dependence of photosynthesis, plant growth responses to elevated atmospheric CO_2 concentrations and their interaction with soil nutrient status.1. General principles and forest ecosystems. Funct Ecol 10:4–32

Lyon B (2004) The strength of El Nino and the spatial extent of tropical drought. Geophys Res Lett 31

Malhi Y, Phillips OL (2004) Tropical forests and global atmospheric change: a synthesis. Philos Trans R Soc Lond B Biol Sci 359:549–555

Malhi Y, Nobre A, Grace J, Kruijt B, Pereira M, Culf A, Scott S (1998) Carbon dioxide transfer over a central Amazonian rain forest. J Geophys Res Atmos 103:31593–31612

Malhi Y, Baker TR, Phillips OL, Almeida S, Alvarez E, Arroyo L, Chave J, Czimczik CI, Di Fiore A, Higuchi N, Killeen TJ, Laurance SG, Laurance WF, Lewis SL, Montoya LMM, Monteagudo A, Neill DA, Vargas PN, Patino S, Pitman NCA, Quesada CA, Salomao R, Silva JNM, Lezama AT, Martinez RV, Terborgh J, Vinceti B, Lloyd J (2004) The above-ground coarse wood productivity of 104 Neotropical forest plots. Glob Change Biol 10:563–591

Nepstad DC, Decarvalho CR, Davidson EA, Jipp PH, Lefebvre PA, Negreiros GH, Dasilva ED, Stone TA, Trumbore SE, Vieira S (1994) The role of deep roots in the hydrological and carbon cycles of Amazonian forests and pastures. Nature 372:666–669

Nepstad D, Lefebvre P, Da Silva UL, Tomasella J, Schlesinger P, Solorzano L, Moutinho P, Ray D, Benito JG (2004) Amazon drought and its implications for forest flammability and tree growth: a basin-wide analysis. Glob Change Biol 10:704–717

Nepstad DC, Tohver IM, Ray D, Moutinho P, Cardinot G (2007) Mortality of large trees and lianas following experimental drought in an Amazon forest. Ecology 88:2259–2269

Nepstad DC, Stickler CM, Soares B, Merry F (2008) Interactions among Amazon land use, forests and climate: prospects for a near-term forest tipping point. Philos Trans R Soc B Biol Sci 363:1737–1746

Nightingale JM, Phinn SR, Held AA (2004) Ecosystem process models at multiple scales for mapping tropical forest productivity. Prog Phys Geog 28:241–281

Notaro M, Vavrus S, Liu ZY (2007) Global vegetation and climate change due to future increases in CO2 as projected by a fully coupled model with dynamic vegetation. J Climate 20:70–90

Paoli GD, Curran LM (2007) Soil nutrients limit fine litter production and tree growth in mature lowland forest of Southwestern Borneo. Ecosystems 10:503–518

Phillips OL, Baker TR, Arroyo L, Higuchi N, Killeen TJ, Laurance WF, Lewis SL, Lloyd J, Malhi Y, Monteagudo A, Neill DA, Vargas PN, Silva JNM, Terborgh J, Martinez RV, Alexiades M, Almeida S, Brown S, Chave J, Comiskey JA, Czimczik CI, Di Fiore A, Erwin T, Kuebler C, Laurance SG, Nascimento HEM, Olivier J, Palacios W, Patino S, Pitman NCA, Quesada CA, Salidas M, Lezama AT, Vinceti B (2004) Pattern and process in Amazon tree turnover, 1976–2001. Philos Trans R Soc Lond B Biol Sci 359:381–407

Phillips OL, Lewis SL, Baker TR, Chao KJ, Higuchi N (2008) The changing Amazon forest. Philos Trans R Soc B Biol Sci 363:1819–1827

Potts M (2003) Drought in a Bornean everwet rain forest. J Ecol 91:467–474

Pregitzer KS, Zak DR, Curtis PS, Kubiske ME, Teeri JA, Vogel CS (1995) Atmospheric CO_2, soil-nitrogen and turnover of fine roots. New Phytol 129:579–585

Ramankutty N, Gibbs HK, Achard F, Defries R, Foley JA, Houghton RA (2007) Challenges to estimating carbon emissions from tropical deforestation. Glob Change Biol 13:51–66

Reijnders L, Huijbregts MAJ (2008) Palm oil and the emission of carbon-based greenhouse gases. J Clean Prod 16:477–482
Ricklefs RE (2001) The economy of nature, 5th edn. W.H. Freeman and Co, New York
Robinson D (2007) Implications of a large global root biomass for carbon sink estimates and for soil carbon dynamics. Proc R Soc B Biol Sci 274:2753–2759
Saleska SR, Miller SD, Matross DM, Goulden ML, Wofsy SC, da Rocha HR, de Camargo PB, Crill P, Daube BC, de Freitas HC, Hutyra L, Keller M, Kirchhoff V, Menton M, Munger JW, Pyle EH, Rice AH, Silva H (2003) Carbon in Amazon forests: unexpected seasonal fluxes and disturbance-induced losses. Science 302:1554–1557
Saleska SR, Didan K, Huete AR, da Rocha HR (2007) Amazon forests green-up during 2005 drought. Science 318:612
Samanta A, Ganguly S, Myneni RB (2010) MODIS Enhanced Vegetation Index data do not show greening of Amazon forests during the 2005 drought. New Phytol 189:11–15
Sampson RN, Apps M, Brown S, Cole CV, Downing J, Heath LS, Ojima DS, Smith TM, Solomon AM, Wisniewski J (1993) Workshop summary statement - terrestrial biospheric carbon fluxes - quantification of sinks and sources of CO_2. Water Air Soil Pollut 70:3–15
Sayer EJ, Powers JS, Tanner EVJ (2007) Increased litterfall in tropical forests boosts the transfer of soil CO_2 to the atmosphere. PLoS One 2:e1299
Schwendenmann L, Veldkamp E (2005) The role of dissolved organic carbon, dissolved organic nitrogen, and dissolved inorganic nitrogen in a tropical wet forest ecosystem. Ecosystems 8:339–351
Soepadmo E (1993) Tropical rain-forests as carbon sinks. Chemosphere 27:1025–1039
Sotta ED, Veldkamp E, Schwendenmann L, Guimaraes BR, Paixao RK, Ruivo M, Da Costa ACL, Meir P (2007) Effects of an induced drought on soil carbon dioxide (CO_2) efflux and soil CO_2 production in an Eastern Amazonian rainforest, Brazil. Glob Change Biol 13:2218–2229
Tsonis A, Elsner J, Hunt A, Jagger T (2005) Unfolding the relation between global temperature and ENSO. Geophys Res Lett 32
USDA (2002) Natural Resources Conservation Service, Soil Survey Division, World soil resources
USDA (2005) Natural Resources Conservation Science, Soil Survey Division, World soil resources
Vandermeer J (1996) Disturbance and neutral competition theory in rain forest dynamics. Ecol Model 85:99–111
Vandermeer JH, Perfecto I (2005) Breakfast of biodiversity: the political ecology of rain forest destruction, 2nd edn. Food First Books, Oakland
Vandermeer J, de la Cerda IG, Boucher D (1997) Contrasting growth rate patterns in eighteen tree species from a post-hurricane forest in Nicaragua. Biotropica 29:151–161
Vandermeer J, Brenner A, de la Cerda JI (1998) Growth rates of tree height Six years after hurricane damage at four localities in Eastern Nicaragua. Biotropica 30:502–509
Veldkamp E, Becker A, Schwendenmann L, Clark D, Schulte-Bisping H (2003) Substantial labile carbon stocks and microbial activity in deeply weathered soils below a tropical wet forest. Glob Change Biol 9:1171–1184
Vitousek PM, Sanford RL (1986) Nutrient cycling in moist tropical forest. Annu Rev Ecol Syst 17:137–167
Watson RT, Noble I, Bolin B (2000) IPCC: special report: land use, change, and forestry
Whigham DF, Olmsted EC, Cano ME, Harmon ME (1991) The impact of hurricane Gilbert on trees, litterfall, and woody debris in a dry tropical forest in northern Yucatan Peninsula. Biotropica 23:434–441
Whigham DF, Dickinson MB, Brokaw NVL (1999) Background canopy gap and catastrophic wind disturbance in tropical forest. In: Walker L (ed) Ecosystems and disturbed ground. Elsevier, The Hague, pp 223–252
Whitmore TC (1989) Changes over twenty-one years in the Kolobangara rain forest. J Ecol 77:409–483
Wilcke W, Hess T, Bengel C, Homeier J, Valarezo C, Zech W (2004) Coarse woody debris in a montane forest in Ecuador: mass, C and nutrient stock. Forest Ecol Manag 205:139–147

Carbon Dynamics in the Temperate Forest

5

Mary L. Tyrrell, Jeffrey Ross, and Matthew Kelty

Executive Summary

Twenty-five percent of the world's forests are in the temperate biome. They include a wide range of forest types, and the exact boundaries with boreal forests to the north and tropical forests to the south are not always clear. There is a great variety of species, soil types, and environmental conditions which lead to a diversity of factors affecting carbon storage and flux. Deforestation is not a major concern at the moment, and the biome is currently estimated to be a carbon sink of 0.2–0.4 Pg C/year, about 37% of the total net terrestrial carbon uptake, disproportionately higher than its representative area, with most of the sink occurring in North America and Europe.

Temperate forests have been severely impacted by human use – throughout history, all but about 1% have been logged-over, converted to agriculture, intensively managed, grazed, or fragmented by sprawling development. Nevertheless, they have proven to be resilient – mostly second growth forests now cover about 40–50% of the original extent of the biome. Although remaining intact temperate forests continue to be fragmented by development, particularly in North America, there is no large-scale deforestation at present, nor is there likely to be in the future. The status of the temperate biome as a carbon reservoir and atmospheric CO_2 sink rests mainly on strong productivity and resilience in the face of disturbance. The small "sink" status of temperate forests could change to a "source" status if the balance between photosynthesis and respiration shifts.

What We Know About Carbon Storage and Flux in Temperate Forests

- Older forests have more carbon stock than younger stands, and mixed species stands in the moist broadleaf and coniferous forest type tend to have higher carbon density than single species stands. Younger stands tend to have higher rates of carbon sequestration, as indicated by net ecosystem productivity (NEP), than mid- or older-aged stands, although the data are highly variable.
- The below ground carbon pool of living biomass (primarily roots), roughly estimated to be 5–10% of total carbon, is much smaller than the above ground pool; however, this is a tenuous conclusion because the below ground biomass carbon pool is the least studied part of the forest carbon budget.
- Soils contain at least half the carbon in temperate forests and possibly as much as two-thirds; this carbon pool appears to be stable under most disturbances, such as logging, wind storms, and invasive species, but not with land

M.L. Tyrrell (✉) • J. Ross
Yale School of Forestry and Environmental Studies, Prospect Street 360, 06511 New Haven, CT, USA
e-mail: mary.tyrrell@yale.edu

M. Kelty
Department of Environmental Conservation, University of Massachusetts Amherst, Amherst, MA, USA

M.S. Ashton et al. (eds.), *Managing Forest Carbon in a Changing Climate*,
DOI 10.1007/978-94-007-2232-3_5, © Springer Science+Business Media B.V. 2012

use change. Huge losses can occur when converting forests to agriculture or development.

- Atmospheric pollution, primarily in the form of nitrogen oxides (NOx) emitted from burning fossil fuels and ozone (O_3), is a chronic stressor in temperate forest regions. Because most temperate forests are considered nitrogen-limited, nitrogen deposition may also act as a growth stimulant (fertilizer effect). Under current ambient levels, nitrogen deposition is most likely enhancing carbon sequestration; however, the evidence regarding long-term chronic nitrogen deposition effects on carbon sequestration is mixed.

What We Do Not Know About Carbon Storage and Flux in Temperate Forests

- Data on mineral soil carbon stocks in temperate forests can only be considered approximations at this time as there is very little research on deep soil carbon (more than 100 cm).
- Global circulation models predict that higher concentrations of atmospheric CO_2 will increase the severity and frequency of drought in regions where temperate forests are found. However, there is a great deal of uncertainty about how drought will affect carbon cycles.
- Little is known about how the interactions between temperature, moisture, available nutrients, pollutants, and light influence key environmental variables, such as drought, to affect ecosystem carbon flows.

What We Think Are the Influences on Carbon Storage and Flux in Temperate Forests

- There is tremendous variability in carbon stocks between forest types and age classes; carbon stocks could easily be lost if disturbance or land use change shifts temperate forests to younger age classes or if climate change or land use change shifts the spatial extent of forest types. On the other hand, if temperate forests are managed for longer rotations, or more area in old growth reserves, then the carbon stock will increase.
- Temperate forests are strongly seasonal, with a well-defined growing season that depends primarily on light (day length) and temperature. This is probably the most important determinant, along with late-season moisture, of temperate forest productivity and hence carbon sequestration.
- On balance, the evidence regarding nitrogen deposition effects on carbon sequestration is mixed. Under current ambient levels, nitrogen deposition is most likely enhancing carbon sequestration. However, under chronic nitrogen deposition, temperate forests may no longer be nitrogen limited, thus the nitrogen "fertilization" effect will be diminished as other resources become constrained.

How We Think the Carbon Status of Temperate Forests Will Change with Changing Climate

- There is evidence of increasing productivity in temperate forests as climate has warmed in the last ~50 years: however, this is confounded by successional dynamics and environmental variables. The atmospheric system has not only experienced changes in temperature, precipitation, and radiation, but in CO_2 concentration and pollutants.
- The few studies that have modeled multi-factor influences on temperate forest net ecosystem productivity or carbon flux have found that combined effects are expected to diminish the effect of CO_2 enrichment alone.
- Natural disturbances, particularly windstorms, ice storms, floods, insect outbreaks, and fire are significant determinants of temperate forest successional patterns. The frequency and intensity of stand-leveling windstorms (hurricanes, tornadoes) is expected to increase under a warmer climate in temperate moist broadleaf and coniferous forest regions, so that fewer stands would reach old-growth stages of development.
- If changing climate alters the frequency and intensity of fires, floods and hurricanes, re-vegetation and patterns of carbon storage will likely be affected, particularly in interior coniferous forests of mountains, the woodlands and pinelands of Mediterranean climates, and coastal forests on eastern sides of continents traditionally exposed to typhoons and hurricanes and floodplains.

1 Introduction

This chapter presents a literature review of carbon dynamics in temperate forests throughout the world. It first characterizes the geographical regions, climates, and forest types within the temperate forest biome. It then describes the carbon stocks that are sequestered within the four forest components: aboveground biomass, belowground biomass, litter layer, and soil. The next section reviews the biotic interactions of carbon uptake through photosynthesis, and carbon loss through respiration and decomposition. The effects of abiotic influences (mostly forest ecosystem disturbances) are also described; they include fire, wind and ice storms, insect outbreaks, nitrogen deposition, ozone pollution, forest management and other land use. These disturbances can greatly affect carbon flux and storage. The final section discusses how changes in climate might impact carbon dynamics through changes in net primary productivity (NPP) and disturbance regimes.

About 25% of the world's forests are in the temperate biome. The majority of this land lies in the Northern Hemisphere, with the southern limit being somewhat above the Tropic of Capricorn (about 30°N). The northern boundary meets the boreal region in an irregular pattern at about 50–55°N. The temperate forest within this zone includes large areas of North America, Europe, and Asia. In North America, there is a broad swath of temperate forest in the east, and a narrower area in the west limited by tall mountain ranges close to the Pacific coast. The interior of the continent consists of prairies, steppes, and mountains. Similarly in Eurasia, a large area of temperate forest covers much of Europe in the west and northeast China, Japan, and Korea in the east, again with a vast non-forested region in the center of the continent.

There is comparatively much less temperate forest in the Southern Hemisphere, simply because the southern continents barely extend below the 30° parallel, where climates would be suitable for temperate forest vegetation. However, there are some important temperate forest regions that occur in southeast Australia (including Tasmania), in New Zealand, on the west coast of Chile, and on the east coast of Argentina (Fig. 5.1; Dixon et al. 1994).

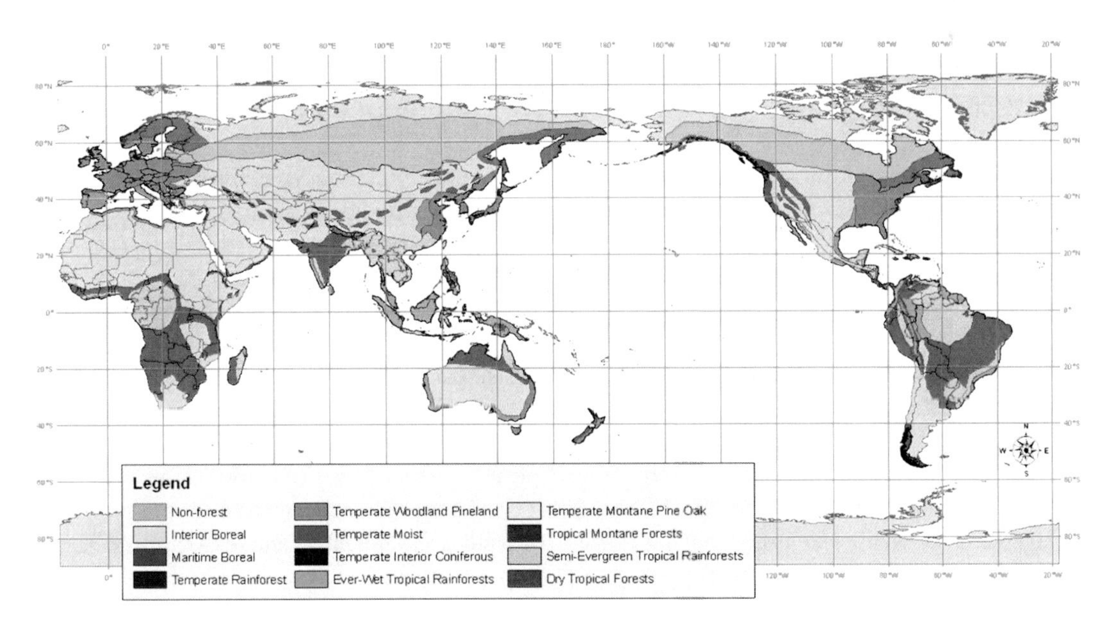

Fig. 5.1 Original extent of boreal, temperate, and tropical forest types of the world prior to land clearing

2 Temperate Forests

In general, temperate forests favor the climatic conditions that characterize the humid mid-latitude regions of western and central Europe, eastern North America, and eastern Asia (Archibold 1995). The northern extent is limited primarily by low winter temperatures (Perry 1994). Climate in these regions exhibits a marked seasonality; it alternates between warm moist summers and winters mild enough to support broad-leaved angiosperms (Perry 1994), at least in the more southerly parts of the biome. The growing season lasts 120–250 days and daily temperatures tend to range from −30° to 30°C, with tree photosynthesis occurring between 5° and 25°C (Martin et al. 2001). Precipitation of between 500 and 1,500 mm tends to be either distributed evenly throughout the year, or peaking in summer, with local variation depending on factors such as latitude, topography, and continental position.

Soils are fertile, often enriched by a decaying litter layer. Soils in Europe are characterized by brown earth, sometimes on calcareous material resembling mollisols. In North America and Asia, alfisols, inceptisols (reflecting the last glaciation), spodosols, and ultisols are common soil types. Soils in the Southern Hemisphere usually consist of highly podsolized material (spodosols) because of high rainfall and granitic substrates (Martin et al. 2001).

The temperate forests can be classified into five major types: *moist broadleaf and coniferous*; *interior coniferous*; *montane oak-pine*; *woodland and pineland*; and *temperate rainforests* (Fig. 5.1).

2.1 Moist Broadleaf and Coniferous Forests

These are mesic, mixed forests with a rich suite of genera, including maple (*Acer*), oak (*Quercus*), hickory (*Carya*), elm (*Ulmus*), linden (*Tilia*), birch (*Betula*), beech (*Fagus*), ash (*Fraxinus*), hemlock (*Tsuga*), and pines (*Pinus*), mostly the soft pine species. Fire plays a relatively minor role, except in the forests dominated by hard pine species on sandy coastal plains. These mixed forests are located in the eastern United States, central Europe, and northeast Asia (including northeast China, Korea and Japan), making up the majority of the temperate forest biome. Northeast Asia has the most diverse tree flora with several genera and families that are not found elsewhere (e.g. *Cunninghamia, Cryptomaria, Phellodendron*).

Soils classified as ultisols underlie much of this area, particularly in the unglaciated parts of eastern North America, and are generally desirable for cultivation because they are usually fertile (though often stony) and require no irrigation because of year-round precipitation. Glaciated regions further north on each continent have either weak soil development (inceptisols) that are often stony and thin to bedrock, or strong organic accumulations and leached upper horizons (spodosols) because cold temperatures and high rainfall inhibit decomposition. These regions are all exposed to both seasonal warm and cold air masses, which cause this forest type to have four distinct seasons. Temperatures vary widely from season to season with annual temperatures averaging about 8°C; precipitation ranges between 750 and 1,500 mm spread fairly evenly throughout the year (Reich and Frelich 2002).

2.2 Interior Coniferous Forests

These forests occur in harsh continental climates and are often found on andisols (volcanic origin) or mountain inceptisols (glacial origin). The principal species are hard pines, spruce (*Picea*), fir (*Abies*), poplar and aspen (*Populus*) and larch (*Larix*). These interior mountain forests are located east of the coastal mountain chain of the U.S. and Canada, and also in Central Asia; they are closely related to interior continental boreal forests in species and forest structure. Soils are young, rocky, often skeletal, and exposed to the extremes of cold winters and dry summers. These forests are fire-adapted, with fire regimes ranging from low intensity, frequent ground fires to high intensity, infrequent stand-replacing fires (McNab and Avers 1994). Precipitation occurs mostly in winter as snow, with the growing season in spring strongly dependent on snow melt.

2.3 Montane Oak-Pine Forests

The mountain ranges of Mexico, Central America, the Himalayas, the Mediterranean region, and Turkey are fire-adapted and relatively dry. These forests are characterized by a diversity of hard pine and evergreen oak species. They grow on elevations ranging from dry, low sites (1,000 m) to moist, high sites (3,000 m). Rainfall occurs mostly in the winter, with groundstory fires occurring in the dry summers. Soils are alfisols or montane-origin inceptisols from the last glaciation. This forest type can be considered a variant of the interior coniferous forests, with the montane oak-pine type being lower in latitude, warmer, and more droughty.

2.4 Woodland and Pineland Forests

These typically fire-adapted forests are located in dry, southern temperate climates, mostly on low elevations along coasts. They include hard pines and oaks in the coastal Mediterranean region and in low elevation areas of Mexico, plus the savannas of Africa and Australia dominated by *Acacia* and *Eucalyptus* species. Soils that are generally classified as alfisols predominate. Such soils are more fertile than ultisols but have dry summers. Annual rainfall is 500–700 mm which falls mostly in winters.

2.5 Temperate Rainforests

These are moist, mesic forests that grow on mountain ranges, usually along western coasts where westerly winds bring high precipitation. Annual rainfall can be greater than 2,000 mm, with year-round precipitation, or with summers somewhat drier. These temperate rainforests include the Pacific Northwest coast of the U.S. and Canada, with the main species being spruce, hemlock, fir, Douglas-fir (*Pseudotsuga*) and western red cedar (*Thuja*); the southwest coastal fringe of Chile and Tierra del Fuego, Argentina, with southern beech (*Nothofagus*); and New Zealand and southeast Australia, with southern beech, eucalyptus, and podocarps (*Podocarpus*). Spodosols and andisols are the predominant soil types. Andisols are volcanic soils that with high precipitation can be very productive. Spodosols are acidic soils associated with bedrock geology, predominantly comprised of minerals such as quartz and silica, and are therefore often nutrient poor.

3 Overview of Forest Cover and Carbon Stocks in Temperate Regions

Temperate forests now cover about 10.4 million km^2 (IPCC 2000; Heath et al. 1993; Dixon et al. 1994), with current forested area estimated to be 40–50% of the original extent (Smith et al. 2009; Bryant et al. 1997). Most of the forest loss occurred in Europe, where many countries had lost more than 90% of their forest cover by the late medieval period (Mather 1990), and China, followed by eastern North America (Malhi et al. 1999; Houghton 1995). Historically, temperate forests have been exploited for timber and charcoal, cleared for agriculture and development, and otherwise heavily impacted by humans (Heath et al. 1993; Nabuurs et al. 2003). Rudel et al. (2005) have described a generalized pattern of forest decline and recovery that has occurred in many countries. With the spread of agriculture and urbanization, forest cover declines until it reaches a turning point; forest cover then begins to increase but this recovery rarely reaches more than 50% of the original forest area. This model appears to apply for many countries in Europe, beginning in the Iron Age (4,000 years BP); forests declined gradually from nearly 100% to only 1–15% of the land area before the turning point, which occurred in various countries anywhere from 1750 to 1950. Now European countries have recovered to anywhere from 10% to more than 50% forest cover, with some Scandinavian countries at 70%. The deforestation of Northeast

China followed a similar general pattern as Europe. The temperate region of China was 50% forested 6,000 years ago, before any substantial deforestation by humans (Ren 2007); forest cover then declined gradually to 25% in the twentieth century. New afforestation programs that were started in 1980 are now increasing forest land area more than 1% annually. In contrast, there was little change in the United States forest cover until the seventieth century. At that time, natural forests covered 46% of the total land area of the current United States, but declined to 30% by about 1930 at the turning point. The decline was not as extreme proportionally as in Europe or China, but U.S. forest cover has increased only a small amount since that time. (FAO 2011).

Currently, the overall temperate forest area is fairly stable at 25% of the global forest, and contains about 11% of the global carbon stock (99–159 Pg carbon) (Heath et al. 1993; Dixon et al. 1994; IPCC 2000). Most countries in Europe and the temperate region of China have increasing forest cover, Australia and North Korea are losing forest cover, and the United States, Japan, South Korea, and New Zealand are stable. Data are not available for the smaller temperate forest areas in South America (FAO 2011). Although stable in area, net productivity is increasing, as most evidence indicates that the temperate forest biome is currently a carbon sink of about 0.2–0.4 Pg of carbon per year (Dixon et al. 1994; Luysssaert et al. 2007; Table 5.1), accounting for 37% of the global carbon uptake (IPCC 2000), well beyond its representative land area, and thus crucial to CO_2 emissions mitigation for the planet. Most temperate forests are currently in "middle age," from 50 to 100 years old, so growth in this biome should continue to be strong in the near term, but less so in the long-term as these forests age.

4 Temperate Forests as Carbon Sinks

Changes in carbon density at a site are generally measured in units of Mg C/ha/year, and are used for comparing the magnitude of carbon fluxes (sink or source) among different sites, as in Table 5.1. Note that positive numbers are sinks for NPP and NEP measurements, whereas negative numbers are sinks for NEE measurements. Net primary productivity (NPP) and net ecosystem productivity (NEP) values are obtained by measuring vegetation and soils; net ecosystem exchange (NEE) values are determined from flux data from eddy covariance towers. At most research sites, only one of the methods is used, making it difficult to compare NEP and NEE data. A study by Curtis et al. (2002) comparing biometric and meteorological methods at five AmeriFlux sites in the northeastern US found no systematic difference between NEP and NEE estimates. Flux measurements (NEE) at these deciduous forests were higher than biometric measurements (NEP) in three sites, similar in one site, and lower in the fifth. On the other hand, ΔC (changes in live biomass and soil) was close to NEE in all but one site. They point out several unquantified pathways of carbon that could account for differences between the two methods, including nonstructural carbohydrate pools and soil leaching.

Most of the European research sites in Table 5.1 were part of the EuroCarboFlux network, with measurements from flux towers. All of these research sites were sinks except for one. The magnitude of the sinks can be seen in the map of hotspots of strong sinks across Europe (Fig. 5.2), with the range of 0–7 Mg C/ha/year. There are 19 sites in Table 5.1 for North America (U.S. and southern Canada). Nearly all (16) sites were sinks, with a similar range of 0–7 Mg C/ha/year; two sites were sources, and one was in balance. One site in Australia which was in a radiata pine (*Pinus radiata*) plantation functioned as a sink. Increasing carbon sinks in China can be inferred from data for plantations showing a mean carbon density of 15.3 Mg C/ha in the 1970s (early in the plantation programs), and increasing to 31.1 Mg C/ha in the 1990s. This represents a sink of approximately 0.8 Mg C/ha/year for that 20-year period (Fang et al. 2001). Natural forests in China had little change in carbon density over that time period.

Successional dynamics and disturbance play an important role in carbon uptake. Younger

Table 5.1 Carbon fluxes in temperate forests, showing carbon density (Mg C/ha/year)

Source		Forest type	Age	Carbon flux (Mg C/ha/year)		
				NPP ("+" = sink)	NEP ("+" = sink)	NEE ("–" = sink)
Barford et al. (2001)	1	Oak-dominated hardwood, Massachusetts, USA	30–100			–2.0
Fahey et al. (2005)	1	Northern hardwood, New Hampshire, USA	70–100	4.83	0 ± 0.2	
Carrara et al. (2003)	1	Mixed conif/decid, Belgium	70			1.1
As reported in Carrara et al. (2003)	1	Deciduous, EuroCarboFlux sites	80–100			–1.77 to –4.6
As reported in Carrara et al. (2003)	1	Coniferous, EuroCarboFlux sites	65–85			–3.19 to –7.2
Knohl et al. (2003)	1	European beech, Germany	Old growth (250)			–4.9
Gough et al. (2008)	1	Mixed northern hardwood, Michigan, USA	6–90 (ave. 85)		1.5	
Hanson et al. (2003)	1	Oak, Tennessee, USA	58–100	7.29 ± 0.69	1.87 ± 0.67	
Malhi et al. (1999)	1	Oak-Hickory, Tennessee, USA	55			–5.85
Valentini et al. (2000)	1	Mixed conifer, France	29		4.3	
Wofsy et al. (1993)	1	Mixed deciduous/hemlock, Massachusetts USA	50–70			–3.7 ± 0.7
Yuan et al. (2008)	1	White pine, Ontario, Canada	55		1.62	
Yuan et al. (2008)	1	Balsam-fir, New Brunswick, Canada	27		5.08	
Hamilton et al. (2002)	1	Loblolly pine, North Carolina, USA	15	7.05	4.28	
Maier and Kress (2000)	1	Loblolly pine, North Carolina, USA	11	5.0 to 12.35	–1.0 to 7.21	
As reported in Schwalm et al. (2010)	1	Loblolly pine, North Carolina, USA	20			–3.5
As reported in Schwalm et al. (2010)	1	White pine, Ontario, Canada	Mature			–1.33
As reported in Schwalm et al. (2010)	1	Mixed deciduous, midwestern USA	Various			–1.32 to –3.46
Law et al. (2003)	2	Ponderosa pine, Oregon, USA	20	2.08	–1.24	
Law et al. (2003)	2	Ponderosa pine, Oregon, USA	70–100	4.0 to 4.85	1.18–1.70	
Law et al. (2003)	2	Ponderosa pine, Oregon, USA	250	3.32	0.35	
Law et al. (1999, 2000)	2	Ponderosa pine, Oregon, USA	Mixed	4.05	2.66	
As reported in Schwalm et al. (2010)	2	Ponderosa pine, USA	int to old			–5.36 to –6.12
As reported in Schwalm et al. (2010)	2	Ponderosa pine, USA	Young			–1.98 to –2.06
Bradford et al. (2008)	2	Subalpine conifer, Rocky Mountains, USA	Mature	2.3 to 3.1		
Jassal et al. (2008)	5	Douglas-fir, Pacific Northwest, USA	58		3.26	
Yuan et al. (2008)	5	Douglas-fir, British Columbia, Canada	55		2.73	
As reported in Schwalm et al. (2010)	5	Douglas-fir, British Columbia, Canada	Mature			–2.44
As reported in Schwalm et al. (2010)	5	Douglas-fir, British Columbia, Canada	Young planatation			0.91
Ryan et al. (1996)		Radiata pine, Australia	20	9.03	2.42	

1 moist broadleaf and coniferous, *2* interior coniferous, *3* montane oak/pine, *4* woodland and pineland, *5* temperate rainforest

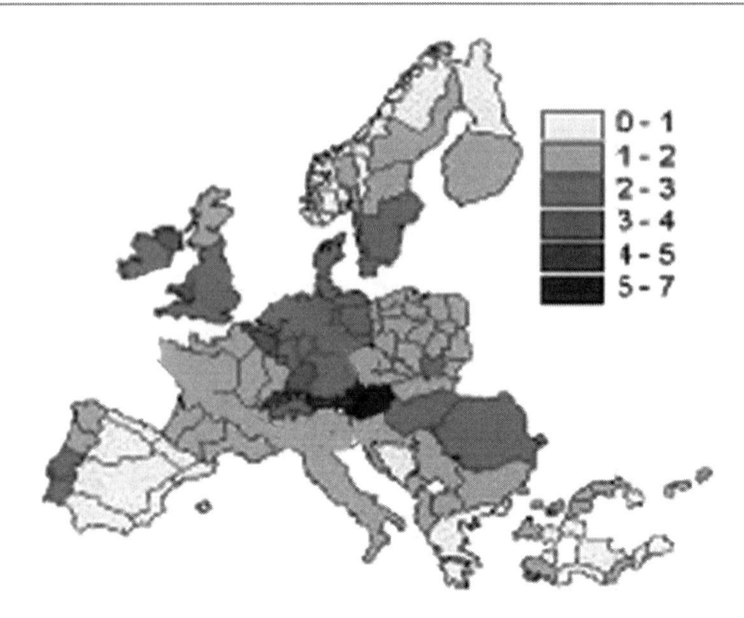

Fig. 5.2 Net ecosystem production (NEP) in European forests, showing carbon density (Mg C/ha/year) (*Source*: Nabuurs et al. (2008). Reprinted with permission)

stands tend to have higher rates of carbon sequestration, as indicated by NPP, than mid- or older-aged stands, although the data are highly variable (Table 5.1). In a meta-analysis of 19 studies from the temperate biome, Pregitzer et al. (2008) found that NPP and NEP both peaked in the 11–30 year age class. This follows the expected pattern of faster growth in young stands, with carbon sequestered via photosynthesis greater than the carbon lost via respiration. In older stands, it is thought that these two processes tend to be more in balance, making them either neutral or a slight sink, although the data in Table 5.2 show older and mid-successional stands having higher NEP and NEE than younger stands in many cases (see Chap. 3, this volume, for a detailed analysis of stand dynamics and carbon). Note, however, that carbon storage is lowest in young stands and highest in old stands, because of the longer period of accumulating carbon through forest production. Older stands can contain up to 2–5 times as much total ecosystem carbon as younger stands (Pregitzer and Euskirchen 2004; Law et al. 2003; Hooker and Compton 2003; Peichl and Arian 2006).

These temperate forest carbon sinks can be attributed largely to both the expansion of forest area in the twentieth century in the U.S., Europe, and China, and to rapid growth of secondary forests in young to mid-successional stages in many areas within those countries. There appears to be an additional factor that may increase forest growth and therefore increase the carbon sink. In a review of 31 published studies on forest productivity, Boisvenue and Running (2006) found that 28 studies showed increases in aboveground net productivity during the twentieth century. Their hypothesis is that increasing carbon dioxide, nitrogen, and other interacting factors are increasing NPP and NEP. This increasing productivity extends across temperate North America, northern Europe, most of central Europe, some parts of southern Europe, and Japan. Most of this sink is occurring in young- to middle-aged stands, with older stands either very small carbon sinks or in some cases, carbon sources, and very young (stand initiation stage) stands acting as carbon sources (Law et al. 2003; Carrara et al. 2003; Malhi et al. 1999; Wofsy et al. 1993).

Table 5.2 Carbon stocks in various pools in temperate forests. The major pools in the temperate biome are aboveground biomass and soil (belowground biomass and litter/CWD are smaller and less often measured, although both can be considerable in some forests)

				Carbon pools (Mg C ha^{-1})					
Source		Forest type	Stand age	Above ground[a]	Below ground	Litter &CWD	Organic soil horizons	Soil	Soil sample depth (cm)
Barford et al. (2001)	1	Oak-dominated hardwood, Massachusetts, USA	30–100	100					
Fahey et al. (2005)	1	Northern hardwood, New Hampshire, USA	70–100	95	25	13	30	127	20+
Bascietto et al. (2004)	1	European beech Germany	70–150	132–177					
Edwards et al. (1989)	1	Oak/hickory, Tennessee, USA	41–83	92–109			15–16	55	100
Fang et al. (2005)	1	All – Japan		27.6	6				
Finzi et al. (1998)	1	Mixed hardwood/hemlock, Connecticut, USA						59–75	15
Gough et al. (2008)	1	Mixed northern hardwood, Michigan, USA	6–90 (ave. 85)	76	23	4		80	Unreported
Hanson et al. (2003)	1	Oak, Tennessee, USA	58–100	108		4	4	64	100
Harris et al. (1975); Edwards et al. (1989)	1	Tulip poplar, Tennessee, USA	41–83	90–96			2–9	97–125	100
Malhi et al. (1999)	1	Oak-Hickory, Tennessee, USA	55	79[a]		7–11	8–27 (incl. roots)	7–55	Unreported
Morrison (1990)	1	Sugar maple, Ontario, Canada	Old growth	104–122[a]			14–16	185–202	100
Ruark and Bockheim (1988)	1	Quaking aspen, Wisconsin, USA	8–66	17–74[a]		4–8	4–9	33–65	60
Yuan et al. (2008)	1	White pine, Ontario, Canada	65	83	17				
Peichl and Arain (2006)	1	White pine, Ontario, Canada	15	40	5			34	Unreported
Peichl and Arain (2006)	1	White pine, Ontario, Canada	30	52	9			30	Unreported
Peichl and Arain (2006)	1	White pine, Ontario, Canada	65	100	19			37	Unreported
Yuan et al. (2008)	1	Balsam-fir, New Brunswick, Canada	27	78	18				
Zhang and Wang (2010)	1	Various hardwood/conifer, Northeastern China	42–59	105		6		161	Unreported
Zhu et al. (2010)	1	Montane conifer/birch, Northeastern China	100+	124	29	14		70	100
Law et al. (2003)	2	Pinus ponderosa, Oregon, USA	20	6	3	12		99	100

(continued)

Table 5.2 (continued)

				Carbon pools (Mg C ha^{-1})					
Source		Forest type	Stand age	Above ground[a]	Below ground	Litter &CWD	Organic soil horizons	Soil	Soil sample depth (cm)
Law et al. (2003)	2	Pinus ponderosa, Oregon, USA	70	53	17	10		76	100
Law et al. (2003)	2	Pinus ponderosa, Oregon, USA	100	102	33	20		102	100
Law et al. (2003)	2	Pinus ponderosa, Oregon, USA	250	134	42	14		64	100
Law et al. (1999, 2000)	2	Pinus ponderosa, Oregon, USA	Mixed	98					
Hamilton et al. (2002)	2	Loblolly pine, North Carolina, USA	15	51	10				
Hooker and Compton (2003)	2	White pine, Rhode Island, USA	10–114	8–183[a]			0–33	58–102	70
Maier and Kress (2000)	2	Loblolly pine, North Carolina, USA	11	11–22	3–7				
Sharma et al. (2010)	3	Montane oak, Garhwal, India	Old growth	115					
Mendoza-Ponce and Galicia (2010)	3	Montane pine, Mexico	12	106	1	7			
Mendoza-Ponce and Galicia (2010)	3	Montane pine, Mexico	30	63	1	8			
Mendoza-Ponce and Galicia (2010)	3	Montane fir, Mexico	75	178	2	2			
de Jong et al. (1999)	3	Montane pine-oak, Mexico		135[a]					
Ordonez et al. (2008)	3	Montane pine-oak, Mexico		92–113	24–29	3–4		93–116	30
Garcia-Oliva et al. (2006)	4	Oak woodland, Navasfrias, Spain	80	34	11	2		103	20
Gower et al. (1992)	5	Rocky mountain Douglas-fir, New Mexico	50	169[a]		8	21	11	30
Smithwick et al. (2002)	5	Fir-Spruce-Cedar, Oregon, USA	150–700	120–628[a]			10–19	37–366	100
Yuan et al. (2008)	5	Douglas-fir, British Columbia, Canada	55	182	37				
Ryan et al. (1996)		Pinus radiata, Australia	20	59	12				

1 moist broadleaf and coniferous, *2* interior coniferous, *3* montane oak/pine, *4* woodland and pineland, *5* temperate rainforest
[a]Total living biomass

When carbon flux data from research sites are combined with forest cover area data, it is possible to scale up carbon flux to a region or nation level. The units at these large scales are generally Pg/year (petragram (Pg) = 10^{15} g). The carbon sink estimates from the 1990s (Dixon et al. 1994) included global and biome level measures. For the forests of the mid-latitude countries (approximately the temperate biome) the United States was a 0.10–0.25 Pg/year carbon sink, Europe was a 0.09–0.12 Pg/year sink, China was a 0.02 Pg/year carbon source, and Australia was in balance. The estimate of the total temperate biome was 0.26 ± 0.09 Pg/year, which provided 35% of the global forest sink (the other 65% of the sink was in the boreal forest biome, whereas the tropical forest biome was a large source of carbon due to deforestation).

The largest change in the temperate forest biome is the switch in China from a carbon source of −0.02 Pg/year in the period of 1949–1980, to a sink of 0.02 by 1998 (Fang et al. 2001). Recent carbon sink estimates for the United States are 0.16 Pg/year derived from forest inventory data (Woodbury et al. 2007), and 0.63 Pg/year derived from AmeriFlux eddy covariance flux data and MODIS satellite observations (Xiao et al. 2011). Recent European estimates vary, but fall within the range of 0.10 to 0.47 Pg/year (Nabuurs et al. 2003; Janssens et al. 2003; de Vries et al. 2006; Papale and Valentini 2003). It should be noted that these numbers are not only for European temperate forests, but also include the boreal forests of Scandinavia. The more recent European data show moderate increases from the 1990s (Dixon et al. 1994), however the most recent US numbers are significanty higher than previously reported.

5 Carbon Pools in Temperate Forests

The focus now shifts to the analysis of carbon that has already been stored within temperate forests. For this review, carbon pool data were analyzed from 30 published studies (Table 5.2) to produce estimates of carbon storage across age classes and forest types. Many of the study sites were in the United States, but others were in Europe, Mexico, Canada, China, Japan, and Australia. Forest carbon storage is often divided into four pools: aboveground biomass, belowground biomass, litter/coarse woody debris, and soil. Most of the published literature is from moist broadleaf and coniferous forests, which by far span the largest geographic area of all temperate forest types (Fig. 5.1). Nonetheless, there are several studies from each of the other types, as shown in Table 5.2. There are 12 studies in which data were collected for all four pools. These studies show that temperate forests stored from 60 to 340 Mg C/ha, with a single outlier of 1,000 Mg C/ha from an old-growth temperate rainforest site in the U.S. Pacific Northwest. Mid-successional stands (40–80 years old) of the moist broadleaf and coniferous forest type contain 100–300 Mg C/ha. The temperate rainforest of similar age (50 years) contains about the same amount (210 Mg C/ha). Thus, these two forest types are similar in carbon storage in their early growth, but the rainforest conifers grow to a much greater age and size resulting in unusually large carbon storage. The study of interior coniferous forest type shows carbon pools of ponderosa pine ranging from 120 to 250 Mg C/ha associated with stand ages of 20 to 100 years. Montane pine-oak on dry sites in Mexico produce 72–235 Mg C/ha (lower value is a 30 year old stand; for higher values age is not available).

5.1 Aboveground Biomass

The data in Table 5.2 indicate that the aboveground carbon pool is the largest pool in temperate forests, and it increases with increasing age of the stand, as forests add woody biomass. Older stands can contain up to 2–5 times as much total ecosystem carbon as younger stands (Pregitzer and Euskirchen 2004; Law et al. 2003; Hooker and Compton 2003; Peichl and Arain 2006). However, soil may actually contain the largest pool of carbon. Measurements of soil carbon are generally made to a depth of 100 cm (or less) and may be missing a significant part of the carbon in deeper soil layers. This will be described further in the soil carbon section.

Tree species composition can be important in forest carbon storage. For example, European beech (*Fagus sylvatica*) and Douglas-fir stands have higher levels of aboveground carbon than other types of similar ages (Bascietto et al. 2004; Yuan et al. 2008; Table 5.2). Mixed-species stands in the moist broadleaf and coniferous forest type tend to have higher carbon density than single species stands (Hanson et al. 2003; Kelty 2006). This may be due to complementarity of tree species characteristics (e.g., deciduous mixed with evergreen species, or tolerant mixed with intolerant species) allowing more complete use of light, water, and nutrients; alternatively it may be due to facilitation effects in which one nitrogen-fixing species produces enriched litter that provides higher nutrient levels to other tree species in the mixture.

5.2 Belowground Biomass

The belowground biomass carbon pool (coarse and fine tree roots and their associated mycorrhizae) is the least studied part of the forest carbon budget, mainly because it is so difficult and labor-intensive to measure the various components of the belowground system. Data indicate that the belowground carbon pool is 5–10% of total ecosystem carbon, and is much smaller than the aboveground carbon pool (Table 5.2). However, this may be understated due to estimation methodologies and under-measurement, particularly of fine root biomass (Vogt et al. 1998).

Trees allocate more biomass to belowground structures when growing on dry sites. Two chronosequences in Table 5.2 indicate that the proportion of aboveground-to-belowground biomass stays relatively constant with stand age. For example, ponderosa pine in Oregon on a semi-arid site has a consistent 24% of total tree biomass in roots from ages 20 to 250 years (Law et al. 2003). White pine in Ontario on a moist site has only 11–16% of total tree biomass in roots across ages 40–100 years (Peichl and Arain 2006). Other species in Table 5.2 show this pattern.

5.3 Litter and Coarse Woody Debris

The litter pool is made up of dead organic matter (leaves, twigs, and other organic debris) on the forest floor that is not completely decomposed and has not yet entered the soil profile. This is the smallest pool, generally less than 10% of total ecosystem carbon. Litter quantity and quality is a function of species composition, and thus varies among forest types. Deciduous forests receive large inputs of litter in the fall, as trees and understory plants senesce; this litter decomposes slowly during the winter months, and quickly in the growing season. Coniferous forests produce less litter, but it generally decomposes more slowly due to high lignin content, so that the litter carbon pool is fairly similar to the deciduous forests. As carbon in the litter pool decomposes, it moves into the atmosphere as respired CO_2 or into the soil in organic carbon compounds. Litter has a rapid turnover compared to most other pools, with the possible exception of fine roots. However, on dry sites there can be substantial litter (10% or more of the total carbon stock) especially in older stands. Some temperate forests can store substantial carbon in the form of large coarse woody debris, which decomposes rather slowly, especially on dry sites.

5.4 Soil Carbon

Carbon generally enters the soil from arthropods mixing litter and soil, from leaching of decomposed litter, and from turnover of fine roots. In forests with earthworm activity, positive effects on carbon storage can occur because of the additional deeper mixing into the soil. Such activity can also have a negative effect when earthworms mix soil layers and expose deeper recalcitrant carbon pools to greater mineralization; in areas with exotic invasive earthworms, a more rapidly decomposing litter layer has caused a sharp decline in soil carbon pools in some sites in North America (Bohlen et al. 2004).

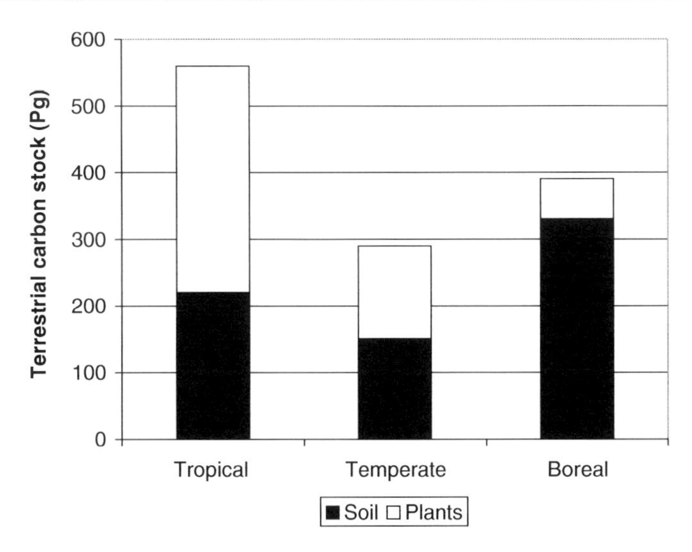

Fig. 5.3 Distribution of world forest carbon stocks by biome and by vegetation and soil (*Source*: Data compiled from Vogt et al. 1998; Eswaran et al. 1995; Goodale et al. 2002; Guo and Gifford 2002)

As indicated by data of Post et al. (1982), greater soil respiration occurs in warm, moist forests than in dry or cool systems. Precipitation, temperature, composition of the microbial community, and nutrient availability are also important in determining soil respiration rates (de Deyn et al. 2008). Hence, northern temperate forests, with cool climates, high precipitation, and extensive mycchorizal fungi associations, should have larger, more stable stores of soil carbon than southern temperate forests. It is difficult to determine whether this is true because there is so much variability in published data. This variability could be due to differences in measurement methodology and depth at which samples are taken, as well as the effects of species and site. Carbon allocation also depends on soil type. Spodosols (infertile/wet–cold) such as those in northern hardwoods contain a greater proportion of carbon in soil (Fahey et al. 2005) versus ultisols (fertile/wet-warm) in more southern hardwood types (Edwards et al. 1989).

A large study of temperate forest biome soils based on over 1,000 samples taken at 0–100 cm soil depth produced an estimate of 60–139 Mg C/ha, with warm moist forests at the lower end and cool moist forests at the upper end of the range (Post et al. 1982). Most soil carbon values in Table 5.2 fit within that range. It is interesting to note that two sites in Table 5.2 had much larger soil carbon density. One is an old-growth sugar maple site with a soil carbon density of 185–202 Mg C/ha; the other is a mixed-conifer temperate rainforest in Oregon with carbon density up to 366 Mg C/ha. These are the kinds of sites that would be expected to have little soil disturbance and therefore greater carbon storage.

Several chronosequence studies have found soil carbon to be relatively stable across age-classes with similar land use histories (Law et al. 2003; Peichl and Arain 2006), although there was a modest increase in soil carbon along a chronosequence of white pine (*Pinus strobus*) stands after agricultural abandonment in Rhode Island, USA (Hooker and Compton 2003). Soil carbon was also found to be stable in an unmanaged mixed pine-hardwood forest in Tennessee, USA (Zhang et al. 2007).

Data on mineral soil carbon stocks in temperate forests can only be considered approximations at this time as there is very little research on deep soil carbon. Current estimates are that soils contain at least half the carbon in temperate forests (Fig. 5.3) and possibly as much as two-thirds (IPCC 2000). Most soil carbon measurements are taken in the top 15–100 cm (Table 5.2). One estimate in an upland oak forest is that the deep mineral soil (1–9 m) contains more carbon (88 Mg C/ha) than the upper 1 m (64 Mg C/ha) (Hanson et al. 2003), which implies that there is a large, relatively stable carbon stock not accounted for in most temperate forest carbon budgets.

6 Biotic Drivers of Carbon Storage and Flux

Carbon flux status depends on the balance among photosynthesis, plant (autotrophic) respiration, and decomposer (heterotrophic) respiration. This balance is highly dependent on temperature, moisture, nutrients, and light, but the interactions among these factors are not well understood (Hanson and Weltzin 2000).

6.1 Photosynthesis and Autotrophic Respiration

Temperate forest net primary productivity (NPP), a measure of the carbon uptake in photosynthesis minus that lost through autotrophic respiration, has variously been found to correlate with: the combined influence of temperature and precipitation in Catalonia Spain (Martinez-Vilalta et al. 2008); spring snowpack depth, summer temperatures, and the Pacific Decadal Oscillation index in the Pacific Northwest, USA (Peterson and Peterson 2001); atmospheric nitrogen deposition in Finland and Russia (Elfving et al. 1996; Ericksson and Karlsson 1996); and length of the growing season in Austria, Belgium, and the Pacific Northwest, USA, among others (Hasenauer et al. 1999; Carrara et al. 2003; Peterson and Peterson 2001).

Many studies have looked at ecosystem responses to temperature variations. There is an optimal temperature (5–25°C) for photosynthesis in trees (Malhi et al. 1999), and in temperate climates it generally occurs from mid-to-late spring until mid-summer. Higher spring temperatures will enhance carbon uptake, as happened at Asian flux sites during a spring high temperature anomaly in 2002 (Saigusa et al. 2008). Most carbon uptake occurs in the spring and early summer, so higher temperatures earlier in the spring will generally increase annual productivity. Other critical factors are moisture, particularly towards the end of the growing season, the timing of the last frost in the spring and first frost in the fall, and summer temperature (higher temperature in summer causes high evapotranspiration, so plants will compensate by closing stomata).

6.2 Length of Growing Season

Temperate forests are strongly seasonal, with a well-defined growing season that depends primarily on light (day length) and temperature. This is probably the most important determinant of temperate forest productivity, or carbon uptake. For example, at one EuroCarboFlux site in Belgium, it has been observed that NEE is highly correlated with the length of the growing season – the forest is a carbon sink in the growing season, between May and August, and a source in the dormant season, from September to April (Carrara et al. 2003).

There is evidence of a longer growing season in North America and Europe over the last 50–100 years. Data from the Long Term Ecological Research site at Hubbard Brook, New Hampshire, USA, indicate that the timing of spring melt has advanced 10–12 days, and green canopy duration has increased by about 10 days since 1958, with significant trends towards an earlier spring (as evidenced by sugar maple leaf-out) (Vadeboncoeur et al. 2006; Richardson et al. 2006). In France and Switzerland, the onset of phenology has advanced considerably in response to spring temperature increases over the last 100 years (Schleip et al. 2008). Higher net carbon uptake in European forests in the spring of 2007, a year of record warm spring temperatures, was attributed to phenological responses to temperature (early bud break in deciduous trees and early release from winter dormancy in conifers) (Delpierre et al. 2009).

The net effects of a longer growing season on carbon sequestration are unclear, and may be confounded by other climate variables such as drought. Satellite observations (combined normalized difference vegetation index data set and climate data) suggest that in mid- to high-latitudes, decreases in carbon uptake during hotter and drier summers offset increased uptake in the spring, thereby reducing or even eliminating the positive benefits of a longer growing season (Angert et al. 2005). However, site-specific ecosystem flux data do not bear this out. A synthesis of data from 21 Fluxnet sites found a positive effect on NEP of earlier spring onset, attributed primarily to temperature, with increased productivity that was not completely offset by later season respiration (Richardson et al. 2010). And carbon flux measurements over an evergreen Mediterranean forest in southern France suggest that increased severity of summer drought did not negatively affect the carbon budget of the ecosystem, and that the annual variability in NEP cannot be fully explained by drought intensity, but is significantly linked with the length of the growing season (Allard et al. 2008). And at two AmeriFlux sites, earlier growing season onset resulted in an increase in net ecosystem productivity both in the spring and over the entire growing season, which the authors suggest could be a result of accelerated nitrogen cycling rates later in the growing season (Richardson et al. 2009).

6.3 Heterotrophic Respiration and Decomposition

Decomposition of organic matter, a form of respiration, emits carbon dioxide, a "flux," from the ecosystem to the atmosphere, primarily from short-lived carbon pools such as soil organic carbon and fine roots (Trumbore 2000). Decomposition and respiration rates in temperate forests depend strongly on soil temperature (Savage et al. 2009; Zhu et al. 2009; Jassal et al. 2007) and moisture (Jassal et al. 2007; Cisneros-Dozal et al. 2007), but also vary with litter quality (Fissore et al. 2009) and nutrient availability (Fahey et al. 2005). Fissore et al. (2009) found that mean residence time of active (acid soluble) soil organic carbon decreased strongly with increasing temperature in 26 deciduous and coniferous forest sites along a 22°C temperature gradient in North America, confirming a positive temperature influence on heterotrophic respiration across forest types. Higher summer temperatures may increase ecosystem respiration because of a direct effect on soil temperature (Yuan et al. 2008). Using MODIS "greenness" data, Potter et al. (2007) found that U.S. forests were largely a sink in 2001, 2003, and 2004, but a source in 2002 when the annual mean temperature was above average in the northeast regions. At the Harvard Forest, Massachusetts, USA, Borken et al. (2006) found that experimental moisture stress caused a decrease in heterotrophic respiration that was not wholly counteracted by increased respiration during natural precipitation levels the following season, resulting in at least a short term net carbon sink.

At the Harvard Forest, Massachusetts, USA, annual CO_2 exchange was found to be particularly sensitive to length of growing season, summer cloud cover, winter snow depth, and drought in summer. The first two regulate photosynthesis, and the latter two affect decomposition and heterotrophic respiration (Goulden et al. 1996). Changes in any of these factors would either increase or decrease the amount of carbon sequestered and stored in the ecosystem. For example, microbial decomposition may be limited by freezing (Goulden et al. 1996), so colder winters, heavier snow packs, or earlier fall freezes will decrease heterotrophic respiration. Higher spring temperatures will bring about earlier leaf-out, and a longer period of spring carbon uptake.

Fine root respiration was found to vary with temperature in soils at Hubbard Brook, New Hampshire, USA, but was much higher for roots in the forest floor than in the soil at all temperatures, attributed to higher nutrient concentration (particularly nitrogen) in root tissues in the forest floor (litter layer) (Fahey et al. 2005). According to Dalal and Allen (2008), elevated CO_2 increases soil respiration rate, possibly due to the enhanced rate of fine root turnover. From the limited evidence on soil respiration and climate variables in temperate forests, it appears that higher temperatures and increased precipitation will increase respiration and hence carbon emissions, whereas lower temperatures and drought conditions will decrease respiration and lower CO_2 emissions.

However, temperature (or any environmental variable) alone cannot explain temperate forest carbon flux dynamics. A few examples bear this out. Hanson et al. (2003) found no relationship of NEP to mean annual temperature in a review of seven studies of U.S. temperate deciduous forests; there was a positive, but not strong, relationship to precipitation. Ecosystem carbon storage was found to increase with altitude in the Great Smoky Mountains (Tennessee, U.S.): it was attributed to decreased respiration at higher elevation due to lower temperatures and higher precipitation (Zhang et al. 2007). Net flux of carbon dioxide, (NEE) at the Harvard Forest, Massachusetts, USA, has been observed to respond quickly to short term changes in climatic conditions such as temperature, precipitation, and snow cover, attributed to changes in rates of decomposition (Barford et al. 2001). Although there was very little response in levels of photosynthesis to changes in environmental variables from summer to summer, there were large shifts observed in annual NEE resulting from brief anomalies in temperature during April and May (Goulden et al. 1996). In a Michigan, USA, northern hardwoods site, high year-to-year fluctuations in carbon storage were observed to correlate with variations in air temperature, whereas respiratory losses were correlated with winter temperatures (Gough et al. 2008). The largest effect was found to be from a combination of high temperature and reduced radiation, which lowered mean annual carbon storage by 28% (Gough et al. 2008).

7 Disturbance and Abiotic Drivers of Carbon Storage and Flux

The net carbon accumulation in forests is heavily dependent on the time elapsed since disturbance (Pregitzer and Euskirchen 2004; Peichl and Arain 2006; Hooker and Compton 2003), because disturbance creates biogeochemical changes (light, temperature, moisture, nutrients) that affect both growth and respiration (Pregitzer and Euskirchen 2004). Fire, drought, windstorms and ice storms, insects and pathogens, nitrogen and other pollutants, and forest management and land use change are the primary natural and human disturbances affecting temperate forests.

7.1 Fire

Although fire plays only a minor role in the moist broadleaf and coniferous forests and in the temperate rainforests, it is a part of the natural disturbance regime in the fire-adapted interior coniferous, montane oak/pine, and woodland and pineland forests. In the United States alone, an average of 2.6 million hectares burned in wildland (both forest and grassland) fires each year between 2001 and 2010, mostly in the west, but also in the southeast and midwest (National Interagency Fire Center 2011), a significant increase over previous decades (Fig. 5.4). According to the reconstruction of fire history by Mouillot and Field (2005), this is still much lower than in the first decade of the twentieth century, when they estimate that fires burned close to 30 million hectares per year. Their analysis shows fires increasing in Europe towards the end of the twentieth century; however, at around 0.5 million hectares per year, the area burned is much smaller than in North America.

Dale et al. (2001) predict a 25–50% increase in burned area throughout the United States over the next 100 years. Understanding the effects of fires on landscape carbon storage over both short-and long-term temporal scales is critical for predicting future changes in both the regional and global carbon budgets (Kasischke et al. 1995). The alteration of net ecosystem production (NEP) varies with time between fires and fire intensity. During a fire, carbon is released to the atmosphere through combustion, creating an immediate CO_2 emission and reduced net primary production (NPP) due to tree mortality. If a new stand becomes established, then the net carbon balance may be zero over a long fire cycle (Kashian et al. 2006). However, net carbon loss to the atmosphere due to increased decomposition and reduced biomass can persist for over a century (Crutzen and Goldhammer 1993).

Short-term effects of fires (from a few years to decades) are important for predicting the Earth's

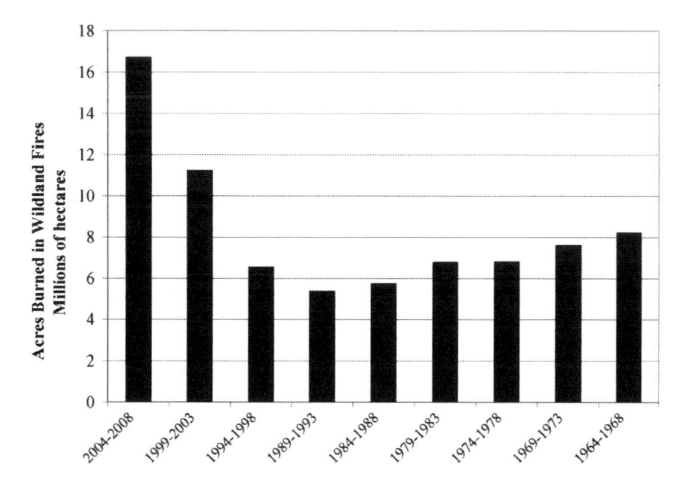

Fig. 5.4 Wildland fires in the United States: Total hectares burned in 5-year periods from 1966 to 2010 (*Source*: Data derived from NIFC (2011). National Interagency Coordination Center, Boise, Idaho. http://www.nifc.gov/fire_info/fires_acres.htm)

carbon balance over the next century because greater fire frequency, extent, or severity will release current carbon stores through combustion resulting in a negative NEP (Kashian et al. 2006). If the burned forest area significantly increases over the next century, these short-term effects will likely influence atmospheric CO_2 concentration (Dale et al. 2001). Short-term effects of fire on carbon storage are regulated by the amount of carbon lost in combustion (Tinker and Knight 2000; Litton et al. 2004), by the rate and amount of regeneration (Kashian et al. 2004; Litton et al. 2004), and by changes in decomposition rates from altered soil conditions and increased woody debris left by the fire (AuClair and Carter 1993; Kurz and Apps 1999).

Long-term effects of fire (over many centuries) on ecosystem carbon balance are regulated by processes that control post-fire regeneration and by fire frequency. If the post-fire stand has poor or no regeneration, forest growth will not replace the carbon lost through combustion and decomposition, and the net carbon storage over a fire cycle will decrease (Kashian et al. 2006). Changing fire frequency will also affect the net carbon storage because the amount of carbon stored in a stand, and the rates of photosynthesis and decomposition, vary with stand age (Kasischke 2000). It is also important to note that more frequent fires will promote a higher proportion of young forests, and these forests tend to store less carbon than older stands because they contain less biomass, even though their rates of production tend to be higher (Ryan et al. 1997). Thus, if changing climate alters the frequency and intensity of fires, re-vegetation and patterns of carbon storage will likely be affected, particularly in interior coniferous forests.

Although fire was not historically as severe a disturbance in moist broadleaf and coniferous forests as in woodland and pinelands and interior coniferous forests, it nonetheless has played an important historical role (Pyne 1982). In eastern North America, Native American burning and lightning resulted in relatively frequent fire in temperate mixed oak and pitch pine (*Pinus rigida*) forests. Fire was also important to drier regions, such as near the prairie–woodland border. These fires had a tremendous impact on the composition and age structure of the forest, since certain species have adaptations, such as thick bark, ability to sprout, and rapid post-fire colonization, that enable them to thrive under such conditions (Reich et al. 1990; Abrams 1992; Kruger and

Reich 1997; Peterson and Reich 2001). Fire frequency regulated the balance between late successional species such as sugar maple, beech (*Fagus grandifolia*), and linden (*Tilia* spp.), and shade-intolerant early successional species such as oak and aspen (*Populus* spp.), which were abundant along the prairie-forest border and areas with sandy soil where fires were most frequent (Grimm 1984; Abrams 1992).

With increasing development following European settlement and expansion, fires in U.S. moist broadleaf and coniferous forests became much less frequent. This was due to cessation of intentional burning, direct suppression of fires, and land use changes that disrupted the contiguity of burnable vegetation across the landscape. Hence, these forests have gradually become increasingly dominated by shade tolerant species such as maple (*Acer* spp.), beech, ash (*Fraxinus* spp.), and linden, with decreased abundance of oaks (Crow 1988; Abrams 1998). In the absence of fire, oaks do not establish well in either shaded understorys or sunlit openings, because they are neither shade tolerant nor fast growing (Reich et al. 1990; Abrams 1992; Kruger and Reich 1997). Hence, a major change in temperate deciduous forests of North America has resulted from the ascendency of fire-intolerant species to a dominant position in these regions. Fire suppression has likely had similar effects in Europe and Asia (Reich and Frelich 2002).

7.2 Drought

Water availability controls tree growth, tree species distribution, and forest composition more than any other perennial factors (Hinckley et al. 1981). Global circulation models predict that increasing concentrations of atmospheric CO_2 will increase the severity and frequency of drought in regions where temperate forests are found (Pastor and Post 1988; Dale et al. 2001). However, there is a great deal of uncertainty about how drought will affect carbon cycles.

Elevated CO_2 generally increases instantaneous water-use efficiency in tree seedlings (Jarvis 1989), but may have negative impacts on other physiological processes. For instance, stomatal closure can occur during a leaf water deficit or by a high internal CO_2 concentration (Hinckley et al. 1981), which may result in an increased resistance to CO_2 uptake (Jarvis 1989). Tschaplinski et al. (1995) found that drought may also slow the growth rate and alter the gas exchange of several tree species growing in an elevated CO_2 atmosphere.

Given a reduced growth rate or altered gas exchange, it is likely that drought may have a negative impact on regional carbon budgets in the short-term (i.e. during the period of the drought or for several years following a drought event), but it is unlikely that it will affect the carbon cycle in the long-term unless there is substantial tree morality as a result of the drought event. Beerling et al. (1996) note that there will be a greater tendency for trees to show drought tolerance in the future and thus, drought may have little consequence on NEP.

7.3 Wind and Ice

Perhaps the most important abiotic disturbance regime in temperate forests is wind and ice, creating a mosaic of gaps of varying sizes that drives successional processes across the landscape (Dale et al. 2001; Nagel and Svoboda 2008). Moist broadleaf and coniferous forests are heavily impacted by wind disturbance, including tornadoes and thunderstorm downbursts in central North America and western Europe and severe low-pressure systems (cyclones and hurricanes) along the eastern Atlantic coast (Dale et al. 2001; Reich and Frelich 2002; Degen et al. 2005; Nagel and Svoboda 2008).

Most windstorms are small- to intermediate-scale events, resulting in gap formation or gap expansion (Worrall et al. 2005), thereby either releasing advance regeneration (accelerating succession) (Webb and Scanga 2001; Uriate and Papaik 2007) or creating conditions for disturbance specialist understory plants to take over the gap (Palmer et al. 2000). Although small-to-intermediate- scale, low intensity windstorms

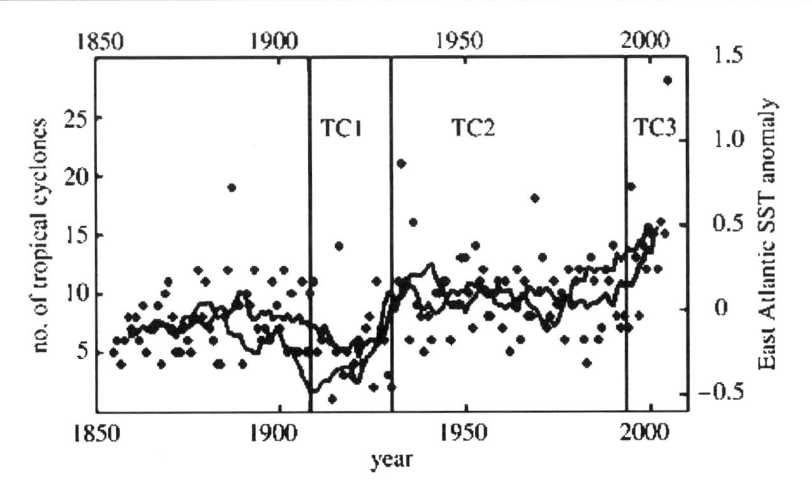

Fig. 5.5 Tropical cyclone occurrence (dots indicate annual totals and the black line is a 9-year running mean) in the North Atlantic together with East Atlantic sea surface temperature (SST) anomalies for the hurricane season (*grey line*) from 1855 to 2005 (*Source*: Holland and Webster (2007). Reprinted with permission)

can change the successional patterns and species composition of forest stands (Hanson and Lorimer 2007; Papaik and Canham 2006; Degen et al. 2005), they may have little impact on total carbon stocks at the landscape scale, particularly if the downed trees and branches are left on the ground. For example, 23 years after an intermediate windstorm in an old-growth beech-fir forest in Slovenia, Nagel et al. (2006) found that, although the basal area of living trees was lower, the basal area of downed logs was higher in areas affected by windthrow.

Hurricanes often create patches of disturbance of intermediate severity across the landscape (McNab et al. 2004; Busing et al. 2009), although intense storms, such as Hurricane Rita along the Gulf Coast in 2005, leave wide swaths of forest damage (Juarez et al. 2008). Carbon moves quickly from the living biomass pool to the dead and downed wood pool (Uriate and Papaik 2007; Busing et al. 2009), reducing NPP and increasing respiration from decomposition, with a net loss of carbon that can last for decades (Fahey et al. 2005; Busing et al. 2009; McNulty 2000). Frequent storms have been shown to depress carbon stocks in southern New England, USA. Maturing second-growth hardwood forests exhibit a decrease in carbon (living and dead aboveground biomass) across a hurricane severity gradient from south (more severe) to north (less severe) (Uriate and Papaik 2007). The authors suggest that in the southernmost areas of New England, storm-free periods were never long enough for the forest stands to reach peak biomass.

The frequency of such stand-leveling winds is expected to increase under a warmer climate, so that fewer stands would reach old-growth stages of development. Thus there would be a decrease in overall carbon sequestration in regions experiencing severe wind storms (Uriate and Papaik 2007). Holland and Webster (2007) looked at 100-year tropical cyclone activity in the eastern Atlantic and concluded that, over the twentieth century, increased storm frequency is related to rises in sea surface temperature; thus, the recent upsurge in tropical cyclones and hurricanes (Fig. 5.5) is likely due in part to global warming. From 1995 to 2007, there were an average of 15 major tropical cyclones (including 8 hurricanes) per year, compared to an average of 9 (5 hurricanes) during the period 1931–1994 (Holland and Webster 2007).

Ice storms are common in moist broadleaf and coniferous forest regions (Goodnow et al. 2008; Changnon 2008), although catastrophic ice storms are rare (Bragg et al. 2003). Injury to trees is widely variable, from minor branch breakage to mortality, and depends on the storm severity, species, and site conditions (Bragg et al. 2003; McCarthy et al. 2006; Boyce et al. 2003). Ice storm damage on the Duke Forest, North Carolina, USA resulted in a transfer of carbon from the living to detrital pools equivalent to 30% of the net ecosystem carbon exchange of the system, with conifers twice as likely to be killed as deciduous trees (McCarthy et al. 2006). Thinned stands had a three-fold increase in carbon transfer to the detritus pool as compared to unthinned stands. Under elevated CO_2 conditions of the Free Air Carbon dioxide Enrichment (FACE) experimental plots, carbon transfer from live biomass to the detritus pool (i.e. storm damage) was significantly less than in the control plots, tentatively suggesting that forests might be less susceptible to ice storm damage in a higher atmospheric CO_2 environment.

7.4 Insects

Conjectures about the effects of climate change on insect populations have been somewhat general to date. It is assumed that disturbance intensity will change across a latitudinal gradient as insect populations extend their ranges to higher latitudes and elevations as temperatures rise, with temperate tree species encountering new non-native insects which migrate much more quickly than trees (Williams and Liebhold 1995; Dale et al. 2001). Such is the case with the mountain pine beetle (*Dendroctonus ponderosae*), which is causing widespread mortality in northwestern North America well beyond its historical range (Kurz et al. 2008). Increased over-wintering survival and higher population growth rates may become more common for insect pests of temperate trees. Therefore, it is important to understand the impacts that larger pest populations – particularly defoliating insects – will have on forests. Nevertheless, few studies have been published that examine the impacts of insect defoliation on the carbon budget in temperate ecosystems.

Large-scale insect infestations can cause high mortality, leading to long-lasting decreases in ecosystem biomass (Knebel and Wentworth 2007). Kurz et al. (2008) predict that the mountain pine beetle outbreak in British Colombia, Canada, will change the 374,000 km^2 affected area from a small carbon sink to a large carbon source throughout the next decade. Great spruce bark beetle (*Dendroctonus micans* Kug.) outbreaks, interacting with climate stress (cold winters and dry summers) led to forest dieback over 10–15 years in Norway spruce (*Picea abies*) plantations in France (Rolland and Lemperiere 2004). Invasive exotic species, such as the European gypsy moth (*Lymantra dispar*) and the emerald ash borer (*Agrilus planipennis*) can severely impact temperate forests in the United States, causing widespread defoliation and/or mortality.

Even major defoliation events in deciduous forests may have only negligible effects on NEP, however; after heavy defoliation from a hurricane (similar in effect to insect defoliation) in Florida, USA, the decline in GPP was offset by a concurrent decline in ecosystem respiration (Li et al. 2007). Defoliation, acting with other environmental variables, can also affect nutrient cycling, because large fluxes of organic matter move from one pool (live biomass) to another (detritis and soil organic matter), changing rates of photosynthesis, decomposition, and critical biochemical parameters such as C:N ratios. Two examples from vastly different ecosystems bear this out. Severe defoliation events, combined with recovery from extreme drought, in the 1960s resulted in dissolved inorganic nitrogen losses from the Hubbard Brook Watershed, in New Hampshire, USA, (Aber et al. 2002), and heavy infestations of bark beetle in ponderosa pine forests in Arizona, USA, did not alter soil respiration rates, but altered nitrogen cycling throughout the growing season, lowering net nitrification rates (Morehouse et al. 2008).

Increased tree mortality on a site will likely result in the release of carbon through increased decomposition of newly dead biomass and in decreased photosynthesis, thus reducing net carbon sequestration on sites with heavy mortality. Repeated defoliation or attacks will only exacerbate this effect and will likely have negative impacts on the regional or global carbon budget. Furthermore, the rate of recovery or presence/absence of regeneration will also determine the amount of carbon being sequestered within a stand that has been heavily defoliated.

7.5 Nitrogen Deposition and Ozone Pollution

7.5.1 Nitrogen

Atmospheric pollution in the form of nitrogen oxides (NOx) emitted from burning fossil fuels may have both positive and negative effects on temperate forest regions (Bouwman et al. 2002; Felzer et al. 2007; Fig. 5.6). NOx disassociates in soil solution as hydrogen ions (H^+) and nitrate (NO_{3-}); the resulting increase in soil pH causes nutrient cations such as calcium (Ca^{2+}) and magnesium (Mg^{2+}) to leach from the soil, and mobilizes toxic cations such as aluminum (Al^{3+}) (Likens and Borman 1995; Driscoll et al. 2001; Puhe and Ulrich 2001). Because most temperate forests are thought to be nitrogen-limited, nitrogen deposition may also act as a growth stimulant (fertilizer effect). On balance, however, the evidence regarding nitrogen deposition effects on carbon sequestration is mixed.

Under current ambient levels, nitrogen deposition is most likely enhancing carbon sequestration, as indicated by data from experimental sites in Europe and the United States. Strong correlations were found between nitrogen levels and photosynthetic capacity (CO_2 absorption capacity) in tree canopies across 11 AmeriFlux sites (Ollinger et al. 2008). It is estimated from EuroFlux data that 10% of net carbon sequestration in Europe is attributed to nitrogen deposition (De Vries et al. 2006). Also, a CO_2 enrichment effect at one FACE site in the southeastern USA was amplified by high levels of soil nitrogen availability (Norby et al. 2005).

Nitrogen acts within a complex of stressors including climate change, drought, insects, diseases, and other air pollutants; therefore, efforts to understand the effect of nitrogen fertilization must be made in the context of these other factors. Thus far, most data come from nitrogen addition experiments where the effects of other factors are intrinsically assumed to be consistent between experimental plots and control plots. The results are inconclusive.

Short-term (one growing season) nitrogen fertilization experiments have produced a decrease in CO_2 emissions, primarily due to decreased soil respiration, in black cherry (*Prunus serotina*) stands (Bowden et al. 2000) and a large increase in NEP in Douglas fir stands (Jassal et al. 2008). Strong responses to nitrogen additions may only be a short-term ecosystem response, however. One-to-three year studies have shown either mixed results (Waldrop et al. 2004) or no detectable change in biomass (Nadelhoffer et al. 1999).

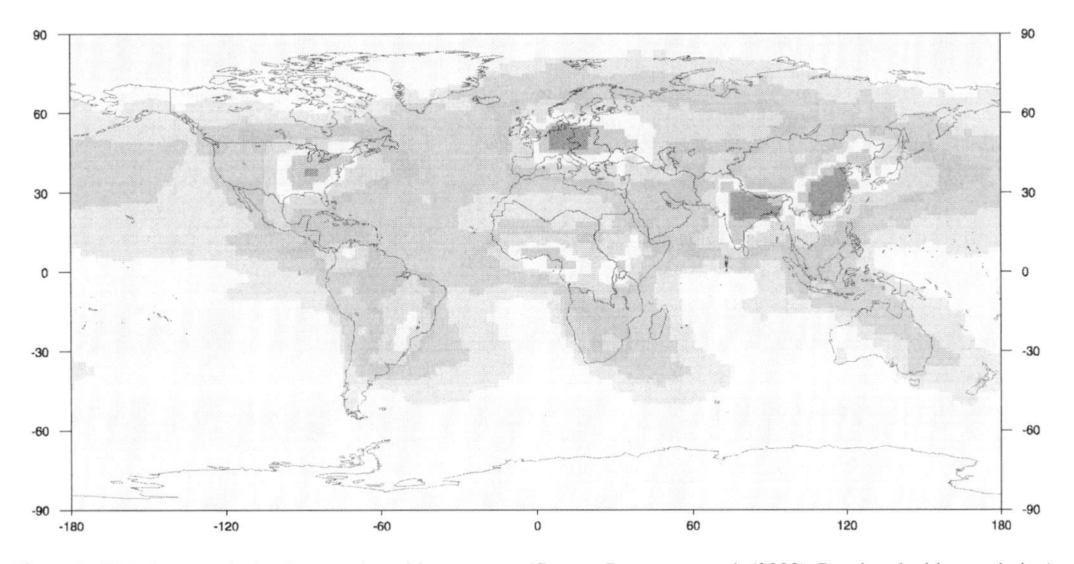

Fig. 5.6 Global atmospheric nitrogen deposition patterns (*Source*: Bouwman et al. (2002). Reprinted with permission)

The reduction of biomass growth in trees with heavy nitrogen fertilization may be a result of unbalanced nutrients within the foliage; this has been observed in several N-fertilization studies (Aber et al. 1998; Magill et al. 2000). Analysis of long-term chronic nitrogen addition experiments in Europe and North America indicate no discernable trend in effects on ecosystem-level carbon sequestration (LeBauer and Treseder 2008; Evans et al. 2008; Pregitzer et al. 2008; Bauer et al. 2004). This leads to the tentative conclusion that under chronic nitrogen deposition, temperate forests may no longer be nitrogen limited; in fact, several of these research sites are leaching nitrogen as nitrate (NO_3).

7.5.2 Ozone

Unlike nitrogen, ozone (O_3) has no known "positive" effects on forests. High levels of ozone cause foliar injury and consequent growth reduction, particularly in conifers, and all other things being equal, carbon sequestration is expected to be lower in forests with high ozone levels (Augustaitis and Bytnerowicz 2008). Ozone is highest in areas with high levels of both sunlight and fossil fuel emissions. This includes most of the temperate forest biome (southwestern and eastern United States, eastern Europe, the Mediterranean, western Asia, and northeastern China (Felzer et al. 2007)). It is projected that 50% of northern hemisphere forests will be affected by toxic levels of ozone by 2100 (Fowler et al. 1999).

Ambient ozone levels have been associated with growth reduction in mature southern pines, particularly loblolly pine (Felzer et al. 2007). In Europe, ozone has been implicated in growth reductions of Aleppo pine (*Pinus halepensis*) in the Mediterranean basin, Swiss stone pine (*Pinus cembra*) in the timberline ecotone of the European mountains (Richardson et al. 2007), and Scots pine in central Europe (Augustaitis and Bytnerowicz 2008). Several pine species in Mexico show ozone-induced damage similar to pines in the western United States (Richardson et al. 2007). In the Great Smoky Mountain National Park, USA, ozone stress is thought to be dampening the potential CO_2 fertilization effect, with carbon stocks increasing only slightly between 1971 and 2001 (Zhang et al. 2007).

7.6 Forest Management and Land Use

Almost all temperate forests have been severely impacted by human use. In Europe and North America, less than 1% of all forests remain in an undisturbed state, free of logging, grazing, deforestation or other intensive use (Reich and Frelich 2002). The largest direct impacts on temperate forests are from conversion to other land uses. Historically, land use change was from forest to agriculture. It is becoming more common for agricultural lands to be shifted back to forests, while forests in other areas are being shifted to urban development.

Management of forest for timber creates a cyclical pattern of carbon release and sequestration, with intensively managed stands storing less carbon than unmanaged forests (Carrara et al. 2003; Gough et al. 2008; Ordóñez et al. 2008; Woodbury et al. 2006). A great variety of wood products are harvested from temperate forests, and these products are important forms of carbon storage. Products have variable turnover times, ranging from the short turnover of biomass wood chips or paper to the much longer turnover of wooden building structures and furniture. (See Chap. 10 of this volume for a detailed discussion of managing temperate forests for carbon.)

8 The Future of Temperate Forests as Carbon Reservoirs: Climate Change Impacts

There is no large-scale deforestation in the temperate forest region at present nor is there likely to be in the future. Forest cover is likely to remain stable because of conservation efforts in the United States, Japan, South Korea, and Europe, and also because reforestation of agricultural land should balance out the loss of forest cover from development and suburban sprawl. Former agricultural lands continue to be planted with pine in the U.S. south (Smith et al. 2009), forest area is expanding in parts of Europe, particularly Spain, Sweden, Norway and Italy (FAO 2011), and recently there have been extensive reforestation efforts with exotic plantations in China (FAO 2011). Although remaining mostly intact, temperate forests continue to be fragmented by development, particularly in North America (Wickham et al. 2008).

The future of the temperate forest biome as a carbon reservoir and atmospheric CO_2 sink rests mainly on its productivity and resilience in the face of changing disturbance regimes in the context of rising atmospheric CO_2. The small sink status (0.2–0.4 Pg C/year) of temperate forests could easily change to a source status if the balance between photosynthesis and respiration shifts. Predictions are that temperatures in temperate regions will increase (IPCC 2007). Warming in Europe and North America is likely to be greatest in the winter, although the Mediterranean and southeastern U.S. are likely to see the largest temperature increases in the summer. Generally this would mean longer growing seasons. Longer and more intense summer heat waves are predicted for East Asia, along with increased precipitation.

There is evidence of increasing productivity in temperate forests as climate has warmed in the last 50 years (McMahon et al. 2010; Xiao et al. 2011; Table 5.1); however, the ability to attribute these changes solely to climate warming is confounded by other factors at work, such as successional dynamics and environmental variables. The atmospheric system has not only experienced changes in temperature, precipitation, and radiation, but also in CO_2 concentration and pollutants, between 1950 and 2005 (Keeling et al. 1995; Innes and Peterson 2001). Current global atmospheric CO_2 is approximately 390 ppm, an increase of about 75 ppm since the 1950s (Tans and Keeling 2011). How forests will respond to rising levels of CO_2 in the long term is still uncertain, but the present overall response is positive – an increase in forest productivity and carbon storage.

What we know about rising levels of atmospheric CO_2 and forest carbon sequestration comes from a few experimental CO_2 enrichment studies in the United States and Europe. A median increase of 23% in net primary production has been recorded across sites exposed to elevated CO_2 (550 ppm) in comparison to control sites (370 ppm) over 1–6 years of Free Air Carbon dioxide Enrichment (FACE) experiments (Norby and Luo 2004). In these fast growing, early successional stands, changes in NPP are related to increased atmospheric CO_2 effects on light energy; these consist of increased light absorption in stands with a lower leaf area index, and increased light use efficiency in those with a higher leaf area index.

Nowak et al. (2004) studied the response of ecosystems to elevated CO_2 using results from FACE experiments in forests across North America and Europe. As expected, leaf CO_2 assimilation and ecosystem primary production increased across all species. The primary production observations, however, are mixed and are overall less than the hypothesized 20% increase in production. Down-regulation of photosynthesis has happened in a number of FACE experiments, but not in all species. Greater nitrogen availability also increased productivity in this study.

Wittig et al. (2005) evaluated GPP of fast-growing Populus species (three years from establishment to canopy closure) in response to elevated CO_2 and found that GPP increased dramatically in the first year but markedly less so in the subsequent years. Hättenschwiler and Körner (2003) similarly found accelerated growth in trees over a 30-year period of elevated CO_2 exposure, with most of the accelerated growth happening at young stages of development. In their 2005 analysis based on FACE data, Körner et al. (2005) found an immediate and sustained enhancement of carbon flux in mature temperate forest trees but, contrary to expectations, found no overall stimulation of growth or litter production after 4 years; hence, forests seem to be "pumping" carbon through faster with no net gain in biomass (NEP). These findings suggest differing responses of trees at different developmental stages.

Interactions between atmospheric CO_2 and carbon sequestration in forests become more complicated when other environmental variables such as drought are involved. Drought stress is expected to outweigh CO_2 enhanced growth in the southern range of Scots pine, based on findings by Martinez-Vilalta et al. (2008) that summer temperature and water availability have been the main climatic drivers of growth over the past 80 years. Hättenschwiler and Körner (2003) suggest, however, that trees exposed to higher CO_2 levels seem to be more tolerant to drought stress, potentially dampening this effect. Körner (2000) concluded that, besides a stimulation of photosynthesis, the most robust findings on plant responses to elevated CO_2 are changes in active tissue quality (wider C:N ratio) and effects on community dynamics. Kozovits et al. (2005) found that the type of competition (intra-specific versus inter-specific) with beech and spruce changed the response of trees to elevated CO_2. DeLucia et al. (2005) found an increase in NPP and NEP in both loblolly pine and deciduous sweetgum forests, but also found an increase in plant respiration that reduced the NPP, more so in the pine than in the deciduous forest. DeLucia et al. (2005) warn that greater allocation to more labile tissues may cause more rapid cycling of carbon back to the atmosphere.

The majority of research has been done using single factor analyses. However, biogeochemical processes and cycles take place in a complex environment of changing climate, increasing atmospheric CO_2 and O_3, nitrogen deposition, and varying land use legacies. The few studies that have modeled multi-factor influences on temperate forest net ecosystem productivity or carbon flux have found that combined effects are expected to diminish the effect of CO_2 enrichment alone. Scenario modeling of the combined influence of CO_2, O_3, temperature, and precipitation by Hanson et al. (2005) produced a 29% reduction in NEE over baseline conditions, even though models of CO_2 enrichment alone yielded substantial increases in NEE. Similarly, both Ollinger et al. (2002) and Zak et al. (2007) found that O_3 significantly dampened the fertilization effect of CO_2 and nitrogen.

Research must continue to further understand stand-level biogeochemical cycling with a focus on large-scale long-term experiments. As the literature shows, there is no clear answer as to whether rising CO_2 concentrations will cause forests to grow faster and store more carbon (Körner et al. 2005). The response to increasing atmospheric CO_2 is confounded by the effects of prior land use and changes in temperature, precipitation, radiation, nitrogen, ozone, and other factors on forest productivity response.

9 Conclusion and Summary Recommendations

Currently the temperate forest biome is a significant part of the net sink for atmospheric carbon dioxide. It combines with the boreal net sink to partially offset the carbon net source of the tropical forests.

- However, the temperate forest sink is small and its current status rests on a tenuous balance between stable forest area, age-class distribution, disturbance regimes (windstorms, fire, insects, management), successional patterns, and the potentially counteracting effects of climate change and levels of atmospheric CO_2 and nitrogen.
- At best, if land use change remains in balance and forest productivity remains high, temperate forests will remain a small carbon sink.

Small changes in forest cover or age-class distribution across the biome would shift temperate forests to being either carbon-neutral or a source of CO_2 emissions, further exacerbating climate change.

- In particular natural disturbances are significant determinants of temperate forest successional patterns and their frequency and intensity is expected to increase under a warmer climate meaning that fewer forests would reach old-growth stages of development.
- If changing climate alters the frequency and intensity of disturbance, patterns of reforestation and carbon storage will be affected, particularly in mountain interiors, woodlands and pinelands of Mediterranean climates, coastal forests and floodplains.

More studies are needed on:

- Mineral soil carbon stocks in temperate forests. They can only be considered approximations at this time as there is very little research on deep soil carbon (more than 100 cm).
- Global circulation models. There is a great deal of uncertainty about how they predict

regional changes in climate. It is expected that the severity and frequency of drought and storms will increase in regions where temperate forests are found. However, there is also a great deal of uncertainty about how drought will affect carbon cycles.

- Interactions. Little is known about how the interactions between temperature, moisture, available nutrients, pollutants, and light influence key environmental variables, such as drought, to affect ecosystem carbon flows.

References

Aber JD, Ollinger SV, Driscoll CT, Likens GE, Holmes RT, Freuder RJ, Goodale CL (2002) Inorganic nitrogen losses from a forested ecosystem in response to physical, chemical, biotic, and climatic perturbations. Ecosystems 5:648–658

Aber J, McDowell W, Nadelhoffer K, Magill A, Berntson G, Kamakea M, McNulty S, Currie W, Rustad L, Fernandez I (1998) Nitrogen saturation in temperate forest ecosystems: hypotheses revisited. Bioscience 48:921–934

Abrams MD (1992) Fire and the development of Oak Forests - in Eastern North-America, Oak distribution reflects a variety of ecological paths and disturbance conditions. Bioscience 42:346–353

Abrams MD (1998) The red maple paradox. Bioscience 48:355–364

Allard V, Ourcival JM, Rambal S, Joffre R, Rocheteau A (2008) Seasonal and annual variation of carbon exchange in an evergreen Mediterranean forest in southern France. Glob Change Biol 14:714–725

Angert A, Biraud S, Bonfils C, Henning CC, Buermann W, Pinzon J, Tucker CJ, Fung I (2005) Drier summers cancel out the CO_2 uptake enhancement induced by warmer springs. Proc Natl Acad Sci USA 102:10823–10827

Archibold O (1995) Ecology of world vegetation. Chapman & Hall, New York

Auclair AND, Carter TB (1993) Forest wildfires as a recent source of CO_2 at Northern latitudes. Can J Forest Res 23:1528–1536

Augustaitis A, Bytnerowicz A (2008) Contribution of ambient ozone to scots pine defoliation and reduced growth in the Central European forests: a Lithuanian case study. Environ Pollut 155:436–445

Barford CC, Wofsy SC, Goulden ML, Munger JW, Pyle EH, Urbanski SP, Hutyra L, Saleska SR, Fitzjarrald D, Moore K (2001) Factors controlling long- and short-term sequestration of atmospheric CO_2 in a mid-latitude forest. Science 294:1688–1691

Bascietto M, Cherubini P, Scarascia-Mugnozza G (2004) Tree rings from a European beech forest chronosequence are useful for detecting growth trends and carbon sequestration. Can J Forest Res 34:481–492

Bauer GA, Bazzaz FA, Minocha R, Long S, Magill A, Aber J, Berntson GM (2004) Effects of chronic N additions on tissue chemistry, photosynthetic capacity, and carbon sequestration potential of a red pine (*Pinus resinosa* Ait.) stand in the NE United States. Forest Ecol Manag 196:173–186

Beerling DJ, Heath J, Woodward FI, Mansfield TA (1996) Drought-CO_2 interactions in trees: observations and mechanisms. New Phytol 134:235–242

Bohlen PJ, Groffman PM, Fahey TJ, Fisk MC, Suarez E, Pelletier DM, Fahey RT (2004) Ecosystem consequences of exotic earthworm invasion of north temperate forests. Ecosystems 7:1–12

Boisvenue C, Running SW (2006) Impacts of climate change on natural forest productivity - evidence since the middle of the 20th century. Glob Change Biol 12:862–882

Borken W, Savage K, Davidson EA, Trumbore SE (2006) Effects of experimental drought on soil respiration and radiocarbon efflux from a temperate forest soil. Glob Change Biol 12:177–193

Bouwman AF, Van Vuuren DP, Derwent RG, Posch M (2002) A global analysis of acidification and eutrophication of terrestrial ecosystems. Water Air Soil Pollut 141:349–382

Bowden RD, Rullo G, Stevens GR, Steudler PA (2000) Soil fluxes of carbon dioxide, nitrous oxide, and methane at a productive temperate deciduous forest. J Environ Qual 29:268–276

Boyce RL, Friedland AJ, Vostral CB, Perkins TD (2003) Effects of a major ice storm on the foliage of four New England conifers. Ecoscience 10:342–350

Bradford JB, Birdsey RA, Joyce LA, Ryan MG (2008) Tree age, disturbance history, and carbon stocks and fluxes in subalpine Rocky Mountain forests. Glob Change Biol 14(12):2882–2897. doi:10.1111/j.1365-2486.2008.01686.x

Bragg DC, Shelton MG, Zeide B (2003) Impacts and management implications of ice storms on forests in the southern United States. Forest Ecol Manag 186:99–123

Bryant D, Nielsen D, Tangley L (1997) The last frontier forests: ecosystems and economies on the edge. World Resources Institute, Washington, DC

Busing RT, White RD, Harmon ME, White PS (2009) Hurricane disturbance in a temperate deciduous forest: patch dynamics, tree mortality, and coarse woody detritus. Plant Ecol 201:351–363

Carrara A, Kowalski AS, Neirynck J, Janssens IA, Yuste JC, Ceulemans R (2003) Net ecosystem CO_2 exchange of mixed forest in Belgium over 5 years. Agric Forest Meteorol 119:209–227

Changnon SA (2008) Space and time distributions of major winter storms in the United States. Natural Hazards 45:1–9

Cisneros-Dozal LM, Trumbore SE, Hanson PJ (2007) Effect of moisture on leaf litter decomposition and its contribution to soil respiration in a temperate forest. J Geophys Res Biogeosci 112:10

Crow TR (1988) Reproductive mode and mechanisms for self-replacement of Northern Red Oak (*Quercus-Rubra*) - a review. Forest Sci 34:19–40

Crutzen PJ, Goldhammer JG (eds) (1993) Fires in the environment: the ecological, atmospheric, and climatic importance of vegetation fires. Wiley, New York

Curtis PS, Hanson PJ, Bolstad P, Barford C, Randolph JC, Schmid HP, Wilson KB (2002) Biometric and eddy-covariance based estimates of annual carbon storage in five eastern North American deciduous forests. Agric Forest Meteorol 113(1–4):3–19

Dalal RC, Allen DE (2008) Greenhouse gas fluxes from natural ecosystems. Aust J Bot 56:369–407

Dale VH, Joyce LA, McNulty S, Neilson RP, Ayres MP, Flannigan MD, Hanson PJ, Irland LC, Lugo AE, Peterson CJ, Simberloff D, Swanson FJ, Stocks BJ, Wotton BM (2001) Climate change and forest disturbances. Bioscience 51:723–734

De Deyn GB, Cornelissen JHC, Bardgett RD (2008) Plant functional traits and soil carbon sequestration in contrasting biomes. Ecol Lett 11:516–531

de Jong BHJ, Cairns M, Haggerty P, Ramírez-Marcial N, Ochoa-Gaono S, Mendoza-Vega J, González-Espinosa M, March-Mifsut L (1999) Land-use change and carbon flux between 1970s and 1990s in the central highlands of Chiapas, México. Environ Manage 23:373–1285

De Vries W, Reinds GJ, Gundersen P, Sterba H (2006) The impact of nitrogen deposition on carbon sequestration in European forests and forest soils. Glob Change Biol 12:1151–1173

Degen T, Devillez F, Jacquemart AL (2005) Gaps promote plant diversity in beech forests (*Luzulo-Fagetum*), North Vosges, France. Ann Forest Sci 62:429–440

Delpierre N, Soudani K, Francois C, Kostner B, Pontailler JY, Nikinmaa E, Misson L, Aubinet M, Bernhofer C, Granier A, Grunwald T, Heinesch B, Longdoz B, Ourcival JM, Rambal S, Vesala T, Dufrene E (2009) Exceptional carbon uptake in European forests during the warm spring of 2007: a data-model analysis. Glob Change Biol 15:1455–1474

DeLucia EH, Moore DJ, Norby RJ (2005) Contrasting responses of forest ecosystems to rising atmospheric CO_2: implications for the global C cycle. Glob Biogeochem Cycles 19:GB3006

Dixon RK, Brown S, Houghton RA, Solomon AM, Trexler MC, Wisniewski J (1994) Carbon pools and flux of global forest ecosystems. Science 263:185–190

Driscoll CT, Lawrence GB, Bulger AJ, Butler TJ, Cronan CS, Eagar C, Lambert KF, Likens GE, Stoddard JL, Weathers KC (2001) Acidic deposition in the northeastern United States: sources and inputs, ecosystem effects, and management strategies. Bioscience 51:180–198

Edwards NT, Johnson D, McLaughlin S, Harris W (1989) Carbon dynamics and productivity. In: Johnson D, VanHook R (eds) Analysis of biogeochemical cycling processes in walker branch watershed. Springer, New York, pp 197–232

Elfving B, Tegnhammar L, Tveite B (1996) Studies on growth trends of forests in Sweden and Norway. In: Spiecker H, Mielikäinen K, Köhl K, Skovsgaard JP (eds) Growth trends in European forests. Springer, Berlin

Ericksson H, Karlsson K (1996) Long-term Changes in Site Index in Growth and Yield Experiements with Norway Spruce (*Picea abies*, [L.] Karst) and Scots Pine (*Pinus sylvestris,* L.). In: Spiecker H, Mielikäinen K, Köhl K, Skovsgaard JP (eds) Growth trends in European forests. Springer, Berlin

Eswaran H, Van den Berg E, Reich PB, Kimble J (1995) Global soil carbon resources. In: Lal R, Kimble JM, Levine E (eds) Soils and global change. Lewis Publishers is an imprint of CRC Press, Boca Raton, pp 27–43

Evans C, Goodale C, Caporn S, Dise N, Emmett B, Fernandez I, Field C, Findlay S, Lovett G, Meesenburg H, Moldan F, Sheppard L (2008) Does elevated nitrogen deposition or ecosystem recovery from acidification drive increased dissolved organic carbon loss from upland soil? A review of evidence from field nitrogen addition experiments. Biogeochemistry 91:13–35

Fahey TJ, Siccama TG, Driscoll CT, Likens GE, Campbell J, Johnson CE, Battles JJ, Aber JD, Cole JJ, Fisk MC, Groffman PM, Hamburg SP, Holmes RT, Schwarz PA, Yanai RD (2005) The biogeochemistry of carbon at Hubbard Brook. Biogeochemistry 75:109–176

Fang JY, Chen AP, Peng CH, Zhao SQ, Ci L (2001) Changes in forest biomass carbon storage in China between 1949 and 1998. Science 292(5525):2320–2322

Fang JY, Oikawa T, Kato T, Mo WH, Wang ZH (2005) Biomass carbon accumulation by Japan's forests from 1947 to 1995. Global Biogeochem Cycles 19:GB2004

FAO (2011) State of the World's forests 2011. Food and Agriculture Organization of the United Nations, Rome

Felzer BS, Cronin T, Reilly JM, Melilloa JM, Wang XD (2007) Impacts of ozone on trees and crops. CR Geosci 339:784–798

Finzi AC, Van Breemen N, Canham CD (1998) Canopy tree soil interactions within temperate forests: species effects on soil carbon and nitrogen. Ecol Appl 8:440–446

Fissore C, Giardina CP, Swanston CW, King GM, Kolka RK (2009) Variable temperature sensitivity of soil organic carbon in North American forests. Glob Change Biol 15:2295–2310

Fowler D, Cape JN, Coyle M, Flechard C, Kuylenstierna J, Hicks K, Derwent D, Johnson C, Stevenson D (1999) The global exposure of forests to air pollutants. Water Air Soil Pollut 116:5–32

Garcia-Oliva F, Hernandez G, Lancho JFG (2006) Comparison of ecosystem C pools in three forests in Spain and Latin America. Ann Forest Sci 63(5):519–523

Goodale CL, Apps MJ, Birdsey RA, Field CB, Heath LS, Houghton RA, Jenkins JC, Kohlmaier GH, Kurz W, Liu SR, Nabuurs GJ, Nilsson S, Shvidenko AZ (2002) Forest carbon sinks in the Northern Hemisphere. Ecol Appl 12:891–899

Goodnow R, Sullivan J, Amacher GS (2008) Ice damage and forest stand management. J Forest Econ 14:268–288
Gough CM, Vogel CS, Schmid HP, Curtis PS (2008) Controls on annual forest carbon storage: lessons from the past and predictions for the future. Bioscience 58:609–622
Goulden ML, Munger JW, Fan SM, Daube BC, Wofsy SC (1996) Exchange of carbon dioxide by a deciduous forest: response to interannual climate variability. Science 271:1576–1578
Gower ST, Vogt KA, Grier CC (1992) Carbon dynamics of rocky-mountain Douglas-Fir - influence of water and nutrient availability. Ecol Monogr 62:43–65
Grimm EC (1984) Fire and other factors controlling the big woods vegetation of Minnesota in the mid-19th century. Ecol Monogr 54:291–311
Guo LB, Gifford RM (2002) Soil carbon stocks and land use change: a meta analysis. Glob Change Biol 8:345–360
Hamilton JG, DeLucia EH, George K, Naidu SL, Finzi AC, Schlesinger WH (2002) Forest carbon balance under elevated CO_2. Oecologia 131:250–260
Hanson JJ, Lorimer CG (2007) Forest structure and light regimes following moderate wind storms: implications for multi-cohort management. Ecol Appl 17:1325–1340
Hanson PJ, Edwards NT, Tschaplinski TJ, Wullschleger SD, Joslin JD (2003) Estimating the net primary and net ecosystem production of a Southeastern upland Quercus forest from an 8-year biometric record. In: Hanson PJ, Wullschleger SD (eds) North American temperate deciduous forest responses to changing precipitation regimes. Springer, New York, Pages 472
Hanson PJ, Weltzin JF (2000) Drought disturbance from climate change: response of United States forests. Sci Total Environ 262:205–220
Hanson PJ, Wullschleger SD, Norby RJ, Tschaplinski TJ, Gunderson CA (2005) Importance of changing CO_2, temperature, precipitation, and ozone on carbon and water cycles of an upland-oak forest: incorporating experimental results into model simulations. Glob Change Biol 11:1402–1423
Harris WF, Sollins P, Edwards NT, Dinger BE, Shugart HH (1975) Analysis of carbon flow and productivity in a temperate deciduous forest ecosystem. In: Reichle D, Franklin J, Goodall D (eds) Productivity of world ecosystems. National Academy of Sciences, Washington, DC, pp 116–122
Hasenauer H, Nemani RR, Schadauer K, Running SW (1999) Forest growth response to changing climate between 1961 and 1990 in Austria. Forest Ecol Manag 122:209–219
Hattenschwiler S, Korner C (2003) Does elevated CO_2 facilitate naturalization of the non-indigenous prunus laurocerasus in Swiss temperate forests? Funct Ecol 17:778–785
Heath LS, Kauppi PE, Burschel P, Gregor HD, Guderian R, Kohlmaier GH, Lorenz S, Overdieck D, Scholz F, Thomasius H, Weber M (1993) Contribution of temperate forests to the worlds carbon budget. Water Air Soil Pollut 70:55–69
Hinckley TM, Teskey RO, Duhme F, Richter H (1981) Temperate hardwood forests. In: Kozlowski TT (ed) Water deficits and plant growth, vol 6, Woody plant communities. Academic, New York
Holland GJ, Webster PJ (2007) Heightened tropical cyclone activity in the North Atlantic: natural variability or climate trend? Philos Trans R Soc A 365:2695–2716
Hooker TD, Compton JE (2003) Forest ecosystem carbon and nitrogen accumulation during the first century after agricultural abandonment. Ecol Appl 13:299–313
Houghton RA (1995) Changes in the storage of terrestrial carbon since 1850. In: Lal R, Kimble J, Levine E, Stewart B (eds) Soils and global change. Lewis Publishers, New York
Innes JL, Peterson DL (2001) Proceedings introduction: managing forests in a greenhouse world -context and challenges. In: Peterson DL, Innes JL, O'Brian K (eds) Climate change, carbon, and forestry in Northwestern North America: Proceedings of a workshop November 14–15 2001, Orcas Island, Washington. General Technical Report PNW-GTR-614, USDA Forest Service
IPCC (2000) IPCC Special Report: land use, land use change, and forestry. Intergovernmental Panel on Climate Change
IPCC (2007) Climate change 2007: Synthesis Report. Intergovernmental Panel on Climate Change
Janssens IA, Freibauer A, Ciais P, Smith P, Nabuurs GJ, Folberth G, Schlamadinger B, Hutjes RWA, Ceulemans R, Schulze ED, Valentini R, Dolman AJ (2003) Europe's Terrestrial biosphere absorbs 7 to 12% of European anthropogenic CO_2 emissions. Science 300:1538–1542
Jarvis PG (1989) Atmospheric carbon-dioxide and forests. Philos Trans R Soc Lond B Biol Sci 324:369–392
Jassal RS, Black TA, Cai TB, Morgenstern K, Li Z, Gaumont-Guay D, Nesic Z (2007) Components of ecosystem respiration and an estimate of net primary productivity of an intermediate-aged Douglas-fir stand. Agric Forest Meteorol 144:44–57
Jassal RS, Black TA, Chen BZ, Roy R, Nesic Z, Spittlehouse DL, Trofymow JA (2008) N2O emissions and carbon sequestration in a nitrogen-fertilized Douglas fir stand. J Geophys Res-Biogeosci 113
Juarez RIN, Chambers JQ, Zeng HC, Baker DB (2008) Hurricane driven changes in land cover create biogeophysical climate feedbacks. Geophys Res Lett 35:5
Kashian DM, Romme WH, Tinker DB, Turner MG, Ryan MG (2006) Carbon storage on landscapes with stand-replacing fires. Bioscience 56:598–606
Kashian DM, Tinker DB, Turner MG, Scarpace FL (2004) Spatial heterogeneity of lodgepole pine sapling densities following the 1988 fires in Yellowstone National Park, Wyoming, USA. Can J Forest Res 34:2263–2276
Kasischke ES (2000) Effects of climate change and fire on carbon storage in North American boreal forests. In: Kasischke ES, Stocks BJ (eds) Fire, climate change, and carbon cycling in the Boreal forest. Springer, New York

Kasischke ES, Christensen NL, Stocks BJ (1995) Fire, global warming, and the carbon balance of Boreal forests. Ecol Appl 5:437–451

Keeling CD, Whorf TP, Wahlen M, Vanderplicht J (1995) Interannual extremes in the rate of rise of atmospheric carbon-dioxide since 1980. Nature 375:666–670

Kelty MJ (2006) The role of species mixtures in plantation forestry. Forest Ecol Manag 233:195–204

Knebel L, Wentworth TR (2007) Influence of fire and southern pine beetle on pine-dominated forests in the Linville Gorge Wilderness, North Carolina. Castanea 72:214–225

Knohl A, Schulze ED, Kolle O, Buchmann N (2003) Large carbon uptake by an unmanaged 250-year-old deciduous forest in central Germany. Agric Forest Meteorol 118:151–167

Korner C (2000) Biosphere responses to CO_2 enrichment. Ecol Appl 10:1590–1619

Korner C, Asshoff R, Bignucolo O, Hattenschwiler S, Keel SG, Pelaez-Riedl S, Pepin S, Siegwolf RTW, Zotz G (2005) Carbon flux and growth in mature deciduous forest trees exposed to elevated CO_2. Science 309:1360–1362

Kozovits AR, Matyssek R, Blaschke H, Gottlein A, Grams TEE (2005) Competition increasingly dominates the responsiveness of juvenile beech and spruce to elevated CO_2 and/or O-3 concentrations throughout two subsequent growing seasons. Glob Change Biol 11:1387–1401

Kruger EL, Reich PB (1997) Responses of hardwood regeneration to fire in mesic forest openings. I. Post-fire community dynamics. Can J Forest Res 27:1822–1831

Kurz WA, Apps MJ (1999) A 70-year retrospective analysis of carbon fluxes in the Canadian forest sector. Ecol Appl 9:526–547

Kurz WA, Stinson G, Rampley GJ, Dymond CC, Neilson ET (2008) Risk of natural disturbances makes future contribution of Canada's forests to the global carbon cycle highly uncerain. Proc Natl Acad Sci USA 105:1551–1555

Law BE, Ryan MG, Anthoni PM (1999) Seasonal and annual respiration of a ponderosa pine ecosystem. Glob Change Biol 5:169–182

Law BE, Sun OJ, Campbell J, Van Tuyl S, Thornton PE (2003) Changes in carbon storage and fluxes in a chronosequence of ponderosa pine. Glob Change Biol 9:510–524

Law BE, Waring RH, Anthoni PM, Aber JD (2000) Measurements of gross and net ecosystem productivity and water vapour exchange of a pinus ponderosa ecosystem, and an evaluation of two generalized models. Glob Change Biol 6:155–168

LeBauer DS, Treseder KK (2008) Nitrogen limitation of net primary productivity in terrestrial ecosystems is globally distributed. Ecology 89:371–379

Li JH, Powell TL, Seiler TJ, Johnson DP, Anderson HP, Bracho R, Hungate BA, Hinkle CR, Drake BG (2007) Impacts of hurricane Frances on Florida scrub-oak ecosystem processes: defoliation, net CO_2 exchange and interactions with elevated CO_2. Glob Change Biol 13:1101–1113

Likens GE, Borman FH (1995) Biogeochemistry of a forested wateshed. Springer, New York

Litton CM, Ryan MG, Knight DH (2004) Effects of tree density and stand age on carbon allocation patterns in postfire lodgepole pine. Ecol Appl 14:460–475

Luyssaert S, Inglima I, Jung M, Richardson AD, Reichsteins M, Papale D, Piao SL, Schulzes ED, Wingate L, Matteucci G, Aragao L, Aubinet M, Beers C, Bernhoffer C, Black KG, Bonal D, Bonnefond JM, Chambers J, Ciais P, Cook B, Davis KJ, Dolman AJ, Gielen B, Goulden M, Grace J, Granier A, Grelle A, Griffis T, Grunwald T, Guidolotti G, Hanson PJ, Harding R, Hollinger DY, Hutyra LR, Kolar P, Kruijt B, Kutsch W, Lagergren F, Laurila T, Law BE, Le Maire G, Lindroth A, Loustau D, Malhi Y, Mateus J, Migliavacca M, Misson L, Montagnani L, Moncrieff J, Moors E, Munger JW, Nikinmaa E, Ollinger SV, Pita G, Rebmann C, Roupsard O, Saigusa N, Sanz MJ, Seufert G, Sierra C, Smith ML, Tang J, Valentini R, Vesala T, Janssens IA (2007) CO_2 balance of boreal, temperate, and tropical forests derived from a global database. Glob Change Biol 13:2509–2537

Magill A, Aber JD, Berntson GM, McDowell WH, Nadelhoffer KJ, Mellilo JM, Steudler PA (2000) Long-term nitrogen additions and nitrogen saturation in two temperate forests. Ecosystem 3:238–253

Maier CA, Kress LW (2000) Soil CO_2 evolution and root respiration in 11 year-old loblolly pine (*Pinus taeda*) plantations as affected by moisture and nutrient availability. Can J Forest Res 30:347–359

Malhi Y, Baldocchi DD, Jarvis PG (1999) The carbon balance of tropical, temperate and boreal forests. Plant Cell Environ 22:715–740

Martin PH, Nabuurs GJ, Aubinet M, Karjalainen T, Vine EL, Kinsman J, Heath LS (2001) Carbon sinks in temperate forests. Annu Rev Energ Env 26:435–465

Martinez-Vilalta J, Lopez BC, Adell N, Badiella L, Ninyerola M (2008) Twentieth century increase of Scots pine radial growth in NE Spain shows strong climate interactions. Glob Change Biol 14:2868–2881

Mather A (1990) Global forest resources. Timber Press, Portland

McCarthy HR, Oren R, Kim HS, Johnsen KH, Maier C, Pritchard SG, Davis MA (2006) Interaction of ice storms and management practices on current carbon sequestration in forests with potential mitigation under future CO_2 atmosphere. J Geophys Res Atmos 111:10

McMahon SM, Parker GG, Miller DR (2010) Evidence for a recent increase in forest growth. Proc Natl Acad Sci USA 107(8):3611–3615. doi:10.1073/pnas.0912376107

McNab WH, Avers PE (1994) Ecological subregions of the United States: Section descriptions. Ecosystem Management Report WO-WSA-5, U.S. Department of Agriculture, Washington, DC

McNab WH, Greenberg CH, Berg EC (2004) Landscape distribution and characteristics of large hurricane-related canopy gaps in a southern Appalachian watershed. Forest Ecol Manag 196:435–447

McNulty SG (2000) Hurricane impacts on US forest carbon sequestration. In: Advances in terrestrial ecosystem: carbon inventory measurements and monitoring conference. Elsevier Sci Ltd, Raleigh, pp S17-S24
Mendoza-Ponce A, Galicia L (2010) Aboveground and belowground biomass and carbon pools in highland temperate forest landscape in central Mexico. Forestry 83(5):497–506. doi:10.1093/forestry/cpq032
Morehouse K, Johns T, Kaye J, Kaye A (2008) Carbon and nitrogen cycling immediately following bark beetle outbreaks in southwestern ponderosa pine forests. Forest Ecol Manag 255:2698–2708
Morrison IK (1990) Organic-Matter and mineral distribution in an Old-Growth Acer-Saccharum forest near the northern limit of its range. Canadian journal of forest Research-Revue canadienne de recherche forestiere 20:1332–1342
Mouillot F, Field CB (2005) Fire history and the global carbon budget: a 1 degrees x 1 degrees fire history reconstruction for the 20th century. Glob Change Biol 11:398–420
Nabuurs GJ, Schelhaas MJ, Mohren GMJ, Field CB (2003) Temporal evolution of the European forest sector carbon sink from 1950 to 1999. Glob Change Biol 9:152–160
Nabuurs GJ, Thurig E, Heidema N, Armolaitis K, Biber P, Cienciala E, Kaufmann E, Makipaa R, Nilsen P, Petritsch R, Pristova T, Rock J, Schelhaas MJ, Sievanen R, Somogyi Z, Vallet P (2008) Hotspots of the European forests carbon cycle. Forest Ecol Manag 256:194–200
Nadelhoffer KJ, Emmett BA, Gundersen P, Kjonaas OJ, Koopmans CJ, Schleppi P, Tietema A, Wright RF (1999) Nitrogen deposition makes a minor contribution to carbon sequestration in temperate forests. Nature 398:145–148
Nagel TA, Svoboda M, Diaci J (2006) Regeneration patterns after intermediate wind disturbance in an old-growth fagus-abies forest in southeastern Slovenia. Forest Ecol Manag 226(1–3):268–278. doi:10.1016/j.foreco.2006.01.039
Nagel TA, Svoboda M (2008) Gap disturbance regime in an old-growth fagus-abies forest in the Dinaric mountains, Bosnia-Herzegovina. Can J Forest Res 38:2728–2737
NIFC (2011) Wildland fire statistics 1960–2010. National Interagency Fire Center. http://www.nifc.gov/fire_info/fires_acres.htm. Accessed 27 Apr 2011
Norby RJ, DeLucia EH, Gielen B, Calfapietra C, Giardina CP, King JS, Ledford J, McCarthy HR, Moore DJP, Ceulemans R, De Angelis P, Finzi AC, Karnosky DF, Kubiske ME, Lukac M, Pregitzer KS, Scarascia-Mugnozza GE, Schlesinger WH, Oren R (2005) Forest response to elevated CO_2 is conserved across a broad range of productivity. Proc Natl Acad Sci USA 102:18052–18056
Norby RJ, Luo YQ (2004) Evaluating ecosystem responses to rising atmospheric CO_2 and global warming in a multi-factor world. New Phytol 162:281–293
Nowak RS, Ellsworth DS, Smith SD (2004) Functional responses of plants to elevated atmospheric CO_2 - do photosynthetic and productivity data from FACE experiments support early predictions? New Phytol 162:253–280
Ollinger SV, Aber JD, Reich PB, Freuder RJ (2002) Interactive effects of nitrogen deposition, tropospheric ozone, elevated CO2 and land use history on the carbon dynamics of northern hardwood forests. Glob Change Biol 8:545–562
Ollinger SV, Richardson A, Martin ME, Hollinger D, Frolking S, Reich PB, Plourde L, Katul G, Munger JW, Oren R, Smith M-L, Paw KT, Bolstad PV, Cook BD, Day M, Martin T, Monson R, Schmid HP (2008) Canopy nitrogen, carbon assimilation, and albedo in temperate and boreal forests: functional relations and potential climate feedbacks. Proc Natl Acad Sci USA 105:19336–19341
Ordonez JAB, de Jong BHJ, Garcia-Oliva F, Avina FL, Perez JV, Guerrero G, Martinez R, Masera O (2008) Carbon content in vegetation, litter, and soil under 10 different land-use and land-cover classes in the central highlands of Michoacan, Mexico. Forest Ecol Manag 255:2074–2084
Palmer MW, McAlister SD, Arevalo JR, DeCoster JK (2000) Changes in the understory during 14 years following catastrophic windthrow in two Minnesota forests. J Veg Sci 11:841–854
Papaik MJ, Canham CD (2006) Species resistance and community response to wind disturbance regimes in northern temperate forests. J Ecol 94:1011–1026
Papale D, Valentini A (2003) A new assessment of European forests carbon exchanges by eddy fluxes and artificial neural network spatialization. Glob Change Biol 9(4):525–535
Pastor J, Post WM (1988) Response of northern forests to CO_2-induced climate change. Nature 334:55–58
Peichl M, Arain AA (2006) Above- and belowground ecosystem biomass and carbon pools in an age-sequence of temperate pine plantation forests. Agric Forest Meteorol 140:51–63
Perry DA (1994) Forest ecosystems. The John Hopkins University Press, Baltimore
Peterson DW, Peterson DL (2001) Mountain hemlock growth responds to climatic variability at annual and decadal time scales. Ecology 82:3330–3345
Peterson DW, Reich PB (2001) Prescribed fire in oak savanna: fire frequency effects on stand structure and dynamics. Ecol Appl 11:914–927
Post WM, Emanuel WR, Zinke PJ, Stangenberger AG (1982) Soil carbon pools and world life zones. Nature 298:156–159
Potter C, Klooster S, Huete A, Genovese V (2007) Terrestrial carbon sinks for the United States predicted from MODIS satellite data and ecosystem modeling. Earth Interact 11(13):11–21
Pregitzer KS, Burton AJ, Zak DR, Talhelm AF (2008) Simulated chronic nitrogen deposition increases carbon storage in Northern Temperate forests. Glob Change Biol 14:142–153

Pregitzer KS, Euskirchen ES (2004) Carbon cycling and storage in world forests: biome patterns related to forest age. Glob Change Biol 10:2052–2077

Puhe J, Ulrich B (2001) Implications of the deposition of acid and nitrogen. In: Puhe J, Ulrich B (eds) Global climate change and human impacts on forest ecosystems. Springer, Berlin, Pages 592

Pyne SJ (1982) Fire in America: a cultural history of wildland and rural fire. Princeton University Press, Princeton

Reich PB, Abrams MD, Ellsworth DS, Kruger EL, Tabone TJ (1990) Fire affects ecophysiology and community dynamics of central Wisconsin oak forest regeneration. Ecology 71:2179–2190

Reich P, Frelich L (2002) Temperate deciduous forests. In: Mooney H, Canadell J (eds) Encyclopedia of global environmental change, vol 2, The earth system: biological and ecological dimensions of global environmental change. Wiley, Chichester, pp 565–569

Ren GY (2007) Changes in forest cover in china during the Holocene. Veg Hist Archaeobot 16(2–3):119–126. doi:10.1007/s00334-006-0075-5

Richardson AD, Black TA, Ciais P, Delbart N, Friedl MA, Gobron N, Hollinger DY, Kutsch WL, Longdoz B, Luyssaert S, Migliavacca M, Montagnani L, Munger JW, Moors E, Piao SL, Rebmann C, Reichstein M, Saigusa N, Tomelleri E, Vargas R, Varlagin A (2010) Influence of spring and autumn phenological transitions on forest ecosystem productivity. Philos Trans R Soc B 365(1555):3227–3246. doi:10.1098/rstb.2010.0102

Richardson AD, Bailey AS, Denny EG, Martin CW, O'Keefe J (2006) Phenology of a northern hardwood forest canopy. Glob Change Biol 12:1174–1188

Richardson AD, Hollinger DY, Dail DB, Lee JT, Munger JW, O'Keefe J (2009) Influence of spring phenology on seasonal and annual carbon balance in two contrasting New England forests. Tree Physiol 29:321–331

Richardson DM, Rundel PW, Jackson ST, Teskey RO, Aronson J, Bytnerowicz A, Wingfield MJ, Proches S (2007) Human impacts in pine forests: past, present, and future. Annu Rev Ecol Evol Syst 38:275–297

Rolland C, Lemperiere G (2004) Effects of climate on radial growth of Norway spruce and interactions with attacks by the bark beetle *Dendroctonus micans* (Kug., Coleoptera: Scolytidae): a dendroecological study in the French Massif Central. Forest Ecol Manag 201:89–104

Ruark GA, Bockheim JG (1988) Biomass, Net primary production, and nutrient distribution for an Age sequence of populus-tremuloides ecosystems. Can J Forest Res 18:435–443

Rudel TK, Coomes OT, Moran E, Achard F, Angelsen A, Xu JC, Lambin E (2005) Forest transitions: towards a global understanding of land use change. Glob Environ Chang 15(1):23–31. doi:10.1016/j.gloenvcha.2004.11.001

Ryan MG, Binkley D, Fownes JH (1997) Age-related decline in forest productivity: pattern and process. Adv Ecol Res 27:213–262

Ryan MG, Hubbard RM, Pongracic S, Raison RJ, McMurtrie RE (1996) Foliage, fine-root, woody-tissue and stand respiration in pinus radiata in relation to nitrogen status. Tree Physiol 16:333–343

Saigusa N, Yamamoto S, Hirata R, Ohtani Y, Ide R, Asanuma J, Gamo M, Hirano T, Kondo H, Kosugi Y, Li SG, Nakai Y, Takagi K, Tani M, Wang HM (2008) Temporal and spatial variations in the seasonal patterns of CO2 flux in boreal, temperate, and tropical forests in East Asia. Agric Forest Meteorol 148:700–713

Savage K, Davidson EA, Richardson AD, Hollinger DY (2009) Three scales of temporal resolution from automated soil respiration measurements. Agric Forest Meteorol 149:2012–2021

Schleip C, Rutishauser T, Luterbacher J, Menzel A (2008) Time series modeling and central European temperature impact assessment of phenological records over the last 250 years. J Geophys Res Biogeosci 113:G04026

Schwalm CR, Williams CA, Schaefer K, Anderson R, Arain MA, Baker I, Barr A, Black TA, Chen GS, Chen JM, Ciais P, Davis KJ, Desai A, Dietze M, Dragoni D, Fischer ML, Flanagan LB, Grant R, Gu LH, Hollinger D, Izaurralde RC, Kucharik C, Lafleur P, Law BE, Li LH, Li ZP, Liu SG, Lokupitiya E, Luo YQ, Ma SY, Margolis H, Matamala R, McCaughey H, Monson RK, Oechel WC, Peng CH, Poulter B, Price DT, Riciutto DM, Riley W, Sahoo AK, Sprintsin M, Sun JF, Tian HQ, Tonitto C, Verbeeck H, Verma SB (2010) A model-data intercomparison of CO_2 exchange across North America: results from the North American carbon program site synthesis. J Geophys Res Biogeosci 115. doi:G00h05.10.1029/2009jg001229

Sharma CM, Baduni NP, Gairola S, Ghildiyal SK, Suyal S (2010) Tree diversity and carbon stocks of some major forest types of Garhwal Himalaya, India. Forest Ecol Manag 260(12):2170–2179. doi:10.1016/j.foreco.2010.09.014

Smith W, Miles P, Perry C, Pugh S (2009) Forest resources of the United States, 2007. US Department of Agriculture, Forest Service, Washington, DC

Smithwick EAH, Harmon ME, Remillard SM, Acker SA, Franklin JF (2002) Potential upper bounds of carbon stores in forests of the Pacific Northwest. Ecol Appl 12:1303–1317

Tans P, Keeling R (2011) Trends in atmospheric carbon dioxide, Mauna Loa, Hawaii. NOAA Earth System Research Laboratory. Available at http://www.esrl.noaa.gov/gmd/ccgg/trends/. Accessed 15 Apr 2011

Tinker DB, Knight DH (2000) Coarse woody debris following fire and logging in Wyoming lodgepole pine forests. Ecosystems 3:472–483

Trumbore S (2000) Age of soil organic matter and soil respiration: radiocarbon constraints on belowground C dynamics. Ecol Appl 10:399–411

Tschaplinski TJ, Stewart DB, Hanson PJ, Norby RJ (1995) Interactions between drought and elevated CO_2 on growth and gas-exchange of seedlings of 3 deciduous tree species. New Phytol 129:63–71

Uriate M, Papaik M (2007) Hurricane impacts on dynamics, structure and carbon sequestration potential of forest ecosystems in Southern New England, USA. Tellus A 59:519–528

Vadeboncoeur M, Hamburg S, Richardson A, Bailey A (2006) Examining climate change at the ecosystem level: a 50-year record from the Hubbard Brook Experimental Forest. Hubbard Brook Ecosystem Study
Valentini R, Matteucci G, Dolman AJ, Schulze ED, Rebmann C, Moors EJ, Granier A, Gross P, Jensen NO, Pilegaard K, Lindroth A, Grelle A, Bernhofer C, Grunwald T, Aubinet M, Ceulemans R, Kowalski AS, Vesala T, Rannik U, Berbigier P, Loustau D, Guomundsson J, Thorgeirsson H, Ibrom A, Morgenstern K, Clement R, Moncrieff J, Montagnani L, Minerbi S, Jarvis PG (2000) Respiration as the main determinant of carbon balance in European forests. Nature 404:861–865
Vogt KA, Vogt DJ, Bloomfield J (1998) Analysis of some direct and indirect methods for estimating root biomass and production of forests at an ecosystem level. In: Box JE (ed) Root demographics and their efficiencies in sustainable agriculture, grasslands and forest ecosystems. Springer, Dordrecht, pp 687–720
Waldrop MP, Zak DR, Sinsabaugh RL, Gallo M, Lauber C (2004) Nitrogen deposition modifies soil carbon storage through changes in microbial enzymatic activity. Ecol Appl 14:1172–1177
Webb SL, Scanga SE (2001) Windstorm disturbance without patch dynamics: twelve years of change in a Minnesota forest. Ecology 82:893–897
Wickham JD, Riitters KH, Wade TG, Homer C (2008) Temporal change in fragmentation of continental US forests. Landscape Ecol 23:891–898
Williams DW, Liebhold AM (1995) Herbivorous insects and global change: potential changes in the spatial distribution of forest defoliator outbreaks. J Biogeogr 22:665–671
Wittig VE, Bernacchi CJ, Zhu XG, Calfapietra C, Ceulemans R, Deangelis P, Gielen B, Miglietta F, Morgan PB, Long SP (2005) Gross primary production is stimulated for three *Populus* species grown under free-air CO2 enrichment from planting through canopy closure. Glob Change Biol 11:644–656
Wofsy SC, Goulden ML, Munger JW, Fan SM, Bakwin PS, Daube BC, Bassow SL, Bazzaz FA (1993) Net exchange of CO_2 in a Midlatitude forest. Science 260:1314–1317
Woodbury PB, Heath LS, Smith JE (2006) Land use change effects on forest carbon cycling throughout the southern United States. J Environ Qual 35:1348–1363
Woodbury PB, Smith JE, Heath LS (2007) Carbon sequestration in the US forest sector from 1990 to 2010. Forest Ecol Manag 241:14–27
Worrall JJ, Lee TD, Harrington TC (2005) Forest dynamics and agents that initiate and expand canopy gaps in *Picea-Abies* forests of Crawford Notch, New Hampshire, USA. J Ecol 93:178–190
Xiao JF, Zhuang QL, Law BE, Baldocchi DD, Chen JQ, Richardson AD, Melillo JM, Davis KJ, Hollinger DY, Wharton S, Oren R, Noormets A, Fischer ML, Verma SB, Cook DR, Sun G, McNulty S, Wofsy SC, Bolstad PV, Burns SP, Curtis PS, Drake BG, Falk M, Foster DR, Gu LH, Hadley JL, Katulk GG, Litvak M, Ma SY, Martinz TA, Matamala R, Meyers TP, Monson RK, Munger JW, Oechel WC, Paw UKT, Schmid HP, Scott RL, Starr G, Suyker AE, Torn MS (2011) Assessing net ecosystem carbon exchange of U.S. terrestrial ecosystems by integrating eddy covariance flux measurements and satellite observations. Agric Forest Meteorol 151(1):60–69
Yuan FM, Arain MA, Barr AG, Black TA, Bourque CPA, Coursolle C, Margolis HA, McCaughey JH, Wofsy SC (2008) Modeling analysis of primary controls on net ecosystem productivity of seven boreal and temperate coniferous forests across a continental transect. Glob Change Biol 14:1765–1784
Zak DR, Holmes WE, Pregitzer KS (2007) Atmospheric CO_2 and O-3 alter the flow of N-15 in developing forest ecosystems. Ecology 88:2630–2639
Zhang QZ, Wang CK (2010) Carbon density and distribution of six Chinese temperate forests. Sci China Life Sci 53(7):831–840
Zhang C, Tian HQ, Chappelka AH, Ren W, Chen H, Pan SF, Liu ML, Styers DM, Chen GS, Wang YH (2007) Impacts of climatic and atmospheric changes on carbon dynamics in the Great Smoky Mountains National Park. Environ Pollut 149:336–347
Zhu BA, Wang XP, Fang JY, Piao SL, Shen HH, Zhao SQ, Peng CH (2010) Altitudinal changes in carbon storage of temperate forests on Mt Changbai, Northeast China. J Plant Res 123(4):439–452. doi:10.1007/s10265-009-0301-1
Zhu JJ, Yan QL, Fan AN, Yang K, Hu ZB (2009) The role of environmental, root, and microbial biomass characteristics in soil respiration in temperate secondary forests of Northeast China. Trees-Struct Funct 23:189–196

Carbon Dynamics in the Boreal Forest

6

Brian Milakovsky, Brent Frey, and Thomas James

Executives Summary

As one of the largest and most intact biomes, the boreal forest occupies a prominent place in the global carbon budget. While it contains about 13% of global terrestrial biomass, its organic-rich soils hold 43% of the world's soil carbon. A growing body of research has attempted to measure how climate influences the processes governing carbon uptake and release, and to predict further changes due to climate change. A review of this body of research produces the key findings outlined below.

Current research on boreal forest carbon pools and the processes that affect them suggest that this biome acts as a weak sink for atmospheric carbon. However, evidence of rapid climate change at northern latitudes has raised concern that the boreal forest could readily shift to a net carbon source if the ecophysiological processes facilitating carbon uptake are sufficiently disrupted. Changes in soil temperatures, respiration rates, and disturbance dynamics (type, extent, and frequency) brought about by climate change or other factors could switch the biome to a net source of carbon. Based on current knowledge, it appears that a warming climate will likely create the conditions for increased carbon release from boreal forests.

The boreal is a large and complex ecosystem and uniform response due to warming is unlikely. Empirical evidence suggests non-linear response, and this will affect forest carbon storage on varied temporal and spatial scales. Furthermore, determining the balance of carbon uptake and release is highly complex, and methods of carbon flux measurement will need to improve for more accurate conclusions of climate change impacts to be made. The following points represent generalizations across all boreal ecosystems.

What We Know About Carbon Storage and Flux in Boreal Forests

- Research indicates that boreal forests across North America and Eurasia have acted as weak sinks for atmospheric carbon in the last century. Storage of carbon in living and dead vegetation and the organic soil pool have generally exceeded carbon release through respiration and combustion. The "sink" status of the boreal forest is largely dependent on factors that keep heterotrophic respiration (release of CO_2 by microbial decomposition of organic matter) lower than carbon uptake through plant growth and accumulation in the soil. Heterotrophic respiration varies with the amount of decaying organic matter, soil moisture, soil temperature, vegetation type, and species/types of decomposers, which in turn

B. Milakovsky • T. James
Yale School of Forestry and Environmental Studies, New Haven, CT, USA

B. Frey (✉)
College of Forestry, Mississippi State University, Starkville, MS, USA
e-mail: bfrey@cfr.msstate.edu

M.S. Ashton et al. (eds.), *Managing Forest Carbon in a Changing Climate*,
DOI 10.1007/978-94-007-2232-3_6,

are influenced by disturbance (particularly fire and insect outbreaks, but also harvesting and ice and wind storms), temperature, precipitation, and duration of thaw.

- The soil carbon pool plays a disproportionately large role in sequestration in boreal forests, frequently constituting the largest pool in the system. In general, carbon accumulation rates in the soil are highest in low-lying, poorly drained sites such as peat bogs or black spruce swamps. More productive, well-drained sites on uplands may produce greater tree growth but store less carbon in the soil pool. Likewise, north-facing aspects that maintain permafrost and cooler temperatures support reduced productivity and higher carbon accumulation rates compared to other topographic positions.
- Studies in Canada have shown that lichens and bryophytes in lowland saturated sites contain upwards of 20% of the above ground carbon. These communities have important effects on how carbon is stored in boreal soils. Thick moss layers limit heat gain from the atmosphere, creating cold and wet conditions that promote the development of permafrost, with limited decomposition, thus are important for carbon storage. These positive feedbacks can be altered by novel disturbance regimes, including severe fire, which alter successional trajectories and increase carbon loss through decomposition and respiration.

What We Do Not Know About Carbon Storage and Flux in Boreal Forests

- Certain regions of the boreal are well studied, including those areas in Canada and Fennoscandia. However, many other regions are underrepresented in global carbon budget projections, and as a result, there is a tremendous amount of uncertainty in estimates of boreal carbon pools.
- There is little quantifiable information about several important carbon pools, including fine root biomass and mycorrhizae, bryophyte and understory layers and coarse woody debris and litter.
- Research is lacking on poorly drained sites, including those found in the larch forests of Siberia, which may be the most vulnerable to soil carbon loss with changes in disturbance regimes and climate.
- Considering the importance of fire in boreal carbon dynamics, there is much that is still not well understood, including extent, frequency, and intensity across the biome; and the interactions among fire intensity, nitrogen, and carbon.

What We Think Are the Major Influences on Carbon Storage and Flux in Boreal Forests

Disturbance

- Increased fire frequency could greatly increase carbon release, especially if it increases the decomposition of "old" carbon from the soil pool by increasing soil temperatures and degrading permafrost. More frequent fires could greatly reduce storage in woody biomass, and cause a concurrent increase in decomposition. Of even greater importance is the enhanced rate of heterotrophic respiration observed after fire. Increased soil temperatures from surface blackening and loss of the insulating bryophyte and litter layers that keep soil respiration low, increased nutrient availability from ash, and carbon inputs from fire-killed trees all contribute to enhanced decomposition rates post-burn. In addition, fire regimes determine the forest age class distribution across the landscape, and influence what vegetation communities develop (with their differing carbon dynamics). On the other hand, an often-overlooked impact of fire is the conversion of woody biomass to charcoal, a very persistent form of carbon that can remain in the soil for centuries. Thus fire may contribute to carbon storage in the soil through charcoal inputs to long-term carbon pool.
- While fire is recognized as the dominant natural disturbance type over much of the boreal forest, secondary disturbances such as insect outbreaks (and "background" insect damage during non-outbreak years) are also critically important. In some circumstances, such as the Canadian boreal and north temperate forests, insects and pathogens annually cause forest volume losses through mortality and growth reductions that are three times the volume lost to fire. Unlike fire, insect damage does not produce a direct emission, but rather exerts its influence through altered rates of decomposition and growth. In

some forest types, insect outbreaks exert the primary influence on age class distribution.

- Drought events have been increasingly implicated as a critical driver of stand dynamics and forest mortality, particularly in the boreal zone. Increased temperatures and extended periods of below-average precipitation have triggered forest dieback and mortality across large areas, with drier regions of the boreal appearing particularly vulnerable. Drought affected regions may also be more vulnerable to insect outbreaks thereby enhancing mortality rates. Resulting massive waves of mortality that have been documented represent a dramatic and sharp increase of carbon in dead standing biomass, with significant consequences for long-term carbon flux.

Age Class Distribution

- The balance of carbon uptake versus respiration loss changes with the stage of stand development in boreal forests, and research indicates that two distinct scenarios may be possible. In the first more frequently observed scenario, a brief period of enhanced post-disturbance (fire or logging) release is followed by a return to sink conditions and, eventually, equilibrium. The "sink" status of boreal forests is thus dependent on a disturbance regime that creates a forest age-class distribution that is skewed towards vigorous, maturing stands. However, other research indicates that decomposition of post-fire detritus may not occur early in stand development, but rather during stand maturation. Such a delayed decomposition response could counteract the high carbon uptake rates observed in maturing stands, making them a weaker sink than traditionally thought.

Climate and Topography

- Extremely high rates of carbon storage are possible in many boreal soils due to insulating bryophyte layers, low temperatures, poor drainage, high moisture content and permafrost formation. Cold and wet conditions slow decomposition rates and allow organic matter to accumulate faster than it is respired away.

How We Think the Carbon Status of Boreal Forests Changes with Changing Climate

- The question of whether moisture availability will decline with climatic warming will probably determine whether warming enhances the boreal carbon sink or turns it into a source. The balance of growth and respiration is significantly influenced by climatic conditions such as temperature, precipitation, and duration of the growing season. Increasing temperatures without concurrent increases in precipitation can cause drought stress, increased respiration, and the loss of carbon from boreal forests. However, if precipitation increases along with temperature, growing conditions could significantly improve and greater carbon uptake could occur. Increasing temperatures in early spring could also increase carbon uptake by lengthening the growing season.
- Sustained increased temperatures could possibly cause the breakdown of permafrost layers in boreal soils. If this occurs, the large stores of carbon bound in these frozen soils could be released.
- It appears that climatic warming is shortening the fire return interval in many boreal forests, speeding up the life cycles of damaging insects, and amplifying drought-driven dieback events. This could result in a large release of carbon, quickly turning the boreal forests from a sink to a source of carbon.
- Peatlands are possibly at greater risk from climate warming than forested areas and there is very little research on these un-forested wetlands, which may hold the majority of the carbon found in the boreal system.
- Over 97% of the total carbon stored in the vast tundra systems to the north of the boreal forest is found in the soil. This has huge implications for the global carbon budget, with the potential for a shifting boreal-tundra border with climate change. It is unclear whether the massive carbon pool in tundra soils would remain intact if converted to a forested biome.

1 Introduction

This chapter reviews the research literature on boreal and sub boreal forests of Eurasia and North America. It first describes the region, the forest

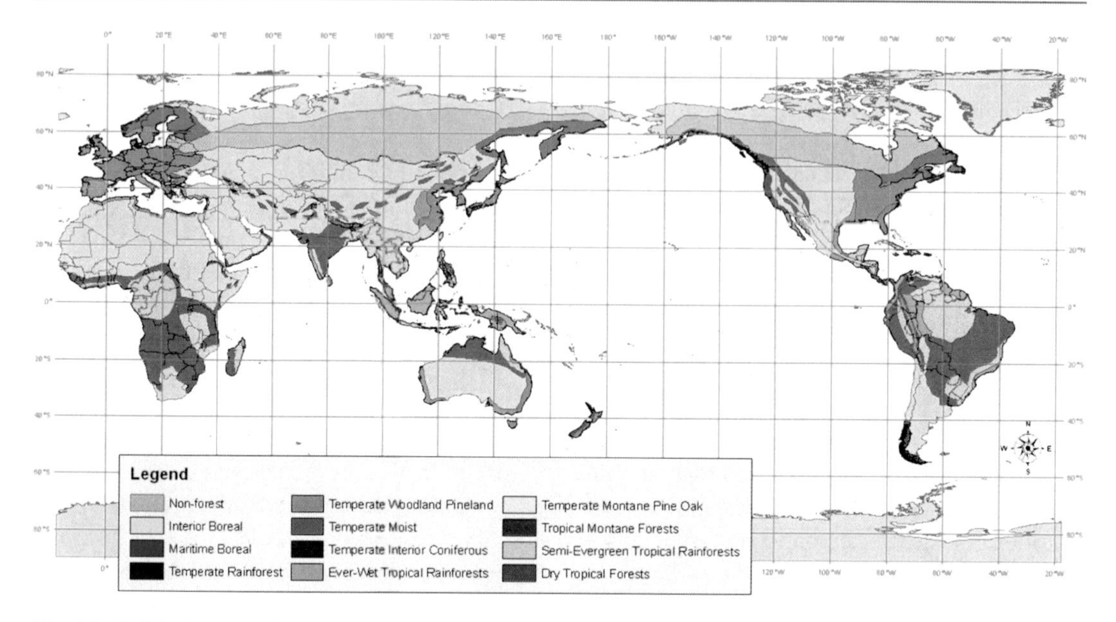

Fig. 6.1 Original extent of boreal, temperate, and tropical forest types of the world prior to land clearing

types, and their climatic variations. It then describes the stocks of carbon within the different components of the forest – above-ground biomass, below-ground biomass, lichens and bryophytes, the litter layer, and the soil. The next part of the chapter is focused on changes among carbon stocks – in particular understanding the biotic interactions of uptake (photosynthesis) and loss (respiration, decomposition); and then how abiotic influences of disturbance (fire, insect outbreaks, drought, forest management) can affect carbon stocks. The chapter highlights areas of carbon forest science that we understand well versus those areas that represent critical gaps in our knowledge and demand further investigation.

1.1 The Boreal Forest System

The boreal forest occupies a vast swath of the northern hemisphere, including much of Canada, Alaska, Fennoscandia, Russia, Mongolia, and northeast China (Fig. 6.1). Its northern limit is close to 68°N in North America and nearly 71°N in Eurasia, north of which tundra vegetation dominates. The southern limit is more variable, blending into temperate mixed forests or grassland and steppe systems, depending on moisture availability (Larsen 1980). Certain temperate forests that border the boreal (such as the Laurentian forest types of eastern North America or the Ussuri Taiga of the Russian Far East) or that occur at high elevations (such as spruce-fir communities in the Rocky Mountains or the Alps) have similar dynamics of carbon storage and release, and much of the research cited in this paper may apply to these regions.

Across their global range, boreal forests share certain key features. Only six tree genera are found as canopy dominants: spruce (*Picea*), fir (*Abies*), pine (*Pinus*), larch (*Larix*), birch (*Betula*), and aspen (*Populus*). Mature stands tend to exhibit very simple structure, dominated by a single stratum of conifers with a well-developed bryophyte layer at ground level (Gower et al. 2001). Understory communities are generally of low diversity (Larsen 1980), but shrub and herb diversity can vary substantially between stands related to overstory composition and soil type (e.g. MacDonald and Fenniak 2007; Légaré et al. 2001). In sub-boreal forests along the southern edge of the zone, aspens and birches may become more dominant, with a concomitant increase in understory diversity (e.g. MacDonald and Fenniak 2007). Boreal landscapes in North America and Eurasia feature vast plains (often the beds of ancient glacial lakes) interspersed with numerous bogs and fens. These plains are bounded by mountain ranges such as the Northern Rockies and the Altai (Fig. 6.1). Soils types vary across the boreal. Higher fertility luvisolic (alfisols) soils are characteristic of some interior regions of the southern boreal, but organic soils (histosols), permafrost soils (gelisols), and heavily leached and nutrient-poor podzols predominate over large areas (Larsen 1980). In lowland areas with sufficient moisture and temperature conditions, large peat deposits form above the mineral soil, sometimes covering many millions of hectares (Gorham 1991).

Differences in climate, moisture availability, and disturbance regimes create distinct zones within the greater boreal continuum. In North America, interior boreal forests characterized by a continental climate occupy the majority of the area. Dominant species assemblages include white spruce (*Picea glauca*), jack pine (*Pinus banksiana*) and spruce-aspen (*Populus tremuloides*) mixedwoods on upland sites, and black spruce (*P. mariana*) with components of Larch (*Larix* sp.) on cold, poorly drained sites. These interior boreal forest types are primarily characterized by a disturbance regime of catastrophic fires. In contrast, maritime influence from the Pacific in the west, and the Atlantic Ocean in the east create moister, more productive conditions in the Cordillarean and Maritime boreal zones, respectively (Apps et al. 1993; Baldocchi et al. 2000) (Fig. 6.1). These forests include a larger component of fir (*Abies* spp.) and cyclical outbreaks of forest insects play a greater role in structuring forest dynamics. In addition, most regions of the North American boreal are heavily influenced by industrial forest use.

In Eurasia, boreal forests west of the Ural Mountains tend to be dominated by Norway spruce (*Picea abies*) and Scots pine (*Pinus sylvestris*), and are significantly influenced by catastrophic fire and industrial forest management practices. The Baltic and White Seas produce a moderating climatic effect for Fennoscandian and northwest Russian forests (Baldocchi et al. 2000), which may explain the higher productivity observed in these areas compared to continental Siberian forests (Schulze et al. 1999) (Fig. 6.1). East of the Urals, a combination of extreme moisture stress and extensive permafrost shifts the competitive advantage to larch species (*Larix* spp.), which are adapted to these difficult growing conditions (Gower and Richards 1990). Large areas of Scots pine are also found in Siberia. A regime of frequent, non-catastrophic ground fires is characteristic of these forests (Harden et al. 1997). Fennoscandian forests have a long history of local forest utilization and commercial forestry (e.g. Berg et al. 2008), while industrial forest use in the Siberian boreal has expanded rapidly in recent years (Archard et al. 2006).

2 Pools of Carbon in the Boreal System

Carbon storage in the boreal forest occurs in distinct but interrelated pools, each of which demonstrates unique response to environmental change. As such, it is very important to address these pools separately before attempting an integrated understanding of boreal carbon dynamics. The major pools are aboveground biomass (ranging from 11% to 59%); soil (ranging from 20% to 85%); and bryophytes/mosses (ranging from 5% to 26%) (Table 6.1). Litter and belowground biomass are much smaller, although the litter pool can be as high as 50% in young Jack pine stands. Belowground biomass is hard to measure and consequently there are limited data for this pool.

2.1 Aboveground Biomass

This pool consists of the live or dead standing biomass of trees, shrubs and herbs. In contrast with tropical and temperate forests, this aboveground pool is usually not the largest in the boreal system but is strongly influenced by site productivity. For example, in relatively productive

Table 6.1 Distribution of carbon among different pools in boreal forests

	Site characteristics			Carbon pools			
Source	Location	Forest type	Age	Aboveground biomass	Bryophytes/ mosses	Litter	Soil
Malhi et al. (1999)	Interior Canada	Black Spruce	115	49.2 Mg/ha (11%)		6.2 Mg/ha (1%)	390.4 Mg/ha (85%)
Goulden et al. (1998)	Interior Canada	Black Spruce Sphagnum site	120	40 ± 20 tons/ha (14%)	45 ± 13 tons/ha (16%)		200 ± 50 tons/ha (70%)
Goulden et al. (1998)	Interior Canada	Black Spruce Pleurozium site	120	40 ± 20 tons/ha (23%)	45 ± 13 tons/ha (26%)		90 ± 51 tons/ha (70%)
Gower et al. (1997)	Interior Canada	Black spruce	115–155	49.2–57.2 Mg/ha (11–12%)			390.4–418.4 Mg/ha (87–88%)
		Aspen	53–67	57.0–93.3 Mg/ha (32–59%)		15.9–19.4 Mg/ha (9–12%)	36.0–97.2 Mg/ha (23–55%)
		Jack pine	25	7.8–12.3 Mg/ha (10–24%)		18.1–40.3 Mg/ha (36–53%)	20.2–28.4 Mg/ha (37–40%)
		Jack pine	65	29.0–34.6 Mg/ha (42–51%)	3.5–5.1 Mg/ha (5–7%)	11.5–14.6 Mg/ha (17–21%)	14.2–25.8 Mg/ha (20–38%)

upland aspen and jack pine sites in central Canada, aboveground vegetation and soil contained roughly equal amounts of carbon. In contrast, in lowland swamps of stunted black spruce (*Picea mariana*), only about 12–13% of the carbon was found aboveground (Gower et al. 1997). Black spruce stands in Manitoba had 40 ± 13 tons carbon ha^{-1} (living and dead biomass), which comprised around 15–23% of total stand carbon depending on whether the sites were saturated swamps or well-drained uplands (Goulden et al. 1998). In southern Siberia, biomass carbon exceeded soil carbon in Scots pine stands, while it was near equal in birch stands, and was exceeded by soil carbon in larch stands (Vedrova et al. 2002). In an interior Canadian black spruce forest, Malhi et al. (1999) reported that aboveground biomass makes up on average around 11% of total stand carbon (see Table 6.1).

Overstory (tree) vegetation appears to dominate the aboveground pool of which approximately 5% may be dead trees (Yarie and Billings 2002). The woody understory comprises a minor component of total forest carbon (Nalder and Wein 1999; Li et al. 2003), and was measured in one study as less than 2% (Wang et al. 2001).

Aboveground productivity in boreal forests is limited by a number of environmental factors, including seasonal distribution of precipitation, timing of soil thaw, soil type, nutrient availability, site aspect, topography, and length of the growing season (Jarvis and Linder 2000; Gower et al. 2001). Many of these factors affect productivity primarily by controlling rates of respiration and decomposition, which will be explained further in the section on "Means of Uptake and Release." One example, nitrogen availability, is often identified as a growth limitation in boreal forests (Bonan and Van Cleve 1992). This limitation is related to very slow decomposition rates, which ties up nitrogen in undecomposed litter (Wirth et al. 2002). Thus, decomposition and its drivers (soil warming, water table depth, forest fire) determine the extent to which nitrogen limits aboveground productivity. In a related way, pollution driven N-deposition in Northern Europe may be increasing aboveground carbon pools in Scandinavia (Mäkipää et al. 1999).

Aboveground carbon storage also appears to differ across forest types. It is greater in mixed woods than pure stands of either deciduous or coniferous trees, perhaps due to the greater foliage mass in stratified mixed stands (Martin et al. 2005). Additionally, aboveground and total net primary production (NPP) are generally higher in deciduous than coniferous stands (Gower et al. 1997, 2001).

Research from the Russian taiga indicates that disturbance and extreme climatic events (i.e. drought) may prevent boreal forests from attaining the maximum density and productivity possible under site conditions (Schulze et al. 1999; Vygodskaya et al. 2002). For instance, southern Siberian forests were kept below the theoretical self-thinning line by frequent ground fires that reduced stand density beyond the levels associated with competition mortality (Schulze et al. 1999). The importance of such events must be considered along with site factors in quantifying the aboveground carbon pool.

2.2 Belowground Biomass

The belowground biomass carbon pool consists of coarse and fine tree roots and their associated mycorrhizae. It is considered one of the most difficult pools to quantify, as labor-intensive destructive sampling is often required to achieve exact figures, and even then measuring fine root mass may not be possible (Table 6.2). Gower et al. (2001) found that the most common bias in estimations of NPP in boreal forests was the exclusion of fine roots and mycorrhizae from the calculation. The few studies that have measured these features show high variability and thus cannot be extrapolated accurately to quantify the belowground carbon pool for the biome.

While precise quantification of belowground biomass is difficult, researchers have been able to identify the approximate proportion of total stand carbon that this pool accounts for (Table 6.1). Data from limited studies show that belowground biomass is highly variable, influenced by such stand and site factors as species composition, stand age, and available moisture. A greater

Table 6.2 Sources of uncertainty in Boreal carbon modeling

	References
Inadequately quantified carbon pools	
Fine root biomass/mycorrhizae	Gower et al. (1997, 2001)
Magnitude of labile soil carbon pool	Rustad and Fernandez (1998); Jarvis and Linder (2000); Bronson et al. (2008)
Bryophyte/understory layers	Gower et al. (2001)
CWD and litter in Russia	Krankina et al. (2002)
Changing allocation patterns within trees	Lapenis et al. (2005)
Poorly understood environmental variables	
Quantifying burned area in Russia	Dixon and Krankina (1993); Conard and Ivanova (1997); Soja et al. (2007)
Recognizing refugia in burned areas	Amiro et al. (2001); Kang et al. (2006)
Fire intensity vs. simply fire occurrence	Wooster and Zhang (2004)
Influence of burn severity on carbon and nitrogen consumption	Balshi et al. (2007)
Accounting for ground vs. crown fires	Wirth et al. (2002)
Changes in insect life cycles	Malmstrom and Raffa (2000)
Possibility of poor post-disturbance stocking	Auclair and Carter (1993); Shvidenko et al. (1997)
Accounting for potential vegetation dieback	Kasischke et al. (1995)
Rates of permafrost degradation	Prokushkin et al. (2005)
Lag time on migration of temperate species into boreal zone	Smith and Shugart (1993)
Quantifying area, depth and bulk density of boreal peatlands	Gorham (1991)
Balance of CO_2 and CH_4 emissions from peatlands	Gorham (1991)
Lack of research on poorly-drained forests	Bond-Lamberty et al. (2004)
Rates of precipitation change	Pastor and Post (1988); Flannigan et al. (1998)
Accuracy of estimation of crown and soil temperatures	Arain et al. (2002)
Varying temperatures of different carbon pools	Lindroth et al. (1998)
Assumption of increased productivity with increased temperature	Briffa et al. (1998); Barber et al. (2000); Wilmking et al. (2004)
Timing of increased temperatures	Lindroth et al. (1998)
Using monthly temperature anomalies as opposed to daily temperature data	Flannigan et al. (1998)
Thresholds in NEP response to climate change	Grant et al. (2006)
Albedo effect of boreal forest cover	Bonan et al. (1992, 1995); Betts (2000); Bala (et al. 2007)
Lack of data on Eurasian larch forests	Gower et al. (2001)

The table summarizes portions of the boreal carbon budget (pools, processes and environmental variables) that are currently poorly understood or quantified, and indicates potential areas for future research on boreal carbon dynamics

percentage of total NPP is allocated to roots in coniferous than in hardwood stands (Bond-Lamberty et al. 2004). One comparative study found that 41–46% of total NPP was allocated to roots in conifer stands but only 10–19% in aspen stands (Gower et al. 1997). However, research in Alaska has shown that hardwood forests can exceed coniferous forests in the production of *fine* roots, which can make up 11–29% of stand biomass (Ruess et al. 1996). Stand age appears to affect the belowground biomass pool by regulating root production. Bond-Lamberty et al. (2004) found that coarse and fine root production peaked at around 70 years in a Canadian black spruce chronosequence, but was 50–70% lower in 151-year-old stands.

Soil moisture limitations may cause trees to allocate more biomass to belowground structures.

Schulze et al. (1999) found that a greater proportion of stand biomass was allocated to roots in Siberian boreal forests than in European Russia or temperate European forests, perhaps due to the extreme moisture deficits that occur in some areas of Siberia. Indeed, increasing aridity across northern Siberia may be causing a shift in allocation from photosynthetic tissues to roots, while increasing moisture in European Russia and southern Siberia is having the opposite effect (Lapenis et al. 2005). Other environmental factors besides moisture could also be at play here: Prokushkin et al. (2005) attributed the high relative allocation of carbon to roots in Siberian forests to low soil temperatures and nutrient availability. It appears that under stressful conditions with low levels of water and nutrients, trees develop larger root systems to access these resources.

2.3 Lichens and Bryophytes

This pool is largely composed of lichens and mosses, which frequently form a dense mat at the ground level in boreal forests. This pool is relatively unique in importance to boreal forests compared to temperate and tropical zones where it is a relatively insignificant component of the carbon budget.

Bryophyte tissues decompose more slowly than woody or non-woody tissue (Turetsky 2003; Turetsky et al. 2010), and thus tend to accumulate between fire events. Soil drainage seems to influence the magnitude of this pool (Turetsky et al. 2005), which is largest in boreal peatlands, where bryophytes are the major vegetation type. In mature lowland black spruce forests, mosses may sequester as much or more carbon than trees, and ten times the amount sequestered by understory vegetation (Harden et al. 1997) (Table 6.1). Czimczik et al. (2006) found that bryophytes made up 20% of total aboveground NPP in black spruce stands. The dominant bryophytes in such saturated sites are *Sphagnum* mosses. In upland spruce sites with better drainage, the moss dominance switches to *Pleurozium* feathermosses, which accumulate significantly less carbon that *Sphagnum* types (Goulden et al. 1998). Moving even further "upland," only 3.2% of stand carbon is stored in mosses in xeric jack pine stands, and in aspen stands the bryophyte pool is even smaller (Nalder and Wein 1999).

Unfortunately, no research on the importance of bryophytes in Eurasian boreal forests was found for this review. Given the circumpolar range of *Sphagnum* and *Pleurozium* species, and the widespread presence of saturated lowland boreal forests in Eurasia, it seems likely that bryophytes also play a large role in that region. Little is also known about the dry lichen communities (often composed of *Cladonia* species) that blanket the floor of xeric conifer woodlands in North America and Eurasia. Despite recognition of their unique importance, lichens and bryophytes remain one of the least studied carbon pools in the boreal forest (Table 6.2).

In addition to their direct role as a carbon pool, bryophyte communities have important effects on how carbon is stored in boreal soils. Thick moss layers (including live mosses and moss-derived organic material) limit heat gain from the atmosphere (Startsev et al. 2007). In black spruce stands, for example, this creates cold and wet conditions near the soil surface that promote the development of permafrost (O'Neill et al. 2002). The limitations on decomposition imposed by such conditions are very important for carbon storage in the soil profile. In white spruce and aspen stands with less-developed bryophyte communities, more rapid transfer of heat, moisture, and oxygen through the soil profile is possible, resulting in warmer and drier subsoil conditions and less stored carbon (O'Neill et al. 2002).

The flammability of different bryophyte communities influences their rates of carbon storage and release. *Pleurozium* mosses dry out completely; consequently, a fire can release the carbon stored therein and expose the soil surface to greater heat and drying. In contrast, *Sphagnum* mosses remain saturated through most of their profile, even during dry seasons. Fires only remove the upper layers, leaving moist lower layers intact to insulate the soil (Harden et al. 1997). In addition, a dense layer of sphagnum moss contributes to higher soil acidity, which facilitates formation of an impermeable soil layer (Bonan and Shugart 1989). This acts as a positive feedback to soil moisture conditions by reducing the movement of moisture through the upper soil horizons, and increases moisture levels near the soil surface. When vigorous, sphagnum moss can even regulate successional trajectories by limiting colonization to species capable of layering, such as spruce (Johnstone et al. 2010). The reduced flammability and decomposition brought about by *Sphagnum* communities contribute to the general trend of greater ground-level and belowground carbon storage in saturated lowland sites than in well-drained uplands. However, this also hints at the potential re-organizing that would take place if fire events in the sphagnum-dominated portions of the boreal were to become more severe (Chapin et al. 2010; Johnstone et al. 2010).

2.4 Litter Layer and Coarse Woody Debris

This pool is made up of dead organic matter that has not decomposed and entered the soil profile. The coarse woody debris component represents an increasingly important element of forests at higher latitudes, and thus may be most at risk of becoming a carbon source under increased warming. Malhi et al. (1999) found that the litter layer composes on average only about 1% of total stand carbon in boreal forests (Table 6.1). The size of this pool is primarily driven by rates of decomposition and disturbance. Disturbances such as fire or insect infestation contribute pulses of dead material to the pool, but fire can also reduce it through direct burning or by raising ground temperatures and stimulating increased decomposition. Also litter and coarse woody debris additions vary by stand type and age (Brassard and Chen 2008). Young post-disturbance stands often have very large litter pools (composed of the dead remains of the previous cohort), which diminish through stand development before increasing again as overstory mortality increases during stand maturation (Goulden et al. 2010). Increased overstory mortality as stands age can thus gradually replenish the supply of litter. As a consequence, coarse woody debris becomes an increasingly significant pool in older-growth stands (Siitonen et al. 2000). In contrast, studies have shown that the forest floor may actually lose carbon as the stand matures, as was identified in Canadian jack pine stands (Nalder and Wein 1999). This sequence of depletion and re-accumulation demonstrates that there is no simple relationship between litter, coarse woody debris, carbon and stand age.

Rates of litter accumulation vary across boreal zones. In Russian boreal forests, these differences may be associated with species composition. Stocks of coarse woody material are greater in Siberia, where rot-resistant larch species predominate, than in pine- and spruce-dominated European Russia (Krankina et al. 2002). Nalder and Wein (1999) found that the density of forest floor carbon was 68% higher in jack pine stands in eastern Canada than in western Canada. The reasons for such differences across the same vegetation community are not entirely clear. Differing site productivity, decomposition rates or fire levels could be involved.

The litter layer also interacts with bryophyte communities to affect soil properties. Like mosses, thick litter layers can insulate the soil, affecting depth of thaw, available moisture and belowground respiration (Bonan et al. 1990). The insulating and moisture-retaining capacity of the forest floor (including both litter and bryophytes) is highest in black spruce forests among all Canadian boreal forest types (Van Cleve et al. 1990). In such stands, the combined litter-bryophyte "ground" layer may store three to four times the carbon held in aboveground biomass (Kasischke et al. 1995).

2.5 Soil Carbon

The soil pool (found below the litter layer, consisting of decomposed organic matter and mineral soil) is the most important in the boreal carbon budget. The amount of carbon held in the soil profile often dwarfs the amount of carbon in forest vegetation (Malhi et al. 1999; Goulden et al. 1998; Kasischke et al. 1995; Wirth et al. 2002), and is a unique feature of the boreal forest (Table 6.1).

Many of the same factors responsible for carbon accumulation in bryophyte and litter layers help explain the prominence of soil carbon: cold, saturated soils have low rates of decomposition allowing carbon-rich organic matter to accumulate in the soil profile faster than respiration losses. Thus, the soil pool is greatest in the coldest, most saturated sites. Unforested wetlands may hold the majority of the carbon found in the boreal system, significantly out of proportion to their position in the landscape (Kasischke et al. 1995; Rapalee et al. 1998). For example, lowland (*Sphagnum* site) black spruce soils contain 200±50 tons carbon ha^{-1}, while upland (*Pleurozium* site) soils contain only 90±20 tons ha^{-1} (Goulden et al. 1998). Soil carbon storage in well-drained (and more productive) aspen and jack pine stands is 2.8–2.9 times less than in saturated black spruce soils, which contain 87–88% of stand carbon (Gower et al. 1997). In contrast, total biomass carbon in the xeric Scots pine stands of Siberia may exceed that of the soil carbon pool (Vedrova et al. 2002; Wirth et al. 2002).

The disproportionately large amount of belowground carbon is even more pronounced in the tundra systems to the north. Over 97% of the total carbon stored in these systems is found in the soil (Billings 1987). If current projections hold, a northward shift in the boreal-tundra ecotone is occurring (Soja et al. 2007), with potentially huge implications for the global carbon budget. It is unclear whether the massive soil pool in tundra sites would remain intact if converted to a forested biome (Kasischke et al. 1995).

The specific location of carbon within the soil profile also varies across time and space. In saturated black spruce sites, soil carbon is often found in the organic horizons or directly below (Goulden et al. 1998; O'Neill et al. 2002), while the majority of soil carbon in upland aspen (92%) and white spruce (82%) stands is found in the mineral soil (O'Neill et al. 2002). Mineral soil carbon typically declines with depth, but the trend varies among soils reflecting prevailing ecosystem processes. For example, in upland larch (*Larix gmelinii*) forests in northeast China, soil carbon concentration decreases relatively rapidly with soil depth across a range of mesic to xeric sites. This may be attributable to pulses of charcoal added to upper layers by recent fires (Wang et al. 2001).

Fires appear to be very important for transferring carbon from vegetation to the soil profile through conversion to charcoal, which is decay-resistant and can reside in the soil 3,000–12,000 years (Deluca and Aplet 2008). While some is transferred into lower soil horizons by cryoturbation (mixing of soil layers by the freeze-thaw process) (Hobbie et al. 2000), the large majority remains above 30 cm in depth, with approximately 70% remaining in the upper 10 cm of the soil profile (Deluca and Aplet 2008). One study estimated that 30% of the biomass killed in a fire enters the soil as charcoal or unburned material, at least half of which may enter the long-term soil pool; the rest is lost to decomposition or re-burning over the next century (Harden et al. 1997). Globally, charcoal additions probably represent about 1% of stored carbon in boreal forest types (Ohlson et al. 2009), but in some forest types may be significantly higher. For example, in the Rocky Mountains, charcoal are estimated to comprise as much as 60% of soil carbon (Deluca and Aplet 2008), while in southern Siberia this figure is 20–24% (Schulze et al. 1999).

3 Biotic Drivers of Uptake and Release

Biosphere-atmosphere carbon flux consists primarily of three processes: photosynthesis, autotrophic respiration (respiration by plants), and heterotrophic respiration (by microbial organisms during decomposition of organic matter). Along with biomass burning, these processes determine the balance between uptake and release of carbon from forests.

3.1 Photosynthesis and Autotrophic Respiration

Plant photosynthesis and respiration processes are coupled, their balance determining net carbon fixation by plants. These two processes are essentially paired because photosynthesis cannot proceed without energetic (respirational) expenditures on the maintenance and production of organs (roots, stems, and leaves) involved in carbon fixation. Carbon uptake by photosynthesis must therefore be paired with carbon loss through autotrophic respiration, which consumes 54–77% of annual net photosynthesis in boreal forests (Ryan et al. 1997). While autotrophic and heterotrophic respiration are often considered together (due to the difficulty of distinguishing them during measurement), only the former is closely paired with photosynthesis. Heterotrophic respiration rates are not necessarily proportional to tree growth (Li et al. 2003; Barr et al. 2007).

The pairing of photosynthesis and autotrophic respiration does not imply that they necessarily respond the same way to environmental stimuli. In one study in a mature Canadian aspen forest, interannual variability of photosynthesis was controlled primarily by growing season length and secondarily by drought, whereas interannual variability in respiration was primarily controlled by drought and secondarily by temperature (Barr et al. 2007). Jarvis and Linder (2000) support the idea that canopy duration (i.e. length of growing season as controlled by spring temperature) is more important in determining total photosynthesis levels than average temperature or soil moisture levels. Indeed, twentieth century increases in spring temperatures attributed to rising atmospheric CO_2 levels may have increased productivity in boreal aspen stands by allowing for earlier leaf out (Chen et al. 1999).

Rising temperatures (especially if encountered in early spring) may stimulate increased photosynthesis, but they also cause a rise in autotrophic respiration. Respiration rates rise faster under rising temperatures than photosynthesis rates, potentially causing carbon release to the atmosphere (Lindroth et al. 1998). Many models of boreal carbon flux assume that respiration responds directly to rising temperature, while photosynthesis is limited by other factors such as light levels, growing season length, and water and nutrient availability. However, in a study of these processes in Canadian peatlands, increasing annual temperature was unexpectedly correlated with increased net carbon uptake, suggesting that photosynthesis may be more responsive than previously thought, and that respiration will not necessarily offset increased carbon uptake in a warming climate (Dunn et al. 2007).

That said, the unexpected results from Dunn et al. (2007) may have been related to the abundant soil moisture available in peatlands. In drier upland forests, soil moisture availability imposes limitations on forest productivity (Chen et al. 1999; Gower et al. 2001; Bond-Lamberty et al. 2007). Rising temperatures unaccompanied by increasing precipitation could cause moisture stress, reducing photosynthesis. But importantly, drought also lowers respiration levels, potentially balancing out the reduced carbon uptake (Barr et al. 2007). The duration and severity of drought is important for several reasons. Mild drought suppresses respiration while photosynthesis remains largely unchanged, whereas severe drought suppresses both, with a dramatic drop in photosynthesis levels as it intensifies (Barr et al. 2007). In addition, drought events will promote species with strong stomatal conductance, such as Scots pine over species less tolerant to arid conditions, including larch (Dulamsuren et al. 2009). In the Mongolian boreal, Dulamsuren et al. (2009) note the competitive advantage of Scots pine under dry conditions, and conclude that a dark conifer for light conifer transition may occur if drought events become more frequent. This will create numerous feedbacks to the carbon budget of these systems (Bonan 2008).

3.2 Heterotrophic Respiration and Decomposition

Heterotrophic respiration, caused by decomposition of organic matter in the soil and litter layers, is the largest source of carbon emissions in the boreal system. Conceptually, decomposition and organic matter accumulation act as opposite influences on the soil and litter carbon pools; if decomposition exceeds organic inputs, there is a net loss of carbon from the system (Harden et al. 1997). Heterotrophic respiration is a large enough component of carbon flux that it might offset not only organic matter accumulation, but also carbon gains from photosynthesis. Indeed, because photosynthesis and autotrophic respiration often rise and fall together, the real determinant of whether a stand is a carbon sink or source may be its rate of heterotrophic respiration.

Certain environmental factors determine this rate. Vegetation type influences respiration rates through the differing qualities of litter produced. For instance, softwood litter decomposes slower than hardwood litter due to its high lignin content (Hobbie et al. 2000), and larch coarse woody material contains chemicals that slow the rate of decay relative to other softwoods (Krankina et al. 2002). Soil moisture exerts an even stronger influence on soil respiration rates (Harden et al. 2000) and needs to be considered along with temperature when simulating ecosystem responses (Krishnan et al. 2008). The high heat capacity of water and thick mats of bryophytes slow the warming of saturated soils. These factors limit baseline respiration rates, and also mitigate large spikes in respiration that follow fires (Harden et al. 1997). This explains the overall trend of higher soil carbon storage in lowland boreal forests than in upland forests. However, the constant saturation that limits release of CO_2 in boreal peatlands also promotes the release of methane (CH_4), an important greenhouse gas. Drying of peatlands would have the opposite result, namely, decreased CH_4, but increased CO_2 emissions (Gorham 1991). This dynamic could become an important element of carbon flux under changing climatic conditions.

Soil temperature may be even more limiting to decomposition rates than soil moisture (O'Neill et al. 2002). Increasing soil temperatures are widely expected to stimulate increased decomposition and respiration rates, but studies suggest that decomposition rates may actually diminish as a consequence of shifts in microbial community structure with soil warming (Allison and Treseder 2008). So responses to increased temperature may not be so easily predicted. Temperature is also important in determining rates of winter respiration, a frequently overlooked process that may make up 20% of yearly respiration (Hobbie et al. 2000). Young deciduous stands that are carbon sinks during the growing season may become sources after senescence due to winter respiration (Pypker and Fredeen 2002; Trofymow et al. 2002). Such respiration appears to take place in deeper soil layers where temperatures remain high enough in the winter to support decomposition (Goulden et al. 1998). The organic matter in these layers is generally composed of much older, less mobile carbon than that which is decomposed in the summertime (Winston et al. 1997; Dioumaeva et al. 2002). The temperature and duration of thaw in these soil layers control the decomposition rate of "old" soil carbon. Whether sustained soil warming associated with climate change would cause significant increases in carbon flux from this long-term pool is unclear (Table 6.2).

Many studies have attempted to quantify how the balance of decomposition and vegetative growth shifts across a post-disturbance chronosequence (Fig. 6.2). Increased respiration after a fire can be a significant source of carbon release. In fact, research has shown that post-fire decomposition may equal (Amiro et al. 2001) or exceed (Auclair and Carter 1993) direct emissions from burning. Fire has a short-term impact on heterotrophic respiration rates by raising soil temperatures, stimulating increased decomposition of soil organic matter (Harden et al. 1997). There is a longer-term respiration response as well, when the trees killed by the fire begin to decompose a few years later. This process can potentially make young post-fire stands a source of carbon despite the vigorous regrowth of trees and mosses (Rapalee et al. 1998; Vedrova et al. 2002; Wirth et al. 2002). At the same time, increased heterotrophic respiration in young post-disturbance stands may be somewhat balanced by a decrease in autotrophic respiration, caused by tree mortality (Wang et al. 2001). Similarly, in a chronosequence of post-harvest stands in central Canada, Li et al. (2003) found that stands younger than 20 years were carbon sources (releasing 193–239 g carbon/m^2 per year), but by 40 years of age had become weak sinks as growth outpaced decomposition. However, a post-fire chronosequence from the same region showed that

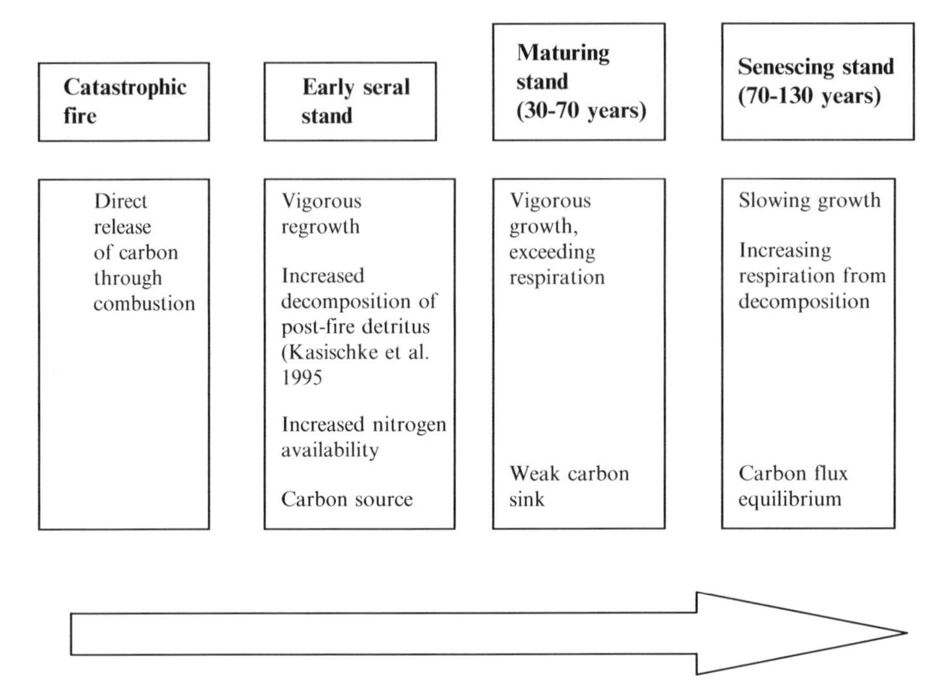

Fig. 6.2 Model of carbon dynamics through stand development in a Canadian black spruce forest (Derived from Litvak et al. (2003) unless otherwise noted in the figure)

significant decomposition of fire-killed litter did not occur in the first few decades and that young stands showed the lowest levels of respiration (Litvak et al. 2003). Czimczik et al. (2006) also did not observe a rise in decomposition in young post-fire stands in Canada. In fact, heterotrophic respiration did not become significant until black spruce dominated the canopy (around 70 years post-fire). Generally, studies suggest an overall pattern of decline in carbon production efficiency as stands age, with older stands tending to show increased carbon losses, in large part due to increased tree mortality (Goulden et al. 2010). These examples demonstrate that disturbance effects on decomposition rates may lag and occur later in stand development, and depend on the type of disturbance. These patterns and lag effects have not been well explored in relation to dieback and tree mortality driven by drought and insects. However, data from the massive mountain pine beetle outbreak in western Canada indicates that forests can rapidly become large net carbon sources in the years following insect attack (Kurz et al. 2008a).

4 Disturbance: Abiotic Drivers of Uptake and Release

Disturbances such as fire, insect and pathogen outbreaks, and logging have important impacts on the boreal carbon budget. Disturbances influence the size of carbon pools by directly destroying (fire) or removing biomass (logging) from the system, and by altering the rates of photosynthesis and respiration as discussed earlier. In fact, disturbance may be the overriding factor in whether or not the boreal forest is a source or sink of carbon. For instance, Kurz et al. (2008c) have estimated that large-scale insect outbreaks have turned Canada's managed forests from a carbon sink to a carbon source. Using Monte Carlo simulations, they predict that this trend will continue due the effects of natural disturbances.

4.1 Fire

The direct emission of carbon to the atmosphere through combustion is a significant component of boreal carbon flux. In upland sites in boreal Canada, Harden et al.'s (2000) model of long-term carbon balance estimated that 10–30% of the annual carbon production has been released as fire emissions, while 40–80% has been released during decomposition and 8–30% fixed as soil carbon. This estimate fits with other observations that increased post-fire decomposition has a greater impact than direct fire emissions (Auclair and Carter 1993; Conard and Ivanova 1997). Quantifying direct emissions is a complicated task, beginning with the process of identifying the area burned in a given year across the vast boreal landscape. Underestimation of burnt area in Russia can significantly bias models, potentially missing a vital source of emissions to the atmosphere (Dixon and Krankina 1993). In contrast, satellite estimation of forest fire extent in Canada overestimated cumulative burned area by approximately 22% because unburned inclusions were not recognized (Kang et al. 2006). These examples demonstrate the difficulty of accurately calculating this component of carbon flux (Table 6.2).

As discussed earlier, fire affects soil properties through changes in temperature and moisture conditions, removal of insulating litter and bryophyte layers, and contribution of decay-resistant charcoal to the soil pool. Fire may also increase nitrogen input from the organic layer to the soil, increasing nitrogen mineralization and vegetation productivity (Kasischke et al. 1995; Johnson and Curtis 2001; Kang et al. 2006). One study in the Canadian boreal demonstrated that deciduous stands are able to respond more rapidly to the increased supply of nitrogen than conifers, due to their faster rate of leaf canopy turnover. Thus, deciduous forests exhibited increased productivity with increasing fire frequency, while the opposite was true of both dry and wet coniferous types (Kang et al. 2006).

Across much of the boreal region, fire exerts a dominant influence on forest age class distribution. Fire-prone landscapes are characterized by a mosaic of age classes, each with differing rates of growth and respiration. Boreal carbon budgets must account for the different patterns of carbon uptake and release that accompany different age class distributions. In Canadian black spruce forests, most of the net biomass accumulation appears to take place from 20 to 70 years after a fire. Stands younger than 20 years lack sufficient leaf area for rapid carbon accumulation and stands older than 70 years are at or near zero carbon balance with the atmosphere (Fig. 6.2). Only a small proportion (9%) of the black spruce stands in central Canada are in the most productive age class (around 36 years old) (Litvak et al. 2003). In boreal Quebec, biomass increased from 27 to 75 years following a fire, and decreased thereafter due to stand degradation. In the Alberta Boreal Plains ecoregion, it took between 15 and 30 years for post-fire stands to attain the same photosynthetic rates as mature areas while biomass continued to increase to at least 60 years of age (Amiro et al. 2000). Kasischke et al. (1995) reported, however, that biomass levels in upland black spruce forests in Alaska and northwest Canada continue to increase for 140–200 years after a fire, before increased overstory mortality sets in.

Such growth rate comparisons across stand age must be paired with rates of post-fire decomposition. In Siberian Scots pine forests, young post-fire stands are sources of carbon, and may take 70 years to reach pre-fire carbon levels (Wirth et al. 2002). Canadian studies also point to high initial rates of decomposition (Li et al. 2003; Litvak et al. 2003), although this trend may not always hold. Using eddy covariance measurements of growing season net ecosystem CO_2 exchange, Litvak et al. (2003) estimated that recently disturbed black spruce stands in Canada are sources of carbon, middle-aged (20–70 years old) stands are sinks, and older (70–130 years old) stands in near balance with the atmosphere. In Siberia, the trajectory is somewhat different: an initial decrease in carbon pools during first 30–40 years after a fire, fairly rapid carbon accumulation over the next 50 years, and lower but steady rates of accumulation in the centuries thereafter (Wirth et al. 2002).

The frequency and intensity of fire determines how forest age classes are distributed in many boreal landscapes (Table 6.3). In boreal forests of North America, Fennoscandia and European Russia, fires have historically been high-intensity and stand-replacing (Harden et al. 2000), and have

Table 6.3 Fire regimes in the boreal forest

Forest type/location	Disturbance type	Return interval (years)	Reference
Pinus sylvestris, NW Russia	Ground fire	20–40	Gromtsev (2002)
Pinus sylvestris, Siberia	Ground fire	25–50	Conard and Ivanova (1997)
Larix sibirica, Siberia	Ground fire	90–130	Conard and Ivanova (1997)
Picea abies, NW Russia	Stand-replacing fire	130–200	Gromtsev (2002)
Dark taiga[a],central Siberia	Stand-replacing fire	400–500	Schulze et al. (2005)
Continental taiga[b], interior Canada	Stand-replacing fire	40–110	Amiro et al. (2000)
Spruce/fir/birch[c], eastern Canada	Stand-replacing fire	136 ± 29	Lesieur et al. (2002)
Boreal/tundra interface[d], NW Canada	Stand-replacing fire	110	Johnson and Rowe (1975)

[a]*Picea obovata, Abies sibirica, Pinus sibirica*
[b]*Picea glauca, P. mariana, Pinus banksiana, P. contorta, Populus tremuloides*
[c]*Picea glauca, P. mariana, Abies balsamea, Betula papyrifera*
[d]*Picea glauca, Pinus banksiana*, muskeg vegetation

a return interval of 40–110 years (Amiro et al. 2000). In Siberia, ground fires that are not stand-replacing are the norm, accounting for about 80% of the area burned. Such fires may burn through Scots pine stands on a short 25–50 year return interval, and larch stands on a 90–130 year interval, leaving many live trees. However, intervals seem to be considerably longer for spruce/fir stands, with fires in this type more likely to be catastrophic (Conard and Ivanova 1997). The total number of fires and the area burned are higher in Siberia than in North America, but the lower intensity of these fires means that more carbon is not necessarily released (Wooster and Zhang 2004). Models that fail to consider that detail can overestimate carbon emissions from Russian forest fires.

Stand-replacing fires have different impacts on carbon dynamics than low-intensity ground fires. The post-fire chronosequences described above tend to occur in catastrophic fire systems, in which the aftermath of fire is nearly always mass mortality and decomposition, and a return to early-successional condition. Ground fires have a more complex result. They can produce uneven-aged communities (Harden et al. 2000), and cause multiple small pulses of mortality and decomposition within the same stand. Rather than causing sudden, complete changes in stand development, ground fires alter competition and productivity levels within the existing cohort. Low-intensity fires in Siberian Scots pine stands result in a 10–20 year growth depression of the surviving trees due to fire damage, followed by 10–15 years of accelerated growth under reduced competition and higher nutrient supply (Schulze et al. 1999). In this forest type, young growth does not appear to necessarily replace the trees lost to ground fires. Instead, low-density stands persist and may never attain the maximum possible stocking (Schulze et al. 1999; Wirth et al. 2002). This "lost" productivity has a significant impact on carbon uptake in Siberian forests; Shvidenko et al. (1997) calculated a 45–50% reduction in forest productivity due to ground fires across large areas of Siberia.

Suppression of forest fires also affects the carbon budget. For example, temperate oak (*Quercus*) forests under fire suppression management had 90% more total ecosystem carbon than those with a frequent fire regime (Tilman et al. 2000). If fire suppression is practiced across a significant portion of the landscape, pools of biomass and litter carbon may exceed estimates for forests under a natural fire regime (Price et al. 1997). However, there is an inherent danger in fire suppression because larger fuel loads may, if ignited, produce much more intense fires than might have occurred in a natural fire regime.

4.2 Insect Outbreaks

While fire is recognized as the dominant natural disturbance type over much of the boreal forest, insect outbreaks (and "background" insect damage during non-outbreak years) are also critically important. Across the Canadian boreal and north temperate forests, insects and pathogens annually cause forest volume losses through mortality and growth reductions that are three times the volume lost to fire. Malstrom and Raffa (2000) found that insects are especially dominant in the moist eastern regions of Canada. Indeed, in the balsam fir (*Abies balsamea*) dominated forests of the Maritime Provinces, cyclical outbreaks of the defoliating insect spruce budworm (*Choristoneura fumiferana*) supplant fire as the primary influence on age class distribution (Baskerville 1975). Unlike fire, insect damage does not produce a direct emission, but rather exerts its influence through altered rates of decomposition and growth (Kurz et al. 2008c).

Kurz et al. (2008c) modeled the impact of spruce budworm and western mountain pine beetle (*Dendroctonus ponderosae*) outbreaks on carbon flux in the Canadian forest. They concluded that these events could switch the region from a carbon sink to a source due to the massive increases in decomposition of dead trees that follow outbreaks. Background levels of insect herbivory are also important. In Fennoscandian and Russian birch (*Betula pubescens*) forests, defoliating insects had a significant effect on leaf area index and net primary production. If certain levels of herbivory are reached, coniferous species may take over the growing space relinquished by damaged birches, speeding stand development and causing related changes in carbon dynamics (Wolf et al. 2008). The combination of drought and defoliating insects can result in significantly reduced production in Canadian aspen forests. If climate change results in an increase in drought and insect outbreaks, closed aspen forests may transition to sparse parklands (Hogg et al. 2002).

4.3 Drought

There is increasing global concern about the potential consequences of altered climate conditions on the extent, duration and severity of drought events and their impacts on forest mortality (Allen et al. 2010). Drought events have been increasingly implicated as a critical driver of stand dynamics and forest mortality, particularly in the boreal zone. Widespread dieback in aspen in western North America (reviewed by Frey et al. 2004) has been attributed to extended periods of unusually severe drought in the region. A function of increased temperatures and periods of below-average precipitation, such events appear to have triggered forest dieback and mortality across large areas, with drier regions of the boreal appearing particularly vulnerable (Hogg et al. 2008). Drought affected regions may also be more vulnerable to insect outbreaks thereby enhancing mortality rates (Frey et al. 2004). Resulting massive waves of mortality that have been documented represent a dramatic and sharp increase of carbon in dead standing biomass (Hogg et al. 2008), with significant consequences for long-term carbon flux.

4.4 Forest Management

Besides its impacts on growth and decomposition rates, the commercial harvest of trees has a direct impact on carbon stocks through the removal of biomass from the forest. The eventual decomposition or combustion of this pool must be considered (refer to Chapter 12 for an analysis of wood products). The greatest difference between timber harvesting and other disturbance types is in the altered contribution it makes to the litter pool compared to fire or insect outbreak. Logging adds litter in pulses that are concentrated around harvest events, and the litter tends to lack stemwood, which is removed from the site for forest products. Intensive site preparation techniques, such as slash burning, can limit this pool even further. Krankina et al. (2002) found that intensively managed European Russian forests had much larger stocks of coarse woody material than unmanaged Siberian forests of similar productivity. In addition, logged stands may maintain higher carbon pools in live biomass compared to post-wildfire stands, where trees are retained in silvicultural activities and their additional beneficial effect on promoting faster regeneration of stand post-disturbance are considered (Seedre and Chen 2010).

Field studies by Martin et al. (2005) suggest that the stand-level impacts of logging on soil carbon dynamics are limited. Harvesting has no consistent effect on carbon levels in soil detritus. Johnson and Curtis (2001) came to a similar conclusion, although they found that whole-tree harvests (as opposed to stem-only harvests that leave tree crowns in the forest) could cause slight decreases in soil carbon. In contrast, Thiffault et al. (2008) observed lower stable C fractions and nutrient retention in soils post-harvest compared to post-wildfire soils of similar age (Thiffault et al. 2008). Furthermore, long-term modeling of managed boreal forests has shown a consistent decline in soil carbon across a 300-year time period compared to forests under a natural disturbance regime (Seely et al. 2002). Long term research plots in managed forests will be necessary to determine if such predictions are accurate.

Timber harvesting is concentrated in certain regions of the boreal forest. Fennoscandia and Maritime Canada are under near-complete management, while vast swathes of interior Canada and Siberia have experienced virtually no logging (although this could change in coming decades). Thus the impacts of forest management on the boreal carbon budget are uneven and difficult to compare with natural disturbances. In south Siberia, the decomposition of logging slash comprised an insignificant proportion of carbon flux to the atmosphere compared to fire emissions and post-fire decomposition (Vedrova et al. 2002). It should also be noted that, unlike natural disturbance, harvesting tends to be concentrated on the most productive portions of the landscape. This could give it an impact out of proportion to area affected (Li et al. 2003).

4.5 Nitrogen-Deposition

Deposition of nitrogen compounds related to pollution has affected several regions, most importantly eastern Europe and Scandinavia. Studies suggest that increased nitrogen-deposition has enhanced productivity in this region (Magnani et al. 2007). While carbon uptake is understood to be highly coupled to nitrogen status, recent findings suggest that increased canopy nitrogen conditions correlate positively with surface albedo, which may represent a further feedback on carbon uptake (Ollinger et al. 2008).

5 Climate Change Impacts on Boreal Carbon Dynamics

The most pressing question is how climate change will affect the carbon balance in the boreal forest. A warming climate could change the productivity/respiration balance, change disturbance regimes, shift forest types, and possibly cause dramatic changes in the extent of the biome itself.

5.1 Increased Productivity Versus Increased Respiration

Much of the uncertainty regarding carbon flux under a changing climate revolves around whether rates of respiration (both autotrophic and heterotrophic) will increase faster than rates of photosynthesis. There is also a question of whether such increased rates will be sustained, or will only constitute a short-term reaction.

Increased CO_2 availability can benefit plant growth, as it is a major constraint on photosynthetic efficiency. Studies have suggested that atmospheric enrichment, or "fertilization" of CO_2 that has been occurring over the past century can enhance growth and may offset increased losses expected from wildfire frequency (Balshi et al. 2007, 2009). Others (e.g. Kurz et al. 2008b) using modeling approaches have suggested that increased productivity is unlikely to offset increased carbon losses due to disturbance.

If climate change results in warmer temperatures in early spring, forest productivity could respond positively thanks to the extension of the growing season (Chen et al. 1999). This could have the greatest effect in deciduous forests due to the stronger response to early-season warmth (Barr et al. 2007). On the other hand, if rising spring temperatures are erratic, they could cause growth reductions by stimulating early de-hardening of tree buds which are then susceptible to frost damage (Hanninen et al. 2005). If rising temperatures come later in the growing season, when moisture stress is a potential problem, then either growth increases could be outstripped by respiration increases (Lindroth et al. 1998), or photosynthesis could actually decrease (Kang et al. 2006). For example, twentieth century decreases in white spruce growth in Alaska have been linked to increased drought stress caused by rising temperatures (Barber et al. 2000). The most common response of trees at the northern Alaskan treeline to increasing temperature is growth reduction, especially on productive sites where competition for moisture is high (Wilmking et al. 2004). Exclusion of such drought impacts from boreal models could potentially skew projections of the carbon budget (Briffa et al. 1998).

Satellite monitoring of boreal forests reveals that productivity declines may be occurring in some regions, perhaps attributed to moisture stress. Goetz et al. (2007) found that more than 25% of boreal forests in Canada that were not recently disturbed showed a decline in productivity with rising global temperatures. Large areas of Siberia showed increased productivity, but this is likely the result of rigorous post-fire regrowth in the wake of many extreme fire seasons.

Thus, whether or not precipitation rises along with temperature has very important consequences for carbon flux (Pastor and Post 1988). If temperature and precipitation increase in tandem, Fennoscandian forests may demonstrate increased productivity (Kellomaki et al. 1997). Predictions of future precipitation changes show strong variation across the boreal system, and even within select ecozones. For instance, while precipitation is expected to increase across most of northern Europe, it is forecasted to decrease in southern Fennoscandia (Flannigan et al. 1998). Similarly, while increased drought stress is modeled for interior Canadian forests, precipitation could rise in maritime eastern Canada (Amiro et al. 2001).

Changing temperature and precipitation regimes will affect decomposition rates in the future. Increasing soil temperatures could increase mineralization and breakdown of organic matter, potentially making more nutrients available for tree growth (Van Cleve et al. 1990). However, the supply of labile nitrogen in the soil may be depleted fairly quickly. In addition, any nitrogen-induced increases may be outweighed by concomitant increases in soil respiration (Bonan and Van Cleve 1992). Soil respiration may be particularly important if a greater proportion of the increased growth goes into roots than aboveground structures (Niinisto et al. 2004). Also, work by Karhu et al. (2010) highlights how responses vary among soil fractions in soil, from labile fractions cycled annually to more recalcitrant material cycled over centuries. Soil organic fractions and sensitivity to warming as estimated by Q_{10} (doubling rates) increases in all soil organic fractions, but most substantially in intermediate fractions. Moreover, the 30–45% increase in carbon loss estimated for soil fractions at current rates of warming would require a 100–120% increase in growth to offset.

However, it is heterotrophic respiration that holds the greatest potential for turning boreal forests from sinks to sources in a warming climate. Bonan and Van Cleve (1992), using models that simulated production and decomposition under warming conditions in Canadian forests, found that respiration increases would balance out photosynthesis gains in black spruce and paper birch (*Betula papyrifera*) forests, and would exceed them in white spruce forests. In a simulation of climatic warming in Finland, gross primary production increased by 12%, but respiration by 22% (Mäkipää et al. 1999). However, climatic simulation in Alaska predicted that increases in heterotrophic respiration would only exceed productivity increases in paper birch stands, while the opposite would be true in white spruce and balsam poplar (*Populus balsamifera*) stands (Yarie and Billings 2002).

Experimental soil warming (+5°C) in north-temperate forests in Maine increased respiration by 25–50% (Rustad and Fernandez 1998). Much of the increase could come from decomposition of deep soil carbon, which currently comprises a small proportion of the whole (Winston et al. 1997; Goulden et al. 1998). In Siberian forests with extreme buildup of organic matter, warming conditions could cause long-term, sustained increases in heterotrophic respiration from humified materials (Dioumaeva et al. 2002). Increased heterotrophic respiration may be limited by certain factors, however. Since the amount of labile organic matter is limited in many boreal soils, respiration rates may tail off after this pool is "burned off" by increased decomposition, (Rustad and Fernandez 1998). In addition, microbial communities in the soil may acclimate to higher temperatures, regulating decomposition rates (Jarvis and Linder 2000; Bronson et al. 2008).

The potential for increases in deep soil decomposition is greatly increased if significant soil thawing and permafrost degradation occurs. This will largely be determined by how a changing climate affects the litter and bryophyte layers that insulate the soil profile. Increasing fire in a warming climate could reduce the thickness of these insulating layers (Harden et al. 2000), and warmer air temperatures would increase the period of time in which there is a positive heat flow from the atmosphere to the ground layer (Kasischke et al. 1995). Both of these factors could cause degradation of permafrost. Camill (2005) found that increasing air temperatures in the latter half of the twentieth century (without an accompanying increase in precipitation) resulted in widespread degradation across the discontinuous permafrost zone of Manitoba. However, drying of the litter layer could reduce decomposition rates (Niinisto et al. 2004), and reduce the layer's thermal conductivity, thereby decreasing the depth of soil thawing (Bonan et al. 1990). If precipitation increased along with temperature, this drying would be prevented and permafrost thaw could increase (Gorham 1991).

The impact of changing temperatures and precipitation is especially hard to understand in boreal peatland systems. On one hand, permafrost degradation and increased heterotrophic respiration are significant possibilities (Hobbie et al. 2000). On the other hand, peat accumulates twice as fast on "collapse scars" as on bogs with intact permafrost (Camill et al. 2001). Thus, the increased productivity of these areas could offset some of the carbon losses. There is also a tradeoff in peatlands between aerobic decomposition (which releases CO_2) and anaerobic decomposition (which releases CH_4). If water tables drop, aerobic decomposition is likely to increase, since waterlogged peat is oxygen-poor, but affected areas could also experience reductions in CH_4 emissions as anaerobic decomposition declines. Under this scenario, it is unclear whether peatlands will become a source or sink. Dried-out peatlands will have accelerated oxidation of organic matter, but reduced emissions of CH_4, whereas waterlogged, collapsed thermokarst basins will accumulate more peat resulting in increased CH_4 emissions (Gorham 1991).

5.2 Changing Disturbance Regimes

Cycles of forest fire and insect outbreak are controlled by weather and the condition of the fuel or host. Both of these factors could be altered by climate change. One possibility is a more rapid build-up of pandemic insect populations as increasing temperatures could cause drought stress in their host tree species as well as shorten insect life cycles. A massive spruce beetle outbreak in Alaska has been attributed to abnormally warm and dry summers since the 1960s (Berg et al. 2006), and similar climatic triggers may be causing the widespread devastation by mountain pine beetle across western North America (Malmstrom and Raffa 2000; Powell and Logan 2005). Indeed, the prospect of future pine beetle and spruce budworm outbreaks caused one model to predict that Canadian boreal forests will be a net source of greenhouse gases in the coming decades (Kurz et al. 2008c).

Climate change may also allow pests that are less cold tolerant to extend their distribution into the boreal zone (Wolf et al. 2008). However, it may also be possible that a warming climate could suppress insect populations under certain conditions. One model predicts that rising temperature without an accompanying rise in precipitation will decrease the area affected by spruce budworm in temperate forests of Oregon (Williams and Liebhold 1995).

There is evidence that fire return intervals have been shortening across the boreal forest during the twentieth century, and this trend could continue (Stocks et al. 1998). Annual area of North American boreal forests burned increased approximately by a factor of three between the 1960s and the 1990s (Kang et al. 2006). One study predicted that Canadian fire return intervals could decline from an average of 150 years to 100–125 years, with significant associated emissions (Kasischke et al. 1995). And just as future rates of photosynthesis and respiration will depend on how precipitation changes in relation to rising temperatures, so too will future fire return intervals (Flannigan et al. 1998; Amiro et al. 2001). It is possible that the most significant impact of rising CO_2 levels in the atmosphere thus far has been an increase in fire frequency, thus altering the boreal forest age-class distribution (Bond-Lamberty et al. 2007).

The potential for altered fire regimes in response to climate change is another topic that will hold implications for the boreal carbon budget. In certain boreal forest-types, climate change is expected to facilitate shorter fire return intervals, which will promote early successional deciduous species (Soja et al. 2007). Because deciduous species accumulate less carbon than spruce stands, a deciduous for coniferous shift in species composition will affect the boreal carbon cycle in many spruce-dominant regions (Kasischke et al. 2010). Additionally, if fire severity changes, more organic matter will be consumed during burn events, subsequently reducing the negative feedbacks associated with Sphagnum moss accumulation and seed germination (Johnstone et al. 2010). Deep thawing would arise in conditions where insulating mosses were removed, and site drainage would likely facilitate drying of the organic layer and subsequently increase fire severity. Newly exposed mineral soils would promote seed germination by different forest species, most likely including light-seeded pioneers (Johnstone et al. 2010).

5.3 Changes in Biome and Forest Type

Some research predicts significant compositional changes within the boreal zone with a changing climate, as well as a shift of its southern border northward with expansion of temperate forests and steppe and invasion of its northern border into the tundra. Some predictions are dramatic: Emanuel et al. (1985) modeled that boreal forests will decrease by 37% if there is a doubling of atmospheric CO_2 concentration. Rising temperatures and degrading permafrost are allowing Siberian pine (*Pinus sibirica*) to invade the understory of larch stands across southern Siberia and Mongolia, and coniferous forests are displacing montane tundra in the mountain ranges of these regions (Soja et al. 2007). In boreal Canada, climate change may make deciduous forest types more competitive (Kasischke et al. 1995), perhaps due to increased fire that favors the hardwood pioneers birch and aspen. A shift to hardwood dominance could change future fire regimes, nutrient dynamics, and even the boreal climate, since the albedo of deciduous forests is higher than coniferous types (Amiro et al. 2006; Goetz et al. 2007). However, caution should be used in predicting major compositional changes through modeling. Models are convenient for parametizing and testing assumptions about complex questions, but the results are only as good as the available data, the assumptions used, and the ability to calibrate and verify the model. Data on feedback between climate and boreal forests are very limited and highly variable, leading to highly variable model results. For example, one model in Alaska predicted that moisture-induced stress would cause the disappearance of existing forest types and their replacement by aspen woodlands (Bonan et al. 1990), but later refinement of the model to include more parameters of biophysical complexity indicated that moisture deficits would likely not reach levels that could cause such widespread mortality (Bonan and Van Cleve 1992).

Compositional changes within the boreal zone could significantly alter carbon dynamics, but conversion of boreal forests to temperate forests, or tundra to boreal forests, could have a greater impact. Such transitions will not be rapid. Rather, the existing community will likely degrade at a faster rate than new vegetation types can invade. During the lag, large CO_2 emissions are possible (Apps et al. 1993). Smith and Shugart (1993) predicted a net carbon loss of 36.6 Pg over a

50–100 year period as other forest types invade the boreal region. The movement of boreal forests into the tundra could greatly increase fuel loads, bringing fire into a system in which it is rare (Kasischke et al. 1995). The impact on soil carbon pools in the tundra is unknown, but concerning. In addition, northward migration of the tree line will change albedo levels in high northern latitudes.

5.4 Albedo Effect

Albedo is not directly related to carbon storage and release; rather, it controls the absorption of heat by the biome. At high northern latitudes, forest cover increases heat absorption because dark conifer crowns have lower albedo (less reflectivity) than light conifers or low, snow-covered tundra vegetation. A growing body of research suggests that light conifer competitiveness is on the wane, and replacement by dark conifers is likely (Kharuk et al. 2009; Lloyd et al. 2011; Shuman et al. 2011). The result of this competitive shift would be a boreal forest that actually exerts a warming influence on regional and global climate, subsequently outweighing their current role as carbon sinks (Betts 2000). The presently high albedo of tundra creates a feedback with the Arctic Ocean, maintaining high levels of sea ice; forest invasion of the tundra zone could alter this interaction, changing dynamics across the entire polar region (Bonan et al. 1995). One modeling exercise that replaced global boreal forests with grass and shrub vegetation predicted a cooling of the earth's climate because of the greater reflectance of these vegetation types (Bala et al. 2007). This research suggests that albedo effects may have a dominant influence on climate at high latitudes. It should be considered, however, that these conclusions are heavily reliant on modeling, and are a relatively recent addition to boreal zone research. At the very least, however, the albedo effect should be considered as a potential balance to any effect that boreal forests may have on slowing climate change through carbon sequestration.

6 Conclusion and Summary Recommendations

Much of the research regarding the impacts of climate change on the boreal carbon budget is based on modeling, and can only predict potential changes.

- Some observations of existing impacts are available, and seem to point toward the potential for greater carbon loss from boreal forests.
- Steadily increasing temperatures across boreal and arctic North America in the past fifty years have been associated with drought-induced growth reductions, permafrost degradation, increased fire frequency, increased soil respiration, and potentially larger outbreaks of insect pests.
- Under increased temperatures, increased respiration associated with rising temperatures seems to outstrip any increases in carbon uptake through growth.
- The possibility of greatly altered carbon dynamics due to permafrost degradation also exists.

However, there is also research suggesting that some of the impacts of climate change may not be as extreme as predicted.

- It is unclear whether increased soil temperatures will cause a sustained increase in carbon release. The pool of labile carbon in the soil may not be large, resulting in only a brief increase in decomposition. While the degradation of permafrost may increase the release of CO_2, it could also result in reduced emissions of CH_4, a potent greenhouse gas.
- Some models also predict an increase in precipitation across much of the boreal zone, which in concert with rising temperatures could cause increased productivity.

Recommendations for further research are necessary particularly on the following topics.

- Understand whether the massive carbon pool in tundra soils would remain intact if converted to a forested biome.
- Concentrate on regions under-represented in global carbon budget projections (e.g. Siberia).

These regions have large uncertainties in estimates of boreal carbon pools.

- Further quantify information about several important carbon pools, including fine root biomass and mycorrhizae, bryophyte and understory layers and coarse woody debris and litter.
- Better understand poorly drained sites, including those found in the larch forests of Siberia, which may be the most vulnerable to soil carbon loss with changes in disturbance regimes and climate.
- Further consider the impacts of fire in boreal carbon dynamics, including extent, frequency, and intensity across the biome; and the interactions among fire intensity, nitrogen, and carbon.

References

Allen CD, Macalady AK, Chenchouni H, Bachelet D, McDowell N, Vennetier M, Kitzberger T, Rigling A, Breshears DD, Hogg EH, Gonzalez P, Fensham R, Zhang Z, Castro J, Demidova N, Lim J-H, Allard G, Running SW, Semerci A, Cobb N (2010) A global overview of drought and heat-induced tree mortality reveals emerging climate change risks for forests. Forest Ecol Manag 259:660–684

Allison SD, Treseder KK (2008) Warming and drying suppress microbial activity and carbon cycling in boreal forest soils. Glob Change Biol 14:2898–2909

Amiro BD, Chen JM, Liu J (2000) Net primary productivity following forest fire for Canadian ecoregions. Can J Forest Res 30:939–947

Amiro BD, Todd JB, Wotton BM, Logan KA, Flannigan MD, Stocks BJ, Mason JA, Martell DL, Hirsch KG (2001) Direct carbon emissions from Canadian forest fires, 1959–1999. Can J Forest Res 31:512–525

Amiro BD, Orchansky AL, Barr AG, Black TA, Chambers SD, Chapin FS, Gouldenf ML, Litvakg M, Liu HP, McCaughey JH, McMillan A, Randerson JT (2006) The effect of post-fire stand age on the boreal forest energy balance. Agric Forest Meteorol 140:41–50

Apps MJ, Kurz WA, Luxmoore RJ, Nilsson LO, Sedjo RA, Schmidt R, Simpson LG, Vinson TS (1993) Boreal forests and tundra. Water Air Soil Pollut 70:39–53

Arain MA, Black TA, Barr AG, Jarvis PG, Massheder JM, Verseghy DL, Nesic Z (2002) Effects of seasonal and interannual climate variability on net ecosystem productivity of boreal deciduous and conifer forests. Can J Forest Res 32:878–891

Archard F, Mollicone D, Stibig H-J, Aksenov D, Laestadius L, Li Z, Popatov P, Yaroshenko A (2006) Areas of rapid forest-cover change in boreal Eurasia. Forest Ecol Manag 237:322–334

Auclair AND, Carter TB (1993) Forest wildfires as a recent source of CO2 at northern latitudes. Can J Forest Res 23:1528–1536

Bala G, Caldeira K, Wickett M, Phillips TJ, Lobell DB, Delire C, Mirin A (2007) Combined climate and carbon-cycle effects of large-scale deforestation. Proc Natl Acad Sci USA 104:6550–6555

Baldocchi D, Kelliher FM, Black TA, Jarvis P (2000) Climate and vegetation controls on boreal zone energy exchange. Glob Change Biol 6:69–83

Balshi MS, McGuire AD, Zhuang Q, Melillo J, Kicklighter DW, Kasischke E, Wirth C, Flannigan M, Harden J, Clein JS, Burnside TJ, McAllister J, Kurz WA, Apps M, Shvidenko A (2007) The role of historical fire disturbance in the carbon dynamics of the pan-boreal region: a process-based analysis. J Geophys Res 112:G02029

Balshi MS, McGuire AD, Duffy P, Flannigan M, Kicklighter DW, Melillo J (2009) Vulnerability of carbon storage in North American boreal forests to wildfires during the 21st century. Glob Change Biol 15:1491–1510

Barber VA, Juday GP, Finney BP (2000) Reduced growth of Alaskan white spruce in the twentieth century from temperature-induced drought stress. Nature 405:668–673

Barr AG, Black TA, Hogg EH, Griffis TJ, Morgenstern K, Kljun N, Theede A, Nesic Z (2007) Climatic controls on the carbon and water balances of a boreal aspen forest, 1994–2003. Glob Change Biol 13:561–576

Baskerville GL (1975) Spruce budworm – super silviculturalist. Forest Chron 51:138–140

Berg EE, Henry JD, Fastie CL, De Volder AD, Matsuoka SM (2006) Spruce beetle outbreaks on the Kenai Peninsula, Alaska, and Kluane National Park and Reserve, Yukon Territory: relationship to summer temperatures and regional differences in disturbance regimes. Forest Ecol Manag 227:219–232

Berg A, Östlund L, Moen J, Olofsson J (2008) A century of logging and forestry in a reindeer herding are in northern Sweden. Forest Ecol Manag 256:1009–1020

Betts RA (2000) Offset of the potential carbon sink from boreal forestation by decreases in surface albedo. Nature 408:187–190

Billings WD (1987) Carbon balance of Alaskan tundra and taiga ecosystems – past, present and future. Quaternary Sci Rev 6:165–177

Bonan GB (2008) Forests and climate change: forcings, feedbacks, and the climate benefits of forests. Science 320:1444–1449

Bonan GB, Shugart HH (1989) Environmental factors and ecological processes in boreal forests. Annu Rev Ecol Syst 20:1–28

Bonan GB, Vancleve K (1992) Soil-temperature, nitrogen mineralization, and carbon source sink relationships in boreal forests. Can J Forest Res 22:629–639

Bonan GB, Shugart HH, Urban DL (1990) The sensitivity of some high-latitude boreal forests to climatic parameters. Climatic Change 16:9–29

Bonan GB, Pollard D, Thompson SL (1992) Effects of boreal forest vegetation on global climate change. Nature 359:716–718

Bonan GB, Chapin FS, Thompson SL (1995) Boreal forest and tundra ecosystems as components of the climate system. Climatic Change 29:145–167

Bond-Lamberty B, Wang CK, Gower ST (2004) Net primary production and net ecosystem production of a boreal black spruce wildfire chronosequence. Glob Change Biol 10:473–487

Bond-Lamberty B, Peckham SD, Ahl DE, Gower ST (2007) Fire as the dominant driver of central Canadian boreal forest carbon balance. Nature 450:89–92

Brassard BW, Chen HYH (2008) Effects of forest types and disturbance on diversity of coarse woody debris in boreal forest. Ecosystems 11:1078–1090

Briffa KR, Schweingruber FH, Jones PD, Osborn TJ, Shiyatov SG, Vaganov EA (1998) Reduced sensitivity of recent tree-growth to temperature at high northern latitudes. Nature 391:678–682

Bronson DR, Gower ST, Tanner M, Linder S, Van Herk I (2008) Response of soil surface CO_2 flux in a boreal forest to ecosystem warming. Glob Change Biol 14: 856–867

Camill P (2005) Permafrost thaw accelerates in boreal peatlands during late-20th century climate warming. Climatic Change 68:135–152

Camill P, Lynch JA, Clark JS, Adams JB, Jordan B (2001) Changes in biomass, aboveground net primary production, and peat accumulation following permafrost thaw in the boreal peatlands of Manitoba, Canada. Ecosystems 4:461–478

Chapin FS, McGuire AD, Ruess RW, Hollingsworth TN, Mack MC, Johnstone Jill F, Kasischke ES, Euskirchen ES, Jones JB, Jorgenson MT, Kielland K, Kofinas G, Turetsky MR, Yarie J, Lloyd A, Taylor DL (2010) Resilience of Alaskas boreal forest to climatic change. Can J Forest Res 40:1360–1370

Chen WJ, Black TA, Yang PC, Barr AG, Neumann HH, Nesic Z, Blanken PD, Novak MD, Eley J, Ketler RJ, Cuenca A (1999) Effects of climatic variability on the annual carbon sequestration by a boreal aspen forest. Glob Change Biol 5:41–53

Conard SG, Ivanova GA (1997) Wildfire in Russian boreal forests – potential impacts of fire regime characteristics on emissions and global carbon balance estimates. Environ Pollut 98:305–313

Czimczik CI, Trumbore SE, Ce MS, Winston GC (2006) Changing sources of soil respiration with time since fire in a boreal forest. Glob Change Biol 12:957–971

DeLuca TH, Aplet GH (2008) Charcoal and carbon storage in forest soils of the Rocky Mountain West. Front Ecol Environ 6:18–24

Dioumaeva I, Trumbore S, Schuur EAG, Goulden ML, Litvak M, Hirsch AI (2002) Decomposition of peat from upland boreal forest: temperature dependence and sources of respired carbon. J Geophys Res Atmos 108:8222

Dixon RK, Krankina ON (1993) Forest-fires in Russia – carbon-dioxide emissions to the atmosphere. Can J Forest Res 23:700–705

Dulamsuren C, Hauck M, Bader M, Oyungerel S, Osokhjargal D, Nyambayar S, Leuschner C (2009) The different strategies of Pinus sylvestris and Larix sibirica to deal with summer drought in a northern Mongolian forest – steppe ecotone suggest a future superiority of pine in a warming climate. Can J Forest Res 39:2520–2528

Dunn AL, Barford CC, Wofsy SC, Goulden ML, Daube BC (2007) A long-term record of carbon exchange in a boreal black spruce forest: means, responses to interannual variability, and decadal trends. Glob Change Biol 13:577–590

Emanuel WR, Shugart HH, Stevenson M (1985) Response to comment: climatic change and the broad-scale distribution of terrestrial ecosystem complexes. Climate Change 7:457–460

Flannigan MD, Bergeron Y, Engelmark O, Wotton BM (1998) Future wildfire in circumboreal forests in relation to global warming. J Vegetation Sci 9:469–476

Frey BR, Lieffers VJ, Hogg EH, Landhausser SM (2004) Predicting landscape patterns of aspen dieback: mechanisms and knowledge gaps. Can J Forest Res 34:1379–1390

Goetz SJ, Mack MC, Gurney KR, Randerson JT, Houghton RA (2007) Ecosystem responses to recent climate change and fire disturbance at northern high latitudes: observations and model results contrasting northern Eurasia and North America. Environ Res Lett 2:045031

Gorham E (1991) Northern peatlands – role in the carbon-cycle and probable responses to climatic warming. Ecol Appl 1:182–195

Goulden ML, Wofsy SC, Harden JW, Trumbore SE, Crill PM, Gower ST, Fries T, Daube BC, Fan SM, Sutton DJ, Bazzaz A, Munger JW (1998) Sensitivity of boreal forest carbon balance to soil thaw. Science 279:214–217

Goulden ML, McMillan AMS, Winston C, Rocha AV, Manies KL, Harden JW, Bond-Lamberty BP (2010) Patterns of NPP, GPP, respiration, and NEP during boreal forest succesion. Glob Change Biol 17:855–871

Gower ST, Richards JH (1990) Larches – deciduous conifers in an evergreen world. Bioscience 40:818–826

Gower ST, Vogel JG, Norman JM, Kucharik CJ, Steele SJ, Stow TK (1997) Carbon distribution and aboveground net primary production in aspen, jack pine, and black spruce stands in Saskatchewan and Manitoba, Canada. J Geophys Res Atmos 102:29029–29041

Gower ST, Krankina O, Olson RJ, Apps M, Linder S, Wang C (2001) Net primary production and carbon allocation patterns of boreal forest ecosystems. Ecol Appl 11:1395–1411

Grant RF, Black TA, Gaumont-Guay D, Kljun N, Barrc AG, Morgenstern K, Nesic Z (2006) Net ecosystem productivity of boreal aspen forests under drought and climate change: mathematical modelling with Ecosys. Agr Forest Meteorol 140:152–170

Gromtsev A (2002) Natural disturbance dynamics in the boreal forests of European Russia: a review. Silva Fenn 36:41–55

Hanninen H, Kolari P, Hari P (2005) Seasonal development of Scots pine under climatic warming: effects on photosynthetic production. Can J Forest Res 35:2092–2099

Harden JW, O'Neill KP, Trumbore SE, Veldhuis H, Stocks BJ (1997) Moss and soil contributions to the annual net carbon flux of a maturing boreal forest. J Geophys Res Atmos 102:28805–28816

Harden JW, Trumbore SE, Stocks BJ, Hirsch A, Gower ST, O'Neill KP, Kasischke ES (2000) The role of fire in the boreal carbon budget. Glob Change Biol 6:174–184

Hobbie SE, Schimel JP, Trumbore SE, Randerson JR (2000) Controls over carbon storage and turnover in high-latitude soils. Glob Change Biol 6:196–210

Hogg EH, Brandt JP, Kochtubajda B (2002) Growth and dieback of Aspen forests in northwestern Alberta, Canada, in relation to climate and insects. Can J Forest Res 32:823–832

Hogg EH, Brandt JP, Michaelian M (2008) Impacts of a regional drought on the productivity, dieback and biomass of western Canadian aspen forests. Can J Forest Res 38:1373–1384

Jarvis P, Linder S (2000) Botany – constraints to growth of boreal forests. Nature 405:904–905

Johnson DW, Curtis PS (2001) Effects of forest management on soil carbon and nitrogen storage: meta analysis. Forest Ecol Manag 140:227–238

Johnson EA, Rowe JS (1975) Fire in the subarctic wintering ground of the Beverly Caribou Herd. Am Midland Nat 94:1–14

Johnstone JF, Hollingsworth TN, Chapin FS, Mack MC (2010) Changes in fire regime break the legacy lock on successional trajectories in Alaskan boreal forest. Glob Change Biol 16:1281–1295

Kang S, Kimball JS, Running SW (2006) Simulating effects of fire disturbance and climate change on boreal forest productivity and evapotranspiration. Sci Total Environ 362:85–102

Karhu K, Fritz H, Inen KHMM, Vanhala P, Jungner HG, Oinonen M, Sonninen E, Tuomi M, Spetz P, Kitunen V (2010) Temperature sensitivity of soil carbon fractions in boreal forest soil. Ecology 91:370–376

Kasischke ES, Christensen NL, Stocks BJ (1995) Fire, global warming, and the carbon balance of boreal forests. Ecol Appl 5:437–451

Kasischke ES, Verbyla DL, Rupp TS, McGuire AD, Murphy K, Jandt R, Barnes J, Hoy EE, Duffy PA, Calef M, Turetsky MR (2010) Alaska's changing fire regime – implications for the vulnerability of its boreal forests. Can J Forest Res 40:1313–1324

Kellomaki S, Karjalainen T, Vaisanen H (1997) More timber from boreal forests under changing climate? Forest Ecol Manag 94:195–208

Kharuk VI, Ranson KJ, Im ST, Dvinskaya ML (2009) Response of *Pinus sibirica* and *Larix sibirica* to climate change in southern Siberian alpine forest-tundra ecotone. Scand J Forest Res 24:130–139

Krankina ON, Harmon ME, Kukuev YA, Treyfeld RF, Kashpor NN, Kresnov VG, Skudin VM, Protasov NA, Yatskov M, Spycher G, Povarov ED (2002) Coarse woody debris in forest regions of Russia. Can J Forest Res 32:768–778

Krishnan P, Black TA, Barr AG, Grant NJ, Gaumont-Guay D, Nesic Z (2008) Factors controlling the interannual variability in the carbon balance of a southern boreal black spruce forest. J Geophys Res 113. doi:10.1029/2007JD008965

Kurz WA, Dymond CC, Stinson G, Rampley GL, Neilson ET, Carroll AL, Ebata T, Safranyik L (2008a) Mountain pine beetle and forest carbon feedback to climate change. Nature 452:987–990

Kurz WA, Stinson G, Rampley G (2008b) Could increased boreal forest ecosystem productivity offset carbon losses from increased disturbance? Philos Trans R Bot Soc 363:2259–2268

Kurz WA, Stinson G, Rampley GJ, Dymond CC, Neilson ET (2008c) Risk of natural disturbances makes future contribution of Canada's forests to the global carbon cycle highly uncertain. Proc Natl Acad Sci USA 105:1551–1555

Lapenis A, Shvidenko A, Shepaschenko D, Nilsson S, Aiyyer A (2005) Acclimation of Russian forests to recent changes in climate. Glob Change Biol 11: 2090–2102

Larsen JA (1980) The boreal ecosystem. Academic Press, Inc, London

Légaré S, Bergeron Y, Leduc A, Paré D (2001) Comparison of the understory vegetation in boreal forest types of southwest Quebec. Can J Bot 79:1019–1027

Lesieur D, Gauthier S, Bergeron Y (2002) Fire frequency and vegetation dynamics for the south-central boreal forest of Quebec, Canada. Can J Forest Res 32:1996–2009

Li Z, Apps MJ, Kurz WA, Banfield E (2003) Temporal changes of forest net primary production and net ecosystem production in west central Canada associated with natural and anthropogenic disturbances. Can J Forest Res 33:2340–2351

Lindroth A, Grelle A, Moren AS (1998) Long-term measurements of boreal forest carbon balance reveal large temperature sensitivity. Glob Change Biol 4:443–450

Litvak M, Miller S, Wofsy SC, Goulden M (2003) Effect of stand age on whole ecosystem CO_2 exchange in the Canadian boreal forest. J Geophys Res Atmos 108(D3):8225

Lloyd AH, Bunn AG, Berner L (2011) A latitudinal gradient in tree growth response to climate warming in the Siberian taiga. Glob Change Biol 17:1935–1945

MacDonald SE, Fenniak TE (2007) Understory plant communities of boreal mixedwood forests in western Canada: natural patterns and response to variable-retention harvesting. Forest Ecol Manag 242:34–38

Magnani F, Mencuccini M, Borghetti M, Berbigier P, Berninger F, Delzon S, Grelle A, Hari P, Jarvis PG, Kolari P, Kowalski AS, Lankreijer H, Law BE, Lindroth A, Loustau D, Manca G, Moncrieff JB,

Rayment M, Tedeschi V, Valentini R, Grace J (2007) The human footprint in the carbon cycle of temperate and boreal forests. Nature 447:848–850

Mäkipää R, Karjalainen T, Pussinen A, Kellomaki S (1999) Effects of climate change and nitrogen deposition on the carbon sequestration of a forest ecosystem in the boreal zone. Can J Forest Res 29:1490–1501

Malhi Y, Baldocchi DD, Jarvis PG (1999) The carbon balance of tropical, temperate and boreal forests. Plant Cell Environ 22:715–740

Malmstrom CM, Raffa KF (2000) Biotic disturbance agents in the boreal forest: considerations for vegetation change models. Glob Change Biol 6:35–48

Martin JL, Gower ST, Plaut J, Holmes B (2005) Carbon pools in a boreal mixedwood logging chronosequence. Glob Change Biol 11:1883–1894

Nalder IA, Wein RW (1999) Long-term forest floor carbon dynamics after fire in upland boreal forests of western Canada. Global Biogeochem Cycles 13:951–968

Niinisto SM, Silvola J, Kellomaki S (2004) Soil CO_2 efflux in a boreal pine forest under atmospheric CO_2 enrichment and air warming. Glob Change Biol 10: 1363–1376

O'Neill KP, Kasischke ES, Richter DD (2002) Environmental controls on soil CO_2 flux following fire in black spruce, white spruce, and aspen stands of interior Alaska. Can J Forest Res 32:1525–1541

Ohlson M, Dahlberg B, Økland T, Brown KJ, Halvorsen R (2009) The charcoal carbon pool in boreal forest soils. Nat Geosci 2:692–695

Ollinger SV, Richardson AD, Martin ME, Hollinger DY, Frolking S, Reich PB, Plourde LC, Katul GG, Munger JW, Oren R, Smith M-L, Paw UKT, Bolstad PV, Cook BD, Day MC, Martin TA, Monson RK, Schmid HP (2008) Canopy nitrogen, carbon assimilation, and albedo in temperate and boreal forests: functional relations and potential climate feedbacks. Proc Natl Acad Sci USA 105:19335–19340

Pastor J, Post WM (1988) Response of northern forests to CO_2-induced climate change. Nature 334:55–58

Powell JA, Logan JA (2005) Insect seasonality: circle map analysis of temperature-driven life cycles. Theor Popul Biol 67:161–179

Price DT, Halliwell DH, Apps MJ, Kurz WA, Curry SR (1997) Comprehensive assessment of carbon stocks and fluxes in a Boreal-Cordilleran forest management unit. Can J Forest Res 27:2005–2016

Prokushkin AS, Kajimoto T, Prokushkin SG, McDowell WH, Abaimov AP, Matsuura Y (2005) Climatic factors influencing fluxes of dissolved organic carbon from the forest floor in a continuous-permafrost Siberian watershed. Can J Forest Res 35:2130–2140

Pypker TG, Fredeen AL (2002) The growing season carbon balance of a sub-boreal clearcut 5 years after harvesting using two independent approaches to measure ecosystem CO_2 flux. Can J Forest Res 32:852–862

Rapalee G, Trumbore SE, Davidson EA, Harden JW, Veldhuis H (1998) Soil carbon stocks and their rates of accumulation and loss in a boreal forest landscape. Global Biogeochem Cycles 12:687–701

Ruess RW, VanCleve K, Yarie J, Viereck LA (1996) Contributions of fine root production and turnover to the carbon and nitrogen cycling in taiga forests of the Alaskan interior. Can J Forest Res 26:1326–1336

Rustad LE, Fernandez IJ (1998) Experimental soil warming effects on CO2 and CH4 flux from a low elevation spruce-fir forest soil in Maine, USA. Glob Change Biol 4:597–605

Ryan MG, Lavigne MB, Gower ST (1997) Annual carbon cost of autotrophic respiration in boreal forest ecosystems in relation to species and climate. J Geophys Res Atmos 102:28871–28883

Schulze ED, Lloyd J, Kelliher FM, Wirth C, Rebmann C, Luhker B, Mund M, Knohl A, Milyukova IM, Schulze W, Ziegler W, Varlagin AB, Sogachev AF, Valentini R, Dore S, Grigoriev S, Kolle O, Panfyorov MI, Tchebakova N, Vygodskaya NN (1999) Productivity of forests in the Eurosiberian boreal region and their potential to act as a carbon sink – a synthesis. Glob Change Biol 5:703–722

Schulze ED, Wirth C, Mollicone D, Ziegler W (2005) Succession after stand replacing disturbances by fire, wind throw, and insects in the dark Taiga of Central Siberia. Oecologia 146:77–88

Seedre M, Chen HYH (2010) Carbon dynamics of aboveground live vegetation of boreal mixedwoods after wildfire and clear-cutting. Can J Forest Res 40: 1862–1869

Seely B, Welham C, Kimmins H (2002) Carbon sequestration in a boreal forest ecosystem: results from the ecosystem simulation model, FORECAST. Forest Ecol Manag 169:123–135

Shuman JK, Shugart HH, O'Halloran TL (2011) Sensitivity of Siberian larch forests to climate change. Glob Change Biol 17:1–15

Shvidenko A, Nilsson S, Roshkov V (1997) Possibilities for increased carbon sequestration through the implementation of rational forest management in Russia. Water Air Soil Pollut 94:137–162

Siitonen J, Martikainen P, Punttila P, Rauh J (2000) Coarse woody debris and stand characteristics in mature managed and old-growth boreal mesic forests in southern Finland. Forest Ecol Manag 128:211–225

Smith TM, Shugart HH (1993) The transient-response of terrestrial carbon storage to a perturbed climate. Nature 361:523–526

Soja AJ, Tchebakova NM, French NHF, Flannigan MD, Shugart HH, Stocks BJ, Sukhinin AI, Varfenova EI, Chapin FS, Stackhouse PW (2007) Climate-induced boreal forest change: predictions versus current observations. Global Planet Change 56:274–296

Startsev NA, Lieffers VJ, McNabb DH (2007) Effects of feathermoss removal, thinning, and fertilization on lodgepole pine growth, soil microclimate and stand nitrogen dynamics. Forest Ecol Manag 240:79–86

Stocks BJ, Fosberg MA, Lynham TJ, Mearns L, Wotton BM, Yang Q, Jin JZ, Lawrence K, Hartley GR, Mason JA, McKenney DW (1998) Climate change and forest fire potential in Russian and Canadian boreal forests. Climatic Change 38:1–13

Thiffault E, Hannam KD, Quideau SA, Pare D, Belanger N, Oh S-W, Munson AD (2008) Chemical composition of forest floor and consequences for nutrient availability after wildfire and harvesting in the boreal forest. Plant Soil 308:37–53

Tilman D, Reich P, Phillips H, Menton M, Patel A, Vos E, Peterson D, Knops J (2000) Fire suppression and ecosystem carbon storage. Ecology 81:2680–2685

Trofymow JA, Moore TR, Titus B, Prescott C, Morrison I, Siltanen M, Smith S, Fyles J, Wein R, CamirT C, Duschene L, Kozak L, Kranabetter M, Visser S (2002) Rates of litter decomposition over 6 years in Canadian forests: influence of litter quality and climate. Can J Forest Res 32:789–804

Turetsky MR (2003) Bryophytes in carbon and nitrogen cycling. Bryologist 106:395–409

Turetsky MR, Mack MC, Harden JW, Manies KL (2005) Spatial patterning of soil carbon storage across boreal landscapes. In: Lovett GM, Jones CG, Turner MG, Weathers KC (eds) Ecosystem function in heterogeneous landscapes. Springer, New York, pp 229–256

Turetsky MR, Mack MC, Hollingsworth TN, Harden JW (2010) The role of mosses in ecosystem succession and function in Alaska's boreal forest. Can J Forest Res 40:1237–1264

Van Cleve K, Oechel WC, Hom JL (1990) Response of black spruce (*picea-mariana*) ecosystems to soil temperature modification in interior Alaska USA. Can J Forest Res 20:1530–1535

Vedrova EF, Shugalei LS, Stakanov VD (2002) The carbon balance in natural and disturbed forests of the southern taiga in central Siberia. J Veg Sci 13:341–350

Vygodskaya NN, Schulze ED, Tchebakova NM, Karpachevskii LO, Kozlov D, Sidorov KN, Panfyorov MI, Abrazko MA, Shaposhnikov ES, Solnzeva ON, Minaeva TY, Jeltuchin AS, Wirth C, Pugachevskii AV (2002) Climatic control of stand thinning in unmanaged spruce forests of the southern taiga in European Russia. Tellus Ser B Chem Phys Meteorol 54:443–461

Wang CK, Gower ST, Wang YH, Zhao HX, Yan P, Bond-Lamberty BP (2001) The influence of fire on carbon distribution and net primary production of boreal *Larix gmelinii* forests in north-eastern China. Glob Change Biol 7:719–730

Williams DW, Liebhold AM (1995) Forest defoliators and climatic-change - potential changes in spatial-distribution of outbreaks of western spruce budworm (*lepidoptera, tortricidae*) and gypsy-moth (*lepidoptera, lymantriidae*). Environ Entomol 24:1–9

Wilmking M, Juday GP, Barber VA, Zald HSJ (2004) Recent climate warming forces contrasting growth responses of white spruce at treeline in Alaska through temperature thresholds. Glob Change Biol 10:1724–1736

Winston GC, Sundquist ET, Stephens BB, Trumbore SE (1997) Winter CO2 fluxes in a boreal forest. J Geophys Res Atmos 102:28795–28804

Wirth C, Schulze ED, Luhker B, Grigoriev S, Siry M, Hardes G, Ziegler W, Backor M, Bauer G, Vygodskaya NN (2002) Fire and site type effects on the long-term carbon and nitrogen balance in pristine Siberian Scots pine forests. Plant Soil 242:41–63

Wolf A, Kozlov MV, Callaghan TV (2008) Impact of non-outbreak insect damage on vegetation in northern Europe will be greater than expected during a changing climate. Climatic Change 87:91–106

Wooster MJ, Zhang YH (2004) Boreal forest fires burn less intensely in Russia than in North America. Geophys Res Lett 31:L20505

Yarie J, Billings S (2002) Carbon balance of the taiga forest within Alaska: present and future. Can J Forest Res 32:757–767

Part II

Measuring Carbon in Forests

Section Summary

The two papers in this section comprise an analysis of the different measurement techniques of carbon in the field and through remote estimation, and with this information global and regional statistics of stored and lost carbon are described. Four categories of methods for measuring forest biomass and estimating carbon are described: (i) forest inventory (biomass); (ii) remote sensing (relationship between biomass and land cover); (iii) eddy covariance (direct measurement of CO_2 release and uptake); and (iv) the inverse method (relationship among biomass, CO_2 flux, and CO_2 atmospheric transport).

Contributors toward organizing and editing this section were: *Mark S. Ashton, Mary L. Tyrrell, Deborah Spalding, and Xuhui Lee*

Measuring Carbon in Forests

7

Xin Zhang, Yong Zhao, Mark S. Ashton, and Xuhui Lee

Executive Summary

Accurate measurement of carbon stocks and flux in forests is one of the most important scientific bases for successful climate and carbon policy implementation. A measurement framework for monitoring carbon storage and emissions from forests should provide the core tool to qualify country and project level commitments under the United Nations Framework Convention on Climate Change, and to monitor the implementation of the Kyoto Protocol.

Currently, there are several methods for estimating forest carbon stocks and flux, ranging from the relatively simple forest biomass inventory to complex, sophisticated experiments and models. Advanced carbon estimation methodologies such as LiDAR and eddy covariance carbon flux experiments may provide reliable, accurate and transparent data and serve as a basis for market tools and international policymaking such as carbon trading, carbon taxes, and reducing emissions credits from deforestation and forest degradation in developing countries (REDD, REDD+). Nevertheless, developing countries, which have limited capacity for data collection and management, need low-cost methodologies with acceptable spatial and temporal resolution and appropriate sampling intensity.

If a standardized verification system across projects, countries, and regions is to ever be attained, policymakers should be aware that there are different basic approaches to measuring forest carbon, which have advantages and disadvantages, and varying degrees of accuracy and precision.

We review the four categories of methods for measuring forest biomass and estimating carbon which are currently in use: i) forest inventory (biomass); ii) remote sensing (relationship between biomass and land cover); iii) eddy covariance (direct measurement of CO2 release and uptake); and iv) the inverse method (relationship among biomass, CO2 flux and CO2 atmospheric transport). These methods all vary in their level of accuracy and the resolution at which data can be obtained. Each technique has its own advantages and disadvantages and there are appropriate circumstances for using each one in measuring CO2 flux and carbon storage for different temporal and spatial scales of evaluation and measurement.

Forest inventory methods are usually direct measures of above-ground biomass accumulation within a forest. They have a long history in development and good data is generally available; however, they are low in time resolution, costly to implement, require technical training and knowledge, are variable in standards for measurement, and are available in only certain regions, mostly developed countries.

X. Zhang • Y. Zhao • X. Lee
Yale School of Forestry and Environmental Studies,
195 Prospect St, New Haven, CT 06511, USA

M.S. Ashton (✉)
Yale School of Forestry and Environmental Studies,
360 Prospect St, New Haven, CT 06511, USA
e-mail: mark.ashton@yale.edu

M.S. Ashton et al. (eds.), *Managing Forest Carbon in a Changing Climate*,
DOI 10.1007/978-94-007-2232-3_7, © Springer Science+Business Media B.V. 2012

Remote sensing methods usually are combined with models that link remote sensing information with CO2 and carbon data (often forest inventory information). Methods can be divided into passive sensing (satellite images, aerial photographs that are characterized by reflected light) and active sensing (radar, LiDAR that emit and receive microwaves or light respectively). Remote sensing is limited by incomplete information, resolution and detection problems, and uncertainties in models that require further development and refinement. Nevertheless, when available at a suitable resolution and spatial scale, it can be the cheapest method of surveying forests.

The eddy covariance method is advanced in its accuracy and resolution, and is a good method for direct measurement of small (hectare-plus) scale CO2 flux; but, it is still restricted by systematic biases, is not accurate in rough topography, and has limited observation sites around the world.

Inverse methods typically are used at continental or global scales. These methods calculate the total sources and sinks, including both anthropogenic and natural, using available atmospheric CO2 concentration data and transportation models. Carbon Tracker is one of the most advanced inverse methods. It was developed by NOAA's Earth Systems Research Laboratory as a system to keep track of carbon dioxide uptake and release at the Earth's surface over time and to continuously improve models and data assimilation methods for higher accuracy and resolution.

What We Do and Do Not Know About Measuring Carbon in Forests

- Forest inventory methods require historical and regional data. Permanent continuous forest inventory (CFI) plots are the best to provide long-term accurate and non-biased assessments. Non-permanent plots can be used but are often biased.
- Most developed countries conduct regular national inventories to evaluate forest health and status. These inventories are therefore a useful data base if biases can be avoided.
- In the past, inventory plots have often been biased toward sampling forests of commercial value. Forests considered degraded or that are now growing back (secondary forest) are often under-represented. Inventories often only include tree species that have commercial value and under-sample small trees.
- Very few inventories account for belowground biomass, litter, and dead wood. Fine spatial-resolution (1–10 m) satellite data have the advantage in providing high resolution details of a specific area. However, disadvantages include a small area of coverage, shadows, and expense in acquisition.
- It is expensive to sample a sufficient number of trees representing the diversity of size and species to generate local allometric equations for use in converting tree data to forest biomass data.
- Medium spatial-resolution (10–100 m) satellite data are the most suitable for regional level above-ground biomass estimation because of better data availability (spatial and temporal), and the lower cost of acquisition and storage. Since spatial resolution is usually sufficient to compare with inventory measurements, this approach is widely used for forests.
- Coarse spatial resolution satellite data (>100 m) are most effective at large national or continental scales. The use at such scales is limited because of the occurrence of mixed pixels, and differences between scale and resolution of forest inventory measurements.
- Aboveground biomass estimation by radar can achieve good accuracy in low and medium density forests, but the relationship between radar backscatter and aboveground biomass weakens when the forest becomes too dense. Its advantage is its ability to penetrate precipitation, cloud cover, and avoid shade/shadow effects from the sun.
- Light Detection and Ranging (LiDAR) is an active remote sensing method, analogous to radar, but using laser light instead of microwaves. The technology needs further development to be widely useful in aboveground biomass estimation.
- Recent technical, financial and logistical (scheduling) problems with the U.S. remote sensing program highlight the need for more countries

or consortiums to provide the international remote sensing community with more options in satellite imagery and Radar/LiDAR data.

- Eddy covariance measurements have been continuously made at certain sites for over 10 years. New observation sites (especially in tropical forest regions), updated models, and remote sensing data will enable eddy covariance methods to continually refine estimates of CO_2 flux from regional to continental scales, making eddy covariance the world's direct tracking system of carbon flux.
- More research needs to be conducted to close the energy budget in eddy covariance measurements and eliminate biases caused by nighttime stratification and complex topography.
- CarbonTracker has emerged as one of the most advanced inverse models currently used for regional and continental inverse estimates of carbon sinks and sources.

1 Introduction

The need to accurately measure the stocks and flux of carbon in forests is urgent given the global consensus that CO_2 emissions have a very strong influence on global warming. Forests are an essential part of the carbon cycle. They are a major terrestrial sink of CO_2, but their land use conversion to agriculture currently accounts for 25% of global carbon emissions. Compared to the combustion of fossil fuel, emissions from land use change are an important issue for developing countries and especially for tropical countries (Houghton and Ramakrishna 1999). Forests are influenced by various anthropogenic and natural disturbances such as fire, disease, insect infestations, harvesting, deforestation, and degradation, all of which can lead to significant carbon emissions. To understand the carbon cycle in the forest, it is important to have valid, cost-effective scientific methods to measure and monitor carbon. Such measures require accuracy and precision in order to have useful data on carbon stocks and flux in forests globally.

Accurate estimation of forest carbon stocks and flux is one of the most important scientific bases for successful policy implementation. Although understanding the methods of measuring the forest carbon cycle may not be a focus of policymakers, it is important that they recognize that there are differences between regions and countries in carbon emission behaviors and carbon storage in forests (and associated land conversion). This understanding will allow them to make better decisions about global and regional resource allocation for measurement capacity, and therefore to optimize adaptation and mitigation strategies for climate change. A measurement framework for monitoring carbon storage and emissions from forests should be the core tool to qualify country and project level commitments under the United Nations Framework Convention on Climate Change (UNFCCC 1997), and to monitor the implementation of the Kyoto Protocol (Brown 2002).

To meet the requirements of the Kyoto Protocol, all Annex I countries[1] must "provide data to establish their level of carbon stocks in 1990 and to enable an estimation of their changes in carbon stocks in subsequent years" (UNFCCC 1997). Developing countries, which have limited capacity in data collection and management, need methodologies with low-cost, acceptable spatial and temporal resolution and appropriate sampling intensity. Furthermore, for the post-Kyoto era, advanced carbon estimation methodologies may provide reliable, accurate, and transparent data and serve as a basis for market tools and international policymaking such as carbon trading, carbon taxes, and reducing emissions credits from deforestation and forest degradation in developing countries (REDD, REDD+).

[1]Annex I Parties to the United Nations Framework Convention on Climate Change (UNFCCC) include the industrialized countries that were members of the OECD (Organisation for Economic Co-operation and Development) in 1992, plus countries with economies in transition (the EIT Parties), including the Russian Federation, the Baltic States, and several Central and Eastern European States.

1.1 Objectives

In this chapter we describe four basic methods of measuring carbon storage and flux in forests: (i) forest inventory; (ii) remote sensing; (iii) eddy covariance; and (iv) the inverse method. These methods are critiqued for their advantages and disadvantages in estimating CO_2 flux and storage. All are evaluated for their accuracy and resolution. In the conclusion section, we describe gaps in data, information, and technologies that need to be addressed if a standardized measurement framework is to be achieved. Recommendations are made on improvements in methodology for more efficient and effective aboveground biomass (AGB) estimation.

1.2 Measuring Carbon

Generally, there are two main approaches to measuring carbon stocks and fluxes in each forest carbon pool: (i) measuring changes in carbon stock, and then inferring a carbon flux under a certain level of confidence; and (ii) measuring carbon flux directly. Generally, biomass, which is readily measured, is widely used to estimate carton stocks using proven formulas for the ratio of carbon to biomass instead of measuring carbon directly, particularly for aboveground carbon (Brown 1997).

Carbon stocks in forests can be classified into five different measurement pools:

- *Aboveground biomass* – Living biomass above the soil, including stem, stump, branches, bark, seeds, and foliage. This category includes live understory.
- *Belowground biomass* – All living biomass of roots greater than a certain defined diameter.
- *Dead wood* – Includes all non-living woody biomass either standing, lying on the ground (but not including litter), or in the soil.
- *Litter* – Includes the litter, humus layers of the soil surface, and all non-living biomass of a certain diameter lying on the ground.
- *Soil organic carbon (SOC)* – Typically includes all organic material in soil to a depth of 1 m, excluding the litter layer and coarse roots of the belowground biomass pool.

2 Forest Inventories and Aboveground Carbon Stock Estimations

Because national forest inventories are commonly available for many countries, different approaches have been developed to estimate above ground biomass (AGB) from inventories. They can be categorized by data source: (i) field measurement; (ii) remote-sensing data; or (iii) ancillary data used in GIS-based modeling (Lu 2006; Wulder et al. 2008). Several approaches to estimating carbon stocks from each of these data sources are shown in Table 7.1.

2.1 Field-Based Methods

The field-based method is usually referred to as an inventory assessment, and can be further classified into volume-to-biomass and diameter-to-biomass approaches. The choice between these approaches is dependent upon the data available and the desired resolution. Generally, the approach of converting timber volume, which is commonly available, to biomass has more uncertainty but requires less detailed data; therefore, this is the most commonly used method. If detailed diameter information and field measurements are available for establishing allometric equations, then the diameter-to-biomass (allometric) approach is generally favored because it is more accurate.

Timber volume data are available for many countries because these data are primarily collected for forest management and revenue accounting. In 1919 (Norway), 1921–24 (Finland), and 1923–24 (Sweden), the Nordic nations started national forest inventories because of the fear that the fuelwood resource would be exhausted (FAO 2000; Brack 2009). Optimally, species, diameter at breast height (DBH), height, site quality, age, increment, and defects are recorded in each inventory dataset (LeBlanc 2009). However, different

Table 7.1 Summary of techniques for above ground carbon stock estimation

Category	Methods	Data used	Characteristics	References
Field measurement methods	Conversion from volume to biomass by biomass expansion factor (BEF)	Volume from sample trees or stands	Individual trees or vegetation stands	Fang et al. (2001); Smith and Heath (2004); Wang et al. (2007a); Woodbury et al. (2007); Wulder et al. (2008)
	Allometric equations	Sample trees	Individual trees	Gehring et al. (2004); Goodale et al. (2002); Jenkins et al. (2003); Zianis and Mencuccini (2004)
Remote sensing methods	Methods based on fine spatial-resolution data	Aerial photographs, IKONOS	Per-pixel level	Thenkabail (2003); Thenkabail et al. (2004a)
	Methods based on medium spatial-resolution data	Landsat TM/ETM+, SPOT	Per-pixel level	Dong et al. (2003) Muukkonen and Heiskanen (2005); Muukkonen and Heiskanen (2007); Cohen and Goward (2004); Lu and Batistella (2005)
	Methods based on coarse spatial-resolution data	IRS-IC WiFS, AVHRR	Per-pixel level	Cross et al. (1991) Laporte et al. (1995)
	Methods based on radar data	Radar	Per-pixel level	Blackburn and Steele (1999); Levesque and King (2003); Sun et al. (2002)
	Methods based on LiDAR Data	LiDAR	Per-pixel level	Anderson et al. (2006); Drake et al. (2003); Lefsky et al. (1999)

Source: Modified from Lu (2006)

countries have various capacities and standards for detailing the inventory information. For example, *Forest Statistics of China 1984–1988* is compiled from more than 250,000 permanent and temporary plots across China, and the technical standard in data collection includes measuring DBH, height, stem volume, age, total area, and site quality (Fang et al. 1998). But in the *National Forest Inventory of Indonesia 1989–1996*, only the number of trees per ha and volume per ha for different diameter classes is available (FAO 2000). In Brazil, very limited data collection is done regionally by consultants, but not by the government or the research academy (Freitas 2006; Wardoyo 2008). It is therefore necessary for some countries to utilize available timber volume data from private company and landowner inventories so as to obtain rudimentary baseline domestic estimates of changes and stocks of standing forest carbon.

2.1.1 Estimating Biomass from Timber Volume

The biomass expansion factor (BEF) is defined as the ratio of all standing aboveground biomass (AGB) to growing stock volume (Mg/m^3) (Fang et al. 2001). It has been developed to estimate aboveground biomass when timber volumes within diameter classes are reported (Brown 2002). Especially for estimating large areas within developing countries that lack detailed information about forest biomass, the BEF is a practical estimate of AGB.

The process of estimating carbon stock by BEF can be simply to use the regression relationships between merchantable plot tree volumes, their annual increments, and estimates of non merchantable volumes, to above ground standing biomass. Estimations of total aboveground biomass from tree volume data is then subsequently expanded to an area based on uniformity of site, stocking and age-class distribution (see Fig. 7.1) (Wulder et al. 2008). BEF varies by different stand density-related factors, such as forest age, site class, stand density, and other biotic and abiotic factors (Brown et al. 1999; Fang et al. 2001). The largest differences are regional and by forest type (see Fig. 7.2) (Brown 2002).

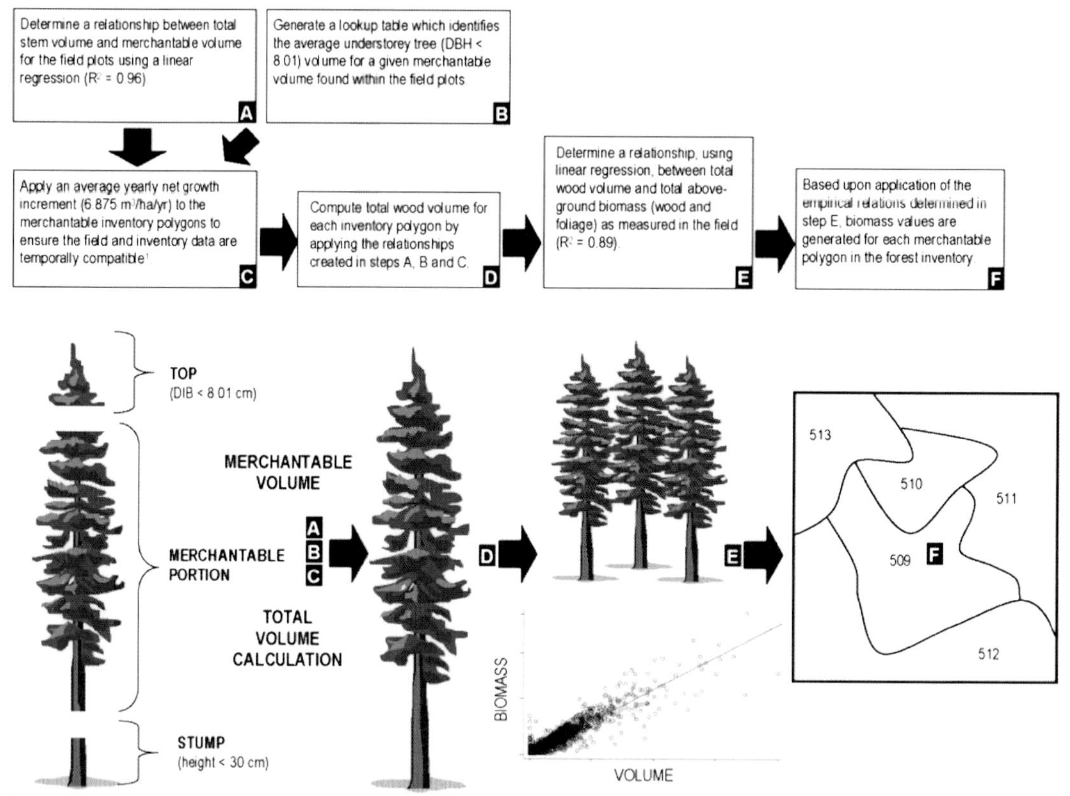

Fig. 7.1 An overview of the process used to estimated biomass from the forest inventory data (*Source*: From Wulder et al. (2008). Reprinted with permission)

2.1.2 Estimating Biomass from Tree Diameter

Compared to the BEF method, allometric equations can provide more precise estimates of aboveground biomass. In the biological sciences, the study of the relationship between the size and shape of organisms is called allometry (Niklas 1994). In the context of biomass estimation, allometry refers to the relationship between individual tree diameters (sometimes with heights) and aboveground biomass for specific species, groups of species, or growth form (Jenkins et al. 2003; Zianis and Mencuccini 2004).

In order to derive an accurate allometric equation for any forest type, an adequate sample of tree sizes and species must be taken. If such data are available at the appropriate scale, the allometric approach can be very accurate. Generally, species groups such as tropical wet-evergreen hardwoods, temperate eastern U.S. hardwoods, pines, and spruces produce highly significant correlations of greater than 0.98 for regressions between diameter at breast height (dbh) and biomass per tree (Brown 1997; Schroeder et al. 1997; Brown et al. 1999; Brown 2002). A study on lianas in Amazon semi-evergreen rain forest showed that a combination of diameter and length is also significantly correlated (R^2=0.91) with biomass (Gehring et al. 2004). This approach is limited, however, by the lack of allometric data for many forest types and regions.

2.1.3 Improvement for Field Based Methods

Estimates of carbon flux from forest inventory measurements require availability of historical

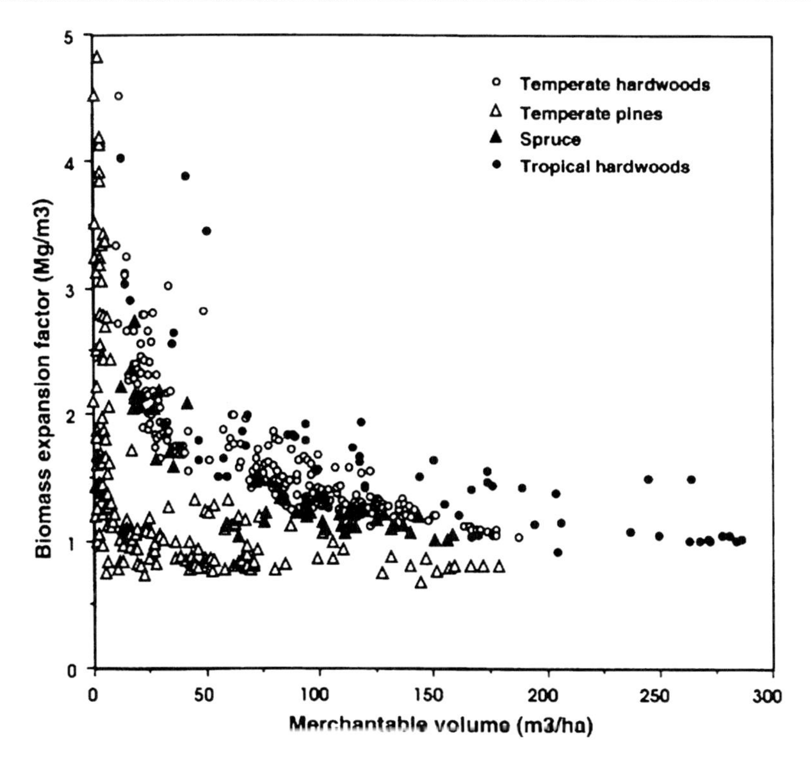

Fig. 7.2 Relationship between BEF for temperate hardwoods, pines and spruce, and tropical hardwoods (*Source*: From Brown (2002). Reprinted with permission)

data at the regional scale. All developed countries conduct regular national inventories (FAO 2000). For the 137 developing countries, 22 have repeated inventories, 54 have a single inventory, 33 have partial forest inventories, and 28 countries have no inventory (Holmgren and Persson 2002). In the U.S., a vast network of permanent sample plots makes up the Forest Inventory and Analysis (FIA) and Forest Health Monitoring (FHM) programs. The FIA program, which has been operating for about 70 years, periodically measures all plots on a state-by-state basis every 5–14 years (Brown 2002; Smith et al. 2002).

Inventory data have several deficiencies that can bring uncertainty, however. First, inventories tend to be conducted in forests that are considered to have commercial value, and the forests that many people depend upon for other values (such as water, recreation, open space, or subsistence) may not be included. Many degraded or semi-deforested open lands, or those regions that are now growing back (secondary forest) are under-sampled or not measured. Often only tree species that have commercial value at the time of the inventory are counted (Brown 1997). This counting bias can bring systematic inaccuracy to the estimation of carbon. Additionally, the assumption that small trees (about 10 cm diameter or less) contribute little to the total forest biomass is not robust according to Schroeder et al. (1997). They concluded that for young hardwood stands in the eastern USA with aboveground biomass less than 50 Mg/ha, trees with dbh of 10 cm or less contain as much as 75% of the biomass of trees with dbh greater than 10 cm.

The cost is high to sample a sufficient number of trees representing a range of size and species in order to generate local allometric equations (Brown 2002). Many developing countries lack funding, staff, and expertise to acquire the data. Additionally, a small number of large diameter trees (>100 cm) and a large number of small diameter trees (<10 cm), which are important to the total biomass, are often missed in a sample for allometry measurements (Brown 1997).

To improve the accuracy and precision of measuring aboveground live tree biomass by inventory methods, Brown (2002) has suggested the following:

- Destructively harvest large diameter trees to establish allometry equations, because they are under-sampled and they have a significant influence on the regression relationship between diameter and biomass.
- Precisely measure small trees (10 cm diameter or less) for temperate hardwood forests (i.e. second growth) or other forest types in which small diameter trees may be significantly underestimated.
- Including height in regression equations can slightly improve the precision, but given the difficulty of measurement, it is not feasible or worth the effort for large areas. The use of remote sensing data can complement tree height data for large-areas, and can improve the precision of allometric regression equations.
- Periodically re-visit the field sites from which the inventory data are derived and modify the allometric equations that may have changed with time and forest growth.

3 Remote Sensing Methods

Inventory data have been used as the basic approach to estimating carbon stock in existing and historical forests worldwide. In recent years, better models and the establishment of more plots have improved accuracy and precision (Smith and Heath 2004). However, sampling intervals are long (5–14 years), so temporal resolution of changes in carbon storage is limited. In addition, gathering inventory data is highly dependent on the capacity of local people to conduct the survey. Assuming that land use change accounts for a significant part of carbon emissions, and that the rate of deforestation is high, remote sensing would appear to be a more suitable method, particularly for use in large and remote forest regions and in developing countries where training on forest inventory procedures is poor.

The remote sensing method monitors forests at different temporal, spatial and spectral resolutions (Patenaude et al. 2005). Several applications of remote sensing for mapping land covers are available and can be categorized as passive (optical) or active (radar).

Optical, or passive, remote sensing technologies include aerial photographs of various kinds (infrared, color, black and white), Normalized Difference Vegetation Index (NDVI) images that are derived from an advanced very high resolution radiometer (AVHRR) sensor, and images from Landsat Thematic Mapper (TM) false color composites and its associates that are at a low resolution (Fig. 7.3). Active remote sensing technologies include radar and LiDar derived images. These can measure structure, detect objects below canopy, and can depict canopy height and stratification (CHM) (Fig. 7.3).

3.1 Optical Remote Sensing

Optical remote sensing captures solar energy reflected by the forest canopy in the visible, near, and middle infrared portion (0.4–2.5 mm) (Patenaude et al. 2005). Optical remote sensing is also called passive remote sensing and can be differentiated from Radar and LiDar methods, which actively emit radiation and then detect the reflectance. The ground sampling distance (GSD) defines the spatial resolution level of the optical remote sensing methods. It can be classified based on degree of resolution into fine, medium, and coarse spatial scales.

3.1.1 Fine Spatial-Resolution Data

Fine spatial-resolution data has a GSD less than 10 m. Aerial photographs (GSD 1.00 m), IKONOS (GSD 0.83 m), and QuickBird (GSD 0.61 m) images are the commonly available fine spatial-resolution data (Lu 2006).

Aerial photographs were widely used in forest surveys starting in the late 1940s, primarily for forest type delineation and stratification, and timber volume estimation (Lu 2006). Since the 1990s, space-borne high spatial-resolution sat-

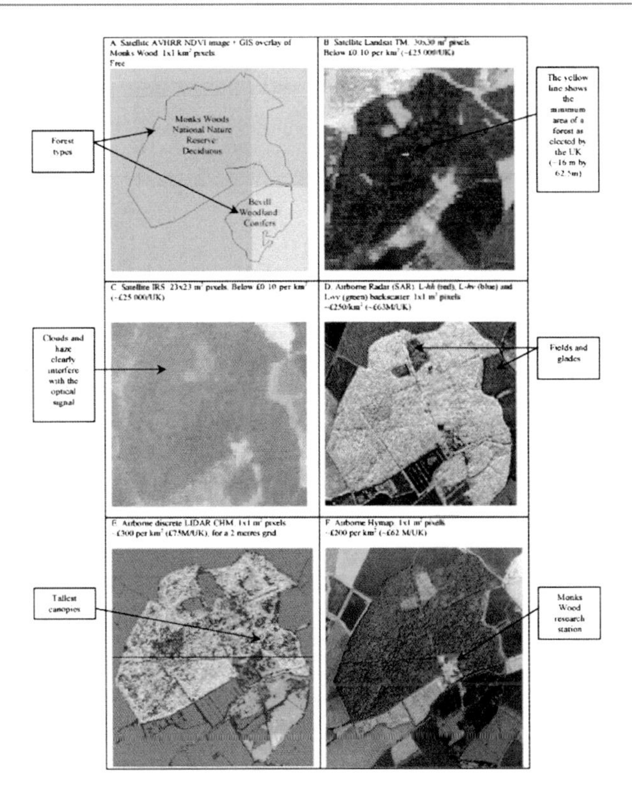

Fig. 7.3 Example of different remote sensing methods on the same site (*Source*: From Patenaude et al. (2005). Reprinted with permission)

ellite images can also be used in biomass estimation as well as in detecting biophysical parameters (height, classification, stand structure). Such images can be used to detect the structural diversity of a forest at a small scale. For example, the IKONOS system, started in September 1999, collects panchromatic data, with a spectral range of 450–900 nm, and four GSD channels of 4 m resolution multi-spectral data (Wulder et al. 2004). Thenkabail et al. (2004b) used multi-date wet and dry season IKONOS images to calculate carbon stock levels of the West African oil palm plantations. It was also used by Thenkabail (2003) to detect small differences in floristic association in the Central African rainforest.

Fine spatial-resolution remote sensing data has the advantage in providing details of a specific area. However, disadvantages include the small area of coverage, preponderance of shadows, and acquisition expense. Therefore, it should mainly be used in small scale projects that are focused on measuring stand-level characteristics (Thenkabail et al. 2004b). Such fine scale resolution can also be useful for the development of reference data for validation or accuracy assessments of medium and coarse scale remote sensing measurements (Lu 2006).

3.1.2 Medium Spatial-Resolution Data

Medium spatial-resolution remote sensing images (10–100 m) are the most suitable for regional level aboveground biomass estimation because of better data availability (spatial and temporal), and the lower cost of acquisition and storage. Since spatial resolution is still good enough to compare with inventory measurements, this approach is widely used for aboveground biomass estimation for various forests (Reese et al. 2002; Tomppo et al. 2002; Foody et al. 2003; Zheng et al. 2004; Muukkonen and Heiskanen 2005, 2007). Landsat Thematic Mapper (TM), Enhanced Thematic Mapper Plus

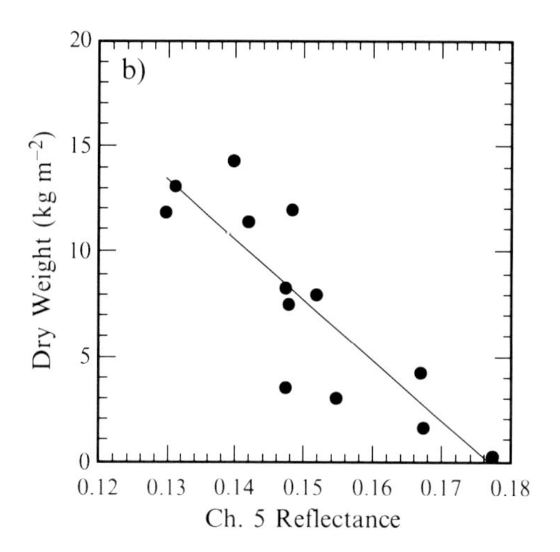

Fig. 7.4 Aboveground biomass of secondary forest versus TM channel 5 reflectance (*Source*: From Steininger (2000). Reprinted with permission)

(ETM+), Multi-Spectral Sensor (MSS), ASTER, AVIRS, and SPOT High Resolution Visible (HRV) are all multispectral sensors commonly used for mapping forest structure and estimating biomass (Muukkonen and Heiskanen 2005).

Landsat has been the most important data source for mapping and remote sensing interpretation. For more than 30 years it has provided appropriate spatial and spectral resolution to detect and characterize forests at an affordable cost (Cohen and Goward 2004). Since 1972, the Landsat program has launched seven satellites. With each launch, sensors have been designed for better spatial and spectral resolution. Landsats 1, 2, 3, and 4 have been decommissioned because better satellites are now available or they had reached the end of their working life. However, due to the failure of Landsat 6 and a defective scan line on Landsat 7, Landsat 5 has been kept running for 24 years and is still widely used for research. The earliest sensor (four-band multispectral scanner sensor – MSS) was deployed on Landsat satellites 1–5. But because of the lower spatial resolution (80 m), and fewer spectral bands of MSS, the TM instrument, and then later the ETM + instrument, which have seven spectral bands and 30 m spatial resolution, are now the primary images used in aboveground biomass estimation (Fig. 7.4).

The Advanced Spaceborne Thermal Emission and Reflection Radiometer (ASTER) was launched in 1999, with three spectral bands in the visible near-infrared region (VNIR), six bands in the shortwave infrared region (SWIR), and five bands in the thermal infrared region (TIR), with 15-, 30-, and 90-m spatial resolution, respectively (Muukkonen and Heiskanen 2005). In spite of its modernity, it is argued that ASTER has relatively narrow SWIR bands 5–8 which are primarily designed for soil and mineral detection, so it is not particularly sensitive to detecting differences among forests (Yamaguchi et al. 1998).

3.1.3 Coarse Spatial-Resolution Data

Overall, coarse spatial-resolution data (greater than 100 m) are most effective at large national or continental scales. However, use at such scales is limited because of the frequent occurrence of mixed-landuse pixels (due to the large pixel size), and differences between scale and resolution of forest inventory measurements and image GSD (Lu 2006). However, the use of fine and medium spatial-resolution data along with coarse spatial-resolution can help estimate aboveground biomass and improve accuracy (Dong et al. 2003; Muukkonen and Heiskanen 2007; Zheng et al. 2007a).

Commonly used coarse spatial-resolution data include NOAA Advanced Very High Resolution Radiometer (AVHRR), Moderate Resolution Imaging Spectroradiometer (MODIS), and SPOT VEGETATION (Table 7.2) (Lu 2006). The AVHRR has collected over 30 years of data and has often been used to assess large areas of forest cover at the scale of a continent (Iverson et al. 1994). For example, for a 1.42 billion ha region of temperate and boreal forest, Dong et al. (2003) used regression analysis between an NDVI dataset, developed from AVHRR at 8 × 8 km resolution, over an 18 year period (1981–1999), and timber volumes from forest inventories to estimate aboveground biomass.

The recent SPOT VEGETATION (VGT) sensor provides imagery with a swath width of 2,250 km and GSD at 1,165 m. Besides the four spectral bands of the SPOT multi-spectral sensor, the Vegetation Instrument has an extra band

Table 7.2 Selected examples of biomass estimation using optical remote sensing data

Datasets	Study area	Techniques	References
IKONOS	West Africa	Empirical regression	Thenkabail et al. (2004b)
Landsat 5	Mauaus, Brazil	Liner and exponential regressions	Steininger (2000)
Landsat 5	Para state and Rondonia state, Brazil	Multiple regression analysis	Lu and Batistella (2005)
SPOT VEGETATION	Canada	Multiple regression and artificial neural network	Fraser and Li (2002)
MODIS, ASTER	Finland	Regression models	Muukkonen and Heiskanen (2007)
Aerial photographs	Suonenjoki, Finland	K nearest-neighbor method and K most similar neighbors	Anttila (2002)
Landsat 5	Sweden	K nearest-neighbor method	Fazakas et al. 1999; Reese et al. (2002)
Landsat TM and IRS-1 C WiFS	Finland and Sweden	K nearest-neighbor method and nonlinear regression	Tomppo et al. (2002)

Source: Modified from Lu (2006)

(0.43–0.47 μm) that is used for the first band (blue) and a 1.65-μm short-wave infrared (SWIR) channel. Fraser and Li (2002) tested the relationship between several values and indexes from VGT and aboveground biomass. The short-wave-based vegetation index (SWVI), in which the SWIR is substituted for the red channels from VGT, has been found to have weak correlation ($R^2 = 0.25$). The other values (red, NIR, SWIR, and NDVI) have either no relation or poor relation with aboveground biomass, and therefore are not useful.

MODIS is a 36-band spectrometer providing a global dataset every 1–2 days with a 16-day repeat cycle. Bands 1 and 2 have GSD at 250 m, bands 3–7 have GSD at 500 m, and bands 8–36 have GSD at 1,000 m. Zheng et al. (2007a) used Landsat 7 ETM+data and field observations to develop an empirical model. After calibration with different sensors, MODIS data were used for model applications at a regional scale. Using a similar approach, Muukkonen and Heiskanen (2007) used ASTER (15×15 m) data to develop regression models with stand forest inventory data volume. MODIS bands 1 and 2 (250×250 m) data were used to estimate stand volume.

3.1.4 Interpretation of Optical Remote Sensing Data

Specific interpretation procedures have been developed to extract information from images. Generally, the procedures are divided into two classes: the traditional approach using parametric methods such as regression models (Holmgren et al. 1997; Steininger 2000), and nonparametric methods such as the k-nearest-neighbor method (k-NN) (Fazakas et al. 1999; Reese et al. 2002) (Table 7.2).

Since coarse spatial resolution data are difficult to couple with forest inventory measurements, researchers usually use fine or medium spatial scale resolution data to link forest inventory data to coarse spatial resolution regional data (Muukkonen and Heiskanen 2007). Regression models differ in variables and equations. Spectral signatures, image textures, and vegetation indexes are among the variables derived from imagery. For example, Lu and Batistella (2005) found that in the Amazon, successional forest is more likely to correlate with a spectral signature, and mature forest is more likely to correlate with texture. Zheng et al. (2007b) showed that leaf area index (LAI), and the normalized difference vegetation index (NDVI) are significant predictors for Chinese fir aboveground biomass, while LAI and stand age can predict 94% of the variation of aboveground biomass.

Regression models include linear, non-linear, multi-, and neural networks. Neural networks in forestry mainly deal with incomplete, disturbed, and noisy datasets (Hanewinkel 2005). The neural network model was used by Steininger (2000)

to develop predictive models of biomass (for example, see Fig. 7.4). Foody et al. (2003) used multiple regression and neutral networks to estimate tropical forest biomass and observed a significant relationship between predicted biomass and that measured from the forest inventories. Other researchers either use ASTER data to estimate aboveground biomass, applying non-linear regression analysis and a neural network approach (Muukkonen and Heiskanen 2005), or fractional textures and semivariance analysis of image fractions integrated with conventional images to establish stepwise multiple regression models to predict forest structure and health (Levesque and King 2003).

Recently, nonparametric methods such as the k-nearest-neighbor method (k-NN) and k most similar neighbor method (k-MSN) have been used to interpret images. In these methods, the prediction is no longer dependent upon the regression of the whole sample space, but on either the weighted mean of neighbors or the distance-weighted mean of most similar neighbors. The accuracy of AGB estimation was tested using the k-MSN method and was deemed acceptable (Anttila 2002). In Sweden, Landsat data was successfully combined with the k-NN method to estimate AGB (Fazakas et al. 1999; Reese et al. 2002).

3.2 Active Remote Sensing: Radar and LiDAR

Unlike optical remote sensing methods using aerial photographs and satellite images that capture the reflectance of solar radiation, Radar and LiDAR systems use their own electromagnetic radiation source independent of solar radiation. Moreover, the microwave portion of the radar wavelength can penetrate precipitation and cloud cover, and avoid shade/shadow effects from the sun (Ranson and Sun 1994; Patenaude et al. 2005). In addition LiDAR can capture detailed stand structure and height, something difficult to achieve by the optical remote sensing method (Table 7.3).

3.2.1 Radar Data

Radio Detection and Ranging (RADAR) systems work by virtue of radiating microwave pulses to subjects and then measuring the returned echo's amplitude (backscatter amplitude) and orientation (polarization). The wavelength emitted in radar is between approximately 1 mm and 1 m. In this range, the C (3.75–7.5 cm), L (15–30 cm), and P (30–100 cm) bands are responsive, respectively, to small structural components (e.g. leaves), large components (e.g. branches), and larger components (e.g. trunks) (Patenaude et al. 2005). Unlike optical remote sensing that detects differences in reflectance of various vegetation and mineral surfaces, radar remotely detects the surface roughness, geometry, and water content of biomass.

There are two types of imaging radar, the earlier Side-Looking Airborne Radar (SLAR) and the later Synthetic Aperture Radar (SAR) (Fig. 7.5). SAR could be air-, space-shuttle-, or satellite-born and is widely used in aboveground biomass estimation. The resolution of SAR is defined in two dimensions: range and azimuth. Unlike the old SLAR radar system, whose azimuth resolution is constrained by antenna length, SAR uses signal processing to increase azimuth resolution by hundreds of times (Canada Centre for Remote Sensing 2008). For transmitting and receiving radiation, the orientation of the electromagnetic wave (polarization) is configured as V for vertical and H for horizontal (e.g. HH is horizontally transmitted and also horizontally received waves, while VH is vertical transmitted and horizontally received radiation). Besides backscatter of amplification in different bands, polarization is also an important characteristic of predicting aboveground biomass. The horizontal and vertical distribution of the target affects the backscattered amplification of the signal (Patenaude et al. 2005).

The interpretations of radar data mainly use regression on different variables. Properly polarized L-band SAR data are among the variables commonly used (Luckman et al. 1998; Castel et al. 2002; Sun et al. 2002).

The L-band HV (LHV) channel of the Shuttle Imaging Radar (SIR-C) data has been shown to be a strong predictor of aboveground biomass (Harrell et al. 1997; Sun et al. 2002). Likewise, the L-band HH SAR channel of the Japanese Earth Resources Satellite 1 (JERS-1) has shown a significant relationship between the backscatter coefficient of JERS-1/SAR data and the stand

Table 7.3 Selected examples of biomass estimation using radar and LiDAR data

Datasets	Study area	Techniques	References
SIR-C	South-eastern USA	Multiple regression analysis	Harrell et al. (1997)
SIR-C	Siberia	Adapted theoretical regression model	Sun et al. (2002)
JERS-1 SAR L band	Ta'pajos, Para' state and Manaus, Amazonas state, Brazil	Forest backscatter regression model	Luckman et al. (1998)
JERS-1 SAR L-band	New South Wales, Australia	Linear regression analysis	Austin et al. (2003)
Airborne laser	Costa Rica	Linear regression, canopy height models	Nelson et al. (1997)
Large-footprint LiDAR	North-east Costa Rica	Multiple regression analysis	Drake et al. (2003)
Small-footprint LiDAR	Piedmont physiographic province of Virginia, south-eastern USA	Measure crown diameter using LiDAR, then estimate biomass using regression analysis	Popescu et al. (2003)

Source: Modified from Lu (2006)

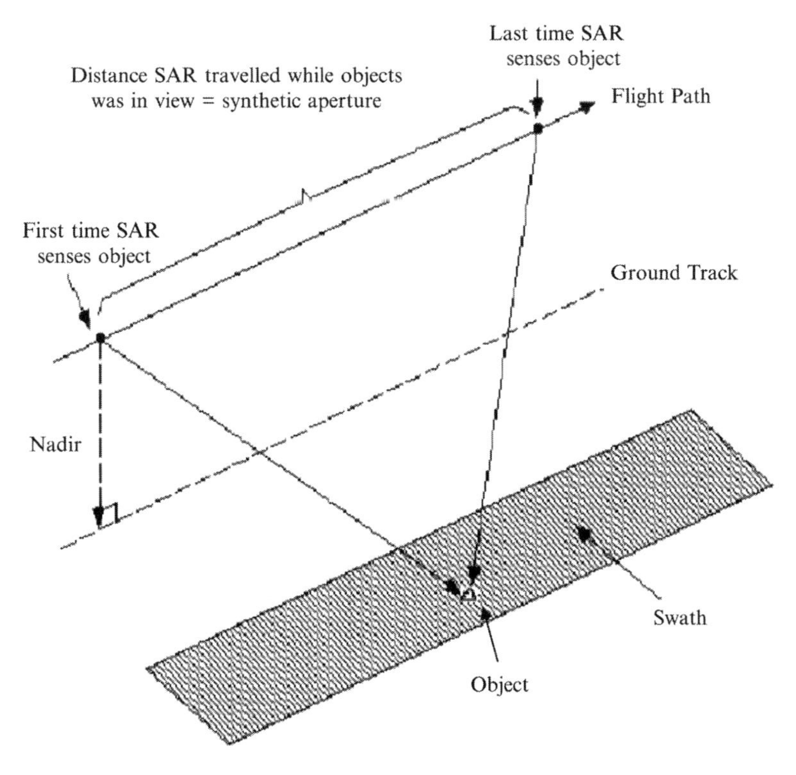

Fig. 7.5 Concept of synthetic aperture (*Source*: From Canada Centre for Remote Sensing (2008). Reprinted with permission from the Government of Canada)

biomass of a pine plantation (Castel et al. 2002). Although low correlations were found between SAR C-band backscatter and aboveground biomass, the addition of C-band HV or HH data can significantly improve estimations (Lu 2006).

Aboveground biomass estimation by radar data can achieve good accuracy in low and medium density forests, but the relationship between radar backscatter and aboveground biomass weakens when the forest becomes too dense, reaching saturation density. Saturation density is correlated with the wavelength of band, polarization, and characteristics of the vegetation canopy and ground conditions (Lu 2006). For example,

Ranson and Sun (1994) found that L, P-band HV data appeared to saturate at 150 tons per hectare in boreal forest, while Luckman et al. (1998) found that the L-band data saturated at 60 tons per hectare in rainforest. This variability can be attributed mainly to density saturation problems rather than real differences in forest type, and emphasizes the importance of being careful when comparing and using biomass estimates derived from different band data and technologies.

3.2.2 LiDAR Data

Laser altimetry, or Light Detection and Ranging (LiDAR), is an active remote sensing method, analogous to radar, but it uses laser light instead of microwaves. The detection principle of LiDAR is similar to that of radar but is different in radiation frequency emitted. A pulse is generated with wavelengths in the visible or near infrared spectrum (900–1,064 nm), and the travel time from the sensor to the target on the ground and back is measured. Unlike optical and radar remote sensing methods, the LiDAR system provides direct information, such as the vertical structure of targets. LiDAR is therefore not actually producing images, so the data need to be converted to aboveground biomass estimations by more sophisticated models. LiDAR measurements are usually taken airborne by aircraft or helicopter (Patenaude et al. 2005).

There are two types of LiDAR systems that are distinguished by the information collected from the return signal: (i) discrete-return devices (DRD); and (ii) waveform recording devices (WRD). DRD can measure one (single-return systems) or a few (multiple-return systems) heights by identifying major peaks. WRD records the time-varying intensity of the returned energy from each laser pulse (Lefsky et al. 2002) (Fig. 7.6). The DRD system has a high spatial resolution (5–90 cm) but provides limited information in stand vertical structure, while the WRD system has a low spatial resolution (10–25 m) but provides enhanced information about the vertical structure of forest.

Similarly to radar, LiDAR data are mainly used in regression models to estimate aboveground biomass. For example, studies by Nelson et al. (1997); Lefsky et al. (2002); Drake et al. (2003) all used regression analyses to estimate aboveground biomass from mean canopy height. Wulder and Seemann (2003) tested the feasibility of using a regression model to spatially extend a LiDAR survey from a sample to a larger area with Landsat TM data. The height measured by LiDAR and correlated with Landsat TM are expected to complement the forest inventory data. At this stage, the regression models still need to be further developed (Wulder and Seemann 2003).

3.3 Improvements for Remote Sensing Methods

Remote sensing is a novel revolutionary technology for aboveground biomass estimation, with unprecedented capability of spatial, temporal, and spectral resolution and potential coverage of remote forest areas. If not restrained by cost, the data can be gathered from anywhere without political or regional restrictions, which overcomes a significant short coming of forest inventory methods for estimating aboveground biomass. Remote sensing data can also complement the conventional inventory data to increase the accuracy of models. However, to improve the utilization of remote sensing data in aboveground biomass estimation, there are several hurdles that need to be overcome.

Patenaude et al. (2005) suggest that the main potential of remote sensing is as a validation tool, rather than as a tool for producing the actual estimate of aboveground biomass, because field measurements are still needed (Fuchs et al. 2009). There are studies that have estimated aboveground biomass and compared results between inventory data and remote sensing data. In both cases MODIS and Landsat TM overestimate aboveground biomass compared with U.S. Forest Inventory Analysis (FIA) (Zheng et al. 2007a; Wulder et al. 2008).

Many direct remote sensing estimations of aboveground biomass still cannot meet an acceptable accuracy without forest inventories. This could potentially be solved with better models, indexes, and instrumentation. An example of this would be further research on the study of effects

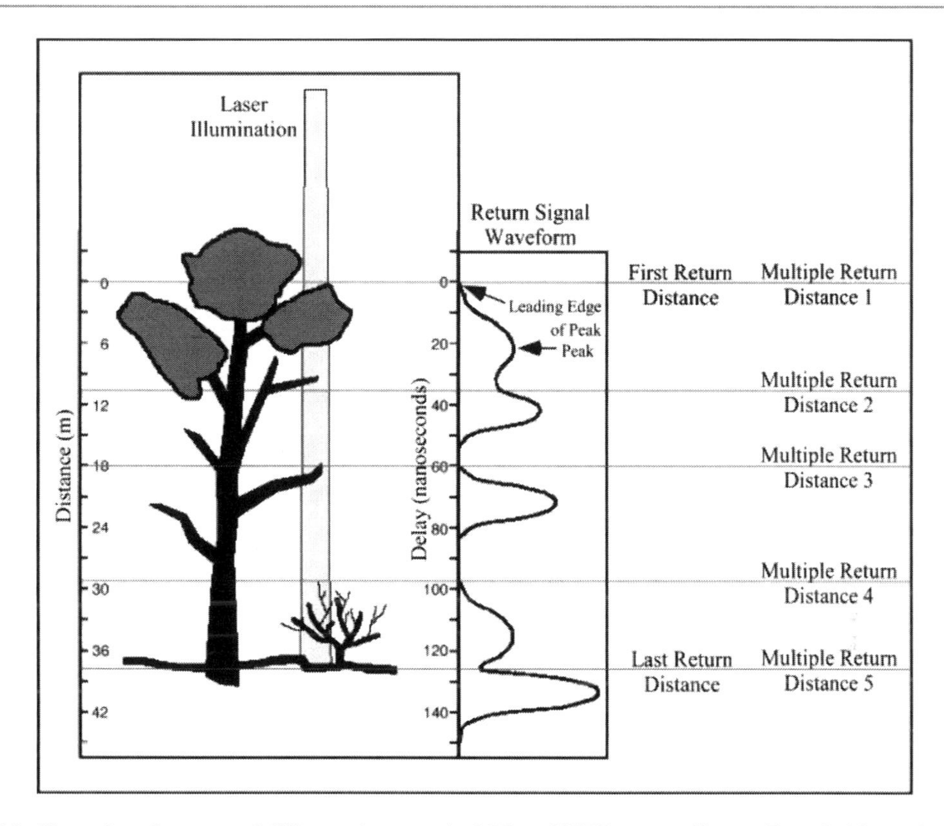

Fig. 7.6 Illustration of conceptual difference between the DRD and WRD system (*Source*: From Lefsky et al. (2002). Reprinted with permission)

of features such as mountains, slopes, and aspects. Such features are a major source of error, and can affect vegetation reflectance, resulting in spurious relationships between aboveground biomass and reflectance. Better estimates of aboveground biomass are always made where land surfaces are flatter.

In the past, remote sensing technology has been dominated by developed nations such as the United Sates. However, this dependence raises the cost and risk of obtaining data worldwide and provides an over-reliance on satellites from a single country's remote sensing program. For example, reliance on the U.S. program has resulted in missed opportunities in data gathering with the failure of Landsat 6, defects in Landsat 7, the delay of LDCM, and the cancellation of vegetation canopy LiDAR. Remote sensing technology in more countries or consortiums is needed to provide the international community with more options in satellite imagery and radar/LiDAR data.

4 Eddy Covariance

4.1 Basic Theory and Advantages

Since the late 1990s, the eddy covariance method has been developed in order to directly measure the uptake and release of CO_2 (CO_2 flux[2]). This method samples three-dimensional wind speed and CO_2 concentration over a forest canopy at a high frequency (around 10 ~20 Hz), and determines the CO_2 flux by the covariance of the vertical wind velocity and CO_2 concentration (Moore 1986; Gash and Culf 1996; Bosveld and Beljaars 2001).

The relationship between (i) CO_2 flux and (ii) the covariance of vertical wind velocity and CO_2 concentration is derived by putting a hypothetical

[2]Flux is the rate of flow of energy or particles across a given surface.

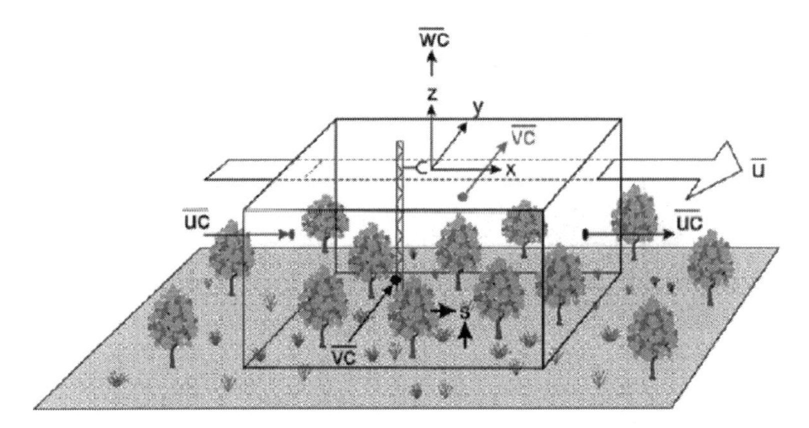

Fig. 7.7 Hypothetic Cartesian control volume over a homogeneous terrain. *V* vertical velocity, *w* horizontal velocity, *u* velocity vector, *c* a constant (*Source*: From Pinnigan et al. (2003). Reprinted with permission)

control volume (box) over a homogeneous canopy (Fig. 7.7). On the upper surface of the "box", three-dimensional wind speeds are recorded in a coordinate system that has the x axis aligned to the averaged wind direction. This assumes that one-dimensional flow (zero mean lateral velocity, zero mean vertical velocity) and stationary flow (no accumulation of CO_2 within the "box") is obtained over a sufficient averaging period (30 min to 1 h). The surface exchange of CO_2 should then be equal to CO_2 exchange at the upper surface of the "box", based on the mass balance within the "box" (Finnigan et al. 2003). By measuring the vertical velocity of CO_2 flow at the height of the upper surface of the "box", the eddy covariance method directly measures CO_2 fluxes over the forest canopy (Lee 2004; Baldocchi and Meyers 1998).

This method is favored because of its high accuracy and appropriate spatial scale. CO_2 flux is usually underestimated by less than 5% during daytime and less than 12% at night. A higher accuracy can be obtained by sampling at a finer temporal and spatial resolution. For example, given normal forest canopy roughness, flat topography, and calm meteorological conditions, an anemometer positioned at 30 m with a sampling interval that is averaged every 30–60 min should provide an accurate estimate of CO_2 flux that covers an area from a hundred meters to several kilometers (Berger et al. 2001).

Eddy covariance measurements have been continuously made at a number of sites for over 10 years (Berger et al. 2001; Haszpra et al. 2005; Su et al. 2008). New observation sites, updated models, and remote sensing data enable the eddy covariance methods to continually refine estimates of CO_2 flux from regional to continental scales (Owen et al. 2007; Sasai et al. 2007; Yang et al. 2007; Yuan et al. 2007).

4.2 Systematic Biases

Since the eddy covariance method is derived from assumptions such as homogeneous canopy, steady environmental conditions, and stationary flow, it suffers from many systematic biases that need to be accounted for.

4.2.1 Energy Imbalance

For eddy covariance measurements, an imbalance exists of about 20% between turbulent energy fluxes (sensible and latent heat that is measured by the eddy covariance system) and available energy (net radiation minus stored energy that are measured separately with radiation sensors and soil heat flux plates) (Wilson et al. 2002; Han et al. 2003; Li et al. 2005). The imbalance can be caused for three reasons: (i) using 30 min as an averaging period in flux estimation filters out low frequency turbulence whose contribution to the flux model is missed (Foken et al. 2006); (ii) flux measurements taken at different heights or across varying topographies represent CO_2 exchange from different source areas, with the result that the source area may not match the representative area separately measured for available energy (Schmid 1997); and (iii) the flux may not be fully detected due to advection or air drainage (Massman and Lee 2002; Hammerle et al. 2007).

Although the CO_2 flux itself is not adversely affected by an energy imbalance, closing the energy budget is important for cross-site comparisons and a better understanding of underestimation and error in CO_2 flux measurement (Wilson et al. 2002).

4.2.2 Nighttime Flux

The boundary layer at nighttime is characterized by low wind speed, thermal stratification, and intermittent turbulence. These characteristics always cause dramatic bias in CO_2 flux estimations (Aubinet et al. 2005; Velasco et al. 2005; Fisher et al. 2007). Vertical and horizontal advection are not negligible, but the correction for advection is usually site-specific (Feigenwinter et al. 2008). Due to thermal stratification, CO_2 concentration builds up within the air layer below the measurement heights, so the storage term can also be significant. But the correction of the storage term is controversial and site-dependent, because CO_2 stored at night might be released in the morning when advection can be negated (Aubinet et al. 2002).

4.2.3 Topography

Over sloping terrain, mathematical rotations of the wind coordinate system are used to meet the basic assumptions of one dimensional flow, but advection is unavoidable (Massman and Lee 2002) and different rotation methods introduce different systematic errors to the estimation (Finnigan 2004). Besides, CO_2 uptake measured at one point may be transported by drainage flows and emitted somewhere else (Sun et al. 1998).

4.3 Data Gaps and Scaling up to Regions and Continents

In addition to the three systematic problems that can lead to bias in estimates, sampling intervals can be interrupted by weather (e.g. heavy rain) and other unforeseen problems such as lightning strikes. A model based on a semi-parametric relationship between net CO_2 flux and environmental conditions, such as light and temperature, can be used to supplement and interpolate between such data gaps (Stauch and Jarvis 2006). Data gaps from eddy covariance measurement exist not only with sampling period (time) but also over area (space). A single eddy covariance measurement can only represent flux over hundreds meters. Multiple observation sites and sophisticated models are required to develop an estimation of regional and global CO_2 budgets.

Since 1998, FLUXNET, a global-scale network for eddy covariance flux measurements, was started to encourage collaboration among flux measurement sites around the globe (Baldocchi et al. 2001) (Fig. 7.8). It supports calibration and comparison of flux measurements among sites and supports collection of vegetation, soil, hydrologic, and meteorological data for each site. Using this network, FLUXNET provides a comprehensive dataset for expanding and scaling up CO_2 flux estimations from a single site to global and regional estimates. However, although the number of FLUXNET tower sites has expanded from around 100 to over 400 in the last decade, most of the sites are located in temperate forest, grasslands, and shrubland, while measurement over some vegetation types such as tropical ever-wet and semi-evergreen rainforest, tropical dry deciduous forest, temperate rain forest, desert, urban areas, and tundra are noticeably under-represented.

Scaling models up to extend flux measurements from single sites to a larger scale involves measurements of two main processes: canopy photosynthesis and ecosystem respiration (Running et al. 1999; Soegaard et al. 2000; Wang et al. 2007b; Baldocchi 2008). Models can be divided into two categories: (i) empirical models which are based on the relationship between CO_2 flux and plant eco-physiological parameters (e.g. photosynthetic light response curves); and (ii) physiological growth models based on stand dynamics (Owen et al. 2007). Both categories of models can be parameterized by eddy covariance measurements, but the parameters can change considerably among different models and different ecosystems.

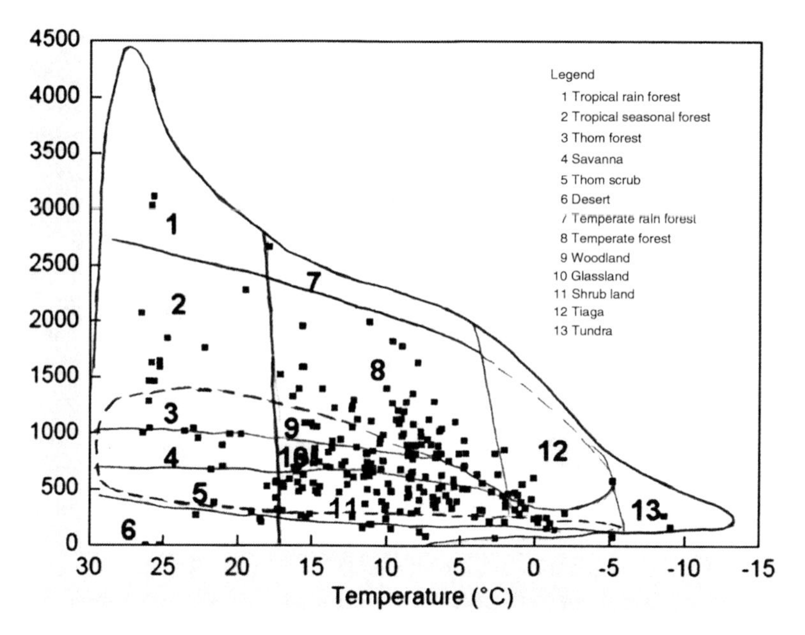

Fig. 7.8 FLUXNET sites in the climate space (*Source*: Site information is from http://daac.ornl.gov/FLUXNET/, biome lines are drawn from Barnes et al. (1998))

Strong relationships between CO_2 uptake and leaf area index have been utilized in the European Arctic region to calculate spatial distribution of Net Ecosystem Exchange (CO_2 flux) based on Landsat TM satellite data (Soegaard et al. 2000). Still others have proposed that net ecosystem exchange may be characterized mainly by non-climatic conditions (e.g. species, age, and site history) (Ball et al. 2007; Luyssaert et al. 2007). In a temperate moist broadleaf and coniferous forest in North Carolina, USA, parameters such as leaf nitrogen concentration and stomatal conductance were measured as inputs to a physiologically based canopy model to estimate gross primary productivity (Luo et al. 2001). Additionally, at observation sites located over heterogeneous landscapes, a footprint model has been used to determine the source area of eddy covariance measurement (Schmid 1997; Soegaard et al. 2000; Chen et al. 2007).

In summary, eddy covariance is a promising method for both CO_2 flux measurements at a regional scale and CO_2 budget estimations at global scales. But more research needs to be conducted to close the energy budget and eliminate biases caused by night time stratification and complex topography. In addition, more sites are needed over various vegetation types that can be calibrated to other sites.

5 Inverse Method

Atmospheric CO_2 concentration can be estimated from sink and source measurements of carbon (forest inventories, flux measurements) combined with transportation models (that model gas movement) using meteorological information. It can also be measured directly. The inverse method has been developed to indirectly calculate sinks and sources of CO_2 from the measured concentration by using the Bayesian inversion technique (Gurney et al. 2002; Rodenbeck et al. 2003). This technique backs out carbon sources and sinks of trace gases including CO_2 through the use of three-dimensional transport models (Gurney et al. 2002) – hence the so-called inverse method. Transportation models and atmospheric CO_2 concentration data therefore determine the accuracy of the inverse method (Patra et al. 2006). Sixteen different transportation models, along with a variety of atmospheric CO_2 datasets, have been used to test, calibrate and estimate regional to

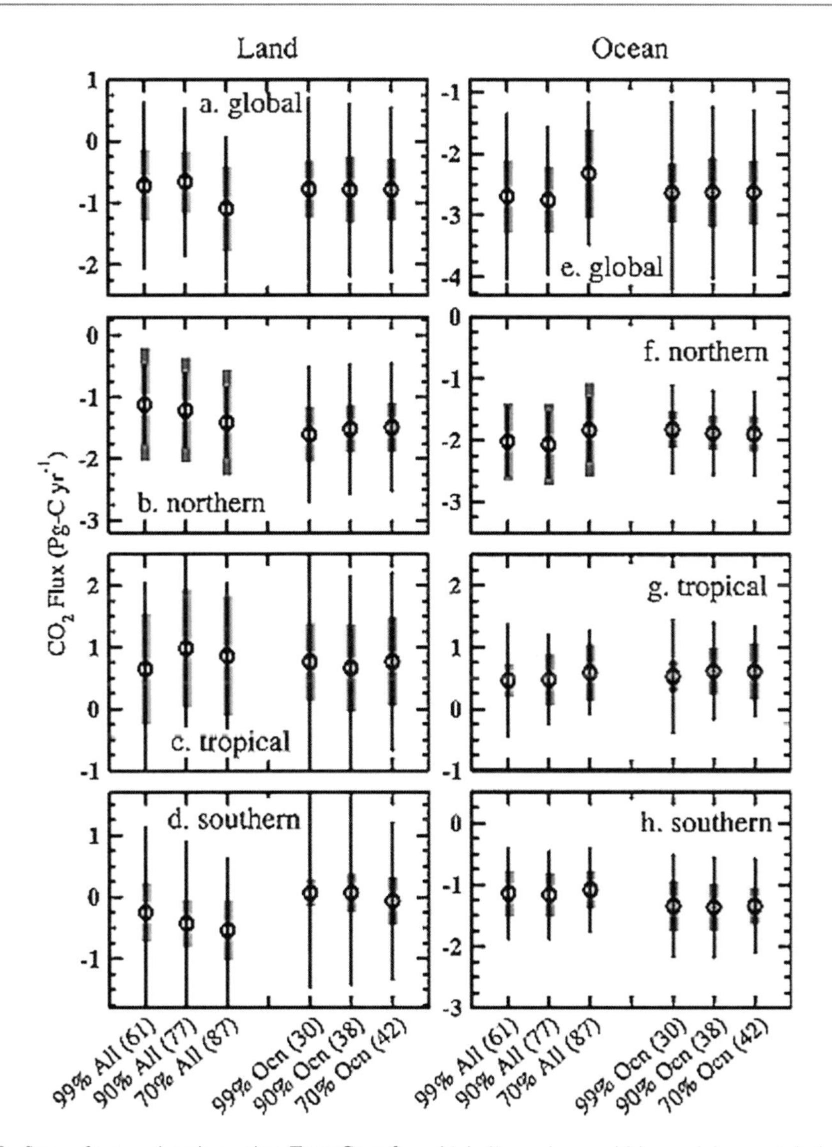

Fig. 7.9 CO_2 fluxes from estimation using TransCom-3 inverse model setup and 16 global transport models. *Black circles* mark the average fluxes obtained from 16 models, *black lines* show between-model uncertainties and *red thick lines* show within-model uncertainties. For each panel, *left part* is derived from 'all site' data; *right part* is derived from 'ocean-only' data (*Source*: From Patra et al. (2006). Reprinted with permission)

continental scale carbon flux (Fig. 7.9). 'Between-model' uncertainties are about 0.51 Pg C per year, and are generally smaller than 'within-model' uncertainties.

The reader should be aware of the following caveats:

1. All models work better over oceans than over land.
2. Different datasets can lead to large differences in estimation. The more sites used in an inverse model, the lower the 'within-model' uncertainty. For example, large uncertainties in the tropical zone data reflect the few observations that are conducted there.
3. Using 'ocean-only' data (excluding the land and coastal measurement sites) instead of 'all

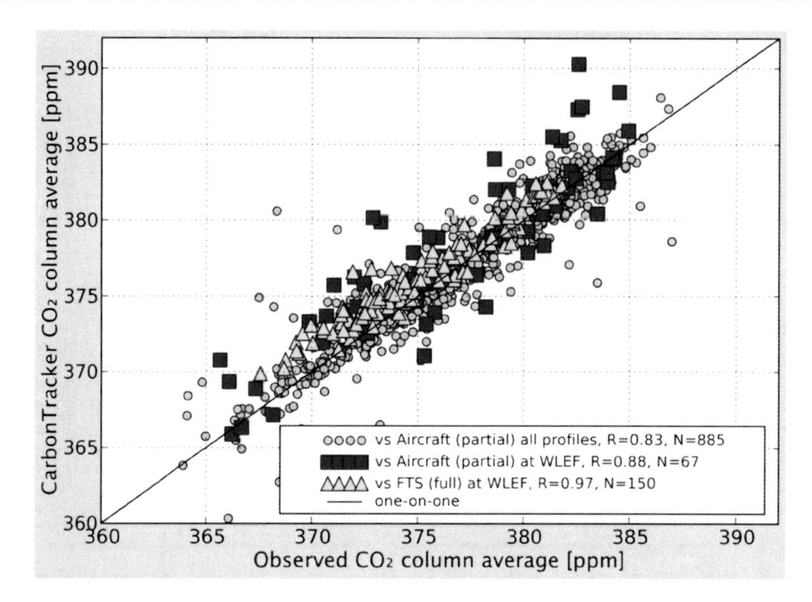

Fig. 7.10 Comparison between column average CO_2 from observations and from CarbonTracker assimilated CO_2 distribution (*Source*: From Peters et al. (2007). Reprinted with permission)

site' data leads to better agreement between models, but the 'within-model' uncertainties increase.

4. Big meteorological or geological events, such as El Nino or a volcanic eruption, bias the data, leading to poor estimation.

With the development of more comprehensive datasets and improved transportation models, CarbonTracker, developed by NOAA's Earth Systems Research Laboratory, has emerged as one of the most advanced inverse models used today (Fig. 7.10). Over the domain covering North America and the eastern Pacific, very good agreement has been achieved between CarbonTracker predictions and real atmospheric measurements (Peters et al. 2007).

CarbonTracker is constrained by about 28,000 flask data points collected by the NOAA ESRL Cooperative Air Sampling Network and continuous CO_2 time series observed at several towers (Peters et al. 2007). Data processing consists of the following steps: (i) develop a 3-dimensional field of atmospheric CO_2 mole fraction around the globe by coupling CO_2 surface exchange models (ocean module, fire module, fossil fuel model and biosphere model) (NOAA 2008) with an atmospheric transport model TM5 (Peters et al. 2004; Krol et al. 2005); (ii) minimize the difference between modeled and observed CO_2 mole fractions by adjusting linear scaling factors which control surface fluxes for large areas; and (iii) build up the history of surface CO_2 exchange at the latitude-longitude resolution of $1° \times 1°$ (Peters et al. 2007).

While measuring CO_2 concentrations, many sites also take measurements for other trace gases (e.g. methane, nitrous oxide, sulfur hexafluoride, carbon monoxide, isotopic ratios of CO_2 and methane). The additional measurements are not only related to climate change, but also can help in source identification of CO_2. Halo-compounds (an organic compound that includes a halogen – e.g. chlorine, fluorine) and hydrocarbons (an organic compound consisting entirely of hydrogen and carbon) have recently been added to the analysis of a subset of air samples along with carbon-14, the best trace for CO_2 emitted through use of fossil fuels.

Although CarbonTracker is an improvement over other inverse models in many aspects, it also suffers from some problems:

1. The accuracy of CarbonTracker depends on the quality and number of observations available. CarbonTracker's ability to accurately

quantify natural and anthropogenic emissions and uptake at regional scales is currently limited by a sparse observational network.
2. Predicted burned area does not match with the observed one in some regions. Methods for dealing with heteroskedastic variables through weighted least squares or nonlinear data transformations increase the influence of low-variance observations while simultaneously decreasing the influence of high variance observations. This is undesirable for estimation (Giglio et al. 2006). Improvements need to be made in the estimation of small burned areas, although they are of less interest compared to the large burns.
3. In the current version of CarbonTracker, relatively small errors in fossil fuel emissions inventories are averaged out by relatively larger errors in other flux emissions (e.g. fires) (Peters et al. 2007).

In order to keep improving this tool for monitoring and predicting the global carbon cycle, all results from CarbonTracker are freely accessible, joint observations are encouraged, and models are updated every year. In addition to the simulated 3-dimensional field of atmospheric CO_2, direct measurement of the 3-dimensional field from satellites is now available (Rayner and O'Brien 2001). The satellite sensors are the Atmospheric Infrared Sounder (AIRS) and the Scanning Imaging Absorption Spectrometer for Atmospheric Cartography (SCIAMACHY) (Buchwitz et al. 2007). In 2008, two dedicated missions called the Orbiting Carbon Observatory (OCO, National Aeronautics and Space Administration) and GoSat (Japanese Space Agency) were launched to quantify CO_2 (Peters et al. 2004). More advanced measurements and more data will improve the performance of CarbonTracker dramatically.

6 Conclusions and Recommendations

The four categories of methods reviewed in this chapter are based on biomass measurement data, remote sensing data, CO_2 flux data (from eddy covariance) and CO_2 concentration data. They all exhibit their own advantages and disadvantages in estimating CO_2 flux and complement each other in different ways (Table 7.4; Fig. 7.11).

Inventory methods quantify biomass accumulation within forests, and are characterized by their long history and adequate data coverage (particularly in developed nations). However,

Table 7.4 Summary of different methods for estimating carbon budgets

Methods	Temporal scale	Spatial scale	Data availability	Uncertainty	Target
Forest inventory	Annual and decades	Regional	Historical data worldwide	1% for growing stock volume, 2–3% for net volume growth and removal, and almost 40% for change in growing a stock volume.	Carbon stock in the forest
Remote sensing	Daily to annual	Regional and global	Start from the end of 1970s	The RMSE for an aggregation area of 510 ha of the unit land area 8.7% for ACIS and 4.6% for world volume.	Carbon stock in the forest
Eddy covariance	Hours to years	Over the course of a year or more	Start from end of 1960s over 400 times worldwide	±50 gCm^{-2} $year^{-1}$ (ideal size)	Net CO_2 exchange across the canopy-atmosphere interface
Inverse method (Carbon Tracker)	Weekly	Global, at 1° × 1° resolution	2000–2006	−1.65 PgCyr (for North American terrestrial biosphere)	Net CO_2 exchange between the terrestrial biosphere and the atmosphere

Source: Compiled from Brown (2002); Patenaude et al. (2005); Lu (2006); Baldocchi (2008); Giglio et al. (2006) and Peters et al. (2007)

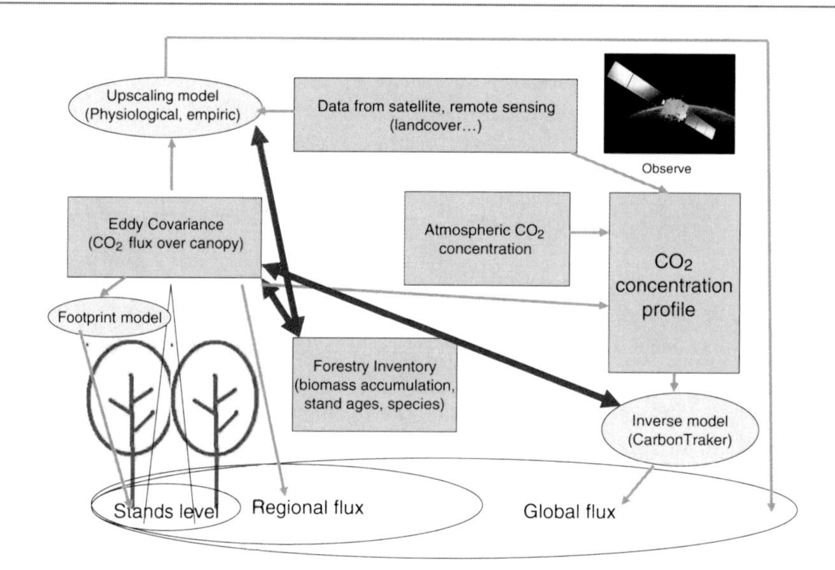

Fig. 7.11 Sketch of current observations and methods system for carbon budget (*red arrow* means inter comparison, *pink arrow* stand for data flow)

they have low time resolution (years) and variable standards of measurement.

Remote sensing methods are most reliable if remote sensing information is jointly used with forest carbon inventories and ecosystem models. However, incomplete information limited by remote sensing techniques and uncertainties in the models require further development.

The eddy covariance method is advanced in its high accuracy and fine temporal resolution (hours), and is a good method for direct measurement of CO_2 flux at the ecosystem scale. However, it is restricted in use by its systematic biases and limited number of observation sites.

Inverse methods are used at continental to global scales. They retrieve the strength of both anthropogenic and non-anthropogenic sources and sinks from atmospheric CO_2 concentration data and transportation models. CarbonTracker is one such inverse model. The data assimilation models in these inverse methods are being improved for higher accuracy and finer spatial resolution.

No single method can meet the accuracy and resolution requirements of all users. A country, user or site will make a choice of method based on the specifics of the circumstance. To accelerate improvements, the user is encouraged to undertake data comparison, collaboration, and assimilation among different methods (Heinsch et al. 2006; Gough et al. 2008). Such improvements should build on a careful synchrony among methods. For example, CO_2 budget estimations from forest inventory are based on biomass accumulation, while CO_2 flux measurements reflect photosynthesis and respiration – usually a 1-year time lag will be found between these two results. In addition, a finer and more comprehensive observation network of CO_2 concentration is required.

References

Anderson J, Martin ME, Smith M-L, Dubayah RO, Hofton MA, Hyde P, Peterson BE, Blair JB, Knox RG (2006) The use of waveform LiDAR to measure northern temperate mixed conifer and deciduous forest structure in New Hampshire. Remote Sens Environ 105:248–261

Anttila P (2002) Nonparametric estimation of stand volume using spectral and spatial features of aerial photographs and old inventory data. Can J Forest Res 32: 1849–1857

National Oceanic and Atmospheric Administration (2008) CarbonTracker. http://www.esrl.noaa.gov/gmd/ccgg/carbontracker/documentation.html. Accessed Apr 2009

Aubinet M, Heinesch B, Longdoz B (2002) Estimation of the carbon sequestration by a heterogeneous forest: night flux corrections, heterogeneity of the site and inter-annual variability. Glob Change Biol 8:1053–1071

Aubinet M, Berbigier P, Bernhofer CH, Cescatti A, Feigenwinter C, Granier A, Grunwald TH, Havrankova

K, Heinesch B, Longdoz B, Marcolla B, Montagnani L, Sedlak P (2005) Comparing CO_2 storage and advection conditions at night at different carboeuroflux sites. Bound Layer Meteorol 116:63–94

Austin JM, Mackey BG, Niel KPV (2003) Estimating forest biomass using satellite radar: an exploratory study in a temperate Australian Eucalyptus forest. Forest Ecol Manag 176:575–583

Baldocchi D (2008) Breathing of the terrestrial biosphere: lessons learned from a global network of carbon dioxide flux measurement systems. Aust J Bot 56:1–26

Baldocchi D, Meyers T (1998) On using eco-physiological, micrometeorological and biogeochemical theory to evaluate carbon dioxide, water vapor and trace gas fluxes over vegetation: a perspective. Agric Forest Meteorol 90:1–25

Baldocchi D, Falge E, Gu LH, Olson R, Hollinger D, Running S, Anthoni P, Bernhofer C, Davis K, Evans R, Fuentes J, Goldstein A, Katul G, Law B, Lee XH, Malhi Y, Meyers T, Munger W, Oechel W, Paw KT, Pilegaard K, Schmid HP, Valentini R, Verma S, Vesala T, Wilson K, Wofsy S (2001) FLUXNET: a new tool to study the temporal and spatial variability of ecosystem-scale carbon dioxide, water vapor, and energy flux densities. Bull Am Meteorol Soc 82:2415–2434

Ball T, Smith KA, Moncrieff JB (2007) Effect of stand age on greenhouse gas fluxes from a Sitka spruce [*Picea sitchensis* (Bong.) Carr.] chronosequence on a peaty gley soil. Glob Change Biol 13:2128–2142

Barnes BV, Zak DR, Denton SR, Spurr SH (1989) Forest ecology. 4th edition, J Wiley & Sons, New York, 345 p

Berger BW, Davis KJ, Yi CX, Bakwin PS, Zhao CL (2001) Long-term carbon dioxide fluxes from a very tall tower in a northern forest: flux measurement methodology. J Atmos Ocean Tech 18:529–542

Blackburn GA, Steele CM (1999) Towards the remote sensing of matorral vegetation physiology relationships between spectral reflectance, pigment, and biophysical characteristics of semiarid bushland canopies. Remote Sens Environ 70:278–292

Bosveld FC, Beljaars ACM (2001) The impact of sampling rate on eddy-covariance flux estimates. Agric Forest Meteorol 109:39–45

Brack C (2009) A brief history of forest inventory. Australian National University, Canberra

Brown S (1997) Estimating biomass and biomass change of tropical forests: a primer. In: FAO Forestry Paper-134, Rome

Brown S (2002) Measuring carbon in forests: current status and future challenges. Environ Pollut 116:363–372

Brown S, Schroeder P, Kern J (1999) Spatial distribution of biomass in forest of the eastern USA. Forest Ecol Manag 123:81–90

Buchwitz M, Schneising O, Burrows JP, Bovensmann H, Reuter M, Notholt J (2007) First direct observation of the atmospheric CO_2 year-to-year increase from space. Atmos Chem Phys 7:4249–4256

Canada Centre for Remote Sensing (2008) GlobeSAR-2 Radar Remote Sensing Training package

Castel T, Guerra F, Caraglio Y, Houllier F (2002) Retrieval biomass of a large Venezuelan pine plantation using JERS-1 SAR data. Analysis of forest structure impact on radar signature. Remote Sens Environ 79:30–41

Chen JM, Chen BZ, Tans P (2007) Deriving daily carbon fluxes from hourly CO_2 mixing ratios measured on the WLEF tall tower: an upscaling methodology. J Geophys Res Biogeosci 112:G01015

Cohen WB, Goward SN (2004) Landsat's role in ecological applications of remote sensing. Bioscience 54:535–545

Cross AM, Settle JJ, Drake NA, Paivinen RTM (1991) Subpixel measurement of tropical forest cover using AVHRR data. Int J Remote Sens 12:1119–1129

Dong JR, Kaufmann RK, Myneni RB, Tucker CJ, Kauppi PE, Liski J, Buermann W, Alexeyev V, Hughes MK (2003) Remote sensing estimates of boreal and temperate forest woody biomass: carbon pools, sources, and sinks. Remote Sens Environ 84:393–410

Drake JB, Knox RG, Dubayah RO, Clark DB, Condit R, Blair JB, Hofton M (2003) Above-ground biomass estimation in closed canopy neotropical forests using LiDAR remote sensing: factors affecting the generality of relationships. Global Ecol Biogeogr 12:147–159

Fang J-Y, Wang GG, Liu G-H, Xu S-L (1998) Forest biomass of China: an estimate based on the biomass-volume relationship. Ecol Appl 8:1084–1091

Fang JY, Chen AP, Peng CH, Zhao SQ, Ci L (2001) Changes in forest biomass carbon storage in China between 1949 and 1998. Science 292:2320–2322

Fazakas Z, Nilsson M, Olsson H (1999) Regional forest biomass and wood volume estimation using satellite data and ancillary data. Agric Forest Meteorol 98:417–425

Feigenwinter C, Bernhofer C, Eichelmann U, Heinesch B, Hertel M, Janous D, Kolle O, Lagergren F, Lindroth A, Minerbi S, Moderow U, Molder M, Montagnani L, Queck R, Rebmann C, Vestin P, Yernaux M, Zeri M, Ziegler W, Aubinet M (2008) Comparison of horizontal and vertical advective CO_2 fluxes at three forest sites. Agric Forest Meteorol 148:12–24

Finnigan JJ (2004) A re-evaluation of long-term flux measurement techniques – part II: coordinate systems. Bound Layer Meteorol 113:1–41

Finnigan JJ, Clement R, Malhi Y, Leuning R, Cleugh HA (2003) A re-evaluation of long-term flux measurement techniques – part I: averaging and coordinate rotation. Bound Layer Meteorol 107:1–48

Fisher JB, Baldocchi DD, Misson L, Dawson TE, Goldstein AH (2007) What the towers don't see at night: nocturnal sap flow in trees and shrubs at two AmeriFlux sites in California. Tree Physiol 27:597–610

Foken T, Wimmer F, Mauder M, Thomas C, Liebethal C (2006) Some aspects of the energy balance closure problem. Atmos Chem Phys 6:4395–4402

FAO (2000) Food and Agriculture Organization of the United Nations, Rome, FAO Forestry Paper, 140 p

Foody GM, Boyd DS, Cutler MEJ (2003) Predictive relations of tropical forest biomass from Landsat TM data and their transferability between regions. Remote Sens Environ 85:463–474

Fraser RH, Li Z (2002) Estimating fire-related parameters in boreal forest using SPOT VEGETATION. Remote Sens Environ 82:95–110

Freitas JVD (2006) Experiences with FRA 2005. Expert Consultation on Global Forest Resource Assessment: towards FRA 2010. Ministry of Environment Brazil

Fuchs H, Magdon P, Kleinn C, Flessa H (2009) Estimating aboveground carbon in a catchment of the Siberian forest tundra: combining satellite imagery and field inventory. Remote Sens Environ 113:518–531

Gash JHC, Culf AD (1996) Applying a linear detrend to eddy correlation data in real time. Bound Layer Meteorol 79:301–306

Gehring C, Park S, Denich M (2004) Liana allometric biomass equations for Amazonian primary and secondary forest. Forest Ecol Manag 195:68–83

Giglio L, van der Werf GR, Randerson JT, Collatz GJ, Kasibhatla P (2006) Global estimation of burned area using MODIS active fire observations. Atmos Chem Phys 6:957–974

Goodale CL, Apps MJ, Birdsey RA, Field CB, Heath LS, Houghton RA, Jenkins JC, Kohlmaier GH, Kurz W, Liu SR, Nabuurs GJ, Nilsson S, Shvidenko AZ (2002) Forest carbon sinks in the northern hemisphere. Ecol Appl 12:891–899

Gough CM, Vogel CS, Schmid HP, Su HB, Curtis PS (2008) Multi-year convergence of biometric and meteorological estimates of forest carbon storage. Agric Forest Meteorol 148:158–170

Gurney KR, Law RM, Denning AS, Rayner PJ, Baker D, Bousquet P, Bruhwiler L, Chen YH, Ciais P, Fan S, Fung IY, Gloor M, Heimann M, Higuchi K, John J, Maki T, Maksyutov S, Masarie K, Peylin P, Prather M, Pak BC, Randerson J, Sarmiento J, Taguchi S, Takahashi T, Yuen CW (2002) Towards robust regional estimates of CO_2 sources and sinks using atmospheric transport models. Nature 415:626–630

Hammerle A, Haslwanter A, Schmitt M, Bahn M, Tappeiner U, Cernusca A, Wohlfahrt G (2007) Eddy covariance measurements of carbon dioxide, latent and sensible energy fluxes above a meadow on a mountain slope. Bound Layer Meteorol 122:397–416

Han IJ, Liu SM, Wang JM, Wang JD (2003) Study on energy balance over different surfaces. In: Geoscience and Remote Sensing Symposium, 2003. IGARSS '03. Proceedings. 2003 IEE International. 5:3208–3210

Hanewinkel M (2005) Neural networks for assessing the risk of windthrow on the forest division level: a case study in southwest Germany. Eur J Forest Res 124:243–249

Harrell PA, Kasischke ES, Bourgeau-Chavez LL, Haney EM, Norman L, Christensen J (1997) Evaluation of approaches to estimating aboveground biomass in southern pine forests using SIR-C data. Remote Sens Environ 59:223–233

Haszpra L, Barcza Z, Davis KJ, Tarczay K (2005) Long-term tall tower carbon dioxide flux monitoring over an area of mixed vegetation. Agric Forest Meteorol 132:58–77

Heinsch FA, Zhao MS, Running SW, Kimball JS, Nemani RR, Davis KJ, Bolstad PV, Cook BD, Desai AR, Ricciuto DM, Law BE, Oechel WC, Kwon H, Luo HY, Wofsy SC, Dunn AL, Munger JW, Baldocchi DD, Xu LK, Hollinger DY, Richardson AD, Stoy PC, Siqueira MBS, Monson RK, Burns SP, Flanagan LB (2006) Evaluation of remote sensing based terrestrial productivity from MODIS using regional tower eddy flux network observations. IEEE Trans Geosci Remote Sens 44:1908–1925

Holmgren P, Persson R, (2002) Evolution and prospects of global forest assessments. In: Perlis A (ed) Unasylva – No. 210 – Forest assessment and monitoring. FAO

Holmgren P, Thuresson T, Holm S (1997) Estimating forest characteristics in scanned aerial photographs with respect to requirements for economic forest management planning. Scand J Forest Res 12:189–199

Houghton RA, Ramakrishna K (1999) A review of national emissions inventories from select non-Annex I countries: implications for counting sources and sinks of carbon. Annu Rev Energy Environ 24:571–605

IPCC (2007) Climate change: synthesis report. In: Pachauri RK, Reisinger A (eds) Contribution of working groups I, II and III to the fourth assessment report of the intergovernmental panel on climate change. IPCC, Geneva, 104 p

Iverson LR, Cook EA, Graham RL (1994) Regional forest cover estimation via remote sensing: the calibration center concept. Landscape Ecol 9:159–174

Jenkins JC, Chojnacky DC, Heath LS, Birdsey RA (2003) National-scale biomass estimators for United States tree species. Forest Sci 49:12–35

Krol M, Houweling S, Bregman B, van den Broek M, Segers A, van Velthoven P, Peters W, Dentener F, Bergamaschi P (2005) The two-way nested global chemistry-transport zoom model TM5: algorithm and applications. Atmos Chem Phys 5:417–432

Laporte N, Justice C, Kendall J (1995) Mapping the dense humid forest of Cameroon and Zaire using AVHRR satellite data. Int J Remote Sens 16:1127–1145

LeBlanc JW (2009) What do we own: understanding forest inventory. University of California Cooperative Extension

Lee XH (2004) A model for scalar advection inside canopies and application to footprint investigation. Agric Forest Meteorol 127:131–141

Lefsky MA, Cohen WB, Acker SA, Parker GG, Spies TA, Harding D (1999) LiDAR remote sensing of the canopy structure and biophysical properties of douglas-fir western hemlock forests. Remote Sens Environ 70:339–361

Lefsky MA, Cohen WB, Parker GG, Harding DJ (2002) LiDAR remote sensing for ecosystem studies. Bioscience 52:19–30

Levesque J, King DJ (2003) Spatial analysis of radiometric fractions from high-resolution multispectral imagery for modelling individual tree crown and forest canopy structure and health. Remote Sens Environ 84:589–609

Li ZQ, Yu GR, Wen XF, Zhang LM, Ren CY, Fu YL (2005) Energy balance closure at ChinaFLUX sites. Sci China Ser D 48:51–62

Lu D (2006) The potential and challenge of remote sensing-based biomass estimation. Int J Remote Sens 27:1297–1328

Lu D, Batistella M (2005) Exploring TM image texture and its relationships with biomass estimation in Rondônia, Brazilian Amazon. Acta Amazonica 35:249–257

Luckman A, Baker J, Honzák M, Lucas R (1998) Tropical forest biomass density estimation using JERS-1 SAR: seasonal variation, confidence limits, and application to image mosaics. Remote Sens Environ 63:126–139

Luo Y, Medlyn B, Hui D, Ellsworth D, Reynolds J, Katul G (2001) Gross primary productivity in duke forest: modeling synthesis of CO_2 experiment and eddy-flux data. Ecol Appl 11:239–252

Luyssaert S, Inglima I, Jung M, Richardson AD, Reichsteins M, Papale D, Piao SL, Schulzes ED, Wingate L, Matteucci G, Aragao L, Aubinet M, Beers C, Bernhoffer C, Black KG, Bonal D, Bonnefond JM, Chambers J, Ciais P, Cook B, Davis KJ, Dolman AJ, Gielen B, Goulden M, Grace J, Granier A, Grelle A, Griffis T, Grunwald T, Guidolotti G, Hanson PJ, Harding R, Hollinger DY, Hutyra LR, Kolar P, Kruijt B, Kutsch W, Lagergren F, Laurila T, Law BE, Le Maire G, Lindroth A, Loustau D, Malhi Y, Mateus J, Migliavacca M, Misson L, Montagnani L, Moncrieff J, Moors E, Munger JW, Nikinmaa E, Ollinger SV, Pita G, Rebmann C, Roupsard O, Saigusa N, Sanz MJ, Seufert G, Sierra C, Smith ML, Tang J, Valentini R, Vesala T, Janssens IA (2007) CO_2 balance of boreal, temperate, and tropical forests derived from a global database. Glob Change Biol 13:2509–2537

Massman WJ, Lee X (2002) Eddy covariance flux corrections and uncertainties in long-term studies of carbon and energy exchanges. Agric Forest Meteorol 113: 121–144

Moore CJ (1986) Frequency-response corrections for eddy-correlation systems. Bound Layer Meteorol 37:17–35

Muukkonen P, Heiskanen J (2005) Estimating biomass for boreal forests using ASTER satellite data combined with standwise forest inventory data. Remote Sens Environ 99:434–447

Muukkonen P, Heiskanen J (2007) Biomass estimation over a large area based on standwise forest inventory data and ASTER and MODIS satellite data: a possibility to verify carbon inventories. Remote Sens Environ 107:617–624

Nelson R, Oderwald R, Gregoire TG (1997) Separating the ground and airborne laser sampling phases to estimate tropical forest basal area, volume, and biomass. Remote Sens Environ 60:311–326

Niklas K (1994) Plant allometry: the scaling of form and process. The University of Chicago Press, Chicago

Owen KE, Tenhunen J, Reichstein M, Wang Q, Falge E, Geyer R, Xiao XM, Stoy P, Ammann C, Arain A, Aubinet M, Aurela M, Bernhofer C, Chojnicki BH, Granier A, Gruenwald T, Hadley J, Heinesch B, Hollinger D, Knohl A, Kutsch W, Lohila A, Meyers T, Moors E, Moureaux C, Pilegaard K, Saigusa N, Verma S, Vesala T, Vogel C (2007) Linking flux network measurements to continental scale simulations: ecosystem carbon dioxide exchange capacity under non-water-stressed conditions. Glob Change Biol 13: 734–760

Patenaude G, Milne R, Dawson TP (2005) Synthesis of remote sensing approaches for forest carbon estimation: reporting to the Kyoto protocol. Environ Sci Policy 8:161–178

Patra PK, Gurney KR, Denning AS, Maksyutov S, Nakazawa T, Baker D, Bousquet P, Bruhwiler L, Chen YH, Ciais P, Fan SM, Fung I, Gloor M, Heimann M, Higuchi K, John J, Law RM, Maki T, Pak BC, Peylin P, Prather M, Rayner PJ, Sarmiento J, Taguchi S, Takahashi T, Yuen CW (2006) Sensitivity of inverse estimation of annual mean CO_2 sources and sinks to ocean-only sites versus all-sites observational networks. Geophys Res Lett 33:L05814

Peters W, Krol MC, Dlugokencky EJ, Dentener FJ, Bergamaschi P, Dutton G, von Velthoven P, Miller JB, Bruhwiler L, Tans PP (2004) Toward regional-scale modeling using the two-way nested global model TM5: characterization of transport using SF6. J Geophys Res Atmos 109:D19314

Peters W, Jacobson AR, Sweeney C, Andrews AE, Conway TJ, Masarie K, Miller JB, Bruhwiler LMP, Petron G, Hirsch AI, Worthy DEJ, van der Werf GR, Randerson JT, Wennberg PO, Krol MC, Tans PP (2007) An atmospheric perspective on North American carbon dioxide exchange: CarbonTracker. Proc Natl Acad Sci USA 104:18925–18930

Finnigan JJ, Clement R, Malhi Y, Leuning R, Cleugh HA (2003) A re-evaluation of long-term flux measurement techniques - part I: averaging and coordinate rotation. Bound Layer Meteorol 107:1–48

Popescu SC, Wynne RH, Nelson RF (2003) Measuring individual tree crown diameter with LiDAR and assessing its influence on estimating forest volume and biomass. Can J Remote Sens 29:564–577

Ranson KJ, Sun G (1994) Mapping biomass of a northern forest using multifrequency SAR data. IEEE Trans Geosci Remote Sens 32:388–396

Rayner PJ, O'Brien DM (2001) The utility of remotely sensed CO_2 concentration data in surface source inversions. Geophys Res Lett 28:175–178

Reese H, Nilsson M, Sandström P, Olsson H (2002) Applications using estimates of forest parameters derived from satellite and forest inventory data. Comput Electron Agric 37:37–55

Rodenbeck C, Houweling S, Gloor M, Heimann M (2003) CO_2 flux history 1982–2001 inferred from atmospheric data using a global inversion of atmospheric transport. Atmos Chem Phys 3:1919–1964

Running SW, Baldocchi DD, Turner DP, Gower ST, Bakwin PS, Hibbard KA (1999) A global terrestrial monitoring network integrating tower fluxes, flask sampling, ecosystem modeling and EOS satellite data. Remote Sens Environ 70:108–127

Sasai T, Okamoto K, Hiyama T, Yamaguchi Y (2007) Comparing terrestrial carbon fluxes from the scale of a flux tower to the global scale. Ecol Model 208:135–144

Schmid HP (1997) Experimental design for flux measurements: matching scales of observations and fluxes. Agric Forest Meteorol 87:179–200

Schroeder P, Brown S, Mo JM, Birdsey R, Cieszewski C (1997) Biomass estimation for temperate broadleaf forests of the United States using inventory data. Forest Sci 43:424–434

Smith JE, Heath LS (2004) Carbon stocks and projections on public forestlands in the United States, 1952–2040. Environ Manage 33:433–442

Smith J, Heath L, Jenkins J (2002) Forest volume-to-biomass models and estimates of mass for live and standing dead trees of U.S. forests. In: General Technical Report NE-298. Northeastern Research Station, p 62

Soegaard H, Nordstroem C, Friborg T, Hansen BU, Christensen TR, Bay C (2000) Trace gas exchange in a high-arctic valley 3. integrating and scaling CO_2 fluxes from canopy to landscape using flux data, footprint modeling, and remote sensing. Global Biogeochem Cycles 14:725–744

Stauch VJ, Jarvis AJ (2006) A semi-parametric gap-filling model for eddy covariance CO_2 flux time series data. Glob Change Biol 12:1707–1716

Steininger MK (2000) Satellite estimation of tropical secondary forest above-ground biomass: data from Brazil and Bolivia. Int J Remote Sens 21:1139–1157

Su HB, Schmid HP, Grimmond CSB, Vogel CS, Curtis PS (2008) An assessment of observed vertical flux divergence in long-term eddy-covariance measurements over two Midwestern forest ecosystems. Agric Forest Meteorol 148:186–205

Sun JL, Desjardins R, Mahrt L, MacPherson I (1998) Transport of carbon dioxide, water vapor, and ozone by turbulence and local circulations. J Geophys Res Atmos 103:25873–25885

Sun G, Ranson KJ, Kharuk VI (2002) Radiometric slope correction for forest biomass estimation from SAR data in the Western Sayani Mountains, Siberia. Remote Sens Environ 79:279–287

Thenkabail PS (2003) Biophysical and yield information for precision farming from near-real-time and historical Landsat TM images. Int J Remote Sens 24:2879–2904

Thenkabail PS, Enclonab EA, Ashton MS, Meer BVD (2004a) Accuracy assessments of hyperspectral waveband performance for vegetation analysis applications. Remote Sens Environ 91:354–376

Thenkabail PS, Stucky N, Griscom BW, Ashton MS, Diels J, Van der Meer B, Enclona E (2004b) Biomass estimations and carbon stock calculations in the oil palm plantations of African derived savannas using IKONOS data. Int J Remote Sens 25:5447–5472

Tomppo E, Nilsson M, Rosengren M, Aalto P, Kennedy P (2002) Simultaneous use of Landsat-TM and IRS-1 C WiFS data in estimating large area tree stem volume and aboveground biomass. Remote Sens Environ 82:156–171

UNFCCC (1997) Kyoto protocol to the United Nations Framework convention on climate change

Velasco E, Pressley S, Allwine E, Westberg H, Lamb B (2005) Measurements of CO_2 fluxes from the Mexico City urban landscape. Atmos Environ 39:7433–7446

Wang S, Chen JM, Ju WM, Feng X, Chen M, Chen P, Yu G (2007a) Carbon sinks and sources in China's forests during 1901–2001. J Environ Manag 85:524–537

Wang YP, Baldocchi D, Leuning R, Falge E, Vesala T (2007b) Estimating parameters in a land-surface model by applying nonlinear inversion to eddy covariance flux measurements from eight FLUXNET sites. Glob Change Biol 13:652–670

Wardoyo (2008) National forestry inventory Indonesia. Ministry of Forestry, Indonesia

Wilson K, Goldstein A, Falge E, Aubinet M, Baldocchi D, Berbigier P, Bernhofer C, Ceulemans R, Dolman H, Field C, Grelle A, Ibrom A, Law BE, Kowalski A, Meyers T, Moncrieff J, Monson R, Oechel W, Tenhunen J, Valentini R, Verma S (2002) Energy balance closure at FLUXNET sites. Agric Forest Meteorol 113:223–243

Woodbury PB, Smith JE, Heath LS (2007) Carbon sequestration in the US forest sector from 1990 to 2010. Forest Ecol Manag 241:14–27

Wulder MA, Seemann D (2003) Forest inventory height update through the integration of LiDAR data with segmented Landsat imagery. Can J Remote Sens 29:536–543

Wulder MA, Hall R, Coops N, Franklin S (2004) High spatial resolution remotely sensed data for ecosystem characterization. Bioscience 54:511–521

Wulder MA, White JC, Fournier RA, Luther JE, Magnussen S (2008) Spatially explicit large area biomass estimation: three approaches using forest inventory and remotely sensed imagery in a GIS. Sensors 8:529–560

Yamaguchi Y, Kahle AB, Tsu H, Kawakami T, Pniel M (1998) Overview of advanced Spaceborne thermal emission and reflection radiometer (ASTER). IEEE Trans Geosci Remote Sens 36:1062–1071

Yang FH, Ichii K, White MA, Hashimoto H, Michaelis AR, Votava P, Zhu AX, Huete A, Running SW, Nemani RR (2007) Developing a continental-scale measure of gross primary production by combining MODIS and AmeriFlux data through support vector machine approach. Remote Sens Environ 110:109–122

Yuan WP, Liu S, Zhou GS, Zhou GY, Tieszen LL, Baldocchi D, Bernhofer C, Gholz H, Goldstein AH, Goulden ML, Hollinger DY, Hu Y, Law BE, Stoy PC, Vesala T, Wofsy SC (2007) Deriving a light use efficiency model from eddy covariance flux data for predicting daily gross primary production across biomes. Agric Forest Meteorol 143:189–207

Zheng D, Rademacher J, Chena J, Crowc T, Breseea M, Moined JL, Ryua S-R (2004) Estimating aboveground biomass using Landsat 7 ETM+data across a managed landscape in northern Wisconsin, USA. Remote Sens Environ 93:402–411

Zheng DL, Heath LS, Ducey MJ (2007a) Forest biomass estimated from MODIS and FIA data in the Lake States: MN, WI and MI, USA. Forestry 80:265–278

Zheng G, Chen JM, Tian QJ, Ju WM, Xia XQ (2007b) Combining remote sensing imagery and forest age inventory for biomass mapping. J Environ Manag 85:616–623

Zianis D, Mencuccini M (2004) On simplifying allometric analyses of forest biomass. Forest Ecol Manag 187:311–332

The Role of Forests in Global Carbon Budgeting

8

Deborah Spalding, Elif Kendirli, and Chadwick Dearing Oliver

Executive Summary

While forests have the capacity to sequester significant amounts of carbon, the natural and anthropogenic processes driving carbon fluxes in forests are complex and difficult to measure. However, since land use change is estimated to be the second largest source of carbon emissions to the atmosphere after the burning of fossil fuels, understanding and quantifying forest carbon sinks and sources is an important part of global carbon budgeting and climate change policy design. Although carbon emissions from land use change have remained fairly steady over the last few decades, there have been significant regional variations within this trend. Specifically, deforestation rates have grown in the tropics, particularly in Asia. In contrast, forests outside the tropics have been sequestering incremental carbon due to CO_2 fertilization and due to forest regrowth on lands that had been cleared for agriculture prior to industrialization.

There are several methods used to measure forest carbon fluxes; these are broadly characterized as top down or bottom up approaches. Top down approaches use atmospheric concentrations of CO_2 as a basis for carbon budgeting. These methods estimate global carbon pools by measuring changes in atmospheric carbon or by using atmospheric transport models to determine regional carbon fluxes across space and time. They can be useful in partitioning global carbon into oceanic and terrestrial biomes. Bottom up approaches include inventory and bookkeeping methods as well as process-based modeling. Forest inventory models require accurate estimates of forest cover and appropriate biomass conversion factors which can be difficult due to lack of comprehensive underlying data and local variations in forest biomass concentrations. Bottom up "bookkeeping" methods are better able to pinpoint the effects of human activity on forest carbon fluxes although they are constrained by a lack of accounting for natural disturbance. More recently, dual-constrained approaches have been used to reconcile top down and bottom up models and to provide full carbon accounting.

What We Do and Do Not Know About Measurement Gaps in Forest Carbon Budgeting

- Knowledge of the amount of carbon stored within each pool and across forest types is limited. Even estimates using broad categories such as carbon in vegetation versus soils vary widely due to a lack of data or assumptions about where carbon is stored within the forest and at what rate carbon is sequestered or released.
- Estimates of forest cover and growing stock are often based on an inadequate number of

D. Spalding (✉)
Working Lands Investment Partners, LLC, New Haven, CT 06510, USA
e-mail: deborah.spalding@yale.edu

E. Kendirli
Consultant, Washington, DC, USA

C.D. Oliver
Yale School of Forestry and Environmental Studies, Prospect Street 360, 06511 New Haven, CT, USA

M.S. Ashton et al. (eds.), *Managing Forest Carbon in a Changing Climate*,
DOI 10.1007/978-94-007-2232-3_8, © Springer Science+Business Media B.V. 2012

field measurement plots, particularly in the tropics. Estimation errors are further magnified by a high degree of heterogeneity in many tropical forests and by the non-normal distribution of carbon pools and fluxes.
- Carbon flux estimates from biological processes in one forest type are often applied across forest types due to a lack of alternative data, despite the fact that biological processes may differ by forest type.
- Natural and anthropogenic disturbances have different impacts on forest carbon cycling over space and time. Carbon flux estimates that do not distinguish by type of disturbance may generate erroneous estimates of disturbance and post-disturbance related carbon fluxes.
- Land use change is a complex and difficult component to quantify in the global carbon budget. The underlying data is often incomplete and may not be comparable across countries or regions due to different definitions of forest cover and land uses. Deforestation rates in the tropics are particularly difficult to determine due to these factors as well as differences in the way land degradation, such as selective logging and fuelwood removals, are accounted for in national statistics.
- Climate change is likely to generate both positive and negative feedbacks in forest carbon cycling. Positive feedbacks may include increased fire and tree mortality from drought stress, insect outbreaks and disease. Negative feedbacks may include increased productivity from CO_2 enrichment. Although the net result from positive and negative climate feedbacks is generally thought to be higher net carbon emissions from forests, the timing and extent of these net emissions are difficult to determine.
- While forests may exhibit greater rates of photosynthesis due to higher levels of CO_2 in the atmosphere, at some point this increased productivity will be inhibited by nutrient limitation. At which point this occurs is likely to differ by region and forest type.
- Temperature increases are likely to have multiple compounding and offsetting impacts which make it difficult to quantify the net impact on carbon cycling. While longer growing seasons may increase carbon uptake in forests, warmer temperatures may lead to increased drought which could offset any increased sequestration from a longer growing season. In addition, warmer temperatures may lead to increased carbon and methane emissions from thawing peatlands.
- The frequency and severity of disturbances are likely to increase. Estimating forest carbon fluxes following disturbance will be difficult if changes in temperature, precipitation, and species composition lead to forest recovery patterns that deviate from history.

Given the uncertainties in forest carbon budgeting, there are several recommendations for policymakers seeking to use carbon budgeting to design forest policy. First, policies should specifically encourage afforestation and discourage deforestation to ensure the ongoing efficacy of the forest carbon sink. Second, models should be selected based on their appropriateness for examining the carbon pools and processes under consideration. Carbon policies should be tested using multiple methodologies to increase accuracy. Third, greater numbers of permanent, long term research plots should be created to improve knowledge of carbon processes and to better estimate carbon fluxes across spatial and temporal gradients. Fourth, countries should be required to adhere to globally accepted methodologies for determining forest cover, land use, and biomass conversion factors. Fifth, regionally specific carbon data should not be extrapolated to other regions and forest types. Finally, policymakers should consider the immediate impacts of policies on forest carbon fluxes as well as the longer term impacts to ensure long term carbon management goals are met.

1 Introduction

While the basic principles of the carbon cycle are well known, there are significant uncertainties surrounding the actual behavior of many of its sinks and sources. This is a particular challenge in forested ecosystems due to the role played by biogeochemistry, climate, disturbance and land use, as well as the spatial and temporal heterogeneity of carbon sequestration across regions and forest types. Nevertheless, as emissions from land use change (largely deforestation) are a significant percentage of the overall global carbon budget, the role of forests continues to be a key component of global carbon policy design.

Forests can act as a sink or a source of carbon under different conditions and across temporal and spatial gradients. Understanding the role of forests in global carbon budgets requires quantifying several components of the carbon cycle, including how much carbon is stored in the world's forests (carbon pools), gains and losses of carbon in forests due to natural and anthropogenic processes (carbon fluxes), exchanges between terrestrial carbon and other sinks and sources, and the ways in which such processes may be altered by climate change.

This chapter will review the current research in global forest carbon budgeting. It will consider the tools used to quantify forest carbon pools and fluxes and their relationship to the global carbon budget. It will demonstrate the complexity of terrestrial carbon sequestration and its interdependence with other components of the carbon cycle by highlighting gaps in knowledge, measurement tools, and models. Finally, it will conclude with some recommendations for future research to better understand forests and their role in global carbon budgeting.

2 The Global Carbon Budget

The basic principles of the global carbon cycle are well known (Fig. 8.1). Still, uncertainties remain surrounding the active processes of the four major reservoirs – the atmosphere, oceans, fossil fuels and terrestrial ecosystems.

Total carbon stored in the world's forests and other terrestrial ecosystems is estimated to be between 500 and 800 billion tones (Adams 2011) and varies across different forest ecosystem types (Fig. 8.2). Similarly, there is significant regional variation in carbon flux estimates due to variable rates of deforestation and afforestation across the globe (Table 8.1a, b).

Contrary to past assumptions, terrestrial ecosystems in aggregate are now believed to be a net carbon sink, although there is still uncertainty regarding the regional allocation of carbon uptake (Stephens et al. 2007). Although the tropical biome has often been viewed as a net carbon source due to widespread deforestation, recent studies suggest it may be neutral, or near

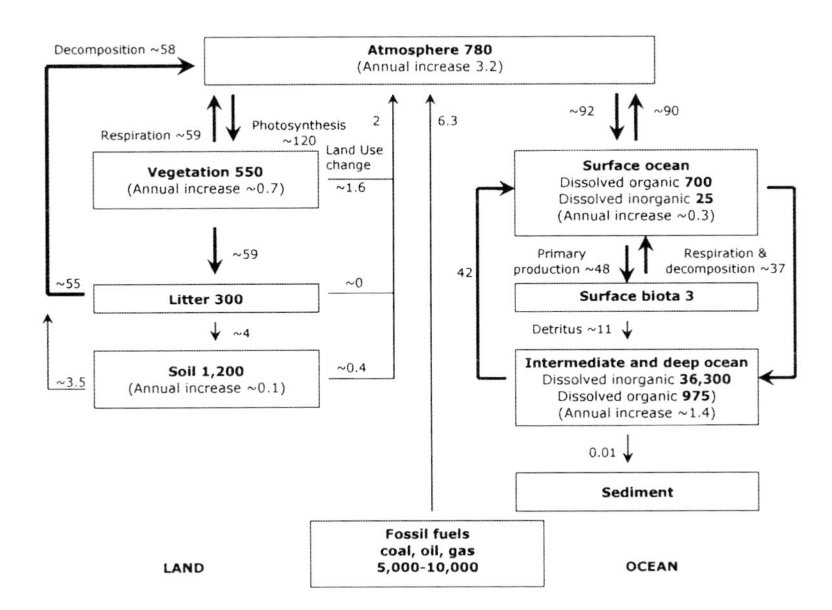

Fig. 8.1 The global carbon cycle (1990s) (*Source*: Houghton (2007))

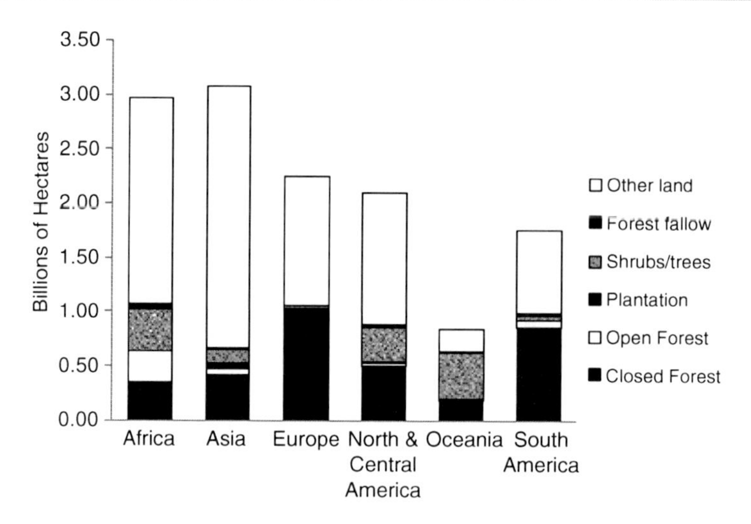

Fig. 8.2 Distribution of forest cover types by continent (*Source*: UNFAO (2000))

Table 8.1 Past changes in land use (**a**) and carbon fluxes from land use changes (**b**)

(a) Average annual rates of tropical deforestation (10^6 ha/year)

	1980s		1990s		
(10^6 ha/year)	FAO (2000)	Defries et al. (2002)	FAO (2000)	Defries et al. (2002)	Achard et al. (2004)
Americas	7.4	4.4	5.2	4.0	4.4
Asia	3.9	2.2	5.9	2.7	2.9
Africa	4.0	1.5	5.6	1.3	2.3
Total	15.3	8.1	16.7	8.0	9.6

(b) Carbon fluxes to the atmosphere from land use change (Pg/C/year)

Region	Total (PgC) 1850–2000	Annual flux 1980–1989	(PgC/year) 1990–1999
Tropical Asia	48	0.88 ± 0.5	1.09 ± 0.5
Tropical America	37	0.77 ± 0.3	0.75 ± 0.3
Tropical Africa	13	0.28 ± 0.2	0.35 ± 0.2
Total tropics	98	1.93 ± 0.6	2.20 ± 0.6
Canada	5	0.03 ± 0.2	0.03 ± 0.2
US	7	(0.12) ± 0.2	(0.11) ± 0.2
Europe	5	(0.02) ± 0.2	(0.02) ± 0.2
Russia	11	0.03 ± 0.2	0.02 ± 0.2
China	23	0.11 ± 0.2	0.03 ± 0.2
Pacific developed	4	0.01 ± 0.2	0.00 ± 0.2
North Africa/Mid East	3	0.02 ± 0.2	0.02 ± 0.2
Total ex tropics	58	0.06 ± 0.5	(−0.02) ± 0.5
Global total	156	1.99 ± 0.8	2.18 ± 0.8

Sources: Houghton (2005b, 2003b)

neutral due to carbon sequestration offsetting most emissions stemming from deforestation (Ramankutty et al. 2007; Stephens et al. 2007). The temperate biome in the Northern Hemisphere is considered a carbon sink although the magnitude of the sink is uncertain (Fan et al. 1998; Liski et al. 2003). The boreal biome is also considered to be a net sink (Myneni et al. 2001) though sub-regions in Canada and Eurasia may be sources.

Table 8.2 Trends in the global carbon budget over time

Global carbon budget	Mean 1959–2006	1970–1999	1990–1999	2000–2006
Economy (kgC/US$)				
Carbon intensity	0.29[a]	0.30	0.26	0.24
Sources (PgC/year)				
Fossil fuel	5.30	5.60	6.50	7.60
Land use change	1.50	1.50	1.60	1.50
Total	6.70	7.00	8.00	9.10
Sinks (PgC/year)				
Atmosphere	2.90	3.10	3.20	4.10
Ocean	1.90	2.00	2.20	2.20
Land	1.90	2.00	2.70	2.80
Airborne fraction	0.43	0.44	0.39	0.45

Source: Canadell et al. (2007)
[a]Data from 1970

The largest source of carbon emissions to the atmosphere is currently believed to be the burning of fossil fuels (Table 8.2). Due to economic growth, population increases and industrialization, fossil fuel emissions have been on a rising trend since 1750 (Raupach et al. 2007). Some portion of these emissions is removed from the atmosphere by oceanic and terrestrial carbon sequestration processes. However, in the last several years, measurements indicate that carbon emissions have grown faster than land and ocean sinks, and that the efficiency of these sinks is declining due to positive feedback mechanisms under global warming (Fung et al. 2005). Some have argued that forest carbon sinks may also diminish as forests which reestablished on abandoned agricultural land in temperate regions continue to mature, and as CO_2 fertilization reaches saturation due to limiting conditions for water and nutrients (Schimel et al. 2001). Thus, while the terrestrial carbon sink remains a key component of the global carbon budget, its long-term role remains uncertain.

The second largest source of emissions results from land use change. Carbon emissions from land use change are thought to driven largely by deforestation in tropical forests of the Americas, Asia, and Africa (Table 8.1; Houghton 2003b, 2005b). However, this has been a fairly recent phenomenon. Until the mid-1930s, there was significant deforestation in North America and Europe. By the mid twentieth century, however, as forests began to regrow on land that had been formerly cleared, forests in temperate North America and Europe began to sequester sufficient amounts of carbon to offset emissions from disturbance and land management practices (Myneni et al. 2001; Goodale et al. 2002). Over the past two decades, net emissions from deforestation in tropical ecosystems have grown approximately 10%, due to increased deforestation in Asia and Africa, and modest reductions in the Americas (DeFries et al. 2002).

In general, deforestation releases carbon to the atmosphere while afforestation removes carbon from the atmosphere. Given the significant loss of terrestrial carbon from land use change/deforestation, it is difficult to reconcile the presence of a net terrestrial sink (Table 8.1). This "missing sink" has been noted across the literature, but its nature is uncertain. It may be due to methodological differences in models used to quantify carbon budgets or due to some factor other than land use change that is not considered in the calculations (Goodale et al. 2002; Houghton 2007). Issues associated with these possible causes will be discussed below.

3 Modeling Global Carbon Budgets

3.1 Top Down Approaches

Although scientists agree on the broad categories of emissions and sources, there are different methods used to quantify carbon sinks and sources, which often produce a wide range of results. Each method accounts for carbon processes in different ways and demonstrates that there remain significant gaps in measurements and knowledge of the terrestrial carbon cycle (see Chap. 7 for more details on measurement methodologies).

Global carbon models generally fall into two methodological categories: top down and bottom up approaches. Top-down approaches rely on observations of atmospheric CO_2 concentrations, changes in concentrations, and atmospheric modeling to infer fluxes from land and ocean sources. They have made a strong contribution in quantifying the source-sink distribution of carbon at global and sub-regional scales (Ciasis et al. 2010; Dolman et al. 2010).

There are two main types of top-down models: measurements of atmospheric CO_2 concentration (through such methods as remote sensing or eddy covariance) and inverse modeling. The first allows for the partitioning of carbon sinks on land and in oceans by measuring changes in atmospheric concentrations of O_2/N_2 alongside measurements of $^{13}C/^{12}C$ ratios (Keeling et al. 1996). Proponents of these models acknowledge there is seasonal variability in measurements and suggest use of multiyear averages to ensure short term seasonal fluxes are not erroneously allocated to long term terrestrial or ocean sinks (Battle et al. 2000). Critics of these models have pointed out that they must be adequately adjusted to account for oceanic outgassing of O_2, otherwise the terrestrial carbon sink will be overstated (Plattner et al. 2002; Manning and Keeling 2006). They argue that the third IPCC report may have inflated the terrestrial carbon sink in the 1990s by 0.2–0.7 PgC/year due to inadequate accounting for outgassing. This would help to explain in part why the third IPCC report reported such a large increase in the net terrestrial carbon sink from the 1980s to the 1990s.

The second top down approach is called *inverse modeling*. Inverse models also examine atmospheric concentrations of CO_2. However, they measure regional distributions of carbon concentrations across space and time and use atmospheric transport modeling to estimate global sources and sinks. Inverse models are heavily influenced by the type of atmospheric transport model used, assumptions about prior regional fluxes, time resolution (annual versus monthly data), spatial resolution (number of source regions), and an ability to reconcile seasonal variations which may impact measurement of carbon fluxes in northern versus southern hemispheres (Schimel et al. 2001; Peylin et al. 2002). This approach has been used to confirm the residual carbon sink across continents.

Limits to inverse modeling include a lack of data to address smaller scales (Ciasis et al. 2010) and to isolate linkages regarding the mechanisms driving carbon fluxes. For example, inverse models typically have a spatial resolution too coarse to account for most inland waters. In riverine systems, carbon in terrestrial materials is transported into oceans, where it is released by air-sea gas exchange. Inverse models may erroneously record the CO_2 outgassing by riverine systems as a "sink" in the terrestrial balance while showing a "source" from the oceanic side, resulting in overestimation of the net terrestrial sink (Aumont et al. 2001; Pacala et al. 2001; Battin et al. 2009).

3.2 Bottom up Approaches

Bottom up approaches are based on stock changes or fluxes at the Earth's surface and infer changes in the atmosphere. Bottom up approaches include inventories, bookkeeping and process-based modeling. Inventory approaches and process based models are often used to produce results at the national or continental levels (Nilsson et al. 2007).

In inventory modeling, data on forest area, timber stocks, and forest growth are converted to biomass estimates to determine the carbon density of the vegetation, and then aggregated to form the forest carbon budget. The robustness of forest inventory methods is a function of accurate estimates of forest cover and appropriate biomass conversion factors. Variability in forest cover and biomass estimates, however, can vary widely as seen in Tables 8.3 and 8.4.

Estimation errors can result from several sources. These include inconsistent definitions

Table 8.3 Carbon stock estimates in plants and soil across biomes

	Area (10^9 ha)		Global carbon stocks (PgC)[a]					
	Dixon et al. (1994); Atjay et al. (1979)	Mooney et al. (2001)	Dixon et al. (1994); Atjay et al. (1979)			Mooney et al. (2001)	IGBP[b]	
Biome			Plants	Soil	Total	Plants	Soil	Total
Tropical forests	1.76	1.75	212	216	428	340	213	553
Temperate forests	1.04	1.04	59	100	159	139[c]	153	292
Boreal forests	1.37	1.37	88[d]	471	559	57	338	395
Tropical savannas and grasslands	2.25	2.76	66	264	330	79	247	326
Temperate savannas and grasslands	1.25	1.78	9	295	304	23	176	199
Deserts and semi deserts	4.55	2.77	8	191	199	10	159	169
Tundra	0.95	0.56	6	121	127	2	115	117
Croplands	1.6	1.35	3	128	131	4	165	169
Wetlands[e]	0.35	–	15	225	240	–	–	–
Total	15.12	14.93[f]	466	2,011	2,477	654	1,567	2,221

Source: Prentice (2001)

[a]Soil carbon values are for the top 1 m, although stores are also high below this depth in peatlands and tropical forests

[b]International Geosphere-Biosphere Programme – Data Information Service, Carter et al. (2000); DeFries et al. (1999)

[c]Estimate likely to be high, being based on mature stand density

[d]Estimate likely to be high due to high Russian forest density estimates including standing dead biomass

[e]Wetlands not recognized in Mooney et al. classification

[f]Total land area includes 1.55 × 109 ha ice cover not listed in this table.

Table 8.4 Estimates of forest area and biomass

	Matthews (2001); Fung et al. (2005)	Houghton (2003a, 1999)
Tropical		
Forest area (10^6 ha)	1,871	2,167
Woody biomass (PgC)	164[a]	288
Biomass per area (kgC/m)	8.8	13.3
Non-tropical		
Forest area (10^6 ha)	1,998	2,659
Woody biomass (PgC)	93[b]	223
Biomass per area (kgC/m)	4.7	8.4
Total		
Forest area (10^6 ha)	3,869	4,827
Woody biomass (PgC)	257	510
Biomass per area (kgC/m)	6.6	10.6

Source: Data from Kauppi (2003)

[a]Tropical pool is esimated as the total (257) minus non-tropical (93) = 164 PgC. All forests of South America and Africa are included in "tropical forest"

[b]88PcC excluding China and 475 PgC esimated for China (Fung et al. 2005). All forests of China, Australia and the US are included in "non-tropical"

of forest cover; incomplete or incorrect data from national sources; inconsistent treatment of areas of sparse tree cover; and accounting for (or omission of) certain types of land, such as recently disturbed forested areas which have temporarily fallen below the threshold canopy cover that has defined it as "forest" (Brown 2002; Kauppi 2003). In addition, inventory model results are dependent upon conversion factors used to translate wood data into carbon. These, in turn, are influenced by whether the model takes into consideration such factors as age class distribution and species composition, and whether it is a natural or managed stand (Goodale et al. 2002; Alexandrov et al. 1999).

Challenges in relying on inventory methods often stem from a lack of adequate data, particularly in the tropics, and the inability to incorporate rapid environmental changes or temporal trends into the major drivers of carbon sequestration (Nilsson et al. 2007). Critics have pointed to a

Table 8.5 Annual carbon fluxes estimated by top down and bottom up models in Pg/C/year (negative values = carbon sink)

	Top down methods		Bottom up methods	
		Inverse calculations		
	O_2 and CO_2	CO_2, $^{13}CO_2$, O_2	Forest inventories	Land-use change
Globe	−0.7 (±0.8)[a]	−0.8 (±0.8)[b]	–	2.2 (±0.6)[c]
Northern mid-latitudes	–	−2.1 (±0.8)[d]	−0.6[e]	−0.03 (±0.5)[c]
Tropics	–	1.5 (±1.2)[f]	−0.6 (±0.3)[g]	0.5–3.0[h]

Source: Houghton (2007)
[a]Plattner et al. (2002)
[b]−1.4 (±0.8) from Gurney et al. (2002), reduced by 0.6 to account for river transport (Aumont et al 2001)
[c]Houghton (2003b)
[d]−2.4 (Gurney et al. 2002), reduced by 0.3 to account for river transport (Aumont et al. 2001)
[e]Forests only, including wood products (Goodale et al. 2002)
[f]1.2 from Gurney et al. (2002), increased by 0.3 to account for river transport (Aumont et al. 2001)
[g]Undisturbed forests (Phillips et al. 1998; Baker et al. 2004)
[h]Fearnside (2000); DeFries et al. (2002); Houghton (2003b); Achard et al. (2004)

lack of adequate accounting for the spatial and temporal heterogeneity of forest carbon components such as belowground biomass, soils, litter, and forest products (Alexandrov et al. 1999; Goodale et al. 2002; House et al. 2003). To address this, many have argued the need for a global network of permanent plots as well as improved remote sensing technology to better understand carbon fluxes from the different components of forested ecosystems (Dixon et al. 1994). Remote sensing is also increasingly being used to address some of the inconsistencies behind national data. The Food and Agriculture Organization of the United Nations (FAO), which publishes the Forest Resource Assessment, recently changed its methodology in the 2010 Assessment to include a global remote sensing survey of forest area as a complement to its historical use of inventory methods (Ridder 2007).

Other bottom up models include "bookkeeping" and "process based" models (Houghton 2005a; Ramankutty et al. 2007). Bookkeeping models track changes in below and above ground carbon stocks through changes in land use. Unlike models based on atmospheric data, bookkeeping models are better able to isolate changes in carbon stocks specifically driven by human activity (such as land clearing) (Houghton 1999), and may provide benefits when viewed alongside top down models. In addition, these models can identify slower carbon releases from residual debris and sequestration from regeneration and regrowth (Achard et al. 2004). However, they exclude emissions from natural disturbances, which can be a significant carbon source (Kurz et al. 1999; Houghton 2003b).

Process based models are specifically designed to incorporate current knowledge regarding ecosystem processes into carbon accounting. In terrestrial ecosystems, these are typically either terrestrial biogeochemical models (TBMs) or dynamic global vegetation models (DVGMs) (Prentice 2001). Unlike top down or bookkeeping models, they seek to measure the interaction between vegetation and the environment (Shvidenko et al. 2010) under a variety of climatic conditions, as well as to diagnose the interannual variation of major carbon fluxes (Nilsson et al. 2007). Process-based models, however, are highly complex and utilize a large number of parameters, while the underlying empirical database is generally limited (Van Oijen et al. 2005). Process-based models do not have a strong ability to adequately assess uncertainties which, combined with limited data, makes it difficult to calibrate the model parameters (Makela et al. 2000; Van Oijen et al. 2005; Zaehle et al. 2005).

Some of the differences between top down and bottom up estimates of carbon fluxes can be seen in Table 8.5.

The different carbon budgeting methods produce a range of results, creating a 'CO_2 accounting gap' across sub-global (continental and smaller) scales from uncertainties in both the data and in the models themselves (House et al. 2003; Gusti and Jonas 2010). In Table 8.5, for instance, inversion models attribute a large amount of carbon uptake to northern latitudes versus land use change-based inventory methods that do not show as large of a sink. Existing global models are still unable to determine carbon sources or sinks with acceptable accuracy at regional and continental spatial scales and at interannual time scales. Generally, temporal patterns are poorly understood at scales greater than a few years (Fung 2000).

To address these concerns, research increasingly uses coupling of top-down and bottom-up models, or a dual-constraint approach (Dargaville et al. 2002). Efforts are being directed towards convergence of top-down/bottom-up methodologies to make up for the 'accounting gap' (Phillips et al. 2009; Gusti and Jonas 2010). A systems approach, 'Full Carbon Accounting' (FCA), estimating all land-based fluxes, whether natural or anthropogenic, may help to reconcile the top-down and bottom-up approaches (Shvidenko et al. 2010). Recently, estimates from top-down and bottom-up methodologies are increasingly converging as model development continues to progress (Malhi et al. 2008).

4 Knowledge and Measurement Gaps in Carbon Budgeting

There continues to be limits to our understanding of carbon cycling processes. Below-ground processes, particularly in forest soils, in biomass, and across forest types are poorly understood (Achard et al. 2004; Heimann and Reichstein 2008) (Chap. 2 of this volume discusses soil measurement issues in detail). For example, root production is a key component of net primary production, yet accurate data on root dynamics is sparse and often inferred from periodic field measurements of live and dead roots or from biomass estimates from allometric equations (Gower et al. 2001; Matamala et al. 2003). Other 'black boxes' include post-disturbance processes in soil on permafrost and nitrogen turnover (Nilsson et al. 2007; McGuire et al. 2009).

Different natural and anthropogenic processes drive carbon cycling in different ways and have varying degrees of influence across forest types (Dixon et al. 1994). In temperate forests, precise deforestation rates may not be as critical as determining the appropriate accounting treatment of residual post-harvest organic matter and carbon in wood products, displacement, and substitution which play a large role in carbon budgets in these regions. In these areas, carbon storage is driven by changes in carbon per unit area as opposed to changes in forested area (Houghton 2005a). In the tropics, however, accurately measuring deforestation is critical since it is a key driver of carbon fluxes.

Most scientists agree that land use change/deforestation in the tropics has caused the tropical biome to be neutral or near neutral in terms of carbon emissions, while outside the tropics, changes in land use (primarily reforestation), coupled with changes in forest age and structure, have resulted in the temperate and boreal biomes becoming a net carbon sink. However, there are significant shortcomings in the data and the way they are appropriated in carbon accounting. This includes a lack of available data and problems with how forest area is defined and classified across regions (Kauppi 2003; Ramankutty et al. 2007; Waggoner 2009). For example, Lepers et al. (2005) pointed to at least 90 different definitions of forests worldwide. Unknown or insufficient precision of measured data also can occur by subjective sampling, bias, deliberate falsification, and inappropriate measurement techniques (Nilsson et al. 2007).

Quantifying carbon fluxes from land use change is also challenging due to the temporal dynamics of land use change. Most scientists agree that northern mid-latitude forests have shown a net carbon accumulation over the last several decades. However, there is less agreement on the processes driving this uptake. This sink is often attributed to secondary forest growth as a result of past land use, such as in the northeastern United States where forests have regenerated following the abandonment of agricultural lands (Barford et al.

2001). Since lands deforested for agriculture were initially a carbon source, today's carbon sink is an "inherited" carbon uptake linked to past carbon emissions. It is important to distinguish the behaviour of sinks driven by past land use from sinks driven by biophysical processes such as CO_2 enrichment or longer growing seasons stemming from climate change (Schimel 2007). While both potentially serve as strong drivers of carbon uptake, their long term trends may be quite different. Studies have shown that carbon sequestration rates are often higher on lands recovering from disturbance (or intensive management) versus long term accumulations on unmanaged, natural landscapes (Schimel et al. 2001). Understanding the underlying mechanisms driving carbon sequestration rates is not only necessary for accurately projecting carbon uptake rates from secondary forests in temperate regions, particularly as they mature, but in projecting long term carbon fluxes in tropical regions where current areas of deforestation may become sinks if cleared lands are converted back to forest.

In addition, emissions from land use change have their own temporal variation. How land is cleared influences when and how much carbon is emitted to the atmosphere. For example, slash and burn clearing for agriculture tends to result in higher emissions in earlier years versus harvests which convert timber to long lived wood products (Ramankutty et al. 2007). Despite the fact that both activities fall into the category of land use change, the carbon fluxes observed over time may be quite different. Estimating future fluxes therefore requires making assumptions about the type of land use change expected to dominate in a particular area.

A central debate revolves around estimates of land use change in the tropics (DeFries et al. 2002; Ramankutty et al. 2007; Malhi 2010). This is largely due to a lack of adequate data although there are challenges posed by using country level data compiled from different underlying methodologies and aggregating them into one statistic. There is also concern that deforestation rates published for the 1980s were actually overstated in national statistics, which has led to underestimations of deforestation in the 1990s (DeFries et al. 2002). Some differences in tropical deforestation rates can be seen in Table 8.1a.

5 Climate Change Feedbacks

Forest ecosystems themselves are subject to local climatic conditions, implying a multitude of climate-ecosystem feedbacks which may amplify or dampen regional and global climate change. Quantifying and predicting these feedbacks are difficult. Positive feedbacks include greater frequency of fire and tree mortality driven by drought, insect outbreaks, and disease, and reduced albedo from less snow and ice cover. Negative feedbacks include increased forest productivity because of CO_2 fertilization, loss of forest cover due to temperature changes, and more frequent and severe disturbances (Lashoff and DeAngelo 1997; Betts 2000). An increase in temperature has different potential effects for the three major forest biomes – tropical, temperate and boreal.

It is generally accepted that the net terrestrial flux from increased temperatures will depend on whether the result is greater photosynthesis (carbon sink) or greater respiration (carbon source). Increased CO_2 concentrations in the atmosphere enhance photosynthesis ("CO_2 fertilization") and thus sequester more CO_2 until photosynthetic uptake of CO_2 becomes saturated. Higher temperatures, however, cause plant respiration to increase more than photosynthesis (Oliver and Larson 1996). There is concern that as global temperatures increase, trees will grow less due to rising respiration rates.

The effect of higher temperatures will directly and indirectly affect all elements of the carbon cycle. For example, warming temperatures observed under climate change are expected to alter seasonal variations, leading to longer growing seasons (Myneni et al. 1997). The net impact of these longer growing seasons on forest carbon storage is still under debate (Angert et al. 2005). Another important driver of tree growth is nitrogen availability, which can determine the magnitude of the CO_2 fertilization effect; at some point, warmer temperatures may result in nitrogen limitation (Oren et al. 2001; Schimel et al. 2001; Janssens et al. 2010).

Changes in climate are also expected to increase the natural disturbance regimes of the world's forests, including fire, drought, disease, and insect/pathogen outbreaks, resulting in both positive and negative feedbacks. Warmer temperatures may also cause compounding disturbance patterns. For example, drought-induced water stress may increase a forest's susceptibility to insect outbreaks. These outbreaks may, in turn, lead to higher fuel loads which increase the probability of stand replacing fires (Dale et al. 2001). On the other hand, carbon uptake by forests following disturbances is often higher than carbon sequestration on natural, unmanaged stands since young stands sequester carbon at faster rates than mature forests (Houghton 2007). Thus, the net effect is not well known.

The health of tropical forests is critical since they store more than one-quarter of the terrestrial carbon (Bonan 2008). As they are particularly vulnerable to a warmer, drier climate (Malhi et al. 2008; Phillips et al. 2009), they may create a positive feedback under climate change that decreases evaporative cooling, releases CO_2, and initiates forest dieback (Betts 2000; Bonan 2008).

The net climate forcing of temperate forests is highly uncertain. Higher albedo with loss of forest cover could offset carbon emissions such that the net climatic effect of temperate deforestation is negligible. Alternatively, reduced evapotranspiration with loss of tree cover could amplify warming. Warmer temperatures may result in increased precipitation in certain regions and decreased fire related disturbance (Bergeron and Archambault 1993). In North America, warmer winters may increase pine beetle infestations in northern areas, creating a positive feedback from the decay of killed trees (Logan et al. 2003; Kurz et al. 2008a, b), but may actually cause decreases in southern areas (Dale et al. 2001).

Developments in the boreal biome may influence the dynamics of atmospheric CO_2 in ways that affect the magnitude of global climate change (Dargaville et al. 2002). Boreal ecosystems store vast amounts of carbon in soils, permafrost, and wetlands. Warmer temperatures may stimulate thaw in carbon rich peatlands, providing positive feedback by releasing carbon and methane (Camill et al. 2001; Heimann and Reichstein 2008). On the other hand, climate forcing from increased albedo as coniferous forests in northern latitudes transition to deciduous forest types may offset the forcing from carbon emission so that boreal deforestation cools the climate. An increase is expected in fire frequency, which is the primary disturbance agent in most boreal forests (Balshi et al. 2007; Bond-Lamberty et al. 2007). The long-term effect of fires on climate will be a balance between post-fire increases in surface albedo, radiative heat emitted during combustion, and increased atmospheric CO_2 from the burned biomass (Bonan 2008).

6 Conclusions and Recommendations

Despite the complexities in measuring forest carbon budgets over time and across different forest types, policies to mitigate climate change must include forests in carbon management strategies. Below are several recommendations to help policy makers navigate the uncertainties:

- Support policies to promote afforestation and discourage deforestation or land degradation since afforestation promotes CO_2 sequestration and deforestation or forest degradation promote release of CO_2 to the atmosphere.
- Select the carbon budget model that best explains the carbon pools and processes under consideration. Dual-constrained models may provide a more comprehensive picture. However, it is important to recognize the constraints on any model used.
- Test the effects of carbon policies against multiple models to ensure against unintended consequences and to better understand the broader impacts of policy design on a wide variety of carbon processes. Accepted methods should be examined periodically to ensure they are appropriate for the current state of knowledge.
- Support the creation of long term research plots to generate better time series data on

land use, forest type, and deforestation rates so that long term processes impacting carbon sinks and sources can be better estimated.

- Require methodological consistency in country level accounting of items such as forest definition, cover type, deforestation rates, and biomass conversion factors. Link funding to compliance with global accounting standards.
- Forest management policies should consider not only the immediate impacts on carbon sequestration, but the longer term effects as well.
- Avoid using region-specific carbon research to make claims about global forest carbon budgets, since model results in one region may not be appropriate in other geographic areas.

Terrestrial carbon cycling is a critical component of the global carbon budget but is probably the least understood and most widely debated of the four main reservoirs. While models exist to measure carbon pools and fluxes in terrestrial ecosystems, no model is able to fully account for all the natural and anthropogenic processes driving carbon fluxes across time and space. When estimating climate change impacts on forest carbon, it is extremely difficult to aggregate the positive and negative feedbacks due to changes in temperature, moisture, and disturbance patterns.

Nevertheless, new forest management strategies must be designed to better optimize terrestrial carbon storage capacity while protecting long term forest carbon sinks from the effects of climate change and human-induced land use change. This will require continued research to better measure global forest carbon fluxes while encouraging a global effort to amass consistent and comprehensive data on the current state of forests across the globe. Although it may be a difficult undertaking, it will be necessary to ensure the ongoing health of the world's forests and their ability to continue as a critical carbon sink.

References

Achard F, Eva HD, Mayaux P, Stibig HJ, Belward A (2004) Improved estimates of net carbon emissions from land cover change in the tropics in the 1990s. Global Biogeochem Cycles 18:12p

Adams J (2011) Estimates of total carbon storage in various important reservoirs. Environmental Sciences Division, Oak Ridge National Laboratory, TN. http://www.esd.ornl.gov/projects/qen/carbon2.html. Accessed March 2011

Alexandrov GA, Yamagata Y, Oikawa T (1999) Towards a model for projecting net ecosystem production of the world forests. Ecol Model 123:183 191

Angert A, Biraud S, Bonfils C, Henning CC, Buermann W, Pinzon J, Tucker CJ, Fung I, Feld CB (2005) Drier summers cancel out the CO_2 uptake enhancement induced by warming springs. Proc Natl Acad Sci USA 102(31):10823–10827

Atjay GL, Ketney P, Duvigneaud P (1979) Terrestrial primary production and phytomass. In: Bolin B, Degens ET, Kempe S, Ketner P (eds) The global carbon cycle. John Wiley & Sons, New York, pp 129–181

Aumont O, Orr JC, Monfray P, Ludwig W, Amiotte-Suchet P, Probst JL (2001) Riverine driven interhemispheric transport of carbon. Global Biogeochem Cycles 15(2):393–405

Baker TR, Phillips OL, Malhi Y, Almeida S, Arroyo L, DiFiore A, Erwin T, Higuchi N, Killeen TJ, Laurance SG, Laurance WF, Lewis SL, Monteagudo A, Neill DA, Vargas PN, Pitman NC, Natalino J, Silva M, Martinez RV (2004) Increasing biomass in Amazonian forest plots. Philos Trans R Soc Lond B Biol Sci 359:353–365

Balshi MS, McGuire AD, Zhuang Q, Melillo JM, Kicklighter DW, Kasischke E, Wirth C, Flannigan M, Harden JW, Clein JS, Burnside TJ, McAllister J, Kurz WA, Apps M, Shvidenko A (2007) The role of historical fire disturbance in the carbno dynamics of the pan-boreal region: a process-based analysis. J Geophys Res 112, G02029-18p

Barford CC, Wofsy SC, Goulden ML, Munger JW, Pyle EH, Urbanski SP, Hutyra L, Saleska SR, Fitzjarrald D, Moore K (2001) Factors controlling long-and short-term sequestration of atmospheric CO_2 in a mid-latitude forest. Science 294(5547):1688–1691

Battin TJ, Luyssaert S, Kaplan LA, Aufdenkampe AK, Richter A, Tranvik LJ (2009) The boundless carbon cycle. Nat Geosci 2:598–600

Battle M, Bender ML, Tans PP, White JWC, Ellis JT, Conway T, Francey RJ (2000) Global carbon sinks and their variability inferred from atmospheric O_2 and $\delta_{13}C$. Science 287:2467–2470

Bergeron Y, Archambault S (1993) Decreasing frequency of forest fires in the southern boreal zone of Quebec and its relation to global warming since the end of the 'Little Ice Age'. Holocene 3(3):255–259

Betts RA (2000) Offset of the potential carbon sink from boreal forestation by decreases in surface albedo. Nature 408:187–190

Bonan GB (2008) Forests and climate change: forcings, feedbacks, and the climate benefits of forests. Science 320:1444–1449

Bond-Lamberty B, Peckham SD, Ahl DE, Gower ST (2007) Fire as the dominant driver of central Canadian boreal forest carbon balance. Nature 450(7166):89–92

Brown S (2002) Measuring carbon in forests: current status and future challenges. Environ Pollut 116:363–372

Camill P, Lynch JA, Clark JS, Adams JB, Jordon B (2001) Changes in biomass, aboveground net primary production, and peat accumulation following permafrost thaw in the boreal peatlands of Manitoba. Can Ecosyst 4:461–478

Canadell JG, LeQuere C, Rapauch MR, Field CB, Buitenhuis ET, Ciais P, Conway TJ, Gillett NP, Houghton RA, Marland G (2007) Contributions to accelerating atmospheric CO_2 growth from economic activity, carbon intensity and efficiency of natural sinks. PNAS 104(47):18866–18870

Carter AJ, Scholes RJ (2000) Spatial global database of soil properties. IGBP Global Soil Data Task CD-ROM, International Geosphere-Biosphere Programme (IGPB), Data Information Systems. Toulouse, France

Ciasis P, Rayner P, Chevallier F, Bousquet P, Logan M, Peylin P, Ramonet M (2010) Atmospheric inversions for estimating CO_2 fluxes:methods and perspectives. Climatic Change 103:69–92

Dale VH, Joyce LA, McNulty S, Neilson RP, Ayres MP, Flannigan MD, Hanson PJ, Irland LC, Lugo AE, Peterson CJ, Simberloff D, Swanson FJ, Stocks BJ, Wotton BM (2001) Climate change and forest disturbances. Bioscience 51(9):723–734

Dargaville R, McGuire AD, Rayner P (2002) Estimates of large-scale fluxes in high latitudes from terrestrial biosphere models and an inversion of atmospheric CO_2 measurements. Climatic Change 55(1–2):273–285

DeFries RS, Field CB, Fung I, Collatz GJ, Bounoua L (1999) Combining satellite data and biogeochemical models to estimate global effects of human-induced land cover change on carbon emissions and primary productivity. Global Biogeochem Cycles 13:803–815

DeFries RS, Houghton RA, Hansen MC, Field CB, Skole D, Townshen J (2002) Carbon emissions from tropical deforestation and regrowth based on satellite observations for the 1980s and 1990s. PNAS 99(22): 14256–14261

Dixon RK, Brown S, Houghton RA, Solomon AM, Trexler MJ, Wisnieski J (1994) Carbon pools and fluxes of global forest ecosystems. Science 263(5144):185–190

Dolman AJ, van der Werf GR, van der Molen MK, Ganssen G, Erisman JW, Strengers B (2010) A carbon cycle science update since IPCC AR-4. Ambio 39(5–6):402–412

Fan S, Gloor M, Mahlman J, Pacala S, Sarmiento J, Takahashi T, Tans P (1998) A large terrestrial carbon sink in north America implied by atmospheric and oceanic carbon dioxide data and models. Science 282:442–446

Fearnside PM (2000) Global warming and tropical land-use change: greenhouse gas emissions from biomass burning, decomposition and soils in forest conversion, shifting cultivation and secondary vegetation. Clim Change 46:115–158

Food and Agriculture Organization of the United Nations (FAO) (2000) FRA 2000: on definitions of forest and forest change. Forest Resources Assessment Programme (FRA). Working Paper 33. Rome

Fung I (2000) Variable carbon sinks. Science 290(5495): 1313

Fung IY, Doney SC, Lindsay K, John J (2005) Evolution of carbon sinks in a changing climate. PNAS 102(32):11201–11206

Goodale CL, Apps MJ, Birdsey RA, Field CB, Heath LS, Houghton RA, Jenkins JC, Kohlmaier GH, Kurz W, Liu S, Naburrs G, Nilsson S, Shvidenko AZ (2002) Forest carbon sinks in the northern hemisphere. Ecol Appl 12(3):891–899

Gower ST, Krankina O, Olson RJ, Apps M, Linder S, Wang C (2001) Net primary production and carbon allocation patterns of boreal forest ecosystems. Ecol Appl 11(5):1395–1411

Gurney KR, Law RM, Denning AS, Rayner PJ, Pak BC, Baker D, Bousquet P, Bruhwiler L, Chen YH, Ciais P, Fung IY, Heimann M, John J, Maki T, Maksyutov S, Peylin P, Prather M, Taguchi S (2004) Transcom 3 inversion intercomparison: model mean resuts for the estimation of seasonal carbon sources and sinks. Global Biogeochem Cycles 18: GB1010-GB1018

Gusti M, Jonas M (2010) Terrestrial full carbon account for Russia: revised uncertainty estimates and their role in a bottom-up/top-down accounting exercise. Climatic Change 103(1):159–174

Heimann M, Reichstein M (2008) Terrestrial ecosystem carbon dynamics and climate feedbacks. Nature 451: 289–292

Houghton RA (1999) The annual net flux of carbon to the atmosphere from changes in land use 1850–1990. Tellus 51B:298–313

Houghton RA (2003a) Why are estimates of the terrestrial carbon balance so different? Glob Change Biol 9: 500–509

Houghton RA (2003b) Revised estimates of the annual net flux of carbon to the atmosphere from changes in land use and land management 1850–2000. Tellus 55B:378–390

Houghton RA (2005a) Aboveground forest biomass and the global carbon balance. Glob Change Biol 11:945–958

Houghton RA (2005b) Chapter 1: Tropical deforestation as a source of greenhouse gas emissions. In: Moutinho P, Schwartzman S (eds) Tropical deforestation and climate change. Amazon Institute for Environmental Research, Washington, DC

Houghton RA (2007) Balancing the global carbon budget. Annu Rev Earth Planet Sci 35:313–347

House JI, Prentice IC, Ramankutty N, Houghton RA, Heimann M (2003) Reconciling apparent inconsistencies in estimates of terrestrial CO_2 sources and sinks. Tellus 55B:345–363

Janssens IA, Dieleman W, Luyssaert S, Subke JA, Reichstein M, Ceulemans R, Ciais P, Dolman AJ, Grace J, Matteucci G, Papale D, Piao SL, Schulze ED, Tang J, Law BE (2010) Reduction of forest soil respiration in response to nitrogen deposition. Nat Geosci 3:315–322

Kauppi P (2003) New low estimate for carbon stock in global forest vegetation based on inventory data. Silva Fennica 37(4):451–458

Keeling RF, Piper SC, Heimann M (1996) Global and hemispheric CO_2 sinks deduced from changes in atmospheric O_2 concentration. Nature 381:218–221

Kurz WA, Apps MJ (1999) A 70 year retrospective analysis of carbon fluxes in the Canadian forest sector. Eco Appl 9(2):526–547

Kurz WA, Stinson G, Rampley GJ, Dymond CC, Neilson ET (2008a) Risk of natural disturbances makes future contribution of Canada's forests to the global carbon cycle highly uncertain. PNAS 105(5):1551–1555

Kurz WA, Dymond CC, Stinson G, Rampley GJ, Neilson ET, Carroll AL, Ebata T, Safranyik L (2008b) Mountain pine beetle and forest carbon feedback to climate change. Nature 452:987–990

Lashoff DA, DeAngelo BJ (1997) Terrestrial ecosystem feedbacks to global climate change. Annu Rev Energy Environ 22:75–118

Lepers E, Lambin EF, Janetos AC, DeFries R, Achard F, Ramankutty N, Scholes RJ (2005) A synthesis of information on rapid land-cover change for the period 1981–2000. Bioscience 55(2):115–124

Liski J, Korotkov AV, Prins CFL, Karjalainen T, Victor DG, Kauppi PE (2003) Increased carbon sink in temperate and boreal forests. Climatic Change 61(1–2): 89–99

Logan JA, Regniere J, Power JA (2003) Assessing the impacts of global warming on forest pest dynamics front. Ecol Environ 1(3):130–137

Makela A, Landsberg J, Ek AR, Burk TE, Ter-Mikaelian M, Agren GI, Oliver CD, Puttonen P (2000) Process based models for forest ecosystem management: current state of the art and challenges for practical implementation. Tree Physiol 20:289–298

Malhi Y (2010) The carbon balance of tropical forest regions, 1990–2005. Curr Opin Environ Sustainab 2(4):237–244

Malhi Y, Roberts JT, Betts RA, Killeen TJ, Wenhong L, Nobre CA (2008) Climate change, deforestation, and the fate of the Amazon. Science 319(5860):169–172

Manning AC, Keeling RF (2006) Global oceanic and land biotic carbon sinks from the Scripps atmospheric oxygen flasks sampling network. Tellus 58B:95–116

Matamala R, Gonzales-Meler MA, Jastrow JD, Norby RJ, Schlesinger WH (2003) Impacts of fine root turnover on forest npp and soil c sequestration potential. Science 302:1385–1387

Matthews E (2001) Understanding the FRA 2000. Forest briefing no. 1, World Resources Institute, Washington, DC, 12p

McGuire AD, Anderson LG, Christensen TR, Dallimore S, Guo L, Hayes DJ, Heimann M, Lorenson TD, MacDonald RW, Roulet N (2009) Sensitivity of the carbon cycle in the arctic to climate change. Ecol Monogr 79(4):523–555

Mooney J, Saugier B, Mooney HA (2001) Terrestrial global productivity. Academic Press, San Diego

Myneni RB, Keeling CD, Tucker CJ, Asrar G, Nemani RR (1997) Increased plant growth in the northern high latitudes from 1981–1991. Nature 386:698–702

Myneni RB, Dong J, Tucker CJ, Kaufmann RK, Kauppi PE, Liski J, Zhou L, Alexeyev V, Hughes MK (2001) A large carbon sink in the woody biomass of northern forests. PNAS 98(26):14784–14789

Nilsson S, Shvidenko A, Jonas M, McCallam I, Thomson A, Balzter H (2007) Uncertainties of a regional terrestrial biota full carbon account: a systems analysis. Water Air Soil Pollut Focus 7(4/5):425–441

Oliver CD, Larson BC (1996) Forest stand dynamics, updateth edn. Wiley, New York

Oren R et al (2001) Soil infertility limits carbon sequestration by forest ecosystems in a CO_2-enriched atmosphere. Nature 411:469–472

Pacala SW, Hurtt GC, Baker D, Peylin P, Houghton RA, Birdsey RA, Heath L, Sundquist ET, Stallard RF, Ciais P, Moorcroft P, Caspersen JP, Shevliakova E, Moore B, Kohlmaier G, Holland E, Gloor M, Harmon ME, Fan SM, Sarmiento JL, Goodale CL, Schimel D, Field CB (2001) Consistent land- and atmosphere-based US carbon sinks estimates. Science 292:2316–2320

Peylin P, Baker D, Sarmiento J, Ciais P, Bousquet P (2002) Influence of transport uncertainty on annual mean and seasonal inversions of atmospheric CO_2 data. J Geophys Res 107(D19): ACH 5–1 – ACH 5–17

Phillips OL, Aragao L, Lewis SL, Fisher JB, Lloyd J, Lopez-Gonzales G, Mahli Y, Monteagudo A, Peacock J, Quesada CA, van der Heijden G, Almeida S, Amaral I, Arroyo L, Aymard G, Baker R, Bánki O, Blanc L, Bonal D, Brando P, Chave J, de Oliveira ATA, Cardozo ND, Czimczik CI, Feldpausch TR, Freitas MA, Gloor E, Higuchi N, Jiménez E, Lloyd G, Meir P, Mendoza C, Morel A, Neill DA, Patino S, Penuela MC, Prieto A, Ramirez F, Schwarz M, Silva J, Silviera M, Thomas AS, ter Steege H, Stropp J, Vasquez R, Zelazowski P, Davila EA, Andelman S, Andrade A, Chao K-J, Erwin T, DiFiore A, Honorio E, Keeling H, Killeen TJ, Laurance WF, Cruz AP, Pitman NCA, Vargas PN, Ramirez-Angulo H, Rudas A, Salamao R, Silva N, Terbrgh J, Torres-Lezama A (2009) Drought sensitivity of the Amazon rainforest. Science 323(5919):1344–1347

Phillips OL, Malhi Y, Higuchi N, Laurance WF, Nunez PV (1998) Changes in the carbon balance of tropical forests: evidence from land-term plots. Science 282:439–442

Plattner GK, Joos F, Stocker TF (2002) Revision of the global carbon budget due to changing air-sea oxygen fluxes. Global Biogeochem Cycles 16(4):12p

Prentice IC (2001) The carbon cycle and atmospheric carbon dioxide. Intergovernmental Panel on Climate Change (IPCC) Third Assessment Report, Chapter 3, Pp 183–239

Ramankutty N, Gibbs HK, Achard F, DeFries R, Foley JA, Houghton RA (2007) Challenges to estimating carbon emissions from tropical deforestation. Glob Change Biol 13:51–66

Raupach MR, Marland G, Clais P, LeQuere C, Canadell JG, Klepper G, Field CB (2007) Global and regional drivers of accelerating CO_2 emissions. PNAS 104(24): 10288–10293

Ridder RM (2007) Global forests resources assessment 2010: options and recommendations for a global remote sensing survey of forests, Forest Resources Assessment Programme, Working Paper 141, 68p

Schimel DS (2007) Carbon cycle conundrums. PNAS 104(47):18353–18354

Schimel DS, House JI, Hibbard KA, Bousquet P, Ciais P, Peylin P, Braswell BH, Apps MJ, Baker D, Bondeau A, Canadell J, Churkina G, Cramer W, Denning AS, Field CB, Friedlinstein P, Goodale C, Heimann M, Houghton RA, Melillo JJ, Moore B III, Murdiyarso D, Noble I, Pacala SW, Prentice IC, Raupach MR, Rayner PJ, Scholes RJ, Steffen WL, Wirth C (2001) Recent patterns and mechanisms of carbon exchange by terrestrial ecosystems. Nature 414:169–172

Shvidenko A, Schepaschenko D, McCallam I, Nilsson S (2010) Can the uncertainty of full carbon accounting of forest ecosystems be made acceptable to policy-makers? Climatic Change 103:137–157

Stephens BB, Gurney KR, Tans PP, Sweeney C, Peters W, Bruhwiler L, Ciais P, Ramonet M, Bousquet P, Nakazawa T, Aoki S, Machida T, Inoue G, Vinnichenko N, Lloyd J, Jordan A, Heimann M, Shibistova O, Langenfelds RL, Steele LP, Francey RJ, Denning AS (2007) Weak northern and strong tropical land carbon uptake from vertical profiles of atmospheric CO_2. Science 316:1732–1735

UNFAO (2000) Global forest resources assessment 2000. Food and agriculture organization of the United Nations, FAO Forestry Paper 140, 481pp

Van Oijen M, Rougier J, Smith R (2005) Bayesian calibration of process-based forest models: bridging the gap between models and data. Tree Physiol 25:915–927

Waggoner PE (2009) Forest inventories: discrepancies and uncertainties. RFF DP 09–29. Resources for the Future, 45p

Zaehle S, Sitch S, Smith B, Hatterman F (2005) Effects of parameter uncertainties on the modeling of terrestrial biosphere dynamics. Global Biogeochem Cycles 19(3), GB2030-16p

Part III

The Management of Carbon in Forests and Forest Products

The following four papers provide a comprehensive synthesis and review of forest carbon management in forests and the life cycle of associated wood products. The papers highlight areas of what is known from recent research, and where the gaps are.

The first three papers discuss the management of forest carbon in existing tropical, temperate, and boreal forests, and in afforestation and reforestation projects. In temperate and boreal forests resiliency treatments (such as fuel reduction thinning and prescribed fire) result in lowered vegetative carbon storage, but they help produce forests that are significantly less susceptible to catastrophic disturbance (with accompanying drastic carbon release). In the tropics, reduced impact logging (RIL) is an important practice to lessen carbon loss, but it is necessary to move beyond RIL to substantially increase carbon storage by developing a more sophisticated silviculture. The largest potential source of carbon sequestration in the tropics is the development of second growth forests on old agricultural lands and plantations established on appropriate sites. However, for all forests, the risk of leakage must be addressed. If carbon sequestration strategies simply displace timber harvests from one forest to another, the ultimate carbon gain is questionable.

The final paper evaluates post-harvest strategies of carbon management. Some studies find that substitution of wood for other construction materials (e.g., steel and concrete) produces net GHG emissions reductions. Substitution effects may be up to 11 times larger than the total amount of carbon sequestered in forest products annually. However, paper products contain significantly more embedded fossil fuel (carbon) energy than wood products, and newer wood products such as oriented strand board and laminated veneer lumber use 80–216% of the energy needed to produce solid sawn lumber. The end-of-use pathways of wood products are also promising. Once discarded, wood products can be burned for energy production, recycled or reused, or put in landfills, where the carbon can remain indefinitely due to anaerobic conditions.

Contributors toward organizing and editing this section were *Mark S. Ashton, Deborah Spalding, Thomas Graedel, Mary L. Tyrrell, and Reid Lifset*

Managing Carbon Sequestration in Tropical Forests

9

Cecilia Del Cid-Liccardi, Tim Kramer, Mark S. Ashton, and Bronson Griscom

Executive Summary

This chapter examines how management methods can be implemented to reduce carbon loss and increase carbon storage in tropical forests. Tropical deforestation and degradation are contributing about 15% of total annual global greenhouse gas emissions. As policy makers work to develop solutions that address climate change, there has been considerable focus on incorporating tropical forests into the overall climate solution. Silvicultural practices will need to be an integral part of reducing carbon loss and improving carbon storage if we are to solve this global challenge while meeting resource needs.

Important Considerations and Trends

- Global climate change negotiations have begun to focus on sustainable forest management as a means to achieving carbon emission reductions, thus presenting opportunities in tropical forest management.
- The most important goal in managing tropical forests for carbon is to conserve standing forests, especially primary forests that are high in carbon.
- Forest carbon storage and uptake vary significantly based on climate, soils, hydrology, and species composition. It is necessary to consider these factors when managing a tropical forest for carbon.
- Reduced impact logging (RIL) is an important practice to lessen carbon loss, but it is necessary to move beyond RIL to substantially increase carbon storage by developing more sophisticated, planned forest management schemes with silvicultural treatments that ensure regeneration establishment, post establishment release, and extended rotations of new stands. Some work on silviculture has been done in the rainforest regions (ever-wet and semi-evergreen), but only in very specific places; almost none has been done in montane or seasonal (dry deciduous) forests.
- Forests can be financially viable compared to other land uses through integration and cultivation of species that provide timber and non-timber products that are stacked (cumulative) and that are compatible with service values – carbon sequestration and water quality.
- To increase forest carbon storage while also meeting societies' resource needs, it is essential to engage in stand- and landscape-level planning aimed at increasing carbon storage.
- Many logged over and second growth forests are ideal candidates for rehabilitation through enrichment planting of supplemental long-lived canopy trees for carbon sequestration.

C.D. Cid-Liccardi • T. Kramer • M.S. Ashton (✉)
Yale School of Forestry & Environmental Studies, 360 Prospect St, New Haven, CT 06511, USA
e-mail: mark.ashton@yale.edu

B. Griscom
Carbon Science Program, The Nature Conservancy, 4245 North Fairfax Drive, Arlington, VA 22203, USA

M.S. Ashton et al. (eds.), *Managing Forest Carbon in a Changing Climate*,
DOI 10.1007/978-94-007-2232-3_9, © Springer Science+Business Media B.V. 2012

- The largest potential source of carbon sequestration in the tropics is the development of second growth forests on old agricultural lands and plantation systems that have proven unsustainable. Every incentive should be provided to encourage this process.

1 Introduction

Tropical forest systems play a large role in the global carbon cycle, with tropical vegetation and soils holding almost 17% of all carbon stored in terrestrial ecosystems (Schlesinger 1997). The large size and distribution of tropical forests make them a significant carbon reservoir (Schimel 1995). Tropical deforestation and degradation account for emissions of about 1.6 Pg C/year, or about 15% of the annual global carbon emissions (van der Werf et al. 2009; Canadella et al. 2007; IPCC 2007). This represents a greater contribution to global climate change than all the planes, trains and automobiles on earth. Estimates for the proportion of these emissions that result directly from forest degradation activities (i.e. logging, fire, fuelwood harvest) range from less than 10% to over 30%, depending upon scale and region of analysis, and methods employed, as reviewed by Griscom et al. (2009).

Improved forest management can reduce these emissions both directly by reducing or reversing the loss of carbon from remaining forests (degradation), or indirectly by improving the incentive to avoid forest conversion (deforestation). In this chapter, we discuss the management options available to reduce or reverse forest degradation and deforestation.

Tropical forests around the world are undergoing a dramatic change, with primary forests (high in biomass and carbon) being converted to agricultural lands or degraded forests (low in biomass and carbon). A significant proportion of this change is occurring as a result of selective logging intensification throughout the tropics (Pinard et al. 1995; Uhl and Kauffman 1990; Verissimo et al. 1992). The spatial extent of logging in the tropics has historically been difficult to quantify, since most logging in tropical forests involves selective tree felling, which is hard to detect using affordable remote sensing data. However, new methods for sub-pixel analysis of low-cost satellite data (i.e. Landsat) are capable of detecting selective logging activity (Asner et al. 2005). Employing these methods, integrated with conventional bookkeeping methods, Asner et al. (2009) conclude that at least 20.3% (3.9 million km^2) of humid tropical forests have been allocated to selective timber harvest. Estimates for emissions from each hectare of forest selectively logged varies, depending upon region and logging practices, from less than 20 Mg C/ha (Keller et al. 2004) to over 100 Mg C/ha (Pinard and Putz 1996).

Logging operations in the tropics rarely use best forest management practices or silvicultural methods and this has resulted in extensive loss of carbon (Putz and Pinard 1993). Less than 1% of tropical forest area is under certified forest management (Siry et al. 2005). The implementation of basic forest management methods throughout the tropics has the potential to considerably reduce carbon loss and increase carbon uptake and storage (Putz et al. 2008a). Efforts to maximize carbon uptake and reduce carbon loss need to be based on site dynamics, well-planned management practices, as well as the application of silvicultural practices that are based on forest type and site characteristics. The high levels of carbon loss that result from deforestation and degradation have led to a considerable focus on incorporating tropical forests into the overall climate mitigation solution (IPCC 2007). Forest management policies and silvicultural practices are and will continue to be an integral part of efforts such as REDD+that are aimed at reducing carbon loss and improving carbon storage.

This chapter will provide an overview of tropical forest management practices and how they can be used to manage tropical forests for carbon. First we will present and outline the major tropical forest biomes and discuss how carbon is related to forest type and the site characteristics of each region. We will then discuss several key concepts that are important to understand when it comes to managing tropical forests for carbon. Then we will discuss the practice of reduced impact logging (RIL) and how it can be applied to reduce the carbon lost through conventional logging practices. After discussing RIL, we will

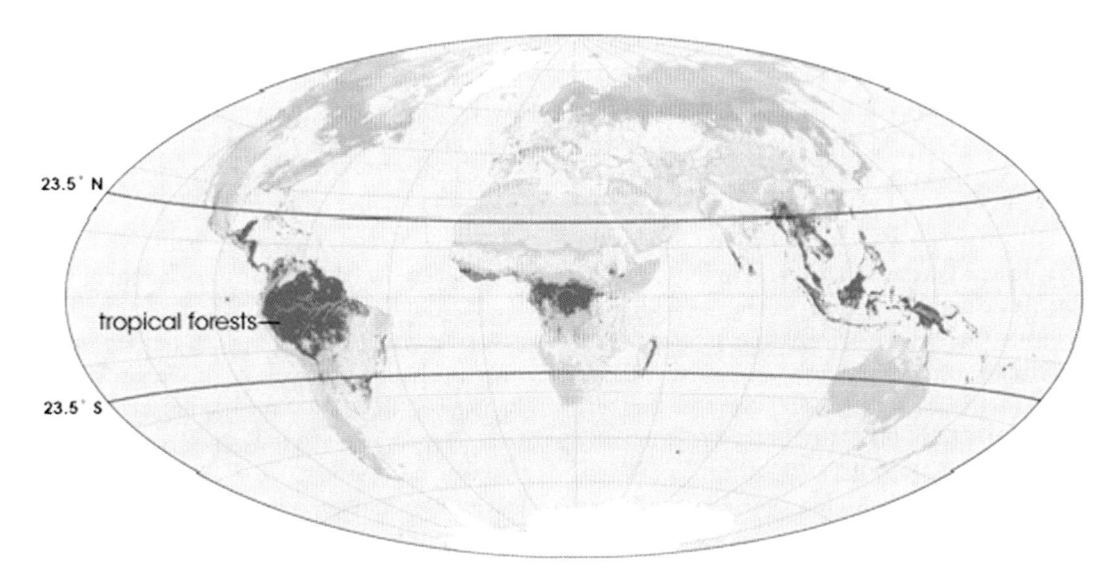

Fig. 9.1 Original extent of boreal, temperate, and tropical forest types of the world prior to land clearing

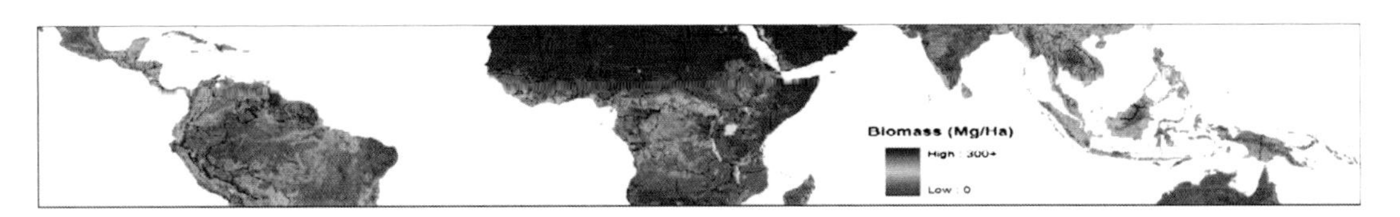

Fig. 9.2 The latest forest biomass carbon map for tropical forests has been developed by integrating optical (MODIS), lidar (GLAS), and RADAR (Alos) satellite data with field sampling. (Reprinted with permission)

shift our focus to management practices and silvicultural treatments that are rarely used now, but which could significantly improve carbon storage and uptake in tropical forests. Finally, we will end with a summary of the key findings and policy implications that will have been outlined within the chapter. It is our hope that this will be instructive for land managers and policy makers who are seeking to better understand the various approaches that are available and appropriate for managing tropical forests for carbon.

2 Tropical Forests of the World

Tropical forests are found throughout the equatorial regions of the world and are broadly categorized by region: Africa, the Neotropics (Central and South America), and Asia (Fig. 9.1). Each region can be divided into three major forest biomes: tropical rainforests (ever-wet, semi-evergreen), tropical montane forests, and tropical seasonal forests (dry deciduous). All biomes are loosely contained between the Tropic of Cancer (23°N) and the Tropic of Capricorn (23°S) and encompass a broad range of regional expressions that vary based on elevation, soil conditions, and regional climatic variations. Carbon uptake and storage vary significantly across biomes, with an average of 200 Mt C/ha in tropical rainforests and 140 Mt C/ha in tropical seasonal forests (Houghton 1999; DeFries et al. 2007). Within each region (Asia, the Neotropics, or Africa) similar biomes can have dramatically different carbon values, which also fluctuate across the landscapes of these regions (Fig. 9.2) (IPCC 2006).

2.1 Tropical Rainforest

Tropical ever-wet and semi-evergreen forests are characterized by more than 80 in. (2,000 mm) of rain annually. These forests have the highest vegetation biomass as well as the largest carbon stocks of all tropical forests (Holzman 2008). The most species rich and structurally diverse forests are around the equatorial latitudes. The greatest expanses of semi-evergreen rainforests are found in the Amazon Basin and the Congo Basin, while the greatest expanses of ever-wet rainforests are on the Southeast Asian islands of Borneo, Sumatra, and New Guinea. The three regional expressions of the tropical rainforests–Neotropical (Central America, the Pacific coast of northern South America, the Amazon, the Caribbean), African (West Africa, Central Africa), and Asian-Pacific regions (South Asia, Indochina, maritime Southeast Asia, Australia/New Guinea) – are each distinct from one another in terms of forest tree composition as well as in carbon levels. These differences are the result of biogeographical origin, climate, soil, and forest structure (Holzman 2008). The greatest similarity exists between the Amazon and Central Africa forests because of a common, but ancient biogeography.

2.1.1 Regional Variations

The neotropical rainforest of Central and South America is the largest and most extensive of the tropical rainforest biomes. A number of tree families are represented in the canopy layer of these forests: Brazil nut (Lecythidaceae), the genera *Tabebuia* (Bignoniaceae), *Anacardium* (Anacardiaceae), and *Vochysia* (Vochysiaceae), and many genera (e.g. *Parkia, Cedrelinga, Dalbergia, Dipteryx*) in the Leguminosae (Meggers et al. 1973). The single most important timber family in the neotropics is mahogany (Meliaceae) with genera such as *Guarea, Swietenia* and *Cedrella* dominating the timber markets. The Neotropical forests, which currently store between 120 and 400 Mt C/ha depending upon species composition, soil, and climate, have historically and continue to experience high levels of deforestation (IPCC 2006).

The African rainforest is smaller in size with less species diversity than the other regions and the forest less dense, with levels of carbon ranging from 130 to 510 Mt C/ha (IPCC 2006). Within the African tropical forest, the canopy layers tend to consist of members of the Caesalpinioideae subfamily of the legume family and include *Gilbertiodendron*, mopane (*Colophospermum mopane*), and senna (*Senna siamea*) (Meggers et al. 1973). However, the most important timber family in this region is again in the Meliaceae, represented by the African mahogany genera (*Entantrophragma, Khaya*).

The Asian-Pacific forest is distinctive due to the presence of the Dipterocarpaceae family of trees that dominates the forest composition. These trees are among the tallest in the tropical rainforest biome and occur in large clumps (Holzman 2008). It is the dominance of the Dipterocarpaceae tree family that gives these forests the highest carbon levels (120–680 Mt C/ha) (IPCC 2006), along with the high carbon peat swamps (>1,000 Mt C/ha) of the region.

2.1.2 Lowland Rainforests

Lowland tropical forests exist below 300 m elevation and constitute the vast majority of tropical ever-wet and semi-evergreen rainforests. These forests have a diversified forest canopy system with the greatest number of commercial tree species, such as the dipterocarps (Dipterocarpaceae), Brazil nut (*Berthollletia excelsa*), and mahogany (*Swietenia, Entantrophragma, Khaya, Cedrella*) species. Lowland tropical forests comprise most of the Amazon and Congo Basins. On their outer margins and along the major river ways, they are being logged and converted at a much faster rate than was predicted because soils are suitable for agriculture and the land more accessible (Fearnside 1993). However, rate of deforestation in Amazonia is less than in SE Asia, but absolute amount of deforestation is higher in Amazonia. Within the lowland rainforests of SE Asia, the peat swamps, where elevated water tables inhibit decomposition, have a substantial storage of organic matter. Draining these peat swamps results in a significant loss of stored carbon (Dixon et al. 1994). Due to emissions from draining peat forests, absolute amount of emissions is very high in SE Asia.

Coastal lowland rainforests are usually located on fertile soils and tend to be extremely workable and in areas of high human influence (Fearnside 1993). Coastal forests usually have flat, deep soils with a sandy component, making them suitable for plantations and tree crop agriculture systems such as oil palm, rubber, and coconut species (Ashton

2003). As a result, many of these forests have been cleared and replaced with tree crop agriculture that requires intensive inputs. These forests will likely remain in this state given their productivity and proximity to markets. As a result, relatively few coastal rainforests still exist. Where they do persist, as in the Chocó rainforest along the Pacific Coast of Panama and Colombia, they hold a large amount of stored carbon (Leigh 1999).

2.1.3 Hill Rainforests

Inland, rainforests with elevations greater than 300 m are much more variable and diverse, with major differences in stored carbon between the broad flat areas and the hilly uplands. These forests are often on marginal lands, in terms of fertility, since the soils often have higher clay content and poor structure (Schimel 1995). The steep slopes in these areas make them less workable and more prone to erosion. When these forests are converted to agriculture and range land, which has occurred in many regions, they are more likely to be abandoned over time and to revert to secondary forests. Often, these forested hilly regions are part of important catchments that provide water for coastal cities and for irrigating crops in coastal lowlands (Ashton 2003). This combination of factors makes these forested regions ideal for long-term carbon management as well as for co-benefits like water.

2.2 Tropical Montane Forests

Tropical montane forests grow above an altitude of 1,000 m. For wet tropical rainforests, an increase in altitude results in changes in forest structure (Vitousek and Stanford 1986). Primarily, these forests become shorter, thicker, and denser with a less developed canopy strata system. They tend to hold less carbon on average than lowland tropical forests as a result of the slower growth rates. Despite this, increased precipitation and decreased decomposition have led to high soil carbon levels, with as much as 61.4 Mt C/ha found in montane forest regions in Ecuador (Rhoades et al. 2000). Regional comparisons show that montane tropical forests in Africa hold between 40 and 190 Mt C/ha, in the Neotropics 60–230 Mt C/ha, and in the Asia-Pacific region, where the highest carbon stocks have been recorded, 50–360 Mt C/ha (Gibbs et al. 2007). South America holds the majority of montane forests because of the Andes, whereas in Africa they are restricted to the upper slopes of the volcanic island mountain systems of East and Central Africa.

Soils in montane forests are usually very fertile, consisting of inceptisols or histosols. As a result, montane forests located adjacent to populated areas often experience a significant loss of forest and soil carbon to intensive agriculture (market gardens – vegetables). Because of the steep slopes, they are also easily eroded. Rhoades et al. (2000) found almost a 24% decrease in soil carbon levels between primary forests and sugar cane fields at high elevations in Ecuador due to soil disturbance and soil erosion. Isolation makes these forests more suitable for carbon reserves because they are less threatened by agricultural conversion and timber extraction. But for the same reason, these forests have limited "additionality".

2.3 Tropical Seasonal Forests

Tropical regions with distinct seasonal rainfall are home to dry deciduous forests. These forests are found in wide bands along the perimeter of the Tropical Rainforest biome towards the margins of the tropical latitudes between 10° and 20° N and S latitudes (Holzman 2008). Seasonal tropical forests primarily occur in South Asia, West and East Africa, northern Australia, the Pacific side of Central America, eastern Brazil, and the southern rim transition region of Amazonia. In seasonal forests there is a distinct cooler and extended dry season and a distinct wet season with precipitation less than 2,000 mm per year unless the climate is strongly monsoonal, where rainfall can be very high but over a short time interval, making most of it surplus. On average these forests tend to be less diverse and more dwarfed in terms of tree size. Fire and large ungulates can play an important role in regulating forest understories in comparison to typical equatorial rainforests. Many of the same families of trees found in tropical rainforests are also found in seasonal tropical forests: however, the species are quite different (Holzman 2008). Trees in the fig family (Moraceae) are widespread throughout all regions, as are trees in the kapok family (Bombacaceae) such as kapok (*Ceiba pentandra*) and palo barrocho (*Chloroleucon chacoense*) trees in the Neotropics and baobab (*Adansonia*) in Africa (Bullock et al. 1995). Many legumes in the subfamilies Mimocaceae (e.g. *Albizia*, *Acacia*) and Fabiaceae (e.g. *Gliricidia*) are common in both the Neotropics and Africa. The seasonally dry miombo woodlands that create an arc around the wet evergreen forests of Central Africa are dominated by *Brachystegia* (Ceaesalpinioideae). Trees in Asian seasonally dry forests are often represented by species in the Combretaceae (*Terminalia*), Verbenacaeae (teak – *Tectona*, *Vitex*), Ebenaceae (ebony – *Diospyros*) and Dipterocarpaceae (sal – *Shorea robusta*) families (Bullock et al. 1995).

As with tropical rainforests, there are distinct regional differences that exist in tropical seasonal forests in terms of species composition, soil quality, and climatic variables, all of which affect the levels of biomass and carbon storage found within each region. This is evident in the regional carbon variations that exist within seasonal tropical forests, with Africa having on average 140 Mt C/ha, the Neotropics 210 Mt C/ha, and the Asian-Pacific region holding the lowest of all, 130 Mt C/ha. (Gibbs et al. 2007)

Under normal climatic conditions, major fires do not appear to be a frequent occurrence in seasonally dry tropical forests (Malaisse 1978; Hopkins and Graham 1983). The most vulnerable dry forests are those adjacent to savanna vegetation because of the flammable grasses and shrubs (Hopkins and Graham 1983). Malaisse (1978) found that local people started most fires in the African miombo (woodland) ecosystems during the dry season to maintain the areas for grazing. When managing a tropical dry forest for carbon, it is vitally important to work with local people to reduce the risk of fire in these forests and develop solutions that work well with local needs (Schwartzman et al. 2000).

Another general characteristic of tropical seasonal forests is that their soils overall are more fertile than in wet tropical regions. Forests of this type have therefore received proportionally greater impact from land conversion to agriculture, with higher human populations that are more dependent on fuelwood from the forest (e.g. West and East Africa; S and E Brazil, Mexico, S. India, Philippines) than in wetter tropical forest biomes (e.g. Central Africa, upper Amazon, Borneo, Sumatra). Many seasonally dry forests are now restricted to the most marginal lands and represent a small fragment of what they once were (see Griscom and Ashton 2011). All of this again emphasizes the importance of site and regional knowledge in managing these forests for carbon.

3 Key Concepts

In order to fully understand how tropical forests are affected by forest management practices, it is important to understand a few key concepts.

3.1 Primary Tropical Forest, Managed Tropical Forest and Second Growth

Primary tropical forests are forests that have attained a great age and exhibit a structural variety that provides higher habitat diversity than forests in other categories. Primary forests usually have multiple horizontal layers of vegetation representing a variety of tree species, age-classes, and sizes (Whitmore 1990; Clark 1996). The forest has proportionately more larger-stature, long-lived trees with high wood densities that promote higher amounts of carbon storage, compared to managed forests and second growth (Thornley and Cannell 2000). In addition, primary forests have large regional variations in structure and floristics that influence carbon storage (Baker et al. 2004; Bunker et al. 2005) We define managed forests as forests where there is a sustained effort to maximize desired long-term social values (timber, carbon, water, biodiversity) with a security in land tenure. A managed forest for timber will on average, comprise smaller more uniformly statured trees that, relative to primary forest, have faster growth with less dense wood (Thornley and Cannell 2000). Compared to managed forests, second growth will be significantly lower in stature, and comprise proportionately higher amounts of very fast growing pioneers with light wood density (Bunker et al. 2005).

We define secondary tropical forests as forests that have grown after a significant disturbance such as fire, logging, or from land clearance for agriculture and will typically lack the large, high carbon, late stage canopy trees. They are often associated with short-term exploitation and insecurity in land tenure. As a result, these forests hold less carbon than primary tropical forests and forests sustainably managed for timber (Brown and Lugo 1990; Chazdon 2008). Secondary forests are common in areas where forests have been cleared for other land uses like agriculture and were later abandoned, as is the case for large areas of Central America and Amazonia. As secondary forests grow, they can exhibit high levels of carbon uptake, acting as significant carbon sinks particularly if the regeneration of the longer-lived species was present and able to compete with the pioneers of the second growth (Brown and Lugo 1990). However in other circumstances second growth can become an arrested shrubland (Whitmore 1990).

3.2 Maximizing Carbon Uptake Versus Maximizing Carbon Storage

Carbon uptake is maximized in tropical forests during the initial stand developmental stages when biomass productivity is at its greatest. The rate of carbon uptake will slow with time as growing space is occupied, and in many systems, it takes a long time before net uptake reaches zero, and accumulation of large woody debris continues even longer. In comparison, maximum carbon storage is achieved in the later stages of stand development when a large amount of carbon is stored in canopy trees (Bunker et al. 2005; Kirby and Potvin 2007). Older forests, with well developed stand structures, will also have higher wood densities, and in some cases, higher soil carbon levels than forests in earlier developmental stages (Kirby and Potvin 2007). Thus in managing tropical forests for carbon, it is important to determine if the forest is going to be managed for maximum carbon uptake, maximum carbon storage, or a combination, since this determines management practices over time (Thornley and Cannell 2000; Bunker et al. 2005; Kirby and Potvin 2007).

3.3 Site and Climatic Factors Limit Productivity and Carbon Storage Potential

Forested landscapes in the tropics vary greatly in terms of biomass productivity and capacity for carbon storage and uptake and, as a result, forest managers will need to take into account all site characteristics across the landscape to assess carbon uptake rates and the carbon storage potential (Baker et al. 2004). These differences can be observed at the regional scale (average carbon biomass estimates given in IPCC 2007). On a more local landscape scale, soil fertility, precipitation levels, and disturbance regimes all greatly influence the maximum amount of biomass and carbon that can exist at a location (Bunker et al. 2005; Gibbs et al. 2007). Tropical forest soils, such as oxisols and ultisols, tend to be deeply weathered and have little to no organic or humus layer (using the USDA (1975) soil classification). In some areas, such as in montane forests, the soils are younger and of volcanic origin, making them fertile and desirable for agriculture. Younger soils, such as inceptisols and entisols, occur on alluvial plains and along rivers or at their ends as deltas and are extremely productive, whereas others are representative of nature's erosive forces (landslides) (Holzman 2008). Tropical forest managers can manipulate forests to adjust carbon uptake levels or manage for species compositions that contain large amounts of carbon, but they will not be able to produce more carbon storage than the site is capable of unless they add fertilizers, add water or take water away, and this is usually too costly for land that is marginal – which the emerging forest carbon market may be restricted to.

3.4 Creating Carbon Additionality Versus Minimizing Carbon Loss

"Additionality" refers to the carbon emissions avoided (or forest carbon remaining as stocks) due to a change in behavior from business-as-usual. In managing tropical forests for carbon, two approaches are possible: (i) create carbon additionality by minimizing the carbon lost from forest management activities or by creating new forests on open lands; and (ii) protecting intact forests thereby protecting stored carbon that would otherwise be released from land clearance or logging.

To create additionality, forest management practices and methods must increase the amount of carbon held within forests when compared to some baseline measure (Lugo et al. 2003). Reforestation, when forests are planted on degraded lands, would be an example of additionality. Another example would be when management practices are implemented that minimize the amount of carbon lost in comparison to "business as usual" logging practices using a set of baseline management conditions. This form of carbon accounting is the basis for using reduced impact forest management practices as a means to minimize carbon loss from current levels.

3.5 Forest Degradation

In order to best determine the appropriate silvicultural treatment to maximize carbon uptake and storage in a forest, it is first necessary to identify the nature of the disturbance and the type of degradation affecting the site. This requires an understanding of forest degradation processes, since so much of the tropical forest biome would now be considered second growth, logged over, or re-growth on old agricultural lands (Ashton et al. 2001a; Chazdon 2008; Asner et al. 2010). Degradation processes can be divided in two categories, structural and functional (Ashton et al. 2001a).

3.5.1 Structural Degradation

Structural degradation is caused by disturbance regimes that alter species composition, structure, and regeneration of a forest. Disturbance regimes that promote structural degradation can be chronic (either bottom up or top down – see Fig. 9.3a, b), or sudden and acute such as temporary and partial land clearance for agriculture (swidden systems) or one time intensive logging (Fig. 9.3c) (Ashton et al. 2001a).

Chronic bottom-up impacts occur when the understory strata of a forest is continually suppressed. As a consequence, the forest structure is simplified because the lower strata lose their ability to successfully regenerate. Examples of such processes are the continuous presence of ungulates and associated herbivory or the intensive cultivation of non-timber forest crops in the understory (Ashton et al. 2001a). The forest becomes impoverished of understory shrubs and tree species as well as seedling regeneration of canopy trees.

Chronic top-down impacts occur when disturbances directly affect the forest canopy. An example would be selective logging with repeated diameter-limit cutting at frequent intervals that progressively removes the tallest trees in the canopy. Here composition and structure shifts from dominance of the large timber tree species to tree species of the subcanopy and the understory (Ashton et al. 2001a).

Forest suffering acute impacts has been partially cleared for agriculture and remains in cultivation for only a short period (less than 5 years) before it is abandoned. After abandonment, the site is colonized by pioneer species and coppice from stumps and root suckers. The biggest shortfall is the shift in composition from late-successional species that have been eliminated from the site because their advance regeneration was eradicated by cultivation to species that are largely pioneers (Fig. 9.3c).

Insert 1. Top Down Disturbance

An example of top-down disturbance is the diameter-limit cuttings that target individual trees in periodic cycles of 10–30 years. At the beginning, the effects of such disturbance can be considered harmless, but over time the canopy will progressively lower in stature and subcanopy tree species and vines will occupy the upper stratum. With the removal of the late-successional canopy trees, the seed source for these species also disappears. This causes loss of advance regeneration and the simplification of forest stratification from the top downward (Ashton et al. 2001a).

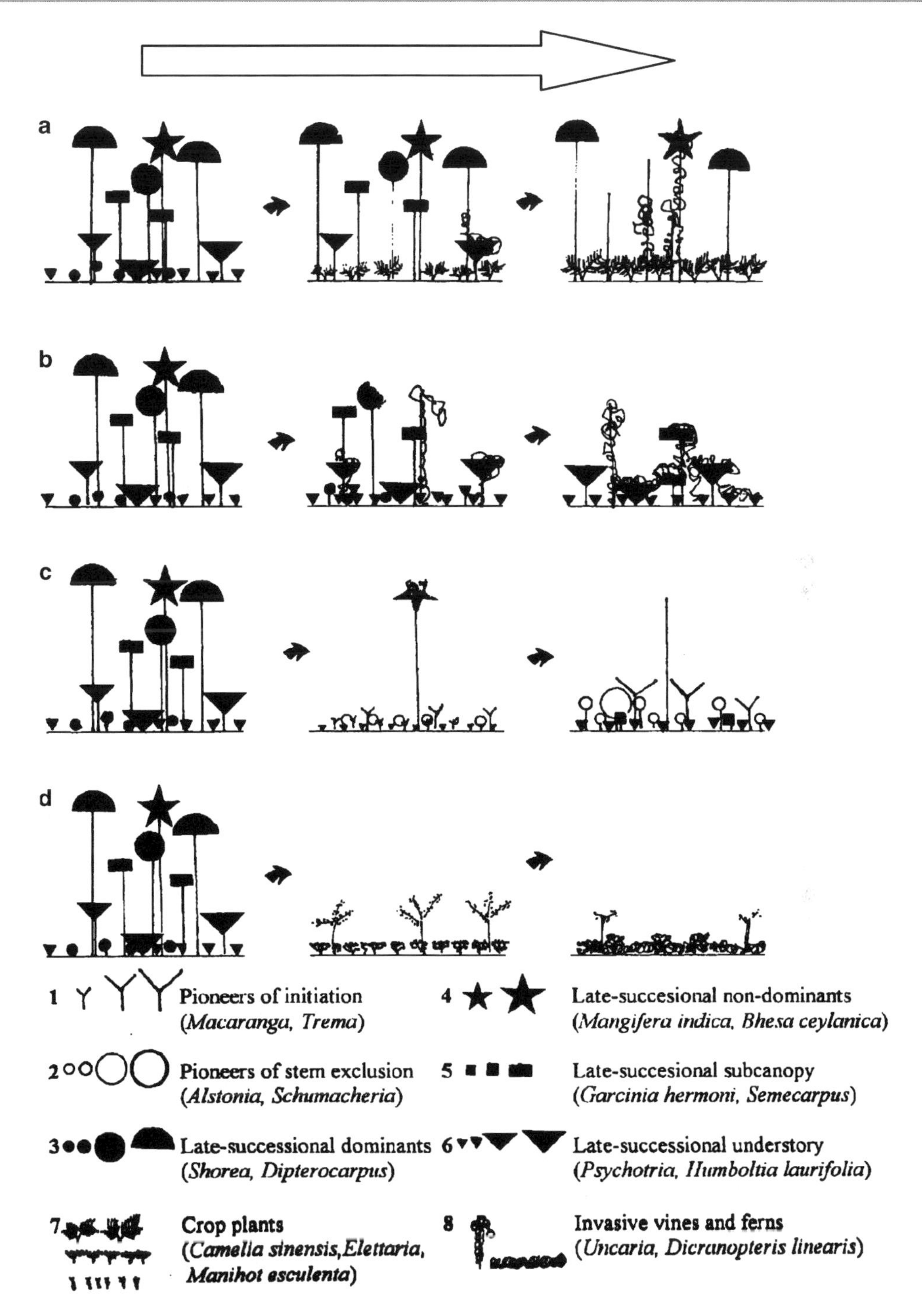

Fig. 9.3 Stand development profiles depicting stand composition and structure for a mature mixed dipterocarp forest (left profiles) that is degraded over a chronosequence by: (**a**) chronic structural bottom-up effects; (**b**) chronic structural top-down effects; (**c**) acute one-time effects from incomplete clearance; and (**d**) acute one-time functional effects from complete clearance for permanent agriculture (*Source*: Ashton et al. 2001a)

3.5.2 Functional Degradation

is caused by acute disturbances that are severe and lethal to the groundstory such that it is eradicated In most cases the soil is intensively turned over with the roots and stumps being removed. Such impacts go beyond shifting forest structure and composition to permanently affecting soil fertility and structure, then altering infiltration, water holding capacity, and therefore subsurface hydrology. The disturbance usually associated with functional degradation is forest clearance for intensive agricultural cultivation or permanent conversion for development (Fig. 9.3d) (Ashton et al. 2001a).

Insert 2. Acute Intensive and Prolonged Disturbance

Permanent conversion of forest to alternative land uses often leads to functional degradation. Of all disturbances, this has the most detrimental effects on soil erosion, hydrological regimes, and edge effects. The majority of tree species' establishment processes (advance regeneration, vegetative sprouting) have been eliminated from the site. After abandonment, the site usually transitions into non-forest composed of fire-prone grasses and ferns. Many of these colonizers are exotic and/or invasive. Once these species have established, they tend to self-perpetuate because of their root networks and their ability to quickly regenerate after fire (Ashton et al. 2001a).

4 Reducing Emissions Through Forest Management

Implementing forest management practices designed to sustain or enhance forest integrity and ecosystem services can generate two forms of carbon additionality: (1) higher average carbon stocks in forests managed for timber, due to reduced emissions and/or improved sequestration associated with timber production activities, and (2) reduced conversion of forests for other land uses. Of these, lower emissions resulting directly from the improved design of logging operations (e.g. no logging in riparian buffers, lower collateral damage to the residual stand) offer the most immediate carbon benefit. Higher sequestration rates due to improved tree growth after harvests, and lower forest conversion for other land uses, are carbon benefits that generally accumulate over longer time periods; however, these may also provide the larger long term source of carbon benefits from improved forest management.

Some studies have found lower deforestation in tropical forests that provide sustained timber production than in strictly protected areas (Hughell and Butterfield 2008; Durán-Medina et al. 2005). This relationship is apparently due to the importance of a local constituency that derives an economic benefit from forest products, especially in the context of many tropical countries where enforcement of protected area boundaries is lax. Ultimately, the incentive to avoid conversion of forests that provide sustained income may prove to be the most important carbon benefit of improved forest management; however, this section will focus on the more direct, and to some extent more immediate, carbon benefits associated with the first form of additionality: higher carbon stocks in existing forests.

There are extensive opportunities in the tropics for reducing emissions and increasing carbon stocks relative to business-as-usual in forests subjected to timber extraction, because a large proportion of tropical forests are exploited for timber (Asner et al. 2009) yet very few tropical forests are sustainably managed (Siry et al. 2005). About 20% of total remaining tropical forest area, or 350 million hectares, are allocated to timber production by countries that are members of the International Tropical Timber Organizations (ITTO 2006). A slightly larger area of tropical forest, 390 million hectares, has recently been identified as under timber production (Asner et al. 2009). Of this total production forest area, roughly equal proportions were identified in Asia/Oceania (45%) and Latin America (41%), while a smaller area was identified in Africa (14%) (Asner et al. 2009). The proportion of remaining tropical forests exploited for timber is likely to increase, in response to the growing demand for timber in domestic and international markets (Kirilenko and Sedjo 2007).

Most timber operations are forest mining operations with no consideration of long term management to sustain or increase forest value, either for timber or other ecosystem services. As such, very few tropical forests are managed using silviculture. A variety of challenges stand in the way of realizing the opportunities to reduce emissions from improved forest management. In many tropical countries destabilizing social phenomena deter investments in basic forest management practices (Uhl et al. 1997). Issues such as land tenure, the lack of environmental regulations, and/or the inability to enforce existing environmental laws exacerbate or encourage short-sighted logging methods and discourage sustainable forestry practices. Other factors such as access to markets and the lack of financial incentives to implement improved forest management practices drive forest management decisions in these often-impoverished areas.

In the exceptional cases where an intent to improve forest management exists, operators are likely to be confronted with a lack of trained experts to implement regionally appropriate forest management techniques, and limited scientific information to inform silvicultural treatments and best operational practices designed for to the species present. In addition, other factors such as access to markets and the lack of financial incentives to implement improved forest management practices drive forest management decisions in these often-impoverished areas. Forest management is in the initial stages of development for many tropical regions.

Notwithstanding our limited knowledge base in the tropics, immediate opportunities for improved forest management are great. Both scientific knowledge and practitioner experience are available for some locations in the tropics, and basic principles of sound silviculture and forest management can be drawn from both tropical and temperate systems.

Generally speaking, very few tropical forests are sustainably managed using silvicultural principles. This is the result of a number of destabilizing social phenomena that deter investments in basic forest management practices (Uhl et al. 1997). One hotspot of research and practitioner experience in tropical forestry can be found in dipterocarp forests of South Asia and the Malay Peninsula where the British colonizers established forestry in the mid-1850s (Ashton et al. 2001a; Ashton 2003), but such forests are small and now restricted because of land conversion. Other tropical regions have made recent advances in research towards improved forest management in the last few decades, including lowland Amazonian Bolivia (BOLFOR project, Peña-Claros et al. 2008a, b), eastern Amazonia (Verissimo et al. 1992; Uhl et al. 1997) and Guyana (De Graaf 1986; De Graaf et al. 2003),

While these examples of research advances provide the foundation for our review here, management practices developed for the tropics have not been designed with the intent of encouraging long-term carbon storage (Putz et al. 2008a). Silvicultural treatments are usually aimed at improving the timber production of commercial species such as mahogany and dipterocarp species, or for various non-timber forest products (Feldpausch et al. 2005). Seldom do these treatments favor carbon sequestration in the short term. The thinning or harvesting of undesired tree species will lower carbon yields over the short term. In the following section we outline the range of reduced impact logging (RIL) methods and silvicultural treatments that are available for tropical forest management, and we consider their implications for achieving emissions reductions and/or net carbon sequestration starting with the most basic and moving towards the more sophisticated.

4.1 Reduced Impact Logging (RIL)

Reduced impact logging (RIL) refers to the use of improved harvesting and forest management practices, in combination with education and training, to reduce avoidable logging damage to residual trees, soils, and critical ecosystem processes (Pinard and Putz 1996). The need for RIL is greatest in the tropics where regulation is usually low and most tree species are non-commercial, thus

Table 9.1 Reduced-impact logging planning and harvesting guidelines condensed from FAO Model Code of Forest Harvesting Practices (Dykstra and Heinrich 1996)

Harvest plan	Formal plan prepared based on timber stock and locations of commercial trees, proposed roads, skid trails, stream crossings, buffer zones, logging unit boundaries
Pre-felling vine cutting	All vines >2 cm DBH to be severed at least 12 months prior to harvesting
Skid trail planning	Skid trails to be located on ridges and designed to minimize skidding distances, skidding on steep slopes, skidding downhill, and stream crossings
Tree felling	Decisions on felling directions based on safety to feller, ease of skidding, and avoidance of damage to harvested tree

subject to unnecessary collateral damage. Under conventional logging practices in the tropics, for every tree harvested, 10–20 other trees are severely damaged by untrained fellers and machine operators working without the aid of detailed maps or supervision (Sist and Ferreira 2007). Carbon lost from this damage can be extensive, with 30–40% of the area often affected by heavy equipment (Chai 1975; Jusoff and Majid 1992). Reduced Impact Logging is standard practice in many parts of the developed world (with important exceptions), but rare in the tropics (Putz et al. 2000), where Siry et al. (2005) report less than 5% of logging operations are certified as sustainably managed.

The use of RIL has the potential to significantly reduce the carbon losses associated with conventional logging. With fewer trees killed or damaged (Johns et al. 1996; Pinard and Putz 1996; Pinard et al. 2000) more carbon remains stored in the living forest. If these residual trees are of higher diameter classes, then a larger amount of carbon will remain sequestered (Johns et al. 1996). Soil carbon is often a significant proportion of the carbon lost due to conventional logging. Forests subject to conventional logging lose much of their silvicultural value due to soil damage (Putz and Pinard 1993). As a result, reducing soil damage is a major emphasis of RIL, especially where logging operations occur on steep slopes and use heavy machines on wet soil. These practices significantly disturb and erode soil and release stored carbon (Putz et al. 2008a).

Potential carbon emissions reductions from improved harvesting practices are often significant. Research in Southeast Asia has shown that RIL areas contain more than 100 Mg more biomass per hectare than conventionally logged areas 1 year after logging (Pinard and Putz 1996). While this is the most dramatic example of carbon benefits from RIL that we are aware of, and few studies specifically quantify carbon, a number of studies have identified 30–50% lower damage using RIL techniques per unit of wood extracted based on metrics like residual tree damage and area logged (Healey et al. 2000; Bertault 1997; Durst and Enters 2001; Pereira et al. 2002; Keller et al. 2004; as reviewed by Griscom et al. 2009). Given the large areas of tropical forest designated as production forests around the world, the implementation of RIL provides an opportunity to avoid emissions of 0.16 gigatons of carbon per year (Gt C y^{-1}), or about 10% of emissions reductions attainable by stopping tropical deforestation, according to Putz et al. (2008a).

Extensive research over the past three decades has provided the scientific grounding for the development of RIL guidelines, outlined in Table 9.1 and condensed from the FAO Model Code of Forest Harvesting Practices (Dykstra and Heinrich 1996). More recently, the Tropical Forest Foundation (TFF) has developed a Standard for RIL (Tropical Forest Foundation 2007) that is the template for RIL training centers located in all three major tropical regions. Specific criteria and indicators have also been developed by TFF for Indonesia (http://www.tff-indonesia.org/en/ril/ril-criteria-and-indicators) as elaborated through a series of manuals on planning, operations, and management (Klassen 2005a, b, 2006a, b). These pre- and post-logging guidelines, standards, and manuals are designed to retain forest biomass and protect the capacity of forests for future regenerative growth, in addition to related social and ecological co-benefits.

4.2 Beyond Reduced Impact Logging: Sustainable Tropical Forest Management

RIL practices can still result in significant loss of stored carbon. To move beyond RIL, and produce greater carbon gains than those obtained from RIL, it is important to look at how silviculture can be used to increase carbon uptake and storage. To do this, forest managers need to carefully take into consideration a landscape's underlying soil and hydrology as well as the disturbances that are acting on the forests (Ashton et al. 2001a; Ashton 2003). In most tropical rain forest regions, silvicultural knowledge is very varied and generally lacking (Ashton et al. 2003) (as previously described). Silvicultural information for montane and dry deciduous forests is particularly limited.

4.2.1 Silvicultural Management and Planning

Scaled Land Use Planning

To effectively store carbon over the long term, it is important for forest managers to delineate forest stands into protected and production stands based on desired forest values, including carbon. The dipterocarp forests of Asia have been heavily logged and the remaining forests are now in the hills and mountains of the region. These areas are recognized by governments as catchments of water supplies for drinking water, fisheries, and agriculture, and are also ideal for the long-term sequestration of carbon, the production of high quality timbers and non-timber forest products, and biodiversity conservation (Ashton 2003; Ashton et al. 2011). Careful stand delineation and planning can help to ensure long-term carbon storage while meeting society's other multiple resource needs. Engaging in landscape scale partitioning of forests into stand management units insures that the maximum amount of carbon can be sequestered without compromising the long-term sustainability of other social values.

The use of stands as the management unit within tropical forests has been largely disregarded in favor of large scale and broadly applied management prescriptions (Appanah and Weinland 1990). Such systems have one diameter limit cutting across a wide landscape of floristic associations (see Ediriweera et al. 2008; Gunatilleke et al. 2006) that do not cater to species differences in growth rates, size classes and site affinities (Hall et al. 2004; Palmiotto et al. 2004a, b). To effectively manage forests for carbon, a unique set of silvicultural treatments should be tailored to the biophysical and social characteristics of each site (Ashton and Peters 1999; Ashton et al. 2001a). Managing stands within the broader context of the landscape allows land managers to identify zones of high carbon value and stands of riparian, wetland and watershed value, as well as areas of high biodiversity. This landscape scale template should reflect an integrated network of stands allocated to production and protection (Ashton 2003) with the focus on maximizing carbon storage within the landscape. Comparative tests of the silvicilture and finances for timber, non-timber and service values between diameter-limit cutting and stand based shelterwood treatments clearly illustrate the inferiority of diameter-limit cutting systems (see Ashton et al. 2001b; Ashton and Hall 2011).

Strategic Harvest Planning

One of the first steps that must be taken to manage a tropical forest for carbon is to develop a long-term strategic harvest plan. Unlike forest management in the temperate developed world, tropical forests are often managed *ad hoc* without long term planning (Uhl et al. 1997). Tenuous ownership rights, abundance of timber resources, and high demand all come together to provide a disincentive to take a long-term forest management perspective (Uhl and Kauffman 1990). A strategic plan for maximizing carbon storage should answer the following question. What type of harvesting must be done and what type of treatments can or should be applied to retain and increase carbon over the long-term? Short term concessions of 10–60 years obviously counter such strategic planning yet they make up the majority of land tenure arrangements on government forest reserves in the tropics (Chomitz 2007).

Changing Rotational Length

A key component of managing tropical forests for maximum carbon storage is the length of harvest rotations (the return time between harvests). The most frequently prescribed logging cycle in tropical forests is 30 years (Sist et al. 2003), but cutting cycles of 60–100 years have been found more likely to sustain timber yields and allow for increased carbon storage (Dykstra and Heinrich 1996; Ashton et al. 2001b; Sist et al. 2003; Ashton 2003). Studies in Brazil by Sist and Ferreira (2007) that focused on timber volume reported that after harvesting 21 m^3/ha from a moist lowland forest, the next planned harvest, 30 years later, would yield only 50% of the first harvest. Despite this study's focus on timber, the results can be extrapolated to biomass and carbon storage.

Little research has been conducted on the effect of harvesting cycles on carbon storage in the tropics. Similar studies based within the temperate regions (Cooper 1983) have found that stands managed for maximum sustained yield store approximately a third of the carbon stored in unmanaged late stage forests (Sohngren and Mendelsohn 2003). Given the increased growth rates and biomass levels of tropical forests and the results based on timber yields, it is reasonable to expect that increased rotation lengths will increase carbon storage in tropical forests (Sohngren and Mendelsohn 2003; Keith et al. 2009).

5 Silvicultural Treatments for Managing Tropical Forests for Carbon

Silvicultural treatments can often be applied within tropical forests to maximize carbon storage, but the degree and intensity with which these interventions need to be applied varies. Where the harvested species are represented by abundant advance regeneration, RIL alone could be sufficient to sustain long-term carbon sequestration and timber yields as long as logging intensities are modest and cutting cycles are long (de Madron and Forni 1997; Sist et al. 2003; Sist and Ferreira 2007; Valle et al. 2007). In other instances, various intensities of silvicultural treatments can be applied both before and after logging operations to promote increased carbon storage as well as to improve the overall health and productivity of the forest. The following section outlines different silvicultural methods and what each one's effect is on forest carbon.

5.1 Stand Level Planning: Regeneration[1]

5.1.1 Species Management Based on Carbon

Tree species that contribute the most to forest carbon storage are often the highly desired commercial timber species (Kirby and Potvin 2007). The selective cutting of these high-carbon timber species dramatically reduces carbon storage within a forest and multiple decades must pass before it is gained back. These findings indicate that efforts to improve carbon storage need to be based on management techniques that promote and encourage long-term regeneration of high-carbon (late successional) species which tend to be slower growing with substantially higher wood densities, and by implication high carbon storage capacities (Baker et al. 2004; Bunker et al. 2005).

Pre-harvest planning combined with species selection allows forest managers to prioritize species, using as criteria (1) the species' overall contribution to carbon storage in the landscape; (2) their relative abundance; and (3) their wood density per capita contributions to carbon storage (Kirby and Potvin 2007). These steps allow forest managers to assess overall forest carbon storage and decide whether silvicultural treatments should be applied and, if so, what treatments will increase carbon storage.

5.1.2 Site Preparation

Many site preparation treatments are applied to improve establishment and growth of the desired species of regeneration or enrichment plantings

[1] Regeneration refers to treatments that prepare the stand for regeneration and the method of regeneration that promotes or excludes different suites of species.

(Smith et al. 1997). Preparation treatments usually precede methods to establish a regenerating stand. When managing tropical rainforest for carbon, preparation treatments that could affect, expose or reduce soil carbon (e.g. scarification of seedbeds, prescribed fire) should be minimized and treatments that free growing space for desired regeneration of tree species (partial removal of clonal understory herbs, shrubs and palms) should be encouraged (Smith et al. 1997).

On the other hand, prescribed burning and scarification (e.g., exposure of mineral soil) may be necessary to encourage regeneration as many important shade-intolerant or light-seeded timber species require such conditions for establishment in this forest type. Both these treatments could have, to some degree, a negative effect on carbon storage if done inappropriately because they could reduce soil organic matter. When considering fire, the main goals are frequent fuel reduction and control of competing vegetation (Smith et al. 1997). On their own, these objectives seem counterintuitive for carbon storage and uptake; however, fuel reduction increases the resilience of fire adapted dry forest and woodlands to more catastrophic fires and competition control will allow the desired regeneration to take hold on the site and occupy the growing space faster. From this perspective, both treatments will have a positive effect on long-term forest carbon storage and uptake.

5.1.3 Reproduction Methods

Methods of reproduction are complex and variable. They broadly comprise two approaches. Episodic methods are characterized and classified by the nature of the regeneration that treatments to the canopy and groundstory promote. They can be characterized as:

1. Clearcut – Seedlings established from buried seedbanks and seed dispersed from outside the stand (e.g. true pioneers dispersed by wind, floods or small birds and bats) (e.g. *Alnus spp., Alstonia spp., Prunus spp., Triplochiton scleroxylon*).
2. Seed Tree – Seedlings established from heavy large seeds. Such species are usually long-lived pioneers that rely upon a within-stand seed source and dispersal but similar seed bed conditions to a clearcut (e.g. *Canarium spp., Cedrella spp., Cordia spp., Entandrophragma spp., Swietenia. macrophylla*).
3. Coppice – Regeneration of sprout origin from the residual parent root of stem. Such species are often understory of subcanopy trees (e.g. *Eugenia spp., Garcinia spp., Psychotria spp., Syzygium spp.*)
4. Shelterwoods – Seedlings established by nearby parent trees and beneath partial shade before being released. (e.g. *Dipterocarpus spp., Calophyllum brasiliense, Pericopsis elata, Shorea spp., Swartzia fistuloides*) (see Smith et al. 1997; Ashton and Hall 2011).

Episodic methods represent a continuum of disturbance regimes that range from catastrophic and lethal whereby all trees die and ground vegetation is destroyed (natures examples – landslide, mudslide, hurricane followed by fire; (Clearcut)) to partial canopy death whereby the groundstory remains largely intact (natures examples – convectional windstorm, drought; (shelterwood)). Such methods largely focus on securing one age class at a time. Shelterwoods are the most inclusive of all regeneration modes depending upon amount of canopy removed but focus on advance regeneration, the most difficult to secure. They can be complex in structure, promoting different species to grow at different rates in intimate mixtures, but do not usually hold onto more than three age-classes (cohorts) at any one time.

The alternative method of regeneration is one that is conceptually applied almost as a continuous series of cuttings that usually represent small canopy openings. This is hard to replicate so treatments are made at periodic intervals at 10–30 years apart. Intervals of time are longer between entry for slower growing stands. The approach is to secure regeneration after each entry in parts of the stand where the openings have been created, and to treat other parts of the stand through selection thinning and liberation where regeneration was established in prior cuttings.

Selection methods tend to promote shade tolerant species and associations, while shelterwoods tend to promote shade intermediate to intolerant species and associations. The main goal of these treatments is to maintain ecosystem structure and function while allowing the regeneration of desirable species (Montagnini and Jordan 2005; Smith et al. 1997). When managing tropical forests for carbon, foresters should seek to increase or maintain forest structure and guild diversity and, by doing so, overall forest resiliency. Stand level planning would allow the forester to use a range of regeneration methods, and their variants. In doing so a forester can cater to changes in species shade tolerances, growth rates and site affinities, and therefore more efficiently manage the standing carbon more compatibly with timber and non-timber production.

5.1.4 Enrichment Planting

Enrichment planting is also known as line, gap, strip or under-planting, depending on the nature of the planting arrangement (Montagnini and Jordan 2005; Smith et al. 1997). Enrichment planting is a method utilized to introduce desirable tree species in degraded forests or stands without affecting the structure or composition already present in the site. In many cases enrichment planting is unnecessary, and can lead to floristic simplification, when only a few species are used, despite evidence of ample well-established natural regeneration.

However, when enrichment does occur, planted species can differ in their rate of growth, shade tolerance, ecological characteristics, and economic value. Choosing one species over another needs to be done paying careful attention to the issue or desired value that is being addressed (Montagnini and Jordan 2005; Ashton et al. 2001a; Ashton 2003).When managing forests for carbon, enrichment planting has the potential to maximize other market values without affecting carbon uptake and storage by also introducing non-timber forest products that comprise herbs, shrubs and smaller sub-canopy trees compatible with the forest successional growth (Ashton et al. 2001b; Ashton 2003). Together, these added values could prevent the conversion of forests into other widespread, low carbon land uses (Montagnini and Jordan 2005).

Insert 3. Non-timber Forest Products Complement Timber Production

A case study in Sri Lanka suggests that rainforest can be managed sustainably for multiple market values. Herbaceous shrubs (i.e. *Cardamomum zeylanicum* – cardamom) can be judiciously cultivated around advance regeneration of the new forest immediately following timber harvesting. In some instances, medicinal vine species such as *Coscinium fenestratum* or climbing palms like *C. zeylanicus* (rattan) can be line-planted along edges and trails and then harvested for their economic value. *Caryota urens* (Fishtail palm) can be under-planted within the regenerating stand and later tapped for sugar when it matures as a sub-canopy tree. The logic behind these plantings is that if lianas and other shrubs and trees grow compatibly with the timber trees, the best option would be to promote and then sequentially harvest them over the time until the timber trees attains maturity. Together with service values for carbon and water, maintaining and managing a tropical rain forest for a diversity of products create greater stacked economic value than land clearance and cultivation for tea, for example. For a private landowner, timber alone cannot compare with the financial rewards of intensive tea cultivation or other agricultural crops; but integrated together with other compatible social values makes it very competitive as a permanent land use (Ashton et al. 2001b).

5.2 Stand Level Planning: Post Establishment[2]

5.2.1 Keeping Track of Stratification Post Regeneration and Post-Establishment

Two stratification processes are a factor in the dynamics of species mixtures (Ashton and Peters 1999; Ashton et al. 2001a). "Static" stratification refers to the late-successional species that occupy distinct vertical strata in the mature forest canopy (i.e. species that will always occupy understory and subcanopy positions at maturity). "Dynamic" stratification refers to the sequential occupation of the canopy strata by species of different successional status (i.e. shorter-lived canopy trees that relinquish their canopy position over time to longer-lived species) (Ashton and Peters 1999; Ashton et al. 2001a; Ashton 2003; Ashton et al. 2011).

If a forest is being managed for carbon uptake and long-term carbon storage, understanding these processes of stand dynamics will provide managers with basic guidelines in the selection of appropriate regeneration methods and thinning treatments for a site. Mixtures that exhibit diverse growth patterns, and differences in shade tolerance and stand development are the best for long-term carbon storage (Ashton et al. 2001a). These stands can have shade tolerant species growing in the understory while the canopy is occupied by late-successional shade-intolerant, high carbon species. In the case of seasonal tropical forests, the lower stratum is often occupied by evergreen species that continue storing carbon even when the deciduous canopy species slow their photosynthetic activity.

Knowing which species are growing upward towards the canopy, which species are overtopped and are of the past canopy ("dynamic"), and which species are remaining in their current stratum ("static") is critical. Keeping track of the "book-keeping" of stratification is therefore a prerequisite for deciding when, where and which silvicultural treatment to use. For example, treatments can: (1) accelerate shade tolerant species into the canopy strata; (2) promote shade tolerant understories to establish; and (3) allow shade intolerant canopy tree species to re-establish in the understory. All require knowledge of the current status of stand condition and stratification process to efficiently promote different aspects of tree growth and hence carbon sequestration and storage.

5.2.2 Release Treatments

After a regenerative disturbance, fast growing pioneer species can occupy the growing space rapidly. It is at this stage that lianas can establish in a site and become harmful competitors of the desired tree species (Ashton et al. 2001a, b). Liana (vine) cutting is an important example of this kind of treatment (Putz et al. 2000, 2008a, b). Active measures to eliminate lianas need to be taken from the beginning of the establishment cycle. The best way of controlling them is to avoid any disturbance to the mineral soil and to maintain the site's growing space fully stocked. Liana removal affects carbon storage because it increases the light available to trees and reduces competition, allowing growth rates and carbon to increase in the stand (Wadsworth and Zweede 2006; Keller et al. 2007; Zarin et al. 2007). The positive benefits of liana removal persist only for about 4 years, requiring repeated treatments over a cutting cycle (Pena-Claros et al. 2008a, b).

Other release treatments that remove older, larger competing trees (liberation) or competing trees of the same cohort that over top the desired species, such as pioneers (cleaning) may also need to be judiciously applied for the successful establishment of the desirable regeneration for the new stand (Smith et al. 1997). These treatments need to be timed correctly, and can be expensive – but can be critical (Ashton 2003; Ashton and Hall. 2011).

5.2.3 Thinning

Thinning occurs naturally (termed "self-thinning") when the number of trees in a stand declines from mortality due to continued competition with other

[2] Post establishment refers to treatments done to the stand after successful establishment of regeneration.

individuals (Smith et al. 1997). The intent of thinning as a management practice is to purposefully regulate and manipulate the distribution of growing space at the stand level to maximize net benefits over the whole rotation before nature does this through self-thinning. Thinning therefore re-allocates growing space to remaining desired trees from competition with undesirable trees (Smith et al. 1997).

The long term objective of thinning is usually to increase the size of the individual tree and/or volume of merchantable wood within a stand. This implies that the initial application of the treatment will result in a loss of standing aboveground carbon because of the reduction in the site's gross carbon volume. The amount of growing space occupied by wood volume (often measured as basal area) of the remaining trees will increase, along with a parallel increase in forest carbon (Smith et al. 1997).

This difference between merchantable wood volume yield and gross biomass production (e.g. carbon) highlights the decisions and tradeoffs between timber and carbon management that land managers will need to make when deciding which silvicultural practices to implement. The goal of many forest managers with both timber and carbon interests is to maintain site merchantable yields while obtaining some baseline long-term carbon storage. This can be accomplished by favoring allocation of growing space to highly valuable timber tree species with high carbon storage (high wood density) that are usually slower-growing and require longer rotations.

6 Management and Policy Implications

6.1 Summary Conclusions

- Tropical forests emit approximately 17% of total annual global greenhouse emissions primarily due to land conversion. Avoiding emissions, and improving sequestration, through improved forest management is an important strategy currently being discussed under REDD+.
- The carbon uptake and storage capacity of a given forest varies greatly depending on the region, forest type, geophysical characteristics, species composition, disturbance regime, site degradation, land tenure, and human use.
- To develop and implement adequate forest management strategies, first it is important to understand that most tropical forests are not sustainably managed, but exploited.
- Implementing stand-level land use delineation, harvest planning, and reduced impact logging techniques can have important effects on increasing tropical forest carbon.
- If the goal is to maximize carbon uptake and storage, along with sustained timber production, more complex silvicultural treatments need to be implemented. This approach will help secure the regeneration of the desired species and the continued vertical stratification of the stand, will increase productivity, and will promote the presence of target species of high economic and carbon sequestration value.
- Successful forest management depends upon site specific silvicultural treatments.
- If appropriate silviculture is achieved, forests will be more resilient to the unpredictability of disturbance and climate change, making them suitable as stable long-term carbon sequestration and storage reservoirs.

6.2 Implications

6.2.1 Areas for Further Investigation

- While abundant literature exists about managing temperate forests and soils for carbon, more research is needed to understand how the application of silvicultural practices affects carbon uptake and storage in tropical forests.
- Future research needs to move beyond reduced impact logging (RIL) and focus on how forested landscapes can be managed for carbon, as well as water, biodiversity, and other ecological values.

6.2.2 Land Managers

- Land managers in tropical forests need to delineate stands and use them as the managing unit within the forest landscape. This would allow them to develop unique silvicultural techniques that are site specific. Stand delineation also helps identify and protect wetlands and riparian corridors, and areas of high diversity.
- Land managers in certain regions with diverse market conditions should not manage tropical forests only for timber production, but to maximize and diversify the services and products they obtain from their forests. This approach will provide an increase in net present value and a possible solution to the problem of exploitation and land conversion.

6.2.3 Policymakers

- Policies need to prioritize the preservation of primary tropical forests since almost all management and silvicultural practices applied to such forests will result in reduced carbon storage levels.
- RIL practices can provide significant carbon benefits; however, they were not designed with carbon in mind. Practices and policies that go beyond RIL can begin to address long-term resource needs as well as maximizing carbon uptake and storage.
- In comparison with data from temperate forests that indicate that some forestry practices have a minimal impact on soil carbon and this pool might not need to be measured all the time, in the tropics, data for soil carbon are lacking.

References

Appanah S, Weinland G (1990) Will the management systems for hill dipterocarp forests stand up? J Trop Forest Sci 3:140–158

Ashton MS (2003) Regeneration methods for dipterocarp forests of wet tropical Asia. For Chron 79:263–267

Ashton MS, Hall JE (2011) The ecology, silviculture, and use of tropical wet forests with special emphasis on timber rich types. In: Gunter S, Mossandl R (eds) Silviculture of tropical forests. Springer, Berlin

Ashton MS, Peters C (1999) Even-aged silviculture in mixed moist tropical forests with special reference to Asia: lessons learned and myths perpetuated. J For 97:14–19

Ashton MS, Gunatilleke CVS, Singhakumara BMP, Gunatilleke I (2001a) Restoration pathways for rain forest in southwest Sri Lanka: a review of concepts and models. For Ecol Manag 154:409–430

Ashton MS, Mendelsohn R, Singhakumara BMP, Gunatilleke CVS, Gunatilleke I, Evans A (2001b) A financial analysis of rain forest silviculture in southwestern Sri Lanka. For Ecol Manag 154:431–441

Ashton MS, Singhakumara BMP, Gunatilleke CVS, Gunatilleke IAUN (2011) Sustainable forest management for mixed-dipterocarp forests: a case study in southwest Sri Lanka. In: Gunter S, Mossandl R (eds) Silviculture of tropical forests. Springer, Berlin

Asner GP, Knapp DE, Broadbent EN, Oliveira PJC, Keller M, Silva J (2005) Selective logging in the Brazillian Amazon. Science 310:480–482

Asner GP, Rudel TK, Aide TM, DeFries R, Emerson R (2009) A contemporary assessment of change in humid tropical forests. Conserv Biol 23(6):1386–1395

Asner GP, Loarie SR, Heyder U (2010) Combined effects of climate and land-use change on the future of humid tropical forests. Conserv Lett 3:395–403

Baker TR, Phillips OL, Malhi Y, Almeida S, Arroyo L, Di Fiore A, Erwin T, Killeen TJ, Laurance SG, Laurance WF, Lewis SL, Lloyd J, Monteagudo A, Neill DA, Patino S, Pitman NCA, Silva JNM, Martinez RV (2004) Variation in wood density determines spatial patterns in Amazonian forest biomass. Glob Change Biol 10:545–562

Bertault JG, Sist P (1997) An experimental comparison of different harvesting intensities with reduced-impact and conventional logging in East Kalimantan, Indonesia. For Ecol Manag 94:209–218

Brown S, Lugo AE (1990) Tropical secondary forests. J Trop Ecol 6:1–32

Bullock SH, Mooney HA, Medina E (1995) Seasonally dry tropical forests. Cambridge University Press, Cambridge/New York, 450p

Bunker DE, DeClerck F, Bradford JC, Colwell RK, Perfecto I, Phillips OL, Sankaran M, Naeem S (2005) Species loss and above-ground carbon storage in a tropical forest. Science 310:1029–1031

Canadella JG, Le Quéré C, Raupacha MR, Field CB, Buitenhuis ET, Ciaisf P, Conway TJ, Gillett NP, Houghton RA, Marland G (2007) Contributions to accelerating atmospheric CO_2 growth from economic activity, carbon intensity, and efficiency of natural sinks. PNAS 104(47):18866–18870

Chai DNP (1975) Enrichment planting in Sabah. Malays For 38:271–277

Chazdon RL (2008) Beyond deforestation: restoring forests and ecosystem services on degraded lands. Science 320:1458–1460

Chomitz KM (2007) At loggerheads: agricultural expansion, poverty reduction, and environment in the tropical forest. A World Bank Report, World Bank, 2007, Washington, DC

Clark DB (1996) Abolishing virginity. J Trop Ecol 12:755–739

Cooper CF (1983) Carbon storage in managed forests. Can J For Res 13:155–166

De Graaf NR (1986) A silvicultural system for natural regeneration of tropical rain forest in Suriname. Ph.D. thesis, Agricultural University Wageningen, Wageningen, pp 250

De Graaf NR, Filius AM, Huesca Santos AR (2003) Financial analysis of sustained forest management for timber: perspectives for application of the CELOS management system in the Brazilian Amazon. For Ecol Manag 177:287–799

De Madron LD, Forni E (1997) Forest management in East Cameroon. Stand structure and logging periodicity. Bois et Forets des Tropiques 294:39–50

DeFries R, Achard F, Brown S, Herold M, Murdiyarso D, Schlamadinger B, de Souza C (2007) Earth observations for estimating greenhouse gas emissions from deforestation in developing countries. Environ Sci Policy 10:385–394

Dixon RK, Brown S, Houghton RA, Solomon AM, Trexler MC, Wisniewski J (1994) Carbon pools and flux of global forest ecosystems. Science 263:185–190

Durán-Medina E, Mas J, Velázquez A (2005) Land use/cover change in community-based forest management regions and protected areas in Mexico. In: Bray D, Marino-Pérez L, Barry D (eds) The community forests of Mexico: managing for sustainable landscapes. University of Texas Press, Austin, pp 213–238, 372 pages

Durst PB, Enters T (2001) Illegal logging and the adoption of reduced impact logging. In: Forest Law Enforcement and Governance: East Asia Regional Ministerial Conference, Denpasar, 2001

Dykstra D, Heinrich R (1996) FAO model code of forest harvesting practice. FAO, Rome, 85 p

Ediriweeera S, Singhakumara BMP, Ashton MS (2008) Variation in light, soil nutrition and soil moisture in relation to forest structure within a Sri Lankan rain forest landscape. For Ecol Manag 256:1339–1349

Fearnside P (1993) Rainforest burning and the global carbon budget: biomass, combustion efficiency, and charcoal formation in the Brazilian Amazon. J Geophys Res 98:16,733–16,743

Feldpausch TR, Jirka S, Passos CAM, Jaspar F, Rhia S (2005) When big trees fall: damage and carbon export by reduced impact logging in southern Amazonia. For Ecol Manag 219:199–215

Gibbs HK, Brown S, Niles JO, Foley JA (2007) Monitoring and estimating tropical forest carbon stocks: making REDD a reality. Environ Res Lett 2:045023 (13pp)

Griscom B, Ganz D, Virgilio N, Price F, Hayward J, Cortez R, Dodge G, Hurd J, Lowenstein FL, Stanley B (2009) The hidden frontier of forest degradation: a review of the science, policy and practice of reducing degradation emissions. The Nature Conservancy, Arlington, 76 pages. Available at: http://www.rainforest-alliance.org/resources/documents/hidden_degradation.pdf. Accessed Oct 2010

Griscom HP, Ashton MS (2011) Restoration of dry tropical forests in Central America: a review of pattern and process. For Ecol Manag 261:1564–1579

Gunatilleke CVS, Gunatilleke IAUN, Esufali S, Harmes KE, Ashton PMS, Burslem DFRP, Ashton PS (2006) Species-habitat associations in a Sri Lankan dipterocarp forest. J Trop Ecol 22:343–356

Hall JS, McKenna JJ, Ashton PMS, Gregoire TG (2004) Habitat characterizations underestimate the role of edaphic factors controlling the distribution of *Entandrophragma*. Ecology 85:2171–2183

Healey JR, Price C, Tay J (2000) The cost of carbon retention by reduced impact logging. For Ecol Manag 139:237–255

Holzman BA (2008) Tropical forest biomes. Greenwood Press, Westport

Hopkins MS, Graham AW (1983) The species composition of soil seed banks beneath tropical lowland rainforests in north Queensland, Australia. Biotropica 15:90–99

Houghton RA (1999) The annual net flux of carbon to the atmosphere from changes in land use 1850–1990. Tellus Ser B Chem Phys Meteorol 51:298–313

Hughell D, Butterfield R (2008). Impact of FSC certification on deforestation and the incidence of wildfires in the Maya Biosphere Reserve. Report. Rainforest Alliance. 18 pages. http://www.rainforest-alliance.org/forestry/documents/peten_study.pdf. Accessed Oct 2010

IPCC (2006) In: Eggleston HS, Buendia L, Miwa K, Ngara T, Tanabe K(eds) IPCC guidelines for national greenhouse gas inventories. Prepared by the National Greenhouse Gas Inventories Programme. Institute for Global Environmental Strategies, Hayama

IPCC (2007) In: Core Writing Team, Pachauri RK, Reisinger A (eds) Climate change 2007: synthesis report. Contribution of working groups I, II and III to the fourth assessment report of the intergovernmental panel on climate change. IPCC, Geneva, 104 pp

ITTO (2006) International tropical timber organization status of tropical forest management 2005. Special edition, Tropical Forest Update 1, 35 p

Johns JS, Barreto P, Uhl C (1996) Logging damage during planned and unplanned logging operations in the eastern Amazon. For Ecol Manag 89:59–77

Jusoff K, Majid NM (1992) An analysis of soil disturbance from logging operation in a hill forest of peninsular Malaysia. For Ecol Manag 47:323–333

Keith H, MacKey BG, Lindenmayer DB (2009) Re-evaluation of forest biomass carbon stocks and lessons from the world's most carbon dense forests. PNAS 106:11635–11640

Keller M, Palace M, Asner GP, Pereira R Jr, Silva JNM (2004) Coarse woody debris in undisturbed and logged forests in the eastern Brazilian Amazon. Glob Change Biol 10:784–795

Keller M, Asner GP, Blate G, McGlocklin J, Merry F, Pena-Claros M, Zweede J (2007) Timber production in selectively logged tropical forests in South America. Front Ecol Environ 5:213–216

Kirby KR, Potvin C (2007) Variation in carbon storage among tree species: implications for the management of a small-scale carbon sink project. For Ecol Manag 246:208–221

Kirilenko AP, Sedjo RA (2007) Climate change impacts on forestry. Proc Natl Acad Sci USA 104:19697–19702

Klassen A (2005a) Planning considerations for reduced impact logging. In: Hasbillah (ed) Prepared for ITTO Project PD 110/01 Rev. 4 (I). Tropical Forest Foundation, Jakarta. http://www.tff-indonesia.org/en/tff-library/technical-procedure-manuals. Accessed Oct 2010

Klassen A (2005b) Operational considerations for reduced impact logging. In: Hasbillah (ed) Prepared for ITTO Project PD 110/01 Rev. 4 (I). Tropical Forest Foundation, Jakarta. http://www.tff-indonesia.org/en/tff-library/technical-procedure-manuals. Accessed Oct 2010

Klassen A (2006a) Planning, location, survey, construction & maintenance for low-impact forest roads. In: Hasbillah (ed) Prepared for ITTO Project PD 110/01 Rev. 4 (I). Tropical Forest Foundation, Jakarta. http://www.tff-indonesia.org/en/tff-library/technical-procedure-manuals. Accessed Oct 2010

Klassen A (2006b) Management considerations for successful implementation of reduced impact logging. In: Hasbillah (ed) Prepared for ITTO Project PD 110/01 Rev. 4 (I). Tropical Forest Foundation, Jakarta. http://www.tff-indonesia.org/en/tff-library/technical-procedure-manuals. Accessed Oct 2010

Leigh EG (1999) Tropical forest ecology: a view from Barro Colorado Island. Oxford University Press, Oxford, 245 p

Lugo AE, Silver WE, Colon SE (2003) Biomass and nutrient dynamics of restored neotropical forest. Water Air Soil Pollut 4:731–746

Malaisse F (1978) The miombo ecosystem. Trop For Ecosyst Nat Resour Res 14:589–606

Meggers BJ, Ayensu ES, Duckworth WD (1973) Tropical forest ecosystems in Africa and South America: a comparative review. Smithsonian Institution Press, Washington, DC

Montagnini F, Jordan CF (2005) Tropical forest ecology: the basis for conservation and management. Springer, Berlin/New York

Palmiotto PA, Davies SA, Vogt KA, Ashton PMS, Vogt DJ, Ashton PS (2004a) Soil related habitat specialization in dipterocarp rain forest tree species in Borneo. J Ecol 92:609–623

Palmiotto PA, Vogt KA, Ashton PMS, Ashton PS, Vogt DJ, Semui H, Seng LH (2004b) Linking canopy gaps, topography and soils in a tropical rain forest: implications for species diversity. In: Losos E, Leigh R (eds) Forest diversity and dynamism: results from the global network of large-scale demographic plots. University of Chicago Press, Chicago, pp 101–125

Pena-Claros M, Fredericksen TS, Alarcon A, Blate GM, Choque U, Leano C, Licona JC, Mostacedo B, Pariona W, Villegas Z, Putz FE (2008a) Beyond reduced-impact logging: silvicultural treatments to increase growth rates of tropical trees. For Ecol Manag 256:1458–1467

Pena-Claros M, Peters EM, Justiniano MJ, Bongers F, Blate GM, Fredericksen TS, Putz FE (2008b) Regeneration of commercial tree species following silvicultural treatments in a moist tropical forest. For Ecol Manag 255:1283–1293

Pereira R Jr, Zweede J, Asner GP, Keller M (2002) Forest canopy damage and recovery in reduced-impact and conventional selective logging in eastern Para, Brazil. For Ecol Manag 168:77–89

Pinard MA, Putz FE (1996) Retaining forest biomass by reducing logging damage. Biotropica 29:278–295

Pinard MA, Putz FE, Tay J, Sullivan TE (1995) Creating timber harvest guidelines for a reduced-impact logging project in Malaysia. J For 93:41–45

Pinard MA, Barker MG, Tay J (2000) Soil disturbance and post-logging forest recovery on bulldozer paths in Sabah, Malaysia. For Ecol Manag 130:213–225

Putz FE, Pinard MA (1993) Reduced-impact logging as a carbon-offset method. Conserv Biol 7:755–757

Putz FE, Dykstra DP, Heinrich R (2000) Why poor logging practices persist in the tropics. Conserv Biol 14:951–956

Putz FE, Sist P, Fredericksen T, Dykstra D (2008a) Reduced-impact logging: challenges and opportunities. For Ecol Manag 256:1427–1433

Putz FE, Zuidema PA, Pinard MA, Boot RGA, Sayer JA, Sheil D, Sist P, Elias, Vanclay JK (2008b) Improved tropical forest management for carbon retention. PLoS Biol 6:1368–1369

Rhoades CC, Eckert GE, Coleman DC (2000) Soil carbon differences among forest, agriculture, and secondary vegetation in lower montane Ecuador. Ecol Appl 10:497–505

Schimel DS (1995) Terrestrial ecosystems and the carbon-cycle. Glob Change Biol 1:77–91

Schlesinger WH (1997) Biogeochemistry: an analysis of global change. Academic Press, Inc, San Diego/London, Illus. Maps, xi+443p

Schwartzman S, Moreira A, Nepstad D (2000) Rethinking tropical forest conservation: perils in parks. Conserv Biol 14:1351–1357

Siry J, Cubbage F, Ahmed M (2005) Sustainable forest management: global trends and opportunities. For Policy Econ 7:551–561

Sist P, Ferreira FN (2007) Sustainability of reduced-impact logging in the Eastern Amazon. For Ecol Manag 243:199–209

Sist P, Fimbel R, Sheil D, Nasi R, Chevallier MH (2003) Towards sustainable management of mixed dipterocarp forests of South-east Asia: moving beyond minimum diameter cutting limits. Environ Conserv 30:364–374

Smith DM, Larson BC, Kelty MJ, Ashton MS (1997) The practice of silviculture: applied forest ecology. Wiley, New York

Sohngen B, Mendelsohn R (2003) An optimal model of forest carbon sequestration. Am J Agric Econ 85: 448–457

Thornley JHM, Cannell MGR (2000) Managing forests for wood yield and carbon storage: a theoretical study. Tree Physiol 20:477–484

Tropical Forest Foundation (2007) TFF Standard for Reduced Impact Logging (TFF-STD-RIL-2006, V1.1.1). Alexandria. http://www.tff-indonesia.org/images/stories/pustaka/others/STANDARD_FOR_RIL.pdf. Accessed Oct 2010

Uhl C, Kauffman JB (1990) Deforestation, fire susceptibility, and potential tree responses to fire in the eastern Amazon. Ecology 71:437–449

Uhl C, Barreto P, Verissimo A, Vidal E, Amaral P, Barros AC, Souza C Jr, Johns J, Gerwing J (1997) Natural resource management in the Brazilian Amazon. Bioscience 47:160–168

USDA (1975) Soil conservation survey-USDA soil taxonomy: a basic system of classification for making and interpreting soil surveys. USDA Agriculture handbook No. 436, US Government Printing Office, Washington, DC

Valle D, Phillips P, Vidal E, Schulze M, Grogan J, Sales M, van Gardingen P (2007) Adaptation of a spatially explicit individual tree-based growth and yield model and long-term comparison between reduced-impact and conventional logging in eastern Amazonia, Brazil. For Ecol Manag 243:187–198

van der Werf GR, Morton DC, DeFries RS, Olivier JGJ, Kasibhatla PS, Jackson RB, Collatz GJ, Randerson JT (2009) CO2 emissions from forest loss. Nat Geosci 2:737–738

Verissimo A, Barreto P, Mattos M, Tarifa R, Uhl C (1992) Logging impacts and prospects for sustainable forest management in an old Amazonian frontier – the case of Paragominas. For Ecol Manag 55:169–199

Vitousek PM, Sanford RL (1986) Nutrient cycling in moist tropical forest. Annu Rev Ecol Syst 17:137–167

Wadsworth FH, Zweede JC (2006) Liberation: acceptable production of tropical forest timber. For Ecol Manag 233:45–51

Whitmore TC (1990) Tropical forest ecology. Oxford University Press, Oxford

Zarin DJ, Schulze MD, Vidal E, Lentini M (2007) Beyond reaping the first harvest: management objectives for timber production in the Brazilian Amazon. Conserv Biol 21:916–925

Managing Carbon Sequestration and Storage in Temperate and Boreal Forests

10

Matthew Carroll, Brian Milakovsky, Alex Finkral, Alexander Evans, and Mark S. Ashton

Executive Summary

If carbon stocks and fluxes in temperate and boreal forests are to be included among efforts to mitigate global climate change, forest managers and policy makers must understand how management affects the carbon budgets in these systems. This chapter examines the effects of management of carbon sequestration, storage, and flux in temperate and boreal forests.

Existing Evidence Reveals the Following Trends

- Drainage of wetlands for increased tree production can result in either net carbon gain or loss, depending on how deep the drainage.
- Silvicultural thinning causes a reduction of the vegetative carbon pool, which recovers over a matter of decades, while the impact on soil carbon is considered limited.
- In certain forest systems, fuels reduction treatments (such as thinning and prescribed fire) result in lowered vegetative carbon storage, but result in forest structures that are significantly less susceptible to stand-replacing disturbance and the commensurate carbon releases from disturbance.
- Regeneration harvests significantly reduce the carbon stocks in vegetation and cause a transient increase in soil respiration, although the annual rate of carbon uptake will be greater in the regenerating stand. Harvested areas often remain net carbon sources for 10–30 years, then return to sinks.
- Carbon sequestration can be increased by extending rotation lengths, especially if maximum biomass productivity has not yet been reached.
- Fertilization can increase carbon storage in vegetation and reduce soil respiration rates, however gains are offset by the carbon released during fertilizer production.

We identified the following key points to consider for carbon storage and sequestration projects in temperate and boreal forests:

- Many forest management activities result in net carbon release and thus cannot demonstrate carbon additionality. Mechanisms should be developed to credit projects that reduce carbon loss, in addition to those that increase carbon gain.
- Where baselines are set for forest carbon project accounting determines which management activities are incentivized.
- The risk of carbon leakage must be addressed. If sequestration strategies simply displace timber harvests from one forest to another, at any geographic scale, carbon gains are neutralized.
- The amount of carbon stored in forest products, emissions from management operations,

M. Carroll • B. Milakovsky • M.S. Ashton (✉)
Yale School of Forestry & Environmental Studies, 360 Prospect St., New Haven, CT 06511, USA
e-mail: mark.ashton@yale.edu

A. Finkral
School of Forestry Northern Arizona University, Flagstaff, AZ 86004, USA

A. Evans
The Forest Guild, Santa Fe, PO Box 519, NM 87504, USA

M.S. Ashton et al. (eds.), *Managing Forest Carbon in a Changing Climate*,
DOI 10.1007/978-94-007-2232-3_10, © Springer Science+Business Media B.V. 2012

and fossil fuel displacement by forest biomass determine whether or not practices like thinning are positive, neutral or negative.
- Many forest management practices have minimal impacts on the soil carbon pool, which is the most difficult pool to measure. Thus, it may be possible that projects involving certain practices could avoid strict quantification of this pool.

Most temperate forests are second growth (Whitney 1996), much of the boreal has recently been cutover, but land conversion is minimal when compared to other regions of the world. Therefore, providing additional carbon storage is a matter of refining silvicultural practices, better quantifying the effects of disturbances, and examining the storage potential of forest products.

1 Introduction

Forests play a major role in the mitigation of climate change, primarily through their ability to assimilate carbon dioxide and sequester it in living tissue, and in their long-term contribution to soil carbon stocks. Temperate and boreal forests are also a significant source of carbon emissions because of wildlife (Wiedinmyer and Neff 2007) and other disturbances (e.g., Zeng et al. 2009). Forest systems cover more than 4.1 billion hectares – approximately one third of the earth's land area (Dale et al. 2001) – and temperate and boreal forests make up roughly 49% of this total. Forests account for 90% of all vegetative carbon in terrestrial ecosystems and assimilate 67% of the total CO_2 absorbed from the atmosphere by all terrestrial ecosystems (Gower 2003).

Whether forests are sinks or sources of terrestrial carbon depends on the balance of processes that cause carbon sequestration (i.e. photosynthesis, peat formation) and release (i.e. increased respiration, forest disturbance). Taken as a whole, the temperate and boreal forest biomes were carbon sinks during the 1980s and 1990s (Schimel et al. 2001), but this may no longer be the case because the Canadian lodgepole pine forests are poised to release massive amounts of carbon as the result of die-off from insect infestations (Kurz et al. 2008). The moist temperate forest sink has been consistently growing with the abandonment of marginal agricultural lands (Houghton et al. 2000), and does not experience the same scale of disturbance-mediated carbon release as in the boreal or inter-mountain forests.

The emphasis on silvicultural practices in boreal and temperate forests is appropriate because increasing forest carbon stocks in these regions is a matter of making adjustments to existing forests and not undergoing radical changes in land use.

2 Boreal and Temperate Forests of the World

Boreal forests comprise the northernmost forest biome of the world, covering much of Alaska, Canada, Fennoscandia, Russia, northern Mongolia and northeast China. Boreal forests are characterized by simple, often single layered stand structure, low tree species diversity (only six genera dominate the entire range: spruce (*Picea*), fir (*Abies*), pine (*Pinus*), larch (*Larix*), birch (*Betula*) and aspen (*Populus*)) and well-developed bryophyte (moss and lichen) communities. Organic-rich peat soils in boreal forests and bogs (histosols or spodosols) are the largest carbon pool in the biome.

Boreal forests can be roughly divided into two major zones – interior continental and maritime (Fig. 10.1). As the name implies, interior continental forests are exposed to cold, dry continental climates. Fire and large-scale insect outbreaks are the dominant disturbance agents. In North America, interior continental boreal forests are dominated by white spruce (*Picea glauca*), Jack pine (*Pinus banksiana*), and aspen (*Populus tremuloides*) in different mixtures. In Eurasia, interior continental forests are found east of the Ural Mountains. Siberian larch (*Larix sibirica*) and Dahurian larch (*Larix gmelinii*), both adapted to extreme cold, drought, and permafrost, cover much of this area.

Maritime boreal forests are found in North America along the Pacific coast (Cordillarean type) and Atlantic coast (Maritime type). In this moderated climate, fir species compose a larger proportion of forest area, and fire gives way to insect outbreaks and commercial harvesting as the primary disturbance agents. Maritime forests are also found in Fennoscandia and northwest Russia near the Norwegian, Baltic and White Seas. Scots pine (*Pinus sylvestris*) and Norway spruce (*Picea abies*)

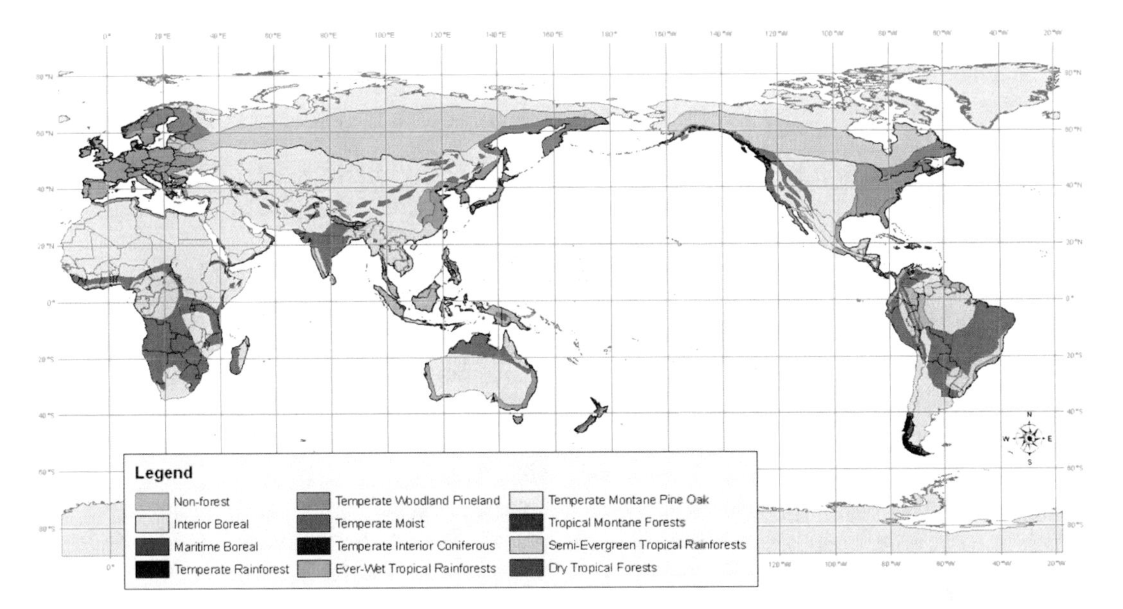

Fig. 10.1 Original extent of boreal, temperate, and tropical forest types of the world prior to land clearing

are the canopy dominants, with a considerable component of aspen and birch. Ground fires, insect outbreaks, and harvesting are major influences.

Temperate forests include a wide range of forest types across the midlatitudes, and the boundaries with boreal forests to the north and tropical forests to the south are subject to interpretation. With a distinct but relatively mild winter, temperate forests are characterized by more diverse climatic conditions and angiosperm species than in the boreal forest type. Generally speaking, the soil carbon pool does not play as large a role here, while the prominence of the vegetative pools increases.

There are five major temperate forest types:

1. *Moist broadleaf and coniferous forests*: mesic, mixed forests with a rich suite of genera, including maple (*Acer*), Oak (*Quercus*), birch (*Betula*), beech (*Fagus*), ash (*Fraxinus*), poplar, aspen (*Populus*), hemlock (*Tsuga),* "soft pines" (*Pinus*), spruce (*Picea*) and fir (*Abies*). Fire plays a relatively minor role in such forests. They are located in the eastern United States and Canada, northern and central Europe, and the Russian Far East. Soils classified as ultisols (USDA 1975) underlie much of this area, particularly in North America, and are generally desirable for cultivation because they are usually relatively fertile (though often stony) and require no irrigation because of precipitation year round.
2. *Interior coniferous forests*: dry, fire-adapted forests in harsh continental mountainous climates, with soils that are inceptisols (glacial non-volcanic) or often andisols (volcanic). "Hard pines" (*Pinus*), spruce, fir and larch predominate. Located in the interior west of the USA and Canada, and in Central Asia, these forest types are closely related to interior continental boreal forests. Soils are young, rocky, often skeletal, and exposed to the extremes of cold winters and dry summers.
3. *Montane oak/pine forests*: *Pinus*- and *Quercus*-dominated systems in mountain ranges of Mexico and Central America, the Himalayas, the Mediterranean and Turkey. They are fire-adapted and relatively dry. Soils are mixed.
4. *Woodland and pineland forests*: Fire-adapted, often open forests in dry, southern climates. They include "hard" pine forests of the U.S. coastal plain, pine and oak in the coastal Mediterranean region, *Acacia-Eucalyptus* savannas of Africa and Australia, and oak woodlands. Soils that are generally classified as alfisols (USDA 1975) predominate. Such soils are more fertile than ultisols but often require partial irrigation because of drier summers. Most forests with alfisols have already been cleared for cultivation, thus this type is restricted to degraded relics.

5. *Temperate rainforests*: Mesic, constantly moist, and often extremely productive forests of mountain ranges along coasts. Spruce, hemlock, Douglas fir (*Pseudotsuga*) and western cedar (*Thuja*) dominate in the Pacific Northwest, the southern beech (*Nothofagus*) in Chile, and southern beech, Eucalypts (*Eucalyptus*) and podocarps (*Podocarpus*) in New Zealand and Australia. Spodosols and andisols are the predominant soil types. Andisols are volcanic soils that with high precipitation can be very productive for pasture. Spodosols are acidic soils associated with bedrock geology that predominantly comprise minerals such as quartz and silica, and are therefore often nutrient poor.

3 The Forest Carbon Cycle

The following concepts pertain to the basic biological dynamics of carbon uptake, storage, and release, and also to important differences in how carbon pools in managed forests are quantified.

3.1 Maximizing Carbon Uptake vs. Maximizing Carbon Storage

Biomass productivity is maximized relatively early in forest development, at the time when annual growth increment dips below the average annual growth increment over the age of the tree or stand. After this point growth slows, and carbon uptake slows along with it. However, while older trees (and stands) may demonstrate reduced uptake rates, the carbon stored within them can greatly exceed that of their younger, perhaps faster-growing, counterparts. Greater pools of soil and litter carbon in older forests may also contribute to this effect, although their pattern is less clear than that of the vegetative pool.

The importance of this difference lies in its management consequences. Managing for productive

Insert 1. Maximizing C Uptake Versus Maximizing C Storage

Aboveground biomass

Litter pool
Belowground biomass
soil pool

These two images demonstrate the contrasting strategies of growing vigorous young forests with high rates of carbon uptake (left), and growing forests to older age classes at which uptake rate is lower, but actual quantities of stored carbon are greate (rights). The downward pointing arrows indicate carbon uptake through photosynthesis, the rates of which are indicated by arrow size. Upward arrows indicate C release through auto- and heterotrophic soil respiration. In the old forest shown on the right, the inputs and outputs are near equilibrium, while on the left, uptake clearly exceeds carbon loss. However, note that the actual size of the aboveground biomass, litter and belowground biomass are considerably larger in the older forest. Importantly, the size of the soil pool does not differ much between the two examples.

young forests promotes maximal carbon uptake, while maintaining old forests and extending rotations leads to larger on-the-ground carbon stocks. In theory, a series of short rotations can sometimes lead to greater total carbon storage than a single long rotation because the stand is growing at a rapid rate for a greater proportion of the time. But each harvest entry is also followed by a release of carbon associated with decomposition.

3.2 Site and Climatic Factors Limit the Carbon Storage Potential of Vegetation

In any given forested site, the maximum potential productivity and carbon storage of vegetation is determined by soil fertility, moisture conditions, and climate. These factors can be regarded as placing a "ceiling" on biomass production. Forest managers can manipulate and re-allocate biomass in different assemblages of species and stand structures. But to create additional carbon storage requires addressing the basic productivity constraints, for instance by fertilizing, irrigating, or draining the site.

A major caveat, however, is that forests may not reach their "biomass ceiling" for hundreds of years, often much longer than the rotations used in conventional forest management (Luyssaert et al. 2008). There are a number of forest management strategies that increase carbon storage (Evans and Perschel 2009). For example, it is often possible to gain carbon benefits simply by growing forests on longer rotations so that they have time to accumulate higher standing volumes (Foley et al. 2009).

3.3 The Carbon Impact of an Activity Changes if the Forest Products Carbon Pool is Included

Thinning results in a reduction of the vegetative carbon pool. It is possible that the residual trees will eventually replace the biomass lost in a harvest, and the pool will equal or exceed its pre-treatment storage. But due to the productivity constraints described above, the pool will never exceed the storage potential of the stand if it had never been thinned. This makes thinning a carbon-negative or at best carbon-neutral activity *unless* the sequestration of carbon within forest products is considered – that is, products are considered to be another "pool" (Eriksson et al. 2007). When the product pool is included, thinning can become carbon-positive because some portion of the harvested carbon will be stored in long-term forest products, while the residual trees are growing at a faster rate and taking up more carbon (e.g. Finkral and Evans 2008).

The inclusion and quantification of the forest products pool in carbon offset programs are topics of much debate and discussion. It is important to recognize the impact that this pool can have on the measurement of the carbon in forest management practices. A comprehensive discussion of the forest products pool is provided in Chapter 12.

3.4 Resiliency: Maximum Carbon Storage at High Risk vs. Reduced Carbon Stocks at Reduced Risk

Forest managers have long recognized that maximizing the density of biomass on a site can be detrimental to forest health. Density-related competition often results in spindly, poorly-formed trees that are not windfirm, are susceptible to insect outbreak, and pose fire risks. On a larger scale, the risk of such disturbances is also increased when a large proportion of the landscape is maintained in dense stands within a limited age class range. Foresters address these concerns by managing for stand- and landscape-level resiliency. Stands are often managed at lower than maximum densities, in order to reduce risk of catastrophic loss. A sacrifice in biomass production is made in order to produce fewer, larger, more vigorous trees.

This principle still applies when carbon uptake and storage is the management goal. Carbon stored in fire-, insect- or windthrow-prone trees and stands is "risky," and some sacrifice in total storage may be necessary to ensure that sequestration is long-term.

3.5 Creating Carbon Additionality vs. Minimizing Carbon Loss

Because of the structure of many carbon offset programs, the primary goal of managing forest carbon is often to create additionality. Certain practices are regarded as reliably "additional," such as afforestation (unless by changing the site a large soil carbon loss is incurred). However, the manipulation of standing forests more commonly results in immediate reductions of carbon pools. Such practices can be adapted in certain ways to reduce their negative carbon impact, such as by leaving more harvest residues or causing less damage to residual trees during harvest. This can result in a form of additionality, compared to business-as-usual management techniques. Activities such as reduced deforestation and reduced impact logging appear additional when compared to such a business-as-usual baseline.

4 Carbon Impacts of Specific Forest Management Practices

4.1 Application of Resiliency

Disturbance plays a vital role in the natural flow of carbon between pools, but as a result of past management practices and a changing climate, many forests in the boreal and temperate regions have become especially susceptible to catastrophic disturbances (Hurteau and North 2009) that release large pulses of carbon into the atmosphere.

Managing for carbon should strive to maximize the amount of stored carbon while minimizing the likelihood of stand-replacing disturbance. This balance is achieved through maximizing forest resiliency, the capacity of a system to absorb disturbance and reorganize while undergoing change so as to retain essentially the same function, structure, and ecosystem services (Folke et al. 2004). This definition works well for carbon purposes because it accounts for a resilient forest's ability to reduce carbon loss from a disturbance and reorganize in such a way that maintains high levels of the desired ecosystem service, carbon sequestration. Here are examples of management responses to four very common disturbances in boreal and temperate forests: fire, wind, insect infestations and climate change.

4.1.1 Fire

Fire is a dominant disturbance agent in many temperate forest regions. In some regions, uncharacteristic fire frequency and intensity is due to changing climactic conditions (Lucas et al. 2007). In many others, the structure of fire dependant temperate forest ecosystems has been altered as a result of a high level of fire suppression over the last 100 years (Covington et al. 1997; Allen et al. 2002; Brown et al. 2004). This has resulted in a buildup of fuels leading to intense fires (Hessburg et al. 2005). Tilman et al. (2000) found that in an oak savannah in Minnesota, when fire was excluded, forests were able to build both above and belowground biomass to levels 90% greater than in forests with frequent ground fires. This sequestered carbon is at high risk of sudden release due to the potential for stand-replacing fire. On such sites, forest managers may choose to balance increased sequestration with increased stability by reducing stem density and fuel loading.

The restoration of more fire-resilient forests is possible and critical (Agee and Skinner 2005). A combination of thinning and burning can build resiliency through the removal of accumulations of biomass fuels at sites. Forests under such management will store less carbon than the maximum possible, but over the long term they may store more than forests that experience stand-replacing fires (Houghton et al. 2000). In the southwestern U.S., a thinning designed to reduce fire risk reduced the total amount of carbon stored in a ponderosa pine stand and turned it into a weak carbon source for a short period following treatment (Dore et al. 2010). Although the carbon sink strength was reduced, the reductions in total stand carbon and gross primary productivity were not as much as in a nearby stand that experienced a high-intensity fire. Furthermore, the thinned stand can continue greater levels of primary production compared to the burned stand (Dore et al. 2010). It is well known that fire severity determines the amount of carbon released during the acute stages of the disturbance. However, some studies indicate that nearly half of the carbon released is lost through the much slower decomposition processes over a period of years (Brown et al. 2004; Hessburg et al. 2005). In fact, some experiments have shown that recently burned and harvested sites are sources of carbon, and that recovery to the same flux as a mature site can take 10 years following a fire (Amiro 2001). Causes of this phenomenon are linked to an increase in soil respiration due to an increase in soil surface temperatures. The complex interactions between fire, soils, vegetation, and site recovery from a disturbance are just beginning to be understood.

Prescribed fire treatments are intended to reduce fuel loads without causing significant mortality to the remaining vegetation. It is important to point out that there is a carbon loss associated with the use of prescribed fire. Surface soils, litter and downed woody material will be carbon sources for some years after the disturbance. Land managers need to weigh these emissions against either a no-action alternative or another silvicultural treatment to determine the best fit for the site. It should be stressed that the carbon loss from a high-intensity fire can be extensive and long-lasting.

Some boreal and temperate forest types, such as lodgepole pine (*Pinus contorta*), have evolved with stand-replacing wildfire. It would thus be misguided to attempt to produce more resilient forest structures – ones "capable of maintaining substantial live basal area after being burned by a wildfire" (Agee and Skinner 2005) – in fire-dependent ecosystems. The autecology of species like *P. contorta* may make stands they dominate inherently more "risky" for carbon sequestration, and inappropriate as sites for long-term storage.

4.1.2 Wind

Unlike fire, the magnitude of carbon loss from a wind disturbance is not so closely linked to stocking density. Wind as a disturbance agent can affect forests through a wide range of magnitude and spatial scales, from a localized downburst damaging a single tree to the large-scale damage caused by hurricanes (McNulty 2002). Over the period 1851–2000 tropical cyclones caused an average carbon release of 25 Tg/y (Zeng et al. 2009). The resilience of trees and understory vegetation to wind disturbance can provide a tight biotic control of ecosystem processes like carbon sequestration, and is based on the structure of the forest prior to the disturbance (Cooper-Ellis et al. 1999). The greater the diversity of functional groups represented in the pre-disturbance forests, the greater capacity the forest has to maintain or recover the ability to sequester carbon in the environment that follows the disturbance (Busing et al. 2009).

4.1.3 Insects/Pathogens

In recent decades there has been no shortage of examples of both native and exotic pests and pathogens causing tree mortality in boreal and temperate forests. Exotic pests and pathogens have great potential to alter forest carbon dynamics (Peltzer et al. 2010; Ayres and Lombardero 2000). Depending upon species-specific characteristics, mixed forests may contribute to ecological stability by increasing resistance and resilience (Larsen 1995). A good example is the mixed hemlock/hardwood forests of the northeastern USA. Hemlock woolly adelgid attacks hemlock trees of all ages and sizes, and infested trees seldom recover (Nuckolls et al. 2008). Carbon effects from the infestation are not surprising; during the first year of infestation, autogenic respiration of CO_2 from roots is reduced although no additional carbon is stored because there is little or no photosynthesis occurring. Decomposition increases as trees die as a result of increased light regimes, leading to increased soil temperatures. Overall the carbon release depends on the size of the infestation and the species mix associated with the hemlock stands. Since most hemlock stands are not single species, or single age class the carbon loss from the ecosystem as a whole is less than in monotypic forest types such as lodgepole pine (Albani et al. 2010; Orwig and Foster 1998). Additionally, large-scale stand-replacing fires are not typical in the eastern US where the hemlock woolly adelgid is found. In the context of carbon sequestration, mixed hemlock/hardwood forests are more resilient to insect infestation than lodgepole pine forests because of their diversity (Schafer et al. 2010).

Insert 2. Managing for Resiliency in Forests Affected by the Mountain Pine beetle

Managing for resiliency in forests affected by the mountain pine beetle

"There are literally several hundred million cubic meters of wood out there in the forests decomposing and releasing carbon dioxide back into the atmosphere," (Kurz et al., 2008) from a massive outbreak of the mountain pine beetle (Dendroctonus ponderosae) across the lodgepole pine (Pinus contorta) forests of interior British Columbia. This infestation and subsequent catastrophic fires in beetle-killed timber are threatening to turn Canada's forests from a carbon sink to a source. It is projected that the region could release 990 million tons to CO_2 *– more than the entire annual emissions reported by Canada in 2005 (Kurz et al., 2008).*

Research has demonstrated that direct management of mountain pine beetle through tree removal, burning or insecticide application is impractical and ineffective. Rather, that alteration of stand structure (age-class distribution, composition and density) has the best chance of minimizing the scale and intensity of the infestations and associated negative carbon flux from these forests (Amman and Logan, 1998). Unfortunately, because of a century long campaign of aggressive fire suppression, and an attempt to maintain a status quo of current stand conditions that goes beyond the natural cycle of regeneration and renewal, there are limited opportunities for appropriate silvicultural treatments.

4.1.4 Climate Change

If climate change alters the distribution, extent, frequency, or intensity of any of these disturbances, large impacts could be expected (Dale et al. 2001). For example, as climate changes, the ability of native and non-native forest pests to establish and spread increases because the range of suitable environment expands. The door opens to insects and pathogens that previously posed less of a risk. Direct effects of climate change on forest pests will likely be increased survival rates due to warmer winter temperatures, and increased developmental rates due to warmer summer temperatures (Hunt et al. 2006). A striking example is in the interior of British Columbia where the mountain pine beetle (*Dendroctonus ponderosae*) infestation is rapidly spreading to the north (Ayres and Lombardero 2000; Peltzer et al. 2010).

The diversity of species in an ecosystem undergoing change appears to be critical for resilience and the generation of ecosystem services (Folke et al. 2004). In this sense, biological diversity provides insurance, flexibility, and a spreading of risk (Duffy 2009). Therefore management should attempt to strive for diverse, mixed species, multiple age class stands, or any combination thereof, for all forest types – simple or complex. It is one important tool that contributes to sustaining the response required for renewing and reorganizing desired ecosystem states after disturbance (Larsen 1995).

Resilience can be influenced at the landscape level by the presence of refugia that escape disturbance and serve an important re-colonization function for surrounding areas. This diversity of species and heterogeneity in the landscape builds integrity, meaning that even if the disturbance causes a change in the stable state of the forest, the new stable state will function in a similar way, providing the same ecosystem services, including carbon sequestration (Perry and Amaranthus 1997).

4.2 The Concept and Application of Thinning

Thinning is a silvicultural practice that lowers stand density through the removal of a portion of the standing volume, often at regular spacing. Thinning clearly impacts the aboveground vegetative carbon pool, and it also affects the litter pool (through the addition of slash and reduction of post-thinning litterfall), and potentially the soil pool (through increased respiration due to increased light and warmth at the soil layer).

Thinning increases the amount of available growing space for residual trees, thereby leaving potential growing space vacant for a period of time immediately post-treatment, resulting in reduced stand carbon storage (e.g., Campbell et al. 2009; Spring et al. 2005; Nilsen and Stand 2008; Balboa-Murias et al. 2006). Importantly, the decrease in stand production does not always scale perfectly with the reduction in stand density. Light-use efficiency of ponderosa pine was almost 60% higher in thinned than unthinned stands (Campbell et al. 2009), perhaps because the trees removed in the treatment were of low vigor and were not using site resources efficiently. Also, if canopy thinning stimulates increased growth in midstory and understory vegetation, reductions in aboveground net primary production can be quickly offset (e.g. thinning in Ohio oak-maple (*Quercus-Acer*) stands, Chiang et al. 2008). However, after thinning, a stimulated shrub layer can also result in net carbon loss if it has lower net primary productivity than the tree layer but similar respiration rates (Campbell et al. 2009).

Different types and intensities of thinning have different impacts on carbon storage. For example, in Allegheny hardwoods, plots thinned from below showed no significant difference in carbon storage from unthinned plots, crown-thinned plots sequestered significantly less carbon, and thinned-from-above plots even less (Hoover and Stout 2007). A pre-commercial thinning in New South Wales increased total stand carbon because all the cut trees remained on the ground (and were sequestered for some time in the litter pool) while the residuals accumulated biomass at a faster rate (McHenry et al. 2006).

Thinning influences litter and soil carbon as well. In general, forest floor carbon declined with increasing thinning intensity in field studies in New Zealand, Denmark, and the USA (Jandl et al. 2007). Litterfall additions to the forest floor and higher ground temperatures stimulated decomposition. However, the impact was moderated by the addition of logging slash to the litter layer, and the fairly rapid return to pre-treatment temperatures in all but the most intensively-thinned plots (Jandl et al. 2007). Increases in CO_2 efflux after thinning have been observed for several years in California

Insert 3 Thinning and the C Balance of a Forest Stand

Thinning and the carbon balance of a forest stand

Flux tower measurements taken in a 40-year-old Scots pine (Pinus sylvestris) stand in southern Finland showed that CO_2 flux did not change after the first commercial thinning. A complex of factors allowed this. A reduction in overstory photosynthesis was balanced by an increase in understory photosynthesis. And while heterotrophic respiration increased with the decomposition of logging slash and roots, this in turn was balanced by a reduction in autotrophic root respiration.

Thus, the "redistribution of sources and sinks is comprehensively able to compensate for the lower foliage area" in the thinned stand.

From Suni et al., 2003

mixed conifers and Ozark oak-hickory (*Quercus-Carya*) stands (Concilio et al. 2005).

The soil pool appears even more buffered from the effects of thinning than the litter pool. Some increase in soil respiration was observed after thinning in Norway spruce, but no significant effects on soil carbon storage could be detected with increasing thinning intensity (Nilsen and Stand 2008). Thinning in South Korean *Pinus densiflora* and German European beech (*Fagus sylvaticus*) forests produced no significant increases in respiration (Dannenmann et al. 2007; Kim et al. 2009). In loblolly pine (*Pinus taeda*) plantations in Virginia, the contribution of logging slash and decaying roots to the soil actually *increased* soil carbon concentration in the 10–40 cm depth 14 years after thinning (Selig 2008).

Thinning thus produces a short term decrease in vegetative and litter carbon pools, and little to no increase in soil respiration. How long this negative impact on carbon storage on-site lasts depends on the intensity and type of thinning, and on how fast residual trees can replace the biomass removed. Whether slash inputs to the litter layer exceed reductions in litterfall also plays a small part in defining when pre-treatment carbon levels are re-attained.

4.3 Site Treatments

4.3.1 Drainage

Drainage is implemented where excessive soil moisture stunts or prohibits the growth of trees. Within the boreal and temperate zones, this practice is most prominent in Fennoscandia, particularly in Finland. Drained peatland forests constitute 18–22% of the total managed area of that country (Minkkinen et al. 2001). Afforestation of drained peatlands has also occurred on a large scale in Great Britain and the coastal mires of the southern United States. These peatland areas are associated with high levels of soil carbon storage, but also with emissions of CH_4 (methane), an important greenhouse gas.

The carbon consequences of land drainage depend on whether the factors that increase sequestration (increased vegetative production, increased litter input, and decreased methane release) exceed the increased respiration caused by oxidation of previously anoxic peat. A critical factor in this balance appears to be how much the water table is lowered in the drainage process. When the water table was lowered from 0–10 cm to 40–60 cm (below the surface) in Finnish mires, CO_2 loss increased 2–3 times and stayed at that rate for at least 3 years (Silvola 1986; Silvola et al. 1996). At this rate, Silvola (1986) found that such mires would switch from a modest carbon sink to a strong carbon source. Similarly, deep drainage of peaty moorlands in Britain for Sitka spruce (*Picea sitchensis*) afforestation would result in sufficient drying such that all but the recalcitrant peat component would decompose resulting in net carbon emissions (Cannell et al. 1993).

In contrast, when the water table in a Finnish mire was only lowered 5–9 cm, emissions barely changed (Silvola et al. 1996). Similarly, afforestation of Irish moorlands did not result in deep drying or oxidation and increased CO_2 release was minimal (Byrne and Farrell 2005). Von Arnold et al. (2005) examined CO_2 and CH_4 efflux (which are usually negatively correlated) in undrained, lightly drained and well-drained (dry) peatlands in Sweden. They found that, from the perspective of minimizing greenhouse gas emissions, the optimal condition was lightly drained peat, because increases in CO_2 efflux were exceeded by the decease in CH_4 efflux. In contrast, both undrained and dry peats were carbon sources to the atmosphere. Importantly, this analysis did not consider the additional sequestration potential of enhanced tree growth and litter production.

When the biomass and litter pools are considered, even greater carbon gains have been recorded in Sweden, Finland and Russian Karelia (Laine and Vasander 1991; Minkinnen and Laine 1998; Sakovets and Germanova 1992). Drained, plowed and afforested peatlands in Scotland were a carbon source for only 4–8 years, at which point increased vegetative productivity switched them to sinks. This effect only increased as the forests matured (Hargreaves et al. 2003).

Thus, drainage of peatlands for increased forest productivity has the potential to be carbon positive or carbon negative, depending on how thorough the drainage is. Shallowly drained sites tend to sequester more carbon than undrained sites because increased tree growth and decreased methane emissions outweigh increased CO_2 emissions. The opposite is true on deeply drained sites.

4.3.2 Fertilization

Tree growth in temperate regions is typically nitrogen-limited. Therefore, nitrogen fertilization is a well-established treatment in this region to increase biomass production. This increased capacity to store carbon is well documented, but must be considered in light of the carbon emissions required to produce and apply the fertilization treatment.

Biomass production is the result of the energy produced by photosynthesis, minus the respiration requirements of the non-photosynthetic plant tissues. Higher fertility increases leaf area, nutrient concentration, and carbon assimilation rates and in turn, improves carbon availability and overall biomass production (Coyle and Coleman 2005). Nitrogen fertilization has been shown to increase biomass production as much as 16 Mg ha^{-1} over 100 years in some intensively managed pine forests in the southeastern United States (Markewitz et al. 2002). On some low fertility sites, nitrogen fertilization can make the difference between the site's being a carbon source or a carbon sink and can lessen the time it takes for a developing stand to go from a source to a sink. The degree of effect that fertilization has depends on the baseline fertility of the site (Maier and Kress 2000).

The fertility of a site can be approximated by determining the nitrogen-use efficiency, a measure of the amount of additional carbon assimilated as a result of the addition of a kg of nitrogen. Nitrogen-use efficiency for carbon sequestration in trees strongly depends on soil nitrogen status as measured by the carbon/nitrogen ratio. Excessive fertilization or appropriate fertilization plus the deposition of anthropogenically elevated levels of atmospheric nitrogen can cause deposition rates to exceed the capacity for nitrogen uptake, and nutrient imbalances can lead to forest decline due to nitrogen saturation (Bauer et al. 2004). The effect of nitrogen saturation is also seen in soils when the biotic component of soil is no longer able to uptake and stabilize the nitrogen in organic compounds. The excess nitrogen is leached out of the soil in the form of nitrates (Magnani et al. 2007).

It has been thought that fertilization decreases soil carbon stocks through an increase in decomposition. However, many recent studies have demostrated that fertilization may increase carbon stocks in the soil. Hagedorn et al. (2001) found that soil organic carbon (SOC) sequestration in fertilized plots was always higher than that in control plots. They and others conclude that fertilization of temperate and boreal forests has high potential to reduce both heterotrophic and autotrophic soil respiration (Pregitzer et al. 2008). Decomposition is slowed as a result of several factors: (i) decreased carbon allocation to mycorrhizae; (ii) direct suppression of soil enzymes responsible for litter degradation; (iii) decreased litter quality; and (iv) decreased growth rates of decomposers. The research highlighting the sequestration of SOC as a result of fertilization is relatively recent and the hypotheses about the mechanisms that drive it are primarily speculation. More research is needed to address this knowledge gap.

Similarly to nitrogen fertilization, temperature can influence soil carbon stocks in the temperate and boreal regions. Temperature can influence nutrient availability and therefore fertility. In the future, therefore, the effect of nitrogen fertilization on soil carbon storage may be offset by the opposite effect of climate change; small increases in temperature will increase the rates of decomposition and nitrogen cycling and the carbon stock of forests may decline due to accelerated decomposition of SOC (Makipaa et al. 1999). This is likely to be a gradual change, but will be most pronounced in the boreal regions where processes are typically more limited by temperature than in temperate regions.

Although nitrogen is limiting in many forests of the temperate and boreal regions, it is not the only fertilization treatment used. In nitrogen-rich sites such as drained peatlands in central Finland or poorly drained loam and clay soils of the upper coastal plain of Georgia, USA, treatments such as additional phosphorus, calcium, potassium or liming are needed to amend critical nutrient levels or pH (Hytönen 1998; Moorhead 1998). In northeastern Oregon and in central Washington where nitrogen is considered limiting, research has shown that the addition of nitrogen and sulfur to Douglas-fir stands produced significant growth response to the nitrogen + sulfur treatment, but not to the nitrogen-alone treatment (Garrison et al. 2000). Similarly, in loblolly pine stands in the coastal plains of Georgia, USA, phosphorus is needed to enhance uptake of nitrogen (Will et al. 2006). Finally, in northwestern Ontario, Canada, the best treatment in terms of total volume increment over that of the control was 151 kg nitrogen ha^{-1} plus 62 kg magnesium ha^{-1}, which produced about 16 m^3 ha^{-1} of extra wood over 10 years (Morrison and Foster 1995).

These examples illustrate the complexities often associated with the correct application of fertilization and amelioration treatments to increase carbon on forested sites. These treatments are site specific; a manager's mastery of the intricacies of the site is essential to increasing the carbon uptake on a site.

It is beyond the scope of this chapter to provide a comprehensive look at the trade-offs between an increase in carbon storage in temperate and boreal forests and the fossil fuel emissions that result from the acquisition, manufacture, transport, and application of fertilizers. Most results indicate that even on the sites where fertilization is most beneficial, the emissions of CO_2 outweigh the carbon sequestered as a result of increased biomass production and SOC stocks (Schlesinger 2000; Markewitz 2006). However, on nitrogen-poor sites, where appropriate, the encouragement of the establishment of nitrogen-fixing plants may be beneficial through natural or artificial seeding (Marshall 2000).

4.4 Concepts and Application of Regenerating Forests

4.4.1 Afforestation and Reforestation

Afforestation and reforestation are silvicultural treatments that typically demonstrate carbon additionality. For example, the average net flux of carbon attributable to land-use change and management in the temperate forests of North America and Europe decreased from a source of 0.06 PgC yr^{-1} during the 1980s to a sink of 0.02 PgC yr^{-1} during the 1990s (Houghton 2003). In the United States this carbon sink is overwhelmingly due to afforestation /reforestation rather than active management or site manipulation (Caspersen et al. 2000). Even though some studies suggest that as forests age the strength of the carbon sink is reduced (and may become a source under certain circumstances), the amount of carbon stored on a forested site is significantly more than any other ecosystem type (Vesterdal et al. 2007).

Land conversion to forests is typically driven by wood demand and not carbon sequestration and it is unlikely that this will change even as carbon markets develop (Eggers et al. 2008).The conversion of land to forests using passive, natural regeneration has been postulated as an option for carbon sequestration because of the low operating costs and potential for co-benefits such as habitat and water quality enhancement (Fensham and Guymer 2009). These co-benefits provide valuable ecosystem services, but proving that the intent of the project was strictly for carbon sequestration (additionality) is complicated. Rules for proving additionality are not well established and/or uniform across carbon offset programs, so landowners planning to invest in afforestation/reforestation for the purpose of capturing market benefits need to make clear that the intent of the project is to sequester carbon.

4.4.2 Regeneration Harvests

Regeneration harvests are silvicultural treatments that remove some or all of the existing forest overstory to release existing regeneration or make growing space available for the establishment of a new cohort. Regeneration harvests alter the aboveground vegetation, with the added potential of affecting the bryophyte, and litter carbon pools; and potentially the mineral soil carbon.

The effect on the vegetative pool depends on the type of regeneration harvest. Uneven-aged treatments such as selection harvesting may have effects similar to thinning in that they only remove a portion of the canopy cover (Laporte et al. 2003;

Harmon et al. 2009). In a comparison of harvest types in Ontario, Canada, carbon storage in northern hardwoods was greater after selection harvesting than clearcutting because vigorous residual trees remained on the site (Lee et al. 2002). Clearcutting has a distinct and stronger effect. A clearcut of old-growth Norway spruce in Finland resulted in a 1/3 reduction in ecosystem carbon (Finer et al. 2003). Whole-tree harvesting on a 100-year rotation was modeled to result in an 81% reduction in biomass carbon compared to uncut forests in boreal China (Jiang et al. 2002).

Harvesting's influence on litter and particularly mineral soil carbon is controversial. An influential study by Covington (1981) in clearcuts at Hubbard Brook Experimental Forest in New Hampshire showed increased decomposition (and hence soil carbon loss) after forest harvest, suggesting that forest floor organic matter declines 50% within 20 years of harvest. A number of studies reinforce this view. In a modeling simulation of the effects of different harvest regimes on carbon stocks in boreal *Larix gmelinii* forests in China, clearcutting was predicted to result in litter and soil carbon loss that was greatest 10–20 years after harvesting, and to slowly recover thereafter (Jiang et al. 2002). A 30-year period of post-harvest soil carbon loss was observed in Nova Scotia red spruce (*Picea rubens*) forests, including from the deep mineral soil (Diochon et al. 2009).

A growing body of research, however, suggests that post-harvest respiration is not as important in the carbon budget as Covington (1981) suggested. A critical re-visit of his study suggested that the loss of organic mass from the forest floor after harvest was due to intermixing into the mineral soil, not increased decomposition (Yanai et al. 2003). If this is true, then the carbon consequences of harvesting are quite different, since organic carbon incorporated into the mineral soil may actually increase total carbon sequestration on the site.

Several comprehensive reviews of harvest effects on soil carbon also indicate limited impact. Depending on the level of slash input and organic matter incorporation into the mineral soil, harvests can result in slightly negative or slightly positive, or often no changes in soil carbon (Johnson 1992; Johnson and Curtis 2001). Conversion of old-growth *Picea* forests in British Columbia to young plantations reduced litter carbon stocks but left mineral soil carbon unaffected (Fredeen et al. 2007). Little or no net loss of forest floor weight was associated with clearcutting or partial cutting in Canadian boreal mixedwoods, perhaps due to rapid return to pre-treatment light and moisture conditions after prolific trembling aspen (*Populus tremuloides*) sprouting (Lee et al. 2002). In both Ontario northern hardwoods (Laporte et al. 2003) and Ozark oak forests (Edwards and Ross-Todd 1983; Ponder 2005; Li et al. 2007), uneven-aged management led to increased soil carbon levels, and clearcutting resulted in no significant change, compared to controls. Rates of both root respiration and microbial respiration may decline after harvest due to tree removal and soil compaction (Laporte et al. 2003). Where increased efflux has been observed, it tends to be small and limited to the uppermost soil layer (such as in a Chilean *Nothofagus pumilio* shelterwood (Klein et al. 2008)), and recovers to pre-harvest conditions after only a few years (aspen clearcuts in Ontario, Canada (Weber 1990)).

Johnson and Curtis (2001) hypothesized that whole tree harvesting could potentially result in soil carbon losses because of the high rates of biomass removal from the site. However, field studies in northern New Hampshire and Maine indicate that this practice results in no reduction in forest floor mass or soil carbon pool relative to uncut areas (Huntington and Ryan 1990; McLaughlin and Philips 2006). Some research suggests that the long-term consequences of management on soil carbon pools will be stronger than the short-term. A 300 year model of Canadian boreal forests shows a consistent decline in soil carbon in managed forests (Seely et al. 2002). Multi-rotation monitoring of managed forests will be necessary to assess the rigor of such models.

As the above studies indicate, there is significant evidence to show that if there is any soil carbon loss following a harvest, it is a short-term component of a site's carbon budget. Mineral soil carbon is usually not affected by harvest, and the loss from litter layers can be offset by slash additions. If the impact on soil carbon is indeed minor, then intensive pre- and post-harvest measurement of soil carbon pools may not be necessary. One of the main criticisms of making soil carbon measurements a low priority is that the research supporting it rarely involves measurement of deep soil carbon. One of the few studies to do so (in a red spruce chronosequence in Nova Scotia) found that younger post-harvest stands had significantly lower carbon storage at the 35–50 cm soil depth (Diochon et al. 2009). Before the conclusion can be made that soil carbon pools are not significantly affected by harvesting, greater attention must be paid to these deep soil layers. A single meta-analysis of impacts of harvesting on mineral soil carbon reveals that soil taxonomy perhaps provides the greatest explanation for susceptibility to mineral soil carbon loss, with ultisols and inceptisols showing a net loss of 7% and 13% respectively, while spodosols and alfisols remained unchanged (Nave et al. 2010). In the same review surface litter horizons were much more sensitive with losses amounting to 30% (hardwood litter had greater carbon loss than coniferous); clearly more studies are needed to substantiate these claims (Nave et al. 2010).

If all the carbon pools, inputs and outputs are considered together, it appears that clearcut stands are carbon sources for the first decade after harvest (thanks to transient increases in respiration), after which they switch to sinks. This pattern holds for boreal forests in British Columbia (Fredeen et al. 2007), Saskatchewan (Howard et al. 2004) and Finland (Kolari et al. 2004), but its applicability in temperate zones is not as clear. Partial regeneration harvests (shelterwoods, selection) appear more site and soil specific. For example many second growth even-aged forests in New England can be managed to increase structural complexity and hence stored carbon by retaining older and larger trees (reserves) within the stand during a regeneration harvest (Keeton 2006; Evans and Perschel 2009; Ashton et al. in press)

4.4.3 Treatment of Harvest Residues

The addition of harvest residues to the litter and soil layers is an important factor in mitigating initial carbon loss from harvested forests. This might suggest a negative carbon influence from removing these residues (and natural litterfall) for utilization, fuel reduction, or site preparation. However, research is mixed. Balboa-Murias et al. (2006) found that logging residues contained 11% of the total biomass carbon stored across a rotation in Spanish radiata pine (*Pinus radiata*) and *P. pinaster* plantations. They thus concluded that residue harvest for biomass burning (a common practice in Spanish forests) would result in reduced ecosystem carbon storage. Piling and burning slash in California clearcuts resulted in soil carbon loss (Black and Harden 1995). Removing harvest residues alone from New Zealand *P. radiata* plantations did not significantly alter soil carbon levels, but removing residuals and the forest floor (i.e. accumulated litterfall) did. In addition, a pattern of increasing soil carbon stocks with increasing residue retention was observed (Jones et al. 2008). In oak forests of Missouri, there was no significant increase in soil respiration between whole-tree harvest and whole-tree harvest + forest floor removal, and both had lower respiration than the control (Ponder 2005). In Australian *Eucalyptus* forests, residue retention had minimal impact on soil carbon levels, but may have some influence if practiced across multiple rotations (Mendham et al. 2003).

It appears that removing logging slash from harvested sites reduces the litter carbon pool, which is important in some forest types. But unless the natural litterfall is also reduced, residue removal has limited impact on soil carbon levels. Moreover the overall carbon impact of biomass removal depends in large part on its utilization such as replacement for fossil fuels (Evans and Finkral 2009) and greenhouse gases produced by its decomposition (Chen et al. 2010).

4.4.4 Changing Rotation Length

Many forests in the temperate and boreal zones are managed on rotations far shorter than the potential age of the species present. Often these rotations are so short that the maximum biomass productivity possible on the site (the "ceiling") is never reached. In a broad review of forest management effects on carbon storage, Cooper (1983) found that, on average, stands managed for maximum sustained yield store only 1/3 of the carbon stored in unmanaged, late successional forests.

Management for a financially optimal rotation results in an even smaller storage.

Research has shown the possibility of creating carbon additionality (in comparison to business-as-usual managed forests) by increasing rotation length. In Chinese boreal forests, Jiang et al. (2002) modeled a variety of rotation lengths and found that 30-year rotations stored only 12% as much carbon as 200-year rotations. In Europe, rotation modeling of spruce and pine forests showed increased carbon storage with increased rotation. This is especially true where stands retain high net primary productivity (NPP) rates even at extended rotations, such as pine plantations in northern Spain (Kaipainen et al. 2004). Further research in Spain supported this finding, although the authors noted that mean annual carbon uptake eventually will decline with increasing rotation as trees become less productive (Balboa-Murias et al. 2006). Jandl et al. (2007) found that lengthening rotations would increase carbon storage until stands reached an advanced developmental stage in which biomass actually began to decline (as observed in some old-growth forests).

As is often the case, the impact of rotation length on soil carbon is complicated. One Finnish study found that soil organic matter was maximized with shorter rotations, because of increased slash inputs to the litter and soil layers (Pussinen et al. 2002). Lengthening rotations in models of wood production in Finland resulted in greater carbon storage when the increase in biomass carbon exceeded the decrease in soil organic matter. This occurred in the case of Scots pine, but not for Norway spruce, suggesting that short rotations are more carbon-positive for the latter

Insert 4. The Principle of Extending Rotations to Sequester More Carbon Per Hectare

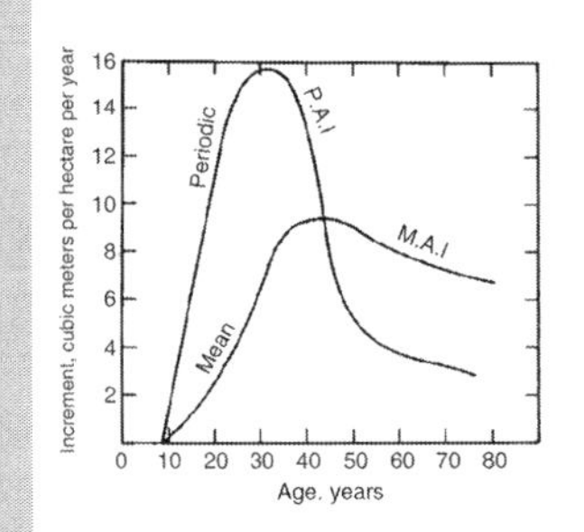

The Principle of Extending Rotations to Sequester More Carbon per Hectare

The figure on the left is an example of how to set timber rotations to maximize stand productivity. When periodic annual increment (PAI) (amount of volume added per hectare in a year) equals mean annual increment (MAI) (the total stand volume divided by stand age) it is time to cut. After this point, the stand will no longer be adding as much volume per year as it has been on average across the rotation.

It is often proposed that carbon credits could be offered to incentivize landowners to extend rotations to the PAI=MAI point. In this way, more biomass will be grown on a hectare over a given time period (add the yield of each rotation) than would have been under "financial maturity" rotations. However, if rotations are extended beyond the maximum productivity point, then total biomass produced per hectare across the time period will be reduced, even though at the end of one such rotation volume/ha is at its greatest level in the example.

It should be kept in mind that this example does not consider the other carbon pools, notably the forest litter. Short rotations add more harvest slash to this pool, but there is also a period of heightened decomposition after each harvest. Research indicates that the soil pool is minimally impacted by forest harvest unless significant soil disturbance takes place.

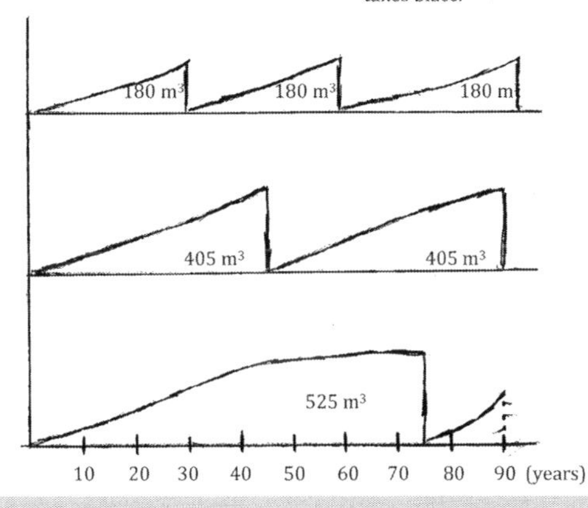

Traditional "financial maturity" rotations of 30 years
Total volume grown per ha across rotations = **540 m³**
Each rotation ends before stand attains max. productivity

Peak productivity rotations (PAI=MAI) of 45 years
Total volume grown per ha across rotations= **810 m³**
Each rotation ends at point of max. productivity

Rotation extended beyond peak productivity (75 years)
Total volume grown per ha across rotations = **540 m³**
Greatest per rotation volume is achieved, but summed rotations produce less volume.

species (Liski et al. 2001). This must be tempered, however, by the increased fossil fuel emissions associated with short-rotation forestry (Liski et al. 2001).

The principle behind lengthening rotations is to bring stands closer to the advanced ages at which maximum biomass is attained. By this same principle, forests that are already in these stages (for instance, old-growth) should be maintained. Harmon et al. (1990) considered the carbon consequences of the conversion of old-growth forests in the Pacific Northwest to managed production forests, finding that it caused a reduction in carbon storage that extended for 250 years, and could probably never be made up for. If forests in this region were managed with rotations of 50, 75 and 100 years, the carbon stored would be at most 38%, 44% and 51%, respectively, of that stored in old-growth (Harmon et al. 1990). Tang et al. (2009) predicted a similar long-term loss in ecosystem carbon with the conversion of Michigan northern hardwoods to younger stand structures. Managing red spruce on 60 year rotations in Nova Scotia would result in the loss of 42% of soil carbon relative to old-growth and 26% relative to 80 year rotations (Diochon et al. 2009). Managed *Eucalyptus* forests in Australia contain only 60% of the aboveground vegetative carbon stored in old-growth.

The key explanation of this discrepancy is the dearth of large (>100 cm in diameter) trees in managed stands. In old-growth rainforest/eucalyptus stands in New South Wales, Australia, such trees make up only 18% of the stems >20 cm, but contain 54% of the vegetative carbon (Roxburgh et al. 2006). These studies suggest, at the least, that when old-growth forests already exist, their maintenance is optimal for carbon sequestration.

5 Management and Policy Implications

5.1 Recommendations for Land Managers

- Relatively few forest management practices can demonstrate true carbon additionality. Afforestation/reforestation usually increases a site's carbon sequestration, unless it results in a significant release of soil carbon (i.e. through intensive site preparation or the oxidation of peat soils). The impact of afforestation/reforestation on soil carbon pools must be carefully monitored.
- Thinning causes a reduction of the vegetative carbon stored on-site, which recovers over a matter of decades (depending on thinning intensity and tree vigor). Thinning's impact on soil carbon appears very limited, as inputs of slash and reduced root respiration seem to make up for reduced litterfall and increased microbial respiration.
- Resiliency treatments (such as fuels reduction thinning and prescribed fire) result in lowered carbon storage on-site and some carbon release from decomposition and combustion. However, they help produce forests that are significantly less susceptible to stand-replacing disturbance (with accompanying carbon releases). Essentially, forest managers using these treatments accept less than maximum short-term carbon storage to ensure long-term and more secure storage.
- Fertilization treatments that improve the nutrient conditions limiting plant growth can increase the vegetative carbon pool (particularly on marginal soils), and increase the soil carbon pool by reducing root and microbial respiration. This must be tempered by consideration of the carbon footprint of fertilizer production, which can match or exceed the additional carbon sequestration.
- Draining of saturated peat soils and subsequent afforestation can cause either a net carbon loss or gain, depending on whether increased tree growth and litterfall and decreased methane release outweigh the increase in respiration from oxidized peat. This may in turn be dependent on the extent to which drainage lowers the peatland water table. Research from drained lands in Finland and the British Isles indicates that net carbon sequestration is possible when the water table remains relatively high after drainage.
- Regeneration harvests significantly reduce the carbon stored on-site, especially even-aged

treatments such as clearcutting. The amount of stored carbon may not rebound for many decades (or centuries, if the pre-harvest stand was in old-growth condition), but the annual rate of carbon uptake will be greater in the regenerating stand. Harvested stands often are net sources of carbon for the first 10–30 years, because of increased litter and soil respiration. They then become net sinks as vegetative growth and litter accumulation exceed respiration.

- Removing harvest residues (slash) for biomass utilization, to reduce fuel levels or to prepare the site for planting, directly reduces the litter carbon pool. The impact on soil carbon is less clear. Treatments that only reduce slash do not result in significant soil carbon loss (over one rotation), but loss occurs if the forest floor (natural litter accumulation) is removed as well.
- Managing stands for maximum sustained yield or financially optimum rotation can result in non-optimal carbon storage. Such rotations are often too short to allow the stand to attain maximum biomass. As such, it is often possible to increase carbon sequestration by extending rotations. This is particularly true on productive sites where high rates of NPP can be sustained through longer rotations. There is a point of diminishing returns, though, when rotations are extended beyond the age of maximum biomass productivity. At some point, it may be possible to store more carbon in a series of short rotations (that maintains the stand in a young, productive stage) than a single longer rotation.
- If old forests *already exist,* however, maintaining them as old forests maximizes carbon storage. Old forests, especially on productive sites, often have very large pools of vegetative carbon in comparison to forests managed on shorter rotations. Soil and litter pools may also be quite large in old-growth forests, and in the boreal, the bryophyte pool as well. The conversion of old-growth to managed forests likely results in a loss of ecosystem carbon that cannot easily be regained.

5.2 Recommendations for Policy Makers

- The concept of carbon additionality is central to carbon credit and offset schemes. It is difficult to demonstrate additionality in most forest management practices. By its nature, forest management often causes reductions in carbon stocks, especially from the vegetative pool. But a contribution can still be made to climate change mitigation by adjusting these practices so as to *minimize carbon release* as opposed to *maximizing carbon sequestration.* The former idea is gaining traction through such mechanisms as offsets for reduced deforestation/degradation and reduced impact logging. If boreal and temperate forests are to be included in a carbon credit and offsets scheme, it will likely be necessary to recognize such contributions, which are potentially more feasible than "traditional" carbon additionality.
- If policy makers choose to include such "reduced carbon release" practices in a credit/offset scheme, they will need to set a baseline that allows these practices to demonstrate additionality. If the baseline is a natural, unmanaged forest, then most forest practices will always appear carbon-negative. But if the baseline is a "business-as-usual" managed forest, then such practices will constitute a creditable improvement over the baseline. Setting baselines is not a purely scientific process; it is an act of policy that determines which forest management activities will be incentivized.
- The practice of extending rotations offers a straightforward biological means of increasing carbon sequestration in existing forests, and thus has become a focus for forest managers participating in carbon offset markets. It has been suggested that carbon offset credits can be used to produce a large-scale dividend of additional carbon sequestration by subsidizing landowners to extend rotations until peak stand productivity (in silvicultural terms, when periodic annual increment and mean annual increment are equal) (Wayburn 2009). In this way, carbon "density" per unit area will

be increased by allowing forests to more closely approach their natural productive potential.

- The well-known market externality of "leakage" complicates the implementation of concept such as extending rotations. If revenues from carbon credits motivate enough landowners to extend rotations, then demand for wood shifts elsewhere. The landowners may well plan to harvest the same (or greater) volume several decades from now, but that does nothing to change the current demand for wood. Mills will be forced to increase the price they pay for roundwood, which will likely motivate landowners not participating in carbon sequestration activities to cut and sell more wood than they otherwise would have (and perhaps *earlier* in the rotation than they planned). Thus, while some landowners delay harvesting in order to accumulate more carbon per forested acre, other landowners will accelerate harvest to fill the gap, neutralizing net carbon gains.
- Another important policy factor is whether to consider forest products as a carbon pool. The choice could well determine whether or not practices like thinning are positive, neutral or negative from a carbon sequestration perspective. If the carbon contained in forest products is "sequestered," then a great many more forestry projects would be eligible for carbon credits and offsets than if that carbon is "released." The designers of offset systems will need to balance the increased measurement and documentation burden of including a forest products carbon pool with the potential to include more projects.
- Many forest management practices have a minimal impact on the soil carbon pool, which is the most difficult to measure. Thus, it may be possible that offsets involving certain forestry practices could go forward without strict quantification of this pool. This would considerably reduce measurement cost. As a rule, quantification would likely be least vital when the practice in question results in minimal soil disturbance.

References

Agee J, Skinner C (2005) Basic principles of forest fuel reduction treatments. For Ecol Manag 211:83–96

Albani M, Moorcroft PR, Ellison AM et al (2010) Predicting the impact of hemlock woolly adelgid on carbon dynamics of Eastern United States forests. Can J For Res 40:119–133

Allen C, Savage M, Falk D, Suckling K (2002) Ecological restoration of southwestern ponderosa pine ecosystems: a broad perspective. Ecol Appl 12:1418–1433

Amiro B (2001) Paired-tower measurements of carbon and energy fluxes following disturbance in the boreal forest. Glob Change Biol 7:253–268

Amman GD, Logan JA (1998) Silvicultural control of mountain pine beetle: prescriptions and the influence of microclimate. Am Entomol 44:166–177

Ashton MS, Frey BF, Koirala R (in press) Growth and performance of regeneration across variable retention shelterwood treatments for an oak-hardwood forest in Southern New England. North J Appl For 00: 000–000

Ayres MP, Lombardero MJ (2000) Assessing the consequences of global change for forest disturbance from herbivores and pathogens. Sci Total Environ 262: 263–286

Balboa-Murias MA, Rodriguez-Soalleiro R, Merino A, Alvarez-Gonzalez JG (2006) Temporal variations and distribution of carbon stocks in aboveground biomass of radiata pine under different silvicultural alternatives. For Ecol Manag 237:29–38

Bauer G, Bazzaz F, Minocha R, Long S, Magill A (2004) N additions on tissue chemistry, photosynthetic capacity, and carbon sequestration potential of a red pine (*Pinus resinosa* Ait.) stand in the NE United States. For Ecol Manag 196:173–186

Black TA, Harden JW (1995) Effect of timber harvest on soil carbon storage at Blodgett experimental forest, California. Can J For Res 25:1385–1396

Brown R, Agee J, Franklin J (2004) Forest restoration and fire: principles in the context of place. Conserv Biol 18:903–912

Busing RT, White RD, Harmon ME, White PS (2009) Hurricane disturbance in a temperate deciduous forest: patch dynamics, tree mortality, and coarse woody detritus. Plant Ecol 201:351–363

Byrne KA, Farrell EP (2005) The effect of afforestation on soil carbon dioxide emissions in blanket peatland in Ireland. Forestry 78:217–227

Campbell J, Alberti G, Martin J, Law BE (2009) Carbon dynamics of a ponderosa pine plantation following a thinning treatment in the northern Sierra Nevada. For Ecol Manag 257:453–463

Cannell MGR, Dewar RC, Pyatt DG (1993) Conifer plantations on drained peatlands in Britain – a net gain or loss of carbon? Forestry 66:353–369

Caspersen JP, Pacala SW, Jenkins JC, Hurtt GC, Moorcroft PR, Birdsey RA (2000) Contributions of land-use history to carbon accumulation in US forests. Science 290(5495):1148

Chen J, Colombo SJ, Ter-Mikaelian MT, Heath LS (2010) Carbon budget of Ontario's managed forests and harvested wood products, 2001–2100. For Ecol Manag 259:1385–1398

Chiang JM, Iverson LR, Prasad A, Brown KJ (2008) Effects of climate change and shifts in forest composition on forest net primary production. J Integr Plant Biol 50:1426–1439

Concilio A, Ma SY, Li QL, LeMoine J, Chen JQ, North M, Moorhead D, Jensen R (2005) Soil respiration response to prescribed burning and thinning in mixed-conifer and hardwood forests. Can J For Res 35: 1581–1591

Cooper CF (1983) Carbon storage in managed forests. Can J For Res 13:155–166

Cooper-Ellis S, Foster DR, Carlton G, Lezberg A (1999) Forest response to catastrophic wind: results from an experimental hurricane. Ecology 80:2683–2696

Covington WW (1981) Changes in forest floor organic-matter and nutrient content following clear cutting in northern hardwoods. Ecology 62:41–48

Covington W, Fulé P, Moore M, Hart S, Kolb T (1997) Restoring ecosystem health in ponderosa pine forests of the southwest. J For 95:23–29

Coyle D, Coleman M (2005) Forest production responses to irrigation and fertilization are not explained by shifts in allocation. For Ecol Manag 208:137–152

Dale VH, Joyce LA, McNulty S, Neilson RP, Ayres MP, Flannigan MD, Hanson PJ, Irland LC, Lugo AE, Peterson CJ, Simberloff D, Swanson FJ, Stocks BJ, Wotton BM (2001) Climate change and forest disturbances. Bioscience 51:723–734

Dannenmann M, Gasche R, Ledebuhr A, Holst T, Mayer H, Papen H (2007) The effect of forest management on trace gas exchange at the pedosphere-atmosphere interface in beech (*Fagus sylvatica* L.) forests stocking on calcareous soils. Eur J For Res 126:331–346

Diochon A, Kellman L, Beltrami H (2009) Looking deeper: An investigation of soil carbon losses following harvesting from a managed northeastern red spruce (*Picea rubens* Sarg.) forest chronosequence. For Ecol Manag 257:413–420

Dore S, Kolb TE, Montes-Helu M, Eckert SE, Sullivan BW, Hungate BA, Kaye JP, Hart SC, Koch GW, Finkral A (2010) Carbon and water fluxes from ponderosa pine forests disturbed by wildfire and thinning. Ecol Appl 20(3):663–668

Duffy JE (2009) Why biodiversity is important to the functioning of real-world ecosystems. Front Ecol Environ 7:437–444

Edwards NT, Ross-Todd BM (1983) Soil carbon dynamics in a mixed deciduous forest following clear-cutting with and without residue removal. Soil Sci Soc Am J 47:1014–1021

Eggers J, Lindner M, Zudin S, Zaehle S, Lisk J (2008) Impact of changing wood demand, climate and land use on European forest resources and carbon stocks during the 21st century. Glob Change Biol 14:2288–2303

Eriksson E, Gillespie AR, Gustavsson L, Langvall O, Olsson M, Sathre R, Stendahl J (2007) Integrated carbon analysis of forest management practices and wood substitution. Can J For Res 37:671–681

Evans AM, Finkral AJ (2009) From renewable energy to fire risk reduction: a synthesis of biomass harvesting and utilization case studies in US forests. Glob Change Biol Bioenerg 1:211–219

Evans AM, Perschel RT (2009) A review of forestry mitigation and adaptation strategies in the Northeast U.S. Clim Change 96:167–183

Fensham RJ, Guymer GP (2009) Carbon accumulation through ecosystem recovery. Environ Sci Policy 12: 367–372

Finer L, Mannerkoski H, Piirainen S, Starr M (2003) Carbon and nitrogen pools in an old-growth, Norway spruce mixed forest in eastern Finland and changes associated with clearcutting. For Ecol Manag 174:51–63

Finkral AJ, Evans AM (2008) The effects of a thinning treatment on carbon stocks in a northern Arizona ponderosa pine forest. For Ecol Manag 255:2743–2750

Foley TG, Richter D, Galik CS (2009) Extending rotation age for carbon sequestration: a cross-protocol comparison of north American forest offsets. For Ecol Manag 259:201–209

Folke C, Carpenter S, Walker B, Scheffer M, Elmqvist T, Gunderson L, Holling CS (2004) Regime shifts, resilience, and biodiversity in ecosystem management. Annu Rev Ecol Evol Syst 35:557–581

Fredeen AL, Waughtal JD, Pypker TG (2007) When do replanted sub-boreal clearcuts become net sinks for CO_2? For Ecol Manag 239:210–216

Garrison MT, Moore JA, Shaw TM, Mika PG (2000) Foliar nutrient and tree growth response of mixed-conifer stands to three fertilization treatments in northeast Oregon and north central Washington. For Ecol Manag 132:183–198

Gower ST (2003) Patterns and mechanisms of the forest carbon cycle. Annu Rev Environ Resour 28:169–204

Hagedorn F, Maurer S, Egli P, Blaser P, Bucher J (2001) Carbon sequestration in forest soils: effects of soil type, atmospheric CO_2 enrichment, and N deposition. Eur J Soil Sci 52:619–628

Hargreaves KJ, Milne R, Cannell MGR (2003) Carbon balance of afforested peatland in Scotland. Forestry 76:299–317

Harmon ME, Ferrell WK, Franklin JF (1990) Effects on carbon storage of conversion of old-growth forests to young forests. Science 247:699–702

Harmon M, Moreno A, Domingo J (2009) Effects of partial harvest on the carbon stores in Douglas-Fir/western Hemlock forests: a simulation study. Ecosystems 12:777–791

Hessburg PF, Agee JK, Franklin JF (2005) Dry forests and wildland fires of the inland Northwest USA: contrasting the landscape ecology of the pre-settlement and modem eras. For Ecol Manag 211:117–139

Hoover C, Stout C (2007) The carbon consequences of thinning techniques: stand structure makes a difference. J For 105:266–270

Houghton R (2003) Revised estimates of the annual net flux of carbon to the atmosphere from changes in land use and land management 1850–2000. Tellus Ser B-Chem Phys Meteorol 55:378–390

Houghton RA, Hackler JL, Lawrence KT (2000) Changes in terrestrial carbon storage in the United States. 2: the role of fire and fire management. Glob Ecol Biogeogr 9:145–170

Howard EA, Gower ST, Foley JA, Kucharick CJ (2004) Effects of logging on carbon dynamics of a jack pine forest in Saskatchewan, Canada. Glob Change Biol 10:1267–1284

Hunt S, Newman J, Otis G (2006) Impacts of exotic pests under climate change implications for Canada's forest ecosystems and carbon stocks. http://www.biocap.ca/rif/report/Hunt_S.pdf. Accessed Nov 2010

Huntington TG, Ryan DF (1990) Whole-tree harvesting effects on soil nitrogen and carbon. For Ecol Manag 31:193–204

Hurteau M, North M (2009) Fuel treatment effects on tree-based forest carbon storage and emissions under modeled wildfire scenarios. Front Ecol Environ e-View 7:409–414

Hytoenen J (1998) Effect of peat ash fertilization on the nutrient status and biomass production of short-rotation Willow on cut-away peatland area. Biomass Bioenerg 15(1):83–92

Jandl R, Linder M, Vesterdal L, Bauwens B, Baritz R, Hagedorn F, Johnson DW, Minkkinen K, Byrne KA (2007) How strongly can forest management influence soil carbon sequestration? Geoderma 137:253–268

Jiang H, Apps MJ, Peng CH, Zhang YL, Liu JX (2002) Modeling the influence of harvesting on Chinese boreal forest carbon dynamics. For Ecol Manag 169:65–82

Johnson DW (1992) Effects of forest management on soil carbon storage. Water Air Soil Pollut 64:83–120

Johnson DW, Curtis PS (2001) Effects of forest management on soil C and N storage: meta analysis. For Ecol Manag 140:227–238

Jones HS, Garrett LG, Beets PN, Kimberly MO, Oliver GR (2008) Impacts of harvest residue management on soil carbon stocks in a plantation forest. Soil Sci Soc Am 72:1621–1627

Kaipainen T, Liski J, Pussinen A, Karjalainen T (2004) Managing carbon sinks by changing rotation length in European forests. Environ Sci Policy 7:205–219

Keeton WS (2006) Managing for late-successional/old-growth characteristics in northern hardwood-conifer forests. For Ecol Manag 235:129–142

Kim C, Son Y, Lee W, Jeong J, Noh N (2009) Influences of forest tending works on carbon distribution and cycling in a *Pinus densiflora S. et Z.* stand in Korea. For Ecol Manag 257:1420–1426

Klein D, Fuentes JP, Schmidt A, Schmidt H, Schulte A (2008) Soil organic C as affected by silvicultural and exploitative interventions in *Nothofagus pumilio* forests of the Chilean Patagonia. For Ecol Manag 255: 3549–3555

Kolari P, Pumpanen J, Rannik U, Ilvesniemi H, Hari P, Berninger F (2004) Carbon balance of different aged Scots pine forests in Southern Finland. Glob Change Biol 10:1106–1119

Kurz WA, Stinson G, Rampley GJ, Dymond CC, Neilson ET (2008) Risk of natural disturbances makes future contribution of Canada's forests to the global carbon cycle highly uncertain. Proc Natl Acad Sci USA 105: 1551–1555

Laine J, Vasander H (1991) Effect of forest drainage on the carbon balance of a sedge fen ecosystem. In: Proceedings of symposium the changing face of fenlands and implications for their future use, Cambridge, 9–11 April 1991. Cited in: Johnson DW (1992) Effects of forest management on soil carbon storage. Water Air Soil Pollut 64: 83–12

Laporte MF, Duchesne LC, Morrison IK (2003) Effect of clearcutting, selection cutting, shelterwood cutting and microsites on soil surface CO_2 efflux in a tolerant hardwood ecosystem of northern Ontario. For Ecol Manag 174:565–575

Larsen JB (1995) Ecological stability of forests and sustainable silviculture. For Ecol Manag 73:85–96

Lee J, Morrison IK, Leblanc JD, Dumas MT, Cameron DA (2002) Carbon sequestration in trees and regrowth as affected by clearcut and partial cut harvesting in a second-growth boreal mixedwood. For Ecol Manag 169:83–101

Li Q, Chen J, Moorhead DL, DeForest JL, Jensen R, Henderson R (2007) Effects of timber harvest on carbon pools in Ozark forests. Can J For Res 37:2337–2348

Liski J, Pussinen A, Pingoud K, Mskipss R, Karjalainen T (2001) Which rotation length is favourable to carbon sequestration? Can J For Res 31:2004–2013

Lucas C, Hennessy K, Mills G, Bathols J (2007) Bushfire weather in southeast Australia: recent trends and projected climate change impacts. Consultancy report prepared for The Climate Institute of Australia, September 2007. Bushfire CRC and Australian Bureau of Meteorology, CSIRO Marine and Atmospheric Research, Aspendale

Luyssaert S, Schulze E-D, Borner A et al (2008) Old-growth forests as global carbon sinks. Nature 455: 213–215

Magnani F, Mencuccini M, Borghetti M et al (2007) The human footprint in the carbon cycle of temperate and boreal forests. Nature 447:848–850

Maier C, Kress T (2000) Soil CO_2 evolution and root respiration in 11 year-old loblolly pine (*Pinus taeda*) plantations as affected by moisture and nutrient availability. Can J For Res 30:347–359

Makipaa R, Karjalainen T, Pussinen A, Kellomaki S (1999) Effects of climate change and nitrogen deposition on the carbon sequestration of a forest ecosystem in the boreal zone. Can J For Res 29: 1490–1501

Markewitz D (2006) Fossil fuel carbon emissions from silviculture: impacts on net carbon sequestration in forests. For Ecol Manag 236:153–161

Markewitz D, Sartori F, Craft C (2002) Soil change and carbon storage in longleaf pine stands planted on marginal agricultural lands. Ecol Appl 12:1276–1285
Marshall V (2000) Impacts of forest harvesting on biological processes in northern forest soils. For Ecol Manag 133:43–60
McHenry MT, Wilson BR, Lemon JM, Donnelly DE, Growns IG (2006) Soil and vegetation response to thinning white cypress pine (*Callitris glaucophylla*) on the north western slopes of New South Wales, Australia. Plant Soil 285:245–255
McLaughlin JW, Philips SA (2006) Soil carbon nitrogen and base cation cycling 17 years after whole-tree harvesting in a low-elevation red spruce (*Picea rubens*)-balsam fir (*Abies balsamea*) forested watershed in central Maine, USA. For Ecol Manag 222:234–253
McNulty SG (2002) Hurricane impacts on US forest carbon sequestration. Environ Pollut 116:S17–S24
Mendham DS, O'Connell AM, Grove TS, Rance SJ (2003) Residue management effects on soil carbon and nutrient contents and growth of second rotation eucalypts. For Ecol Manag 181:357–372
Minkkinen K, Laine J (1998) Long-term effects of forest drainage on the peat carbon stores of pine mires in Finland. Can J For Res 28:1267–1275
Minkkinen K, Laine J, Hokka H (2001) Tree stand development and carbon sequestration in drained peatland stands in Finland – a simulation study. Silva Fennica 35:55–69
Moorhead (1998) Fertilizing pine plantations: a county agent's guide for making fertilization recommendations. Georgia Cooperative Extension Services, College of Agriculture and Environmental Sciences and the Warnell School of Forest Resources, The University of Georgia, Athens
Morrison IK, Foster NW (1995) Effect of nitrogen, phosphorus and magnesium fertilizers on growth of a semimature jack pine forest, northwestern Ontario. For Chron 71:422–425
Nave LE, Vance ED, Swanston CW et al (2010) Harvest Impacts on Soil Carbon Storage in Temperate Forests. For Ecol Manag 259:857–866
Nilsen P, Stand LT (2008) Thinning intensity effects on carbon and nitrogen stores and fluxes in a Norway spruce (*Picea abies* (L.) Karst.) stand after 33 years. For Ecol Manag 256:201–208
Nuckolls A, Wurzburger N, Ford C, Hendrick R (2008) Hemlock declines rapidly with hemlock woolly adelgid infestation: Impacts on the carbon cycle of southern Appalachian forests. Ecosystems 12:179–190
Orwig DA, Foster DR (1998) Forest response to the introduced hemlock woolly adelgid in southern New England, USA. J Torrey Bot Soc 125:60–73
Peltzer DA, Allen RB, Lovett GM et al (2010) Effects of biological invasions on forest carbon sequestration. Glob Change Biol 16:732–746
Perry D, Amaranthus M (1997) Disturbance, recovery, and stability. In: Kohm K, Franklin J (eds) Creating a forestry for the twenty-first century: the science of ecosystem management. Island Press, Washington, DC, pp 32–55
Ponder F (2005) Effect of soil compaction and biomass removal on soil CO_2 efflux in a Missouri forest. Commun Soil Sci Plant Anal 36:1301–1311
Pregitzer K, Burton A, Zak D, Talhelm A (2008) Simulated chronic nitrogen deposition increases carbon storage in northern temperate forests. Glob Change Biol 14: 142–153
Pussinen A, Karjalainen T, Makippa R, Valsta L, Kellomaki S (2002) Forest carbon sequestration and harvests in Scots pine stand under different climate and nitrogen deposition scenarios. For Ecol Manag 158:103–115
Roxburgh SH, Wood SW, Mackey BG, Woldendorp G, Gibbons P (2006) Assessing the carbon sequestration potential of managed forests: a case study from temperate Australia. J Appl Ecol 43:1149–1159
Sakovets VV, Germanova NI (1992) Changes in the carbon balance of forested mires in Karelia due to drainage. Suo (Helsinki) 43:249–252
Schäfer KVR, Clark KL, Skowronski N et al (2010) Impact of insect defoliation on forest carbon balance as assessed with a canopy assimilation model. Glob Change Biol 16:546–560
Schimel DS, House JI, Hibbard KA, Bousquet P, Ciais P, Peylin P, Braswell BH, Apps MJ, Baker D, Bondeau A, Canadell J, Churkina G, Cramer W, Denning AS, Field CB, Friedlingstein P, Goodale C, Heimann M, Houghton RA, Melillo JM, Moore B, Murdiyarso D, Noble I, Pacala SW, Prentice IC, Raupach MR, Rayner PJ, Scholes RJ, Steffen WL, Wirth C (2001) Recent patterns and mechanisms of carbon exchange by terrestrial ecosystems. Nature 414:169–172
Schlesinger W (2000) Carbon sequestration in soils: some cautions amidst optimism. Agric Ecosyst Environ 82: 121–127
Seely B, Welham C, Kimmins H (2002) Carbon sequestration in a boreal forest ecosystem: results from the ecosystem simulation model, FORECAST. For Ecol Manag 169:123–135
Selig MF (2008) Soil carbon and CO_2 efflux as influenced by the thinning of loblolly pine (*Pinus taeda* L.) plantations on the Piedmont of Virginia. For Sci 54:58–66
Silvola J (1986) Carbon-dioxide dynamics in mires reclaimed for forestry in eastern Finland. Ann Bot Fennosc 23:59–67
Silvola J, Alm J, Ahlholm U, Nykanen H, Martikainen PJ (1996) CO_2 fluxes from peat in boreal mires under varying temperature and moisture conditions. J Ecol 84:219–228
Spring D, Kennedy J, MacNally R (2005) Optimal management of a flammable forest providing timber and carbon sequestration benefits: an Australian case study. Aus J Agric Resour Econ 49:303–320
Suni T, Vesala T, Rannik U, Keronen P, Marranen T, Sevanto S, Gronholm T, Smolander S, Kulmala M, Ojansuu R, Ilvesniemi H, Uotila A, Makela A, Pumpanen J, Kolari P, Beringer F, Nikinmaa E, Altimir N, Hari P (2003) Trace gas fluxes in a boreal forest remain unaltered after thinning. PowerPoint presentation. www.boku.ac.at/formod/Monday/T_Suni.ppt

Tang J, Bolstad PV, Martin JG (2009) Soil carbon fluxes and stocks in a great lakes forest chronosequence. Glob Change Biol 15:145–155

Tilman D, Reich P, Phillips H, Menton M, Patel A, Vos E, Peterson D, Knops J (2000) Fire suppression and ecosystem carbon storage. Ecology 81:2680–2685

USDA (1975) Soil conservation survey-USDA soil taxonomy: a basic system of classification for making and interpreting soil surveys. USDA Agriculture handbook No. 436, US Government Pringing Office, Washinton, DC

Vesterdal L, Rosenqvist L, Van Der Salm C, Hansen K, Groenenberg B-J, Johansson MB (2007) Carbon sequestration in soil and biomass following afforestation: experiences from oak and Norway spruce chronosequences in Denmark, Sweden and the Netherlands. In: Heil Gerrit W, Muys Bart, Hansen Karin (eds) Environmental effects of afforestation in north-western Europe. Springer, New York, pp 19–51

Von Arnold K, Weslien P, Nilsson M, Svensson BH, Klemedtsson L (2005) Fluxes of CO_2, CH_4, and NO_2 from drained coniferous forests on organic soils. For Ecol Manag 210:239–254

Wayburn L (2009) Forests in the United States' climate change policy. Presentation at Yale University, New Haven, 30 Mar 2009

Weber MG (1990) Forest soil respiration after cutting and burning in immature aspen ecosystems. For Ecol Manag 31:1–14

Whitney G (1996) From coastal wilderness to fruited plain: a history of environmental change in temperate north America. Cambridge University Press, New York

Wiedinmyer C, Neff J (2007) Estimates of CO_2 from fires in the United States: implications for carbon management. Carbon Balance Manag 2:10

Will RE, Markewitz D, Hendrick RL, Meason DF, Crocker TR, Borders BE (2006) Nitrogen and phosphorus dynamics for 13-year-old loblolly pine stands receiving complete competition control and annual N fertilizer. For Ecol Manag 227:155–168

Yanai RD, Currie WS, Goodale CL (2003) Soil carbon dynamics after forest harvest: an ecosystem paradigm reconsidered. Ecosystems 6:197–212

Zeng H, Chambers JQ, Negrón-Juárez RI et al (2009) Impacts of tropical cyclones on U.S. Forest tree mortality and carbon flux from 1851 to 2000. Proc Natl Acad Sci 106:7888–7892

11 Managing Afforestation and Reforestation for Carbon Sequestration: Considerations for Land Managers and Policy Makers

Thomas Hodgman, Jacob Munger, Jefferson S. Hall, and Mark S. Ashton

Executive Summary

Forest management of planted and natural secondary forests for carbon sequestration, applied in the appropriate contexts, presents many opportunities for climate change mitigation and adaptation.

In climate change policy discussions, planted and natural secondary forests are placed in the category of afforestation and reforestation (A/R) projects. Temperate regions currently contain most of the existing planted and naturally regenerating forests. However, establishment of new forests is fastest in the tropics, especially Southeast Asia and Latin America.

We Identify Two Key Success Factors for A/R Projects in General, and for Carbon Sequestration in Particular

- *Site selection.* In order to manage A/R projects for carbon sequestration successfully, managers must select appropriate sites. Selecting the right site can result in forests that are both productive and efficient at sequestering carbon. In particular, it is important to understand how new forests will affect soil carbon reserves. On inappropriate sites, A/R projects can result in losses of soil carbon that are in conflict with the objective of sequestering carbon. In addition, newly established forests can affect water quantity, water quality, and biodiversity. While opportunity exists for carbon sequestration projects, carbon should not supplant all other forest values. Rather, managers should treat carbon as one of many management objectives for forests.
- *Species selection.* Selecting species that are appropriate for site conditions and management objectives is necessary for a successful A/R project. Mixed-species forests, containing species that occupy different ecological niches on the same site, have the potential to store more biomass, and therefore carbon. Single-species forests are less complex to manage, and often benefit from years of research and phenotypic selection, resulting in high growth rates and carbon sequestration. Therefore, while mixed-species forests have great potential, the extensive research and knowledge regarding single species forests often leads to more certain timber production and carbon sequestration.

There Are Some Common Forest Management Practices That Effect Forest Carbon Sequestration

- *Site preparation.* Generally, site preparation increases root and tree growth, improving biomass production. However, site preparation can cause loss of soil carbon and inherently involves significant fossil fuel emissions.

T. Hodgman • J. Munger • M.S. Ashton (✉)
Yale School of Forestry & Environmental Studies, 360 Prospect St, New Haven, CT 06511, USA
email: mark.ashton@yale.edu

J.S. Hall
Smithsonian Tropical Research Institute, Balboa, Ancon, Republic of Panama

M.S. Ashton et al. (eds.), *Managing Forest Carbon in a Changing Climate*, DOI 10.1007/978-94-007-2232-3_11, © Springer Science+Business Media B.V. 2012

- *Fertilization.* When managers supply the proper nutrients to a forest in the proper amounts, fertilization increases carbon sequestration. Fertilization also results in greenhouse gas emissions due to the fertilizer production and application process. Alternatives to fertilizer include planting nitrogen-fixing species in A/R projects.
- *Irrigation.* Irrigation can dramatically increase forest growth rates, but may be prohibitively expensive or impractical.
- *Herbicides.* Controlling competing vegetation with herbicides produces the best results when applied as part of site preparation. After A/R projects fully occupy a site, there is little benefit to carbon sequestration from herbicides.
- *Thinning.* Selectively harvesting individual trees, commonly called thinning, always has a negative short-term impact on forest carbon stocks. However, thinning improves timber quality and tree vigor and can reduce the risk of a reversal of carbon sequestration due to fire, windthrow, insect infestations and disease.
- *Harvesting.* Forest managers can increase carbon stocks by reducing logging impacts on residual trees and the forest floor. Increasing rotation lengths and retaining logging slash on site can also increase carbon stocks.

There Are Some Key Afforestation and Reforestation Implications for Forest Managers and Policy Makers

- Afforestation of sites that have historically not supported forests generally has adverse affects on forest values other than carbon sequestration. Policy makers should consider whether their incentives for forest carbon should promote this type of activity.
- Forest managers should consider using nitrogen-fixing species in place of fertilizers. This can result in reduced emissions from fertilizer production and increased forest biomass.
- Thinning, while reducing short-term carbon sequestration, is an important management technique to reduce the risk of forest loss, improve long-term carbon sequestration, and improve timber quality.
- A/R activities often involve site preparation and/or soil disturbances, which affect soil carbon sequestration. Soil carbon often represents a significant portion of total ecosystem carbon; therefore, policy makers should include soil carbon pools in A/R carbon legislation to avoid unintended carbon emissions.
- Policy makers should consider how to incentivize or protect other ecosystem services besides carbon to help ensure that unintended negative side effects of A/R projects do not ensue.
- Large, industrial, single-species plantations developed by institutional investors dominate A/R projects. Policy makers should seek ways to make native and mixed species plantations economically competitive with single species systems because, in some cases, they offer additional carbon storage and reduced risk of carbon loss from pests and disease.
- Additional research is needed on the management of native species in tropical countries.
- Long-term carbon sequestration studies of A/R projects are lacking. It is important to monitor existing projects as they progress into older forests.

1 Introduction

Deforestation and forest management account for an estimated 17.3% of global greenhouse gas (GHG) emissions (IPCC 2007). As a result, forests and forest management are receiving significant attention in both domestic and international climate change policy discussions (Angelsen 2008; Broekhoff 2008). The Clean Development Mechanism (CDM) of the Kyoto Protocol currently includes afforestation and reforestation[1] (A/R) projects; however, only three such projects

[1] For the purposes of this paper, we define reforestation as planting or natural regeneration of forest on land that previously supported forest (i.e. planting trees on cropland, which supported forest prior to land clearance). We define afforestation as planting trees on land that has never previously supported forest (i.e. planting trees on steppe or pampas grassland ecosystems that do not naturally support forest).

have been approved by the CDM board as of May 2009 (UNFCCC 2009). In addition to A/R projects, policy makers are now considering including carbon credits from Reduced Emissions from avoided Deforestation and forest Degradation (REDD) under a successor to the Kyoto Protocol. Various voluntary carbon registries and sub-national level programs also include forestry to some degree (for more on the topic of global policy, see Chap. 17, this volume).

While forestry has received much attention as a low cost source of emission reductions, a key success factor for forest carbon projects is high-quality management. Land managers will need to understand both the science of how forests grow and sequester carbon, and the communities and people associated with forests. In this chapter, we review the silviculture (the science of managing forests) of afforestation and reforestation projects as it relates to carbon sequestration.

First, we present some of the trends in planted and secondary forests across the globe. We then discuss key concepts for reforestation and afforestation projects. Next, we review the suitability of different sites for A/R projects. Assuming a site is suitable for A/R, we then discuss species selection. Finally, we review how some of the most common silvicultural treatments affect the carbon balance of A/R projects. Throughout, we illustrate the management of A/R projects for carbon sequestration with two case studies: reforestation with Eucalyptus spp. in Brazil, and Acacia spp. in Indonesia.

It is our hope that this chapter will be instructive for foresters managing A/R projects for carbon sequestration under different circumstances, and help policy makers develop appropriate and effective forest carbon offset legislation.

1.1 Global Afforestation/Reforestation Trends

To understand the trends in afforestation and reforestation on a global scale, first it is helpful to define different types of forests. Primary forests are those forests that have never been cleared and have developed under natural ecological processes. Secondary forests are those forests that have regenerated by natural processes following the clearance of primary forests or a change in land use, for example, to agriculture, and then abandonment and reversion back to forest. Plantations are forests that humans have planted either on landscapes that once supported primary forest or on land that did not previously support forest. In this review, reforestation and restoration efforts with the aim of establishing a forest primarily for biodiversity values are considered plantations. Plantations may be established using native or exotic species, or a combination of both. Afforestation projects are always plantations, while reforestation projects may be plantations or secondary forests.

1.2 Historic Patterns of Forest Cover Based Upon Site Productivity

Variation in soil quality and productive capacity drives the distribution of land use across the globe. The inherent productive capacity of land has led to common processes of land colonization and abandonment in areas experiencing afforestation and reforestation. This phenomenon has been identified by (Mather and Needle 1998) as the "Forest Transition," which they characterize as an adjustment of agriculture to site quality and inherent productivity.

In the early stages of a human colonization, vast areas of forest are cleared for agriculture. As a society industrializes and urbanizes, marginal lands are abandoned and agriculture is concentrated on the most productive sites. Agricultural abandonment typically follows one of two pathways: scarcity of employment in rural areas leading to migration to urban areas, or scarcity of forest products to meet demand. These pathways result in different types of secondary forests. Scarcity of employment (currently in Europe and the Mediterranean) results in more naturally regenerating forests, while scarcity of forest products (currently in SE Asia and Latin America) results in more intensively managed plantations (Rudel et al. 2005). This phenomenon has been observed throughout the temperate regions, and similar processes are beginning in the tropics (Rudel et al. 2002).

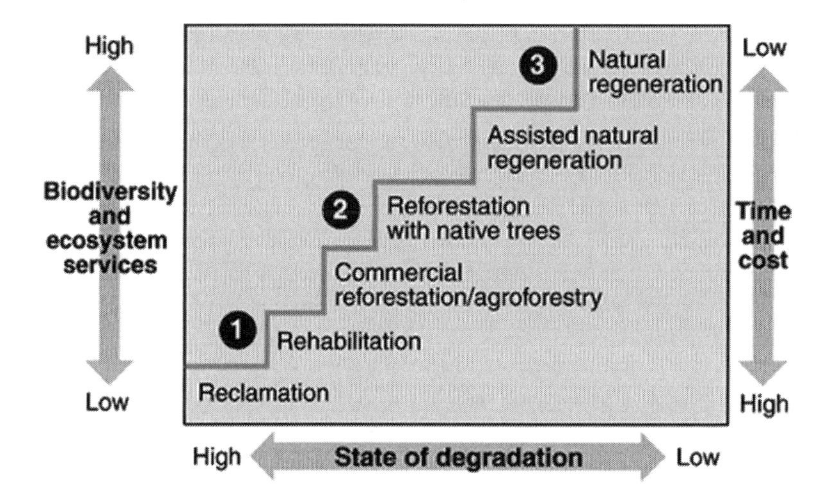

Fig. 11.1 Restoration staircase of previously forested landscapes (From Chazdon 2008. Reprinted with permission)

Table 11.1 Total global forest areas, 1990–2005 (000s hectares)

	Annual rate of change				
	1990	2000	2005	1990–2000	2000–2005
Total forest area	4,077,291	3,988,610	3,952,025	(8868)	(7317)
Total primary forest area	1,397,585	1,373,536	1,337,764	(2405)	(7154)
Total plantation area[a]	102,636	126,938	139,772	2,430	2,567
Other forest – including secondary	2,577,070	2,488,136	2,474,489	(8893)	(2729)

[a]Plantation Area in this table only includes planted exotic species. Table 11.2 includes native and exotic planted forests
Source: Derived from FAO (2006a). Authors' analysis

Depending on access to markets and the economics of competing land uses, deforestation and reforestation may occur simultaneously in the same region (Sloan 2008). In Panama, for example, reforestation has begun in many parts of the country, while deforestation continues in others. International forestry companies and investors are responsible for most of the reforestation occurring in Panama (Sloan 2008). As well-capitalized forestry firms convert pasture to plantations, ranchers and farmers move to new frontiers and continue deforestation.

As marginal agricultural lands are abandoned, they often transfer to pasture and then to forest, or directly to forest. This creates an opportunity for A/R as countries industrialize. The specifics of a given A/R project depend on site quality and access to markets – more intensive silviculture is practiced on the more productive abandoned land, while natural regeneration is often more practical on low productivity and remote sites.

If land has been degraded, natural regeneration is often impossible or impractically slow (Ashton et al. 2001; Holl and Aide 2010; see Fig. 11.1 from Chazdon 2008). Infrastructure, roads, access to international timber markets and human capital in the form of professional foresters, all make intensive forest management more economical. If these elements are absent, it is more likely that A/R will take the form of natural regeneration.

1.3 Current Patterns in Forest Cover

Global primary forest area has declined from 1,397 million hectares in 1990 to 1,337 million hectares in 2005, or a loss of approximately four million hectares of primary forest per year (FAO 2006a). The rate of primary forest loss is accelerating, and accounts for the majority of global forest losses from 2000 to 2005 (Table 11.1).

Table 11.2 Total planted forest area. Includes exotic and native species (000s ha)

	Total planted forests[a]		
	Area		
Region	1990	2000	2005
Africa	13,783	14,371	14,838
Asia	100,896	114,820	131,984
Europe	68,400	76,328	79,394
North and Central America	14,758	26,084	29,050
Oceania	2,447	3,491	3,865
South America	9,157	11,462	12,215
Total World	209,441	246,558	271,346

[a]Includes planted native species and planted exotic species
Source: Derived from FAO (2006b)

In contrast, plantations, of both native and exotic species, compose an increasingly large proportion of global forest area. Global plantation area increased from 209 million hectares in 1990 to 271 million hectares in 2005, equating to 4.1 million hectares of new plantations per year (Table 11.2). While these rates of primary forest loss and new plantation establishment are similar in magnitude, it should not be inferred that primary forest is being converted directly to plantation forests, although this may be true in some regions.

The loss of primary forests is especially disturbing from a global carbon balance perspective. Primary forests have been shown to contain more carbon than the secondary forests, plantations, agriculture, agroforestry systems and pastures that replace them (Montagnini and Nair 2004; Gibbs et al. 2007; Kirby and Potvin 2007). Therefore, if reducing greenhouse gas emissions through forest management is an objective of society, the first priority should be to minimize the loss of intact primary forest.

Although the exact area is unknown, naturally regenerating secondary forests compose a significant portion of the forests under the A/R umbrella (FAO 2006b). Given the young age of many of the planted and naturally regenerating forests that have established globally since 1990, they are likely sequestering large amounts of CO_2 from the atmosphere (see, e.g. Kraenzel et al. 2003; Piotto et al. 2010; Asner et al. 2010). Ironically, while afforestation projects will sequester carbon, they may be counterproductive in high latitudes and only marginally effective in temperate regions in mitigating global warming (Bala et al. 2007).

Planted forests are following different trends in different regions of the world. Asia has the largest area of planted forests, followed by Europe and the Americas (Table 11.2, Figs. 11.2 and 11.3). The FAO classifies planted forests by their primary purpose (production or protection) and species. Pinus (pine species) is by far the most commonly planted genus. Acacia, Eucalyptus and Cunninghamia (Asian fir) also represent large components of global planted forests (Table 11.3). Acacia, Eucalyptus and Tectona (teak) are tropical species, while the other commonly planted genera are temperate species. This suggests that while A/R is becoming more prevalent in tropical countries, there is still much more land area of A/R in temperate regions. We do not present the distribution of species by region here, but it is available in FAO's Global Planted Forests Thematic Study (FAO 2006b).

The extent and high growth rate of planted forests has generated interest in using A/R projects as a means of carbon sequestration. While the rate of establishment of A/R forests has increased in recent years, there are still large areas suitable for A/R projects. The IPCC estimates that the potential exists for 345 million hectares of new plantations and agroforests (Cannell 1999). In addition, many policy makers and foresters point to the positive effects A/R can

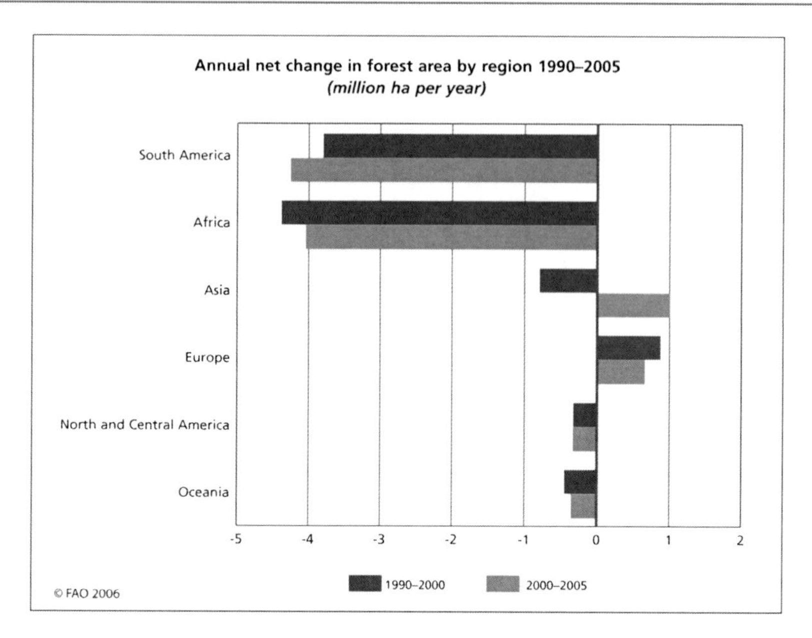

Fig. 11.2 Annual net change in forest area by region 1990–2005 (millions of hectares per year) (From FAO 2006a. Reprinted with permission)

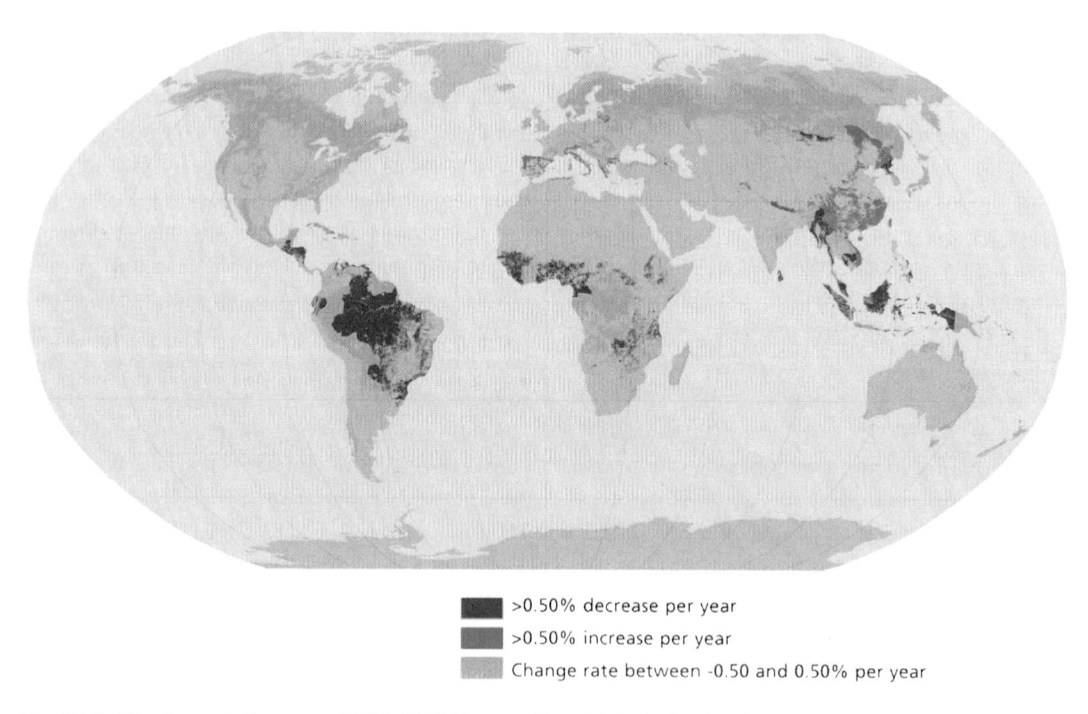

Fig. 11.3 Net change in forest area 2000–2005 (Source: From FAO 2006a. Reprinted with permission)

have on ecosystem services such as water, soil quality and biodiversity (Plantinga and Wu 2003; Cusack and Montagnini 2004; Schoeneberger 2005; Carnus et al. 2006).

On the other hand, many ecologists, soil scientists and foresters have raised concerns over certain A/R projects that may cause loss of soil carbon (Farley et al. 2004; Hirano et al. 2007) or

Table 11.3 Global plantation area by species in 2006

	Plantation area (1,000 ha)	
Genus	Productive	Protective
Acacia	7,357	1,554
Eucalyptus	11,981	1,693
Cunninghamia	15,393	770
Picea	6,284	867
Pinus	46,067	8,802
Populus	4,241	4,949
Tectona	5,819	20
All Others	44,794	29,775
Total	**141,936**	**48,430**

Source: From FAO (2006b)

reduced stream flow and water yield (Scott and Lesch 1997; Farley et al. 2005; Wang et al. 2008). Soil carbon stocks directly affect the carbon balance of A/R projects; therefore, they should be included in a carbon offset program. The impacts on other ecosystem services, while not directly related to carbon sequestration, are tradeoffs that land managers and policy makers will need to evaluate (Chisholm 2010).

Before reviewing the literature on appropriate locations for A/R projects, we first introduce some key concepts that are essential for successful A/R implementation and management.

2 Key Concepts

2.1 Forest Ecosystem Carbon Stocks and Flows

Globally, terrestrial ecosystems, including forests, are estimated to sequester between 1.8 and 3 billion tons of carbon dioxide (CO_2) annually (Dixon et al. 1994; Canadell and Raupach 2008). To manage forests of any type for carbon sequestration, it is important to understand how carbon is stored and cycled through a forest ecosystem. Very broadly speaking, carbon is present in four pools: above ground biomass, below ground biomass, dead woody debris, and soil carbon. Above ground biomass includes all tree and plant parts including the tree stem, branches, and leaves. Below ground biomass includes both coarse and fine plant roots. Dead woody debris includes decaying biomass on the forest floor such as leaves, branches and entire trees. Soil carbon includes the organic matter incorporated into the soil itself. Carbon flows between these sinks and the atmosphere in a complex manner, described in more detail by Dixon et al. (1994) and (Malhi et al. 1999). A more in-depth review of forest stand dynamics in relation to carbon sequestration is provided in Chap. 3 of this volume.

2.2 Additionality

A common principle underlying carbon offset projects and protocols is additionality over a specified baseline. In other words, to be awarded offsets, a project must demonstrate that the carbon it sequesters is beyond what would have happened in the absence of the project. This principle applies to A/R projects as well, and is particularly important when considering site selection for A/R projects.

2.3 Degraded Forests

The term "degraded" is used loosely in describing forests impacted by human activity or management, and is used to justify converting land to plantations. Here we distinguish between structural and functional degradation, and suggest that structurally degraded forests are still functioning forests and therefore should not be eligible for A/R funding. In contrast, functionally degraded sites are no longer able to successfully support trees and are therefore legitimate sites for restoration and A/R.

Structural degradation usually entails small scale but continuous chronic site disturbance that alters the species composition or structure of the forest canopy. Functional degradation is usually the result of acute, one-time disturbances, which alter the site productivity and physical characteristics of the soil (Ashton et al. 2001).

Heavily logged primary forests, with little remaining valuable timber, are often regarded as structurally degraded. While merchantable timber may be lacking, many cutover forests continue to grow and serve as carbon sinks and wildlife habitat. Furthermore, recently cutover forests are not non-forested land and therefore should not be eligible for A/R funding because it will create perverse incentives to high-grade natural forests, classify them as degraded, and then replace them with plantation forests. Rather, structurally degraded forests should be considered under REDD or Improved Forest Management methodologies. In contrast, functional degradation alters forest sites to such a degree that trees can no longer grow on the site and active reforestation is often the most efficient and practical means to restore forest to the site (Parrotta 1990, 1991; Lamb et al. 2005). Examples of functionally degraded sites are those that have been used for surface mining, or intensive agriculture and pasture. Such sites when abandoned have lost the capacity to naturally re-vegetate to forest because: (i) no viable seed source for natural regeneration exists; (ii) the hydrology and the fertility of surface soil horizons have been altered to an extent that cannot allow seed germination or establishment of trees; or (iii) degradation has allowed opportunistic vegetation to colonize that is maintained by new cyclical disturbances (e.g. fire).

2.4 Risk Aversion

In addition to carbon sequestration, forests should be managed to minimize the risk of carbon loss through disturbance. Depending upon site productivity, managers can assume different levels of risk in their management strategy. High productivity sites support shorter rotations and encourage managers to practice more intensive, expensive management. On high productivity sites, there is less chance of disturbance over the short rotations and the pay-off is greater at the end of the rotation. Even if there is a disturbance, the rotations are short enough that one can easily replant and start over.

Forests on marginal lands grow more slowly and therefore landowners must wait longer to derive a useful product. Longer rotations expose forest stands to disturbance (disease, fire, wind throw) for greater periods, making the loss of some or the entire timber crop more likely. In addition, long rotations result in lower rates of return, all else being equal, because cash flows are realized further in the future.

Therefore, managers generally practice less capital-intensive silviculture as they move to progressively less productive sites. This is supported by the land use trends that can be observed on the landscape and in the theoretical models of forest management (Mather and Needle 1998). Natural regeneration and passive management for carbon sequestration may be more appropriate on marginal lands, although land managers will need to conduct their own economic and silvicultural analyses for their specific site.

Insert 1. Management Intensity – *Acacia mangium*

Intensive management of *A. mangium* plantations on rich sites in Indonesia (fluvisols) has been found to maintain high production levels of carbon and/or timber over successive rotations without significant loss of nutrients (Mackensen and Folster 2000). *A. mangium* plantations on poorer sites (arenosols, acrisols, ferralsols), however, cause nutrient losses that threaten the long-term productivity of the site and that can only be compensated for with expensive investments in fertilizer. Thus, site productivity needs to be considered in deciding how intensively to manage a site.

2.5 Promoting Resiliency Against and Preparing for Disturbances

Practicing sound silviculture to promote resiliency and minimize risk of major disturbances is key when managing for carbon sequestration. A major stand-clearing disturbance such as a fire can release most or all of a forest's aboveground carbon stocks, reversing any carbon sequestration benefit. Protecting a stand against disturbance can involve management practices that reduce a stand's aboveground carbon stocking, such as thinning. Management that slightly reduces carbon stocks in the short-term is worthwhile when it helps avoid the types of disturbances that can wipe out a forest carbon offset project.

While forest fire fighting capabilities have long been developed in most temperate countries, most tropical countries are woefully unprepared to fight fires (see, e.g. Cochrane 2002); a fire protection plan is essential in many areas. Thinning stands in fire-prone regions is one of the most effective means for reducing the risk of catastrophic fire (Finkral and Evans 2008). We discuss the specifics of how thinning affects forest carbon sequestration in greater detail below. Maintaining a mix of species can also be effective in lessening the potential damage from disturbances that target a particular species, such as insect outbreaks (Jandl et al. 2007).

3 Beyond Above-Ground Carbon Storage: Soil, Water, and Biodiversity

The addition of trees to a non-forested site will increase above-ground carbon storage in almost all cases. A/R projects vary in how they impact soil carbon, water, and biodiversity, and adverse impacts to any of these forest values must be considered when deciding where to site an A/R project. Ideally, A/R projects will provide carbon benefits as well as economic and ecological benefits — increases in biodiversity, water quality – and the decision to proceed will be straightforward. Managers will face difficult choices when A/R run the risk of having negative impacts on other ecological values, even while providing carbon additionality.

Insert 2. Risk: *Acacia mangium*

Acacia mangium accounts for approximately 80% of short-rotation plantations in Indonesia. Incidence of heartrot fungi in these stands is as high as 46.7% in some regions of the country (Barry et al. 2004). Root rot is also prevalent (as high as 28.5% incidence) in southeast Asian *A. mangium* plantations, particularly in second and third rotations. Root rot, however, was found less often in former grasslands than in lowland former rainforest. Further, waiting for 2 months between harvesting and replanting was found to reduce the incidence of root rot (Irianto et al. 2006). Using a mixed species approach can help diversify the investment in reforestation, so that if a plantation becomes heavily infected, not all trees are lost. Forest fire poses an additional risk in Indonesian *Acacia mangium* stands. The high litter fall produced by *A. mangium* combined with dry conditions and Imperata grassland understories has caused significant losses of forest to fire (Saharjo and Watanabe 2000).

A. mangium has been promoted and planted in Latin America and early growth and survival across a rainfall gradient in Panama found the species to consistently out-compete most native species (Park et al. 2010; Breugel et al. 2010). However, at one site in Panama 18.3% of the trees suffered a discoloring fungal infection (Hall, unpublished data). In addition, because this species has been found to be highly invasive (Daehler 1998), any decision to plant it outside its native range should be carefully considered.

3.1 Soil Carbon

While most A/R projects will increase aboveground carbon stocks (because trees tend to store more carbon than other types of land cover, namely, shrubs, grasses, or crops), they will not necessarily increase soil carbon stocks. In grasslands, carbon accumulates in the soil each year as grasses die and decompose. If a forest replaces grassland, the tilling and site preparation necessary to plant new trees exposes the soil carbon to increased levels of oxygen. This speeds up soil carbon decomposition rates, and carbon dioxide emissions.

A similar phenomenon occurs if peat is drained to improve site conditions to plant forests. Peat is generally very moist, and creates an oxygen-poor environment where decomposition happens very slowly. Draining peat increases the oxygen levels in peat soils, resulting in faster decomposition and carbon dioxide emissions (Jaenicke et al. 2008).

Given that approximately 75% of terrestrial carbon is stored in soils (Paul et al. 2002), it is vitally important to monitor soil carbon as well as aboveground carbon for A/R projects. Afforestation projects – projects on land that has never previously supported forests — run the biggest risk of causing large soil carbon releases. A/R projects may also cause changes to other environmental services such as water runoff and biodiversity. Managers should weigh these potentially negative changes against the benefits of carbon sequestration in deciding whether to initiate an A/R project.

3.1.1 Agricultural Land

Abandoned agricultural land is perhaps the most common land cover type for A/R projects. Agricultural land is found across a wide range of ecological settings and can encompass land used for crops as well as for pasture, making it difficult to generalize about its suitability for A/R Projects. In this section, we attempt to differentiate between some of the different types of agricultural land, and to assess their suitability for A/R projects.

In general, afforestation of cropland has been found to increase soil carbon content in the long-term, following an initial decrease. In contrast, afforestation of pastures has been shown to slightly decrease soil carbon (Paul et al. 2002). However, these overall trends vary by region and forest type.

Completing a meta analysis of the effects of land use change on soil carbon stocks, Guo and Gifford (2002) identified patterns that they suggest should be treated as hypotheses rather than conclusions. Overall they found that converting pasture to plantations reduced soil carbon on average by 10%; however, this trend was driven by studies of conversion to conifer as opposed to broadleaf plantations (65 vs 18 studies). Indeed, while variability exists, they found no significant change when broadleaf species were planted on pastures.

In this same analysis Guo and Gifford (2002) found only six studies of the conversion of pasture to secondary forest that met their study criteria, with a range from over 40% loss of soil carbon to a gain of approximately 5% (mean, approximately 20% loss). In a study of soil organic carbon (SOC) during natural secondary succession on pasture in Panama, Neumann-Cosel et al. (2011) found significantly lower SOC under pastures converted from forest and grazed for more than 20 years than under forests. Further, 15 years of secondary succession was insufficient to note a significant increase in SOC. These authors highlight the different trends found in relation to SOC under tropical secondary succession in recent studies. For example, of 10 recent studies reviewed of work undertaken in Latin America, five studies could not identify a trend in SOC with age of recovery. Neumann-Cosel et al. (2011) point to the importance of soil type, soil mineralogy, and texture in helping to determine SOC. Van der Kamp et al. (2009) attribute the vast differences in soil carbon accumulation found in Sumatra under transition from Imperata grassland to secondary forest as opposed to their study of soil carbon during secondary succession based in East Kalimantan to soil properties.

For studies comparing the transition of crop land to plantations (29) and secondary forest (9),

Guo and Gifford (2002) found all studies to show an increase in soil carbon. With both crop and pastureland, management intensity affects how soil carbon stocks change (Guo and Gifford 2002). Letting a secondary forest grow on former crop or pasture land often results in greater soil carbon levels than a plantation because there is less soil disturbance (Guo and Gifford 2002).

3.1.2 Afforestation or Reforestation

Afforestation or reforestation affects pastureland (potentially arable grassland) soil carbon stocks in various ways depending on specific site conditions. One key source of variability is precipitation (Guo and Gifford 2002). Afforestation on arid or semi-arid grasslands can create carbon additionality, although care must be taken to select species that are efficient in their water use. In Inner Mongolia, poplar (*Populus spp.*) and Mongolian pine (*Pinus sylvestris var. mongolica*) have been used to afforest semi-arid grasslands. Soil carbon under poplar plantations recovers to pre-afforestation stocks by age 15, while soil carbon under Mongolian pine persists below the pre-afforestation grassland levels after 30 years. Although soil carbon decreased under Mongolian pine, significant increases in aboveground and belowground (root) carbon stocks resulted in both pine and poplar stands being net carbon positive (Hu et al. 2008).

Afforestation of pasture land in the Patagonian semi-arid steppe has also resulted in net carbon sequestration (Laclau 2003a; Nosetto et al. 2006). The grass-shrub steppe of Patagonia stores approximately 95.5 Mg C/ha, predominantly as soil carbon (Laclau 2003a). Afforestation with exotic Ponderosa pine (*Pinus ponderosa*) resulted in no loss of soil carbon and significant gains in aboveground biomass after 14 years. Naturally regenerated native cypress (*Austrocedrus chilensis*) stands, with an average age of 45 years, showed significant increases in soil carbon (Laclau 2003a) and even greater total carbon storage than Ponderosa pine. The different average stand ages in this study make comparison of the rates of carbon storage between the exotic ponderosa pine and native cypress difficult. However, given that cypress regenerates naturally in the steppe ecosystem, it is likely a more efficient method of long-term carbon storage than planted Ponderosa pine.

Afforestation of grasslands in wetter climates has greater potential to release large amounts of soil carbon. For instance, in the Ecuadorian highlands, radiata pine (*Pinus radiata*) has been used to afforest grasslands on carbon-rich, volcanic soils in a relatively wet climate (Farley et al. 2004). The wet, oxygen poor soils store large amounts of carbon. When these wet grasslands are drained and exposed to oxygen, rapid decomposition of soil carbon occurs. In 25 year old plantations of radiata pine in the Ecuadorian highlands, carbon stocks were reduced in the first 10 cm of soil from 5 kg/m^2 under native grasslands to 3.5 kg/m^2. Soil carbon content decreased at greater depths as well (Farley et al. 2004). In contrast, reforestation of grasslands that were once tropical forest has the potential to increase soil carbon storage. In Indonesia, imperata grasslands now cover 8.5 million ha of what was once primary forest (van der Kamp et al. 2009). Secondary forest growth has the potential to store 61.7 tons/ha (East Kalimantan) to 219 tons/ha (Sumatra) of carbon as compared to imperata grassland baselines of 39.64 tons/ha (East Kalimantan) and 47 tons/ha (Sumatra) (van der Kamp et al. 2009).

3.1.3 Peatland

While A/R on peatland is not as common as on agricultural land, the large amounts of carbon stored in peatland soil warrants discussion. The carbon-rich peatlands of Southeast Asia have increasingly become a target for drainage and conversion to plantations.

The impacts of afforestation on peatlands depends upon the depth of peat (and hence the amount of drainage required) as well as the climate. In colder climates such as the UK or Scandinavia, shallow peats requiring less drainage result in lower levels of soil carbon release, and afforestation projects may be net positive (Hargreaves et al. 2003; Byrne and Farrell 2005; Byrne and Milne 2006). For instance, afforestation of peatland in Britain in peat less than

35.5 cm deep resulted in an increase in aboveground biomass that could compensate for the loss of soil carbon of 50–100 g C/m^2/year (Cannell et al. 1993). In deeper peat in Britain, where carbon release from drainage can reach 200–300 g C/m^2/year, aboveground biomass did not compensate for the loss in soil carbon (Cannell et al. 1993). Peatlands in Southeast Asia are deeper and store considerably more carbon than those in the UK or Scandinavia. Indonesian peatlands have been estimated to store 55 Gt C (Jaenicke et al. 2008). One recent study of carbon release from a drained peat swamp estimated an average of 313–602 g C/m^2/year released over three years (Hirano et al. 2007). Although peatlands only comprise 3% of the world's land area, they store one-third of the world's soil carbon (Rydin and Jeglum 2006). Given these high levels of soil carbon release, policy makers and managers need to look closely at afforestation projects on tropical peatlands to ensure additionality.

3.2 Water

Management of forests for water has long been part of the tool kit for watershed protection such that it should not come as a surprise to foresters and policy makers interested in carbon sequestration that planting trees – be it for afforestation or reforestation – will alter the hydrological cycle (Bruijnzeel 2004).

Afforestation on grassland and shrubland can alter the hydrology of a system by decreasing runoff and increasing transpiration. This can be particularly problematic in drier locations where water limitation is an issue. Globally, grassland and shrubland afforestation have been found to reduce annual runoff by as much as 44% and 31%, respectively, for up to 20 years after afforestation (Farley et al. 2005). Fast-growing species demand more water and will induce greater water flow reductions (Bruijnzeel et al. 2005). Studies conducted in South America have demonstrated significant decreases in water levels, both in the drier steppe (in Patagonia, using Ponderosa Pine and native Cypress), as well as in the wetter pampas (in Argentina, using *Eucalyptus camaldulensis*; Engel et al. 2005; Licata et al. 2008). In addition, it was found that afforestation of the Argentine pampas with *E. camaldulensis* acidifies the soil, reduces the soil cation exchange capacity (Jobbagy and Jackson 2003), and results in soil and groundwater salinization (Jobbagy and Jackson 2004). Therefore, while afforestation may enhance short-term carbon sequestration, it can also alter soil chemistry in ways that can significantly reduce future productivity and impact groundwater quality.

Reforestation, like afforestation, can also reduce overall water flow due to the demand for water by trees (Bruijnzeel et al. 2005). Whereas afforestation reduces water flow to levels that the site may not be adapted for, reforestation may ultimately reduce water flow to levels similar to when the site was previously forested (Bruijnzeel et al. 2005); however, planting fast growing trees can reduce stream flow below natural forest conditions as vigorously growing stands exceed the water use of the original mature forest (Malmer et al. 2010). Further, reforestation can have other positive side effects in relation to water values. Deforestation has been shown to increase the risk and severity of flooding (Bradshaw et al. 2007), and reforestation has been proposed as a means of reducing flood risk. Nevertheless, the forest flood mitigation effect observed in small watersheds can disappear at larger spatial scales (Wilk and Hughes 2002; Malmer et al. 2010).

Another potential benefit of reforestation in tropical areas with a seasonal climate is the potential to recreate the "sponge effect" – or improve groundwater recharge during the wet season and thus increase dry season stream flow as water is released into the stream (Ilstedt et al. 2007; Malmer et al. 2010; Stallard et al. 2010). Although this is commonly perceived as a well understood phenomenon outside the community of hydrologists studying this effect (Malmer et al. 2010), there is little empirical evidence to support this and the processes are poorly understood (Bruijnzeel 1989, 2004; Scott et al. 2005). Ilstedt et al. (2007) recently completed a meta analysis where they evaluated evidence for improved infiltration, a key process in groundwater recharge, with reforestation in the tropics. While these authors were only able to identify 14 studies that met their methodological and statistical criteria for inclusion in the analysis, all studies showed improved infiltration with reforestation. More recent studies of infiltration under plantations and secondary succession in Latin America have also found improved infiltration over relatively short periods of time (Hassler et al. 2010; Zimmermann et al. 2010); however, further research needs to be conducted to understand the process of groundwater recharge as well as the variation with climate, soils, geology, and species used in reforestation (Malmer et al. 2010).

3.3 Biodiversity

The effect of A/R on biodiversity has been a hotly debated topic. The specific effects of any given project depend heavily on its historical context and location within the broader landscape. Plantation forests almost always provide more suitable habitat for forest species than agricultural land (Brockerhoff et al. 2008) while those planted using ecological restoration techniques provide biodiversity and other ecosystem services (Benayas et al. 2009) . Planted forests can also enhance the matrix between remnant natural forest patches, which has multiple benefits:

- edge effects on natural forests are decreased;
- planted forests (depending upon choice of tree species) facilitate dispersal between natural forest patches;
- forest generalist species often use resources provided by planted forests; and,

Insert 3. Panama Canal Watershed Case Study

The Panama Canal Watershed consists of a landscape mosaic including large protected forests and a variety of other land uses. The Panama Canal Authority (ACP) manages water for drinking, hydroelectric power generation, and ship passages within the Canal where 2.6 billion m^3 of water are used for ship transits (Stallard et al. 2010). However, the Panama Canal Watershed is located in an area of seasonal rainfall where rare but severe dry seasons limit the ACP's ability to provide water for all of its uses and can even lead to reductions in Canal cargo transits.

Studies of paired experimental catchments – where stream flow of a forested catchment have been compared to that of a vegetation mosaic catchment that includes large areas of active and abandoned pastures – have shown pronounced differences in stream flow between seasons and land uses (Condit et al. 2001; Ibanez et al. 2002). Even during a relatively wet dry season, dry season flow was reduced in the mosaic as compared to the forested catchment (Stallard et al. 2010), thus indicating the sponge effect is not limited to severe dry seasons. The extent to which forest soils absorb water during the wet season and subsequently release it during the dry season is of intense interest in the Panama Canal Watershed. The sponge effect, how it changes with land use, and tradeoffs inherent to maximizing water, carbon storage, and biodiversity values is under study by the "Agua Salud Project" – a long term initiative undertaken by the Smithsonian Tropical Research Institute, the ACP and other partners in Panama (Stallard et al. 2010).

- plantation forests can reduce harvesting pressure on and habitat loss in existing natural forests.

In some cases, A/R increases plant and animal diversity, particularly on degraded lands. Imperata grasslands have replaced large areas of primary forest in Indonesia (van der Kamp et al. 2009). Reforestation of these non-native grasslands with *A. mangium* increased arthropod diversity, although not as much as a naturally regenerating secondary forest (Maeto et al. 2009). Agroforests and silvopastoral systems can increase carbon sequestration (Kirby and Potvin 2007; Andrade et al. 2008) and are important for biodiversity conservation (Schroth et al. 2004; Gibbons et al. 2008). Shade grown coffee (Greenberg et al. 1997) and cacoa (Bael et al. 2007) have been studied for their biodiversity values, while a number of authors have studied the importance of native trees for enhancing seed dispersal and understory regeneration (see, e.g., Tucker and Murphy 1997; Jones et al. 2004; Zamora and Montagnini 2007). Indeed, the strategic spatial juxtaposition of plantations and other land management strategies that enhance tree cover across the landscape can dramatically increase the biodiversity value at the landscape scale in rural areas by increasing habitat and improved connectivity between protected areas and other forest patches (Harvey et al. 2006, 2008).

However, at the landscape scale, A/R is not always desirable from a biodiversity perspective. In the case of afforestation, planted forest replaces a natural habitat type (i.e. grassland). If the species that are native to a region depend on grassland ecosystems, afforestation will reduce habitat available for these species and be detrimental to landscape scale biodiversity (e.g. bird diversity following afforestation in South Africa, (Allan et al. 1997)). In addition, the modeled impacts of afforestation on a South African fynbos site using radiata pine projected a large loss of plant and insect biodiversity (Garcia-Quijano et al. 2007).

Insert 4. Biodiversity – Eucalyptus plantations

The effect of eucalyptus plantations on biodiversity has been examined in the Brazilian Amazon using forest birds as a biodiversity indicator. A study in the northeast Amazon estimated bird species richness using point count estimates. Primary forest (106.5 species) had greater bird species diversity than secondary forest (70 species), which in turn had greater diversity than eucalyptus plantations (50 species) (Barlow et al. 2007a). Eucalyptus plantations contained almost no species in common with primary forest, and contained very few habitat specialists. Primary forests also contained greater butterfly diversity than secondary forests and eucalyptus plantations, although eucalyptus plantations contained a higher number of individuals (Barlow et al. 2007b). Since reforestation takes place on land that is not forested, the appropriate baseline against which to measure reforestation is pastureland or agricultural land. In São Paulo State, Brazil, Blue-winged Macaws used eucalyptus plantations as habitat, but never used pastureland, coffee plantations or rubber plantations (Evans et al. 2005).

Retaining forest strips in the northeastern Amazon, whether riparian or upland, that extend into and through the eucalyptus plantation matrix greatly increases bird diversity (Hawes et al. 2008). Riparian and upland forest strips were found to have species assemblages that closely reflected continuous primary forest. This suggests that while eucalyptus plantations themselves do not contribute significantly to biodiversity conservation, managers can design plantations to contain riparian and upland reserves that do provide significant biodiversity conservation benefits.

4 Management Objective, Site, and Species Selection

4.1 Management Objective

When designing a project or managing a piece of land it is important to clearly define the overall management and secondary objectives while also considering associated impacts. Soil carbon, water, and biodiversity have been addressed above and could be included within the management objectives or considered for associated impacts. Ecosystem services often interact (Bennett et al. 2009) where some can be bundled together (Raudsepp-Hearne et al. 2010). While carbon sequestration can be the primary management objective of an A/R project, it often serves as a secondary objective to timber production, ecological restoration, or livelihood projects (Hall et al. 2010; Garen et al. 2011). Two livelihood related types of projects for which carbon sequestration can play a significant role include Agroforestry and Silvopastoral systems.

4.1.1 Timber Production and Ecological Restoration

Planting trees on pastures, agricultural, or grasslands for both timber production and ecological restoration result in the conversion of vegetation cover into forests of different levels of complexity. As discussed below, carbon sequestration in timber plantations can depend upon management intensity but should result in significant carbon sequestration, often more so should land be left fallow and succession allowed to occur naturally. For example, a study of 20 year old teak plantations in Panama found an average carbon sequestration of 120 t/ha. In contrast, preliminary analysis of aboveground biomass in 12–15 year old secondary forest at a site nearby found an average carbon sequestration of approximately 40 t/ha (80 t/ha dry biomass, Neumann-Cosel et al. 2010); however, plant species diversity is extremely high (Breugel et al. 2011). The objective of ecological restoration is to recreate a forest that is similar in diversity and structure to the forest that existed on the site before conversion to other uses. Such efforts employ ecological principals to foster recruitment and also typically include a large number of species in the initial planting (Tucker and Murphy 1997; Elliot et al. 2003; Rodrigues et al. 2010). When successful, they will sequester a significant amount of carbon; however, inappropriately designed projects can lead to stand stagnation (Rodrigues et al. 2009).

4.1.2 Agroforestry

For the purposes of this review, agroforestry refers to a number of different practices of growing trees on agricultural lands, including alley cropping, riparian buffer strips, forest farming, and wind breaks. While agroforestry systems generally do not sequester as much carbon as primary forests, secondary forests or plantations, they can provide a means of integrating forest carbon sequestration into agricultural production (Montagnini and Nair 2004). The inclusion of these systems by smallholders in the tropics could produce significant carbon sequestration. Estimates of carbon storage in agroforestry systems ranges greatly, from 0.29 to 15.21 Mg C/ha/year, depending upon site productivity (Nair et al. 2009). Agroforestry systems can also increase soil carbon storage (Haile et al. 2008; Takimoto et al. 2008). Agricultural crops and trees grown together can provide complementary carbon storage benefits, similar to mixed-species plantations. However, like site preparation associated with establishment of tree plantations, conversion of land, such as a pasture, into an agroforestry system that requires soil tillage will usually reduce total soil carbon, even with row plantings of trees (Guo and Gifford 2002).

4.1.3 Silvopastoral Systems

Murgueitio et al. (2010) describe silvopastoral systems, as comprising different agroforestry arrangements that combine fodder plants, such as grasses and leguminous herbs, with shrubs and trees for animal nutrition and complementary uses. Silvopastoral systems can vastly improve economic and ecological sustainability (Murgueitio et al. 2010) while also sequestering carbon (Andrade et al. 2008) and providing biodiversity values (Harvey et al. 2004). A study by

Sharrow and Ismail (2004) found that a silvo-pastoral system in western Oregon sequestered more carbon than a pure plantation or pasture, which they attribute to the complementary nature of the pasture's soil carbon storage with the trees' biomass storage.

4.2 Site Selection

One of the most important management decisions for a successful afforestation/reforestation project is selecting an appropriate site. Foresters working in temperate regions have long recognized the relationship between site and tree growth as evidenced by the use of site index (Avery and Burkhart 1994). Early growth of 49 tropical species grown in a rainfall-precipitation matrix in Panama underscore the relationship between site and growth for the species studied (Breugel et al. 2010). Degraded tropical soils pose a particular challenge to reforestation efforts as loss of nutrients and soil structure can be sufficiently elevated as to require significant site intervention in order to obtain successful establishment and growth (Ashton et al. 2001). In addition, site selection can impact associated ecosystem services as discussed above.

4.3 Species Selection

A small number of genera comprise much of the global plantation area (FAO 2000). Single-species plantations of exotic species such as eucalyptus, pine, acacia, and teak have several advantages that make them popular: they are fast-growing species with known markets; a large body of knowledge exists on their silviculture; and, growing only one species makes for less complicated silviculture than growing multiple species. They can clearly sequester large quantities of carbon in a relatively short period of time (see, e.g., Kraenzel et al. 2003), particularly when grown on appropriate sites and/or with intensive management (Forrester et al. 2010; Toit et al. 2010).

However, there are situations where alternatives to use of exotics in single-species plantations can generate increased carbon sequestration. Mixed-species plantations can potentially increase carbon storage over single-species plantations through the integration of nitrogen-fixing trees, and the use of trees with complementary growth patterns (Ashton and Ducey 1997). Mixed-species plantations can also reduce the risk of damage from pests and disease, which is important for ensuring the permanence of carbon storage. There is a growing body of research indicating that viable alternatives to single-species plantations exist and can potentially sequester more carbon (Erskine et al. 2006; Forrester et al. 2006; Nichols et al. 2006; Parrotta 1999; Piotto et al. 2009), while also reducing the risk of carbon loss through disturbance.

4.3.1 Mixed Species

Within a forest or stand tree interactions can be characterized as competitive, complementary or facilitative (Kelty 2006; Forrester et al. 2006). Although managed single-species plantations can grow well, owing to their similar morphologies and resource acquisition strategies, individual tree interactions are competitive. Trees of different species compete for resources but interactions may also be complementary or facilitative. For example, species with different rooting morphologies can acquire water at different soil depths such that, while they use more water at the stand level, their acquisition strategies are complementary such that they do not compete to the extent they would if they acquired water at the same depth. An example of a facilitative interaction could be where a nitrogen-fixing species increases ecosystem nitrogen through litterfall such that more nitrogen is available for growth of individuals of other species. Mixed-species plantations can increase carbon storage over single-species plantations by incorporating species that facilitate the growth of others and those with complementary light, nutrient, and water requirements and/or acquisition strategies. Mixed-species plantations can also reduce the risk of carbon loss from pest and disease outbreaks.

Nitrogen-fixing species: Nitrogen-fixing or leguminous species are typically from the Fabaceae or

Leguminosae family, and host rhizobia bacteria on their roots that can convert nitrogen gas, N_2, into biologically available nitrogen, NO_3 or NH_4. Growing nitrogen-fixing species such as *Albizia spp*. in combination with conventional plantation species can increase productivity. Nitrogen is often limiting in tropical plantations, and increasing the available nitrogen can increase biomass production and carbon storage Binkley et al. (1992; Parrotta 1999; Balieiro et al. 2008). For instance, the benefits of nitrogen-fixing species on eucalyptus plantation productivity have been researched extensively in Hawaii. Binkley et al. (1992) found that planting 34% eucalyptus and 66% albizia maximized total biomass on volcanic soils in Hawaii. Not only can total biomass be maximized with the introduction of nitrogen-fixing species, but growth rates of the primary timber species (eucalyptus, e.g.) can be increased as well (DeBell et al. 1997), which can increase revenue from wood products. In addition, high growth rates can be sustained longer into eucalyptus rotations (Binkley et al. 2003). Forrester et al. (2005) attributed benefits to eucalyptus growth from the addition of nitrogen-fixing acacia not only to increases in available nitrogen but also to increased rates of nitrogen and phosphorous cycling (also see Siddique et al. 2008).

The benefits from nitrogen-fixers vary based upon soil properties (Boyden et al. 2005). If the supply of other nutrients is limited, nitrogen fixers will not necessarily enhance productivity. Also, many leguminous nitrogen-fixers do not grow well in acidic soil conditions (Binkley et al. 1992).

Competition versus facilitation: Using mixtures of species can increase plantation productivity if the species are complementary in their use of resources (Kelty 2005; Carnus et al. 2006). That is, if species have different requirements for light and nutrients, and different growth rates, competition between species may be less intense than within a single species, and total biomass growth on the site can be increased (Ashton et al. 1998; Forrester et al. 2005). Mixtures of complementary species have been found to maintain productivity at higher densities than single-species plantations (Amoroso and Turnblom 2006).

Mixed-species stands can also be less productive than monocultures if the species used are too similar in their requirements. Chen et al. (2003) studied different combinations of mixed-conifer species plantations in British Columbia and found no combinations that were superior to single-species plantations. They suggested that strategic selection of shade tolerant and intolerant species mixtures might have produced better results in the mixed-species stands. Performance of mixed species plantations in central Oregon was also found to vary depending on species composition and the initial spacing of trees (Garber and Maguire 2004).

Risk aversion: In addition to potential increases in carbon sequestration, mixed-species stands can reduce the risk of significant pest and disease outbreaks, which can release stored carbon (Montagnini and Porras, 1998). Jactel et al. (2005) concluded, based upon a meta-analysis of single vs. mixed species stands, that damage from insect outbreaks was significantly higher in single-species stands. Mixed-species stands are also less vulnerable to fungal pathogens (Pautasso et al. 2005). In a mixed-species forest, even if one species is attacked by pests or pathogens, another species can replace it and continue to sequester carbon and provide other forest values. Piotto et al. (2010) offer an excellent example from the humid zone of Costa Rica. These authors describe 100% mortality for five species when grown in monocultures. One species, *Calophyllum brasiliense*, showed excellent growth at 13 years of age. Redondo-Brenes and Montagnini (2006) highlighted its carbon sequestration potential and calculated a rotation length of 18.5 years. However, all trees in monoculture plots died at 15 years of age while mixed species plots in the experiment survived (Piotto et al. 2010) In addition, the passive dispersal of disease is slower in mixed species stands (Pautasso et al. 2005).

4.3.2 Native Species Versus Exotics

Overall, the handful of major exotic plantation species used globally (*Acacia mangium*, *Eucalyptus* spp., *Pinus caribbaea*, *Tectona grandis*) have some significant advantages for carbon sequestration: fast growth rates, a large body of knowledge on how to successfully manage them, existing wood markets, and less complex silvicultural knowledge than is required to grow a native or mixed-species plantation.

Native species also have the potential to be equally if not more productive than exotic species and may be better suited to maintaining the long-term productivity of a site. However, long-term silvicultural research, as well as development of markets, is necessary in order to improve the viability of native species plantations. Studies comparing native species to exotic species have shown that some native species have similar or superior growth rates compared to exotic species. Currently, the primary disadvantage of native species plantations in the tropics is the lack of research and knowledge of their silviculture and wood properties compared with conventional exotic species. However, this is starting to change, particularly in Central America (Carnevale and Montagnini 2002; Hooper et al. 2002; Wishnie et al. 2007; Hall et al. 2010; van Breugel et al. 2010) and Asia (Otsamo et al. 1997; Shono et al. 2007; Thomas et al. 2007). Markets for native species wood products are not as well developed as markets for traditional plantation species, meaning that land owners can be more certain of investment returns for exotic species (Streed et al. 2006).

Teak (*Tectona grandis*) is a highly valuable exotic species grown throughout Central America for which markets are well developed. It can exhibit high growth rates (van Breugel et al. 2010) and sequester vast amounts of carbon during a 20 year rotation (Kraenzel et al. 2003), particularly when grown on fertile loamy soils in areas of moderate rainfall. However, Piotto et al. (2004), found *Schizolobium parahyba* to be competitive to teak at 68 months in the Nicoya Peninsula of Costa Rica, a finding also found after 4 years of growth at a relatively dry and fertile site in Panama (Hall et al. 2010), Hall et al. (2010) also found 5 year growth of *Dalbergia retusa* at a dry, infertile site in Panama to rival that of teak whereas both *Terminalia amazonia* and *Vochysia guatemalensis* preformed as well or better than teak at a wet, infertile site in Panama (Hall et al. unpublished data). Heavy erosion can occur with the onset of rains when teak is planted on inappropriate sites (Carnus et al. 2006). In one study in Costa Rica, teak also resulted in lower soil organic carbon levels than native species on land converted from pasture (Boley et al. 2009).

In a study comparing over 15 years of growth in monocultures and mixtures in the humid zone of Costa Rica, Piotto et al. (2010) found the mixture of *Terminalia amazonia*, *Dipteryx panamensis,* and *Virola koschnyi* to achieve the highest aboveground biomass as compared to monocultures and other mixtures; however, while exhibiting a markedly higher biomass, it was not statistically significantly higher than *T. amazonia* monoculture. Depending upon wood density, fast growing species can sequester carbon quickly in the early years, while slow growing species accumulate more carbon in the long term.

In another study in Costa Rica, carbon storage was shown to vary considerable, depending on stand management. Nine to twelve year old single species plantations of *Terminalia amazonia* and *Dipteryx panamensis* contained 55.1–79.1 and 36.9–91.0 Mg C/ha, respectively, in Sarapiqui, but only 27.5 and 36.5–44.4 Mg C/ha, in San Carlos (Redondo-Brenes 2007). This difference was best explained by stand density, as stands that had silvicultural thinnings stored more carbon than those that had not due to the adverse effects of stand density on tree vigor.

5 Managing Afforestation/Reforestation for Carbon Sequestration

In this section, we review the carbon balance of A/R forest management in terms of the most common silvicultural treatments that forest managers employ. For each treatment, we present general information regarding how the treatment affects forest carbon balances, with more in-depth case studies on eucalyptus and acacia management. A variety of silvicultural treatments are available to improve tree growth and carbon storage in forests, as well as to minimize risk from catastrophic disturbance. Below we have summarized how silvicultural treatments can influence carbon sequestration in the context of A/R projects.

It should be emphasized that having a knowledgeable manager to oversee an A/R project is more important than any particular silvicultural practice. Growing healthy, well-formed trees through sound forestry practices will produce carbon additionality as well as merchantable timber. This point is particularly important since A/R projects often require a combination of carbon credits and timber sales in order to be economically feasible.

5.1 Site Treatments

5.1.1 Pre-planting/Site Prep

Site preparation includes a variety of operations such as stump removal, mowing, disking, excavating planting pits, ripping, subsoiling, ploughing and control of competing vegetation. Site preparation has the potential to affect carbon sequestration in three ways. First, site preparation increases the ease with which trees establish and begin growth, accelerating carbon sequestration. Tilling or cultivating the soil prior to planting of eucalyptus increases root growth, uptake of nutrients and water, and initial growth rates (de Moraes Goncalves et al. 2002). The degree to which soils are cultivated prior to planting depends on the specific structure of the soil. Second, site preparation that disturbs the soil exposes soil carbon to oxygen in the atmosphere, which increases CO_2 emissions from soil organic carbon (SOC) decomposition (Jandl et al. 2007). Finally, site preparation is one of the most energy-intensive operations associated with A/R management, and results in significant CO_2 emissions (Table 11.4). Fossil fuel emissions from site preparation can be reduced if mowing is used instead of disking in clearing/cleaning operations, and if furrowing and ridging are performed instead of ripping and subsoiling (Dias et al. 2007; Table 11.4).

Table 11.4 Carbon dioxide emissions from typical site preparation, stand tending, and infrastructure establishment operations

Operation	CO_2 specific emissions (kg CO_2 ha^{-1})
Stump removal (with digger)	324
Clearing/cleaning	
Mowing	97
Disking	128
Soil scarification	
Excavating planting pits	70
Ripping	235
Subsoiling	117
Ploughing	86
Furrowing and ridging	68
Terrace construction	689
Soil loosening (disking)	39
Selection of coppice stems (with chainsaw)	21
Precommercial thinning (with chainsaw)	27
Infrastructure establishment	
Road building	67
Road maintenance	23
Firebreak building	9
Firebreak maintenance	2

From Dias et al. (2007). Reprinted with permission

5.1.2 Fertilization

One of the limiting resources to tree growth and carbon sequestration is nutrient availability. When certain nutrients are unavailable, trees cease to grow at optimal rates. Fertilization is one silvicultural tool available to land managers to increase the biomass production of A/R projects. Applying fertilizer to a stand can increase its growth rates, and hence, the speed at which it sequesters carbon. Intelligent application of fertilizer requires knowledge of the site and the species in order to know which particular nutrient is limiting growth. As a general rule, phosphorous tends to be limiting in tropical sites while nitrogen tends to be limiting in temperate climates.

Several studies have shown that fertilizer increases aboveground carbon storage in forests (Shan et al. 2001; Sampson et al. 2006; Coyle et al. 2008; Luxmoore et al. 2008). Coyle et al. (2008) measured increases in belowground biomass from fertilization in sweetgum, while Gower et al. (1992) observed reduced litterfall, and reduced mass and production of fine roots. Other studies found that fertilization had no significant impact on soil carbon (Shan et al. 2001; Sartori et al. 2007; Luxmore et al. 2008).

It is widely accepted that fertilizer increases the rate of above and below ground biomass production; however, Markewitz (2006) raised concerns about greenhouse gas emissions associated with fertilizer application. Including fertilizer production, packaging, transport and application in forest carbon budgeting results in 1.48 t of C emissions per ton of nitrogen fertilizer application (Markewitz 2006). Therefore, while fertilizer may increase carbon sequestration in forest biomass, there are large emissions costs associated with fertilizer application. These should be considered when evaluating the net carbon balance of plantation systems.

In some cases, inter-planting of leguminous trees and ground covers (e.g. *Desmodium spp., Pueraria spp.*) can be more beneficial to productivity and aboveground carbon sequestration than fertilization (Ashton et al. 1997). For instance, in a study of reforestation on eroded pastureland in Costa Rica, inter-planted leguminous species increased productivity in a native species plantation while fertilizer had no effect (Carpenter et al. 2004). Nevertheless, the use of modest amounts of fertilizer may be necessary to improve establishment success. In a species selection trial that included over 35,000 trees, Breugel et al. (2010) attribute the impressive 2 year survival rates to the addition of fertilizer upon planting.

Insert 5. Site Preparation – Eucalyptus

In Brazil, four general soil types have been identified where eucalyptus is planted: sandy, loamy, oxidic and kaolinitic. Soil cultivation can be restricted to the planting holes in well-structured and well-drained soils (sandy or loamy soils), while more intensive site preparation is necessary on compacted or cohesive soils (kaolinitic, oxidic soils) (de Moraes Goncalves et al. 2002).

5.1.3 Irrigation

Irrigation can also enhance tree growth and aboveground carbon sequestration by providing additional water in moisture limited environments (Gower et al. 1992; Coyle et al. 2008). However, irrigation is a relatively expensive silvicultural treatment; therefore, its cost can only be justified by a high increase in productivity.

Insert 6. Irrigation – Eucalyptus

Irrigation in *E. globulus x urophylla* stands in Bahia, Brazil significantly increased plantation growth. Aboveground net primary productivity (ANPP) increased by 18% in irrigated stands in a historically wet year, and by 116% in a normal rainfall year (Stape et al. 2008). The majority of ANPP is concentrated in the bole, suggesting significant gains in timber production with irrigation. In terms of carbon, net ecosystem productivity (ANPP plus below ground NPP, litter and soil carbon fluxes) increased with irrigation from 2.3 to 2.7 kg C/m^2/year in the wet year, and from 0.8 to 2.0 kg C/m^2/year in the normal year. In terms of efficiency of carbon production, each additional 100 mm of water contributed 0.075 kg C/m^2/year in wet years and 0.125 kg C/m^2/year in dry years. This suggests that irrigation most efficiently increases net carbon sequestration in dry years.

5.1.4 Understory Elimination/ Herbicides

Understory elimination can improve biomass growth of the over-story trees but also removes biomass from the understory. In the Southeast U.S., understory-elimination and application of herbicide in pine plantations has increased aboveground carbon stores, while at the same time causing net primary production and soil carbon to decrease (Shan et al. 2001; Sarkhot et al. 2007; Sartori et al. 2007). In a 3 year study comparing growth of teak and *T. amazonia* under different herbicide and cleaning regimes of the exotic, large statured grass, *Saccharum spontaneum* in

Panama, Craven et al. (2008) found regular understory cleaning essential to reduce mortality and improve growth of the species tested. They found that only by cleaning the understory seven times a year could they match the growth obtained by annual herbicide application and cleaning two times a year. Further, while there was no statistically significant difference in growth between annual herbicide application with cleaning four times a year and annual herbicide treatment and cleaning four times a year for teak, both basal diameter and height were significantly higher for *T. amazonian* with the latter as compared to the former treatment.

5.2 Thinning

Forest thinning, or the selective removal of trees, is a silvicultural technique used to manipulate the spacing between individual trees in a forest stand and improve the growth of the remaining individuals. Thinning causes an immediate loss of carbon from the forest, unless carbon stored in wood products is considered sequestered under carbon offset policies. Multiple studies have measured the reduction in aboveground carbon from thinning, with heavier thinnings resulting in greater reductions (Balboa-Murias et al. 2006; Nilsen and Strand 2008; Campbell et al. 2009). A/R forests respond to thinning in a common fashion across various sites and species: thinning increases the biomass, thus the carbon content, of individual trees, while reducing the stand level carbon stock (Sayer et al. 2001; Eriksson 2006; Munoz et al. 2008; Campbell et al. 2009). In a *Eucalyptus nitens* stand in Los Alamos, Chile, stands thinned to 400 stems/ha contained 333 tons/ha of biomass, significantly less than the 437 tons/ha present in stands with 1,100 stems/ha (Munoz et al. 2008). Impacts of thinning on soil carbon are inconclusive, although Selig et al. (2008) measured an increase in soil carbon from thinning in southeastern U.S. loblolly pine plantations.

As stated above, carbon sequestration is often only one of many management objectives of A/R projects, and while thinning reduces stand level carbon stocks, it can positively affect other stand attributes. Thinning re-allocates growing space to the remaining individuals in the stand, improving their growth rates and quality. This results in higher quality sawtimber and generally increases the economic returns at the end of the rotation. If carbon offsets are awarded for harvested wood products or fuel switching from fossil fuels to renewable biomass energy, the carbon balance of thinning operations may become more favorable (Eriksson 2006).

The risk reduction benefits of thinning should be an important consideration in carbon storage projects. Thinning can provide protection against the risk of a major disturbance such as fire, which could cause massive carbon release from the system. Finkral and Evans (2008) found that thinning an over-stocked ponderosa pine forest in Arizona resulted in a net release of 3,114 kg C/ha in aboveground carbon (assuming no storage in wood products). In the event of a stand replacing forest fire, however, this thinned stand is predicted to release 2,410 kg C/ha less than an un-thinned stand experiencing the same intensity fire, although a stand replacing fire is much less likely in the thinned stand (also see Dore et al. 2010) .

5.3 Harvesting

Harvesting will inherently release some amount of carbon from a forest due to fossil fuels used by vehicles, soil carbon lost through respiration and erosion, and above-ground carbon lost from trees removed from the forest (although this carbon may continue to be stored for long periods of time depending upon whether the wood is being used in long-lived products). There is a growing body of literature, however, on strategies for minimizing carbon loss from forests during harvesting. See Chap. 9, this volume, for a more in-depth discussion of reduced impact logging in the tropics.

5.3.1 Rotation Length

Longer rotations can increase the total carbon stored in a forest as trees continue to add biomass (Paul et al. 2002), and will delay the point at which carbon is released during harvest. However, many A/R projects have the additional objective of producing harvestable timber. Lengthening rotations can cause tension between the dual goals of maximizing timber value and storing additional carbon. Also, lengthening rotations can increase the risk of disturbance, such as fire, if the forest is allowed to become over-stocked (Laclau 2003b).

5.3.2 Importance of Harvest Residue

Whole-tree harvesting has become more common as biofuel markets develop. Leaving residual woody debris in the forest is important for minimizing carbon loss to the system at the time of harvest (Kim et al. 2009) and also helps protect against nutrient leaching and erosion, which helps prevent loss of carbon from the system (Mendham et al. 2003). Stem-only harvesting can produce higher carbon stocks than whole-tree removal harvesting (Jones et al. 2008). Whole-tree removal in turn maintains higher carbon stocking than whole-tree removal that also removes the litter and dead woody debris on the forest floor.

6 Management and Policy Implications

6.1 Recommendations for Land Managers

- Land managers should clearly define their management objective and seek strategies that maximize other goods and services, while also taking care to eliminate unnecessary adverse effects.
- Afforestation may result in above-ground carbon additionality, but can result in adverse impacts on other ecosystem values such as water, biodiversity, and soil carbon. Managers should seek to minimize these impacts.
- Reforestation generally results in carbon additionality, and adverse impacts to other ecosystem values are less likely because the land has naturally supported forest in the past.
- Land managers should consider using nitrogen-fixing and other species that enhance nutrient cycling and litter production in mixtures to reduce fertilizer inputs and increase biomass production.
- Thinning increases the value of timber and reduces the risk of catastrophic disturbances in a stand, but reduces stand level biomass and carbon. We believe thinning should be used as a risk mitigation strategy for A/R carbon projects, despite the lower carbon stocks that will result.
- There is a growing body of research suggesting that native species plantations can be competitive with exotic species from a growth and yield perspective. Land managers should explore opportunities to implement native species silvicultural systems, due to their positive co-benefits.

6.2 Recommendations for Policy Makers

- Soil carbon is an important component of forest ecosystem carbon stocks. Excluding soil carbon from carbon legislation may result in projects that look additional on paper, but are not additional due to extensive losses of soil carbon that can occur when soil is disturbed, and with changes in hydrology.
- Solely focusing on carbon sequestration, to the exclusion of other forest values (water supply, biodiversity, nutrient depletion), may result in undesired consequences of A/R projects. A "no negative side effects" policy is important for A/R policy.
- Policy makers should consider how to incentivize the international timberland investment community to use native species. Large institutional investors and international companies are responsible for many A/R projects, and they currently do not use mixed species on a large scale.
- While A/R projects are important, primary forests hold even more carbon than A/R forests. Reducing deforestation of primary forests should be a top priority for mitigating climate change through forestry activities.

6.3 Recommendations for Further Investigation

- More research is still needed into the impacts of A/R projects on water, biodiversity and other ecological values across the many different ecosystems where A/R projects occur.
- While there are a number of studies addressing changes to soil carbon on former crop and pasture land, more research is still needed on the long-term impacts of A/R projects on soil carbon across some of the other ecosystems where A/R projects occur.
- Much less research has been done on managing native species plantations than on managing the major exotic species. More long-term research is still needed to reduce the uncertainty associated with native species plantations.

References

Allan DG, Harrison JA, Navarro RA, vanWilgen BW, Thompson MW (1997) The impact of commercial afforestation on bird populations in Mpumalanga province, South Africa – insights from bird-atlas data. Biol Conserv 79:173–185

Amoroso MM, Turnblom EC (2006) Comparing productivity of pure and mixed Douglas-fir and western hemlock plantations in the Pacific Northwest. Can J For Res 36:1484–1496

Andrade HJ, Brook R, Ibrahim M (2008) Growth, production and carbon sequestration of silvopastoral systems with native timber species in the dry lowlands of Costa Rica. Plant Soil 308:11–22

Angelsen A (2008) Moving ahead with REDD: issues, options and implications. CIFOR, Bogor

Ashton PMS, Ducey MJ (1997) The development of mixed plantations as successional analogs to natural forests. In: Landis D, South DB (eds) The national proceedings: forest and conservation nursery associations 1996. Gen. Tech. Rep. PNW – 389, U.S.D.A. Forest Service

Ashton PMS, Samarasinghe SJ, Gunatilleke IAUN, Gunatilleke CVS (1997) Role of legumes in release of successionally arrested grasslands in the central hills, Sri Lanka. Restor Ecol 5:36–43

Ashton PMS, Gamage S, Gunatilleke I, Gunatilleke CVS (1998) Using Caribbean pine to establish a mixed plantation: testing effects of pine canopy removal on plantings of rain forest tree species. For Ecol Manag 106:211–222

Ashton MS, Gunatilleke CVS, Singhakumara BMP, Gunatilleke I (2001) Restoration pathways for rain forest in southwest Sri Lanka: a review of concepts and models. For Ecol Manag 154:409–430

Asner GP, Powell GVN, Mascaro J, Knapp DE, Clark JK, Jacobson J, Kennedy-Bowdoin T, Balaji A, Paez-Acosta G, Vivtoria E, Secada L, Valqui M, Hughes RF (2010) High-resolution forest carbon stocks and emissions in the Amazon. Proc Natl Acad Sci USA. 107:16738–16742 www.pnas.org/cgi/doi/10.1073/pnas.1004875107

Avery TE, Burkhart HE (1994) Forest measurements. McGraw-Hill, New York

Bala G, Caldeira K, Wickett M, Phillips TJ, Lobell DB, Delire C, Mirin A (2007) Combined climate and carbon-cycle effects of large -scale deforestation. Proc Natl Acad Sci USA 104:6550–6555

Balboa-Murias MA, Rodriguez-Soalleiro R, Merino A, Alvarez-Gonzalez JG (2006) Temporal variations and distribution of carbon stocks in aboveground biomass of radiata pine and maritime pine pure stands under different silvicultural alternatives. For Ecol Manag 237:29–38

Balieiro FD, Pereira MG, Alves BJR, de Resende AS, Franco AA (2008) Soil carbon and nitrogen in pasture soil reforested with eucalyptus and guachapele. Revista Brasileira De Ciencia Do Solo 32:1253–1260

Barlow J, Mestre LAM, Gardner TA, Peres CA (2007a) The value of primary, secondary and plantation forests for Amazonian birds. Biol Conserv 136:212–231

Barlow J, Overal WL, Araujo IS, Gardner TA, Peres CA (2007b) The value of primary, secondary and plantation forests for fruit-feeding butterflies in the Brazilian Amazon. J Appl Ecol 44:1001–1012

Barry KM, Irianto RSB, Santoso E, Turjaman M, Widyati E, Sitepu I, Mohammed CL (2004) Incidence of heartrot in harvest-age *Acacia mangium* in Indonesia, using a rapid survey method. For Ecol Management 190:273–280

Benayas JMR, Newton AC, Diaz A, Bullock JM (2009) Enhancement of biodiversity and ecosystem services by ecological restoration: a meta-analysis. Science 325:1121–1124

Bennett EM, Peterson GD, Gorden LJ (2009) Understanding relationships among multiple ecosystem services. Ecol Lett 12:1394–1404

Binkley D, Dunkin KA, Debell D, Ryan MG (1992) Production and nutrient cycling in mixed plantations of eucalyptus and albizia in Hawaii. For Sci 38:393–408

Binkley D, Senock R, Bird S, Cole TG (2003) Twenty years of stand development in pure and mixed stands of *Eucalyptus saligna* and nitrogen-fixing *Facaltaria moluccana*. For Ecol Manag 182:93–102

Boley JD, Drew AP, Andrus RE (2009) Effects of active pasture, teak (*Tectona grandis*) and mixed native plantations on soil chemistry in Costa Rica. For Ecol Manag 257:2254–2261

Boyden S, Binkley D, Senock R (2005) Competition and facilitation between Eucalyptus and nitrogen-fixing Falcataria in relation to soil fertility. Ecology 86:992–1001

Bradshaw CJA, Sodhi NS, Peh KSH, Brook BW (2007) Global evidence that deforestation amplifies flood risk and severity in the developing world. Glob Change Biol 13:2379–2395

Brockerhoff EG, Jactel H, Parrotta JA, Quine CP, Sayer J (2008) Plantation forests and biodiversity: oxymoron or opportunity? Biodivers Conserv 17:925–951

Broekhoff D (2008) Creating jobs with climate solutions: how agriculture and forestry can help lower costs in a low-carbon economy. Testimony Before Senate Subcommittee on Rural Revitalization, Conservation, Forestry, and Credit of the United States Senate Committee on Agriculture, Nutrition, and Forestry. 21 May 2008

Bruijnzeel LA (1989) (De)forestation and dry season flow in the tropics: a closer look. J Trop For Sci 1:145–161

Bruijnzeel LA (2004) Hydrological functions of tropical forests: not seeing the soil for the trees? Agric Ecosyst Environ 104:185–228

Bruijnzeel LA, Gilmour DA, Bonell M, Lamb D (2005) Conclusion – forests, water and people in the humid tropics: an emerging view. In: Forests, water and people in the humid tropics. Cambridge University Press, Cambridge/New York, pp 906–925

Byrne KA, Farrell EP (2005) The effect of afforestation on soil carbon dioxide emissions in blanket peatland in Ireland. Forestry 78:217–227

Byrne KA, Milne R (2006) Carbon stocks and sequestration in plantation forests in the Republic of Ireland. Forestry 79:361–369

Campbell J, Alberti G, Martin J, Law BE (2009) Carbon dynamics of a ponderosa pine plantation following a thinning treatment in the northern Sierra Nevada. For Ecol Manag 257:453–463

Canadell JG, Raupach MR (2008) Managing forests for climate change mitigation. Science 320:1456–1457

Cannell MGR (1999) Growing trees to sequester carbon in the UK: answers to some common questions. Forestry 72:237–247

Cannell MGR, Dewar RC, Pyatt DG (1993) Conifer plantations on drained peatlands in Britain – a net gain or loss of carbon. Forestry 66:353–369

Carnevale NJ, Montagnini F (2002) Facilitating regeneration of secondary forests with the use of mixed and pure plantations of indigenous tree species. For Ecol Manag 163:217–227

Carnus JM, Parrotta J, Brockerhoff E, Arbez M, Jactel H, Kremer A, Lamb D, O'Hara K, Walters B (2006) Planted forests and biodiversity. J For 104:65–77

Carpenter FL, Nichols JD, Sandi E (2004) Early growth of native and exotic trees planted on degraded tropical pasture. For Ecol Manag 196:367–378

Chazdon RL (2008) Beyond deforestation: restoring forests and ecosystem services on degraded lands. Science 320:1458–1460

Chen HYH, Klinka K, Mathey AH, Wang X, Varga P, Chourmouzis C (2003) Are mixed-species stands more productive than single-species stands: an empirical test of three forest types in British Columbia and Alberta. Can J For Res 33:1227–1237

Chisholm RA (2010) Trade-offs between ecosystem services: water and carbon in a biodiversity hotspot. Ecol Econ 69:1973–1987

Cochrane MA (2002) Spreading like wildfier: tropical forest fires in Latin America and the Caribbean. United Nations Environment Programme, 97 pp

Condit R, Robinson WD, Ibanez R, Aguilar S, Sanjur A, Martinez R, Stallard RF, Garcia T, Angehr GR, Petit L, Wright SJ, Robinson TR, Heckadon S (2001) The status of the Panama Canal watershed and its biodiversity at the beginning of the 21st century. Bioscience 51:389–398

Coyle DR, Coleman MD, Aubrey DP (2008) Above- and below-ground biomass accumulation, production, and distribution of sweetgum and loblolly pine grown with irrigation and fertilization. Can J For Res 38:1335–1348

Craven D, Hall J, Verjans JM (2008) Impacts of herbicide application and mechanical cleanings on growth and mortality of two timber species in Saccharum spontaneum grasslands of the Panama Canal Watershed. Restor Ecol. doi:10.1111/j.1526-100X.2008.00408.x

Cusack D, Montagnini F (2004) The role of native species plantations in recovery of understory woody diversity in degraded pasturelands of Costa Rica. For Ecol Manag 188:1–15

Daehler CC (1998) The taxonomic distribution of invasive angiosperm plants: ecological insights and comparison to agricultural weeds. Biol Conserv 84:167–180

de Moraes Goncalves JL, Stape JL, Laclau JP, Smethurst P, Gava JL (2002) Silvicultural effects on the productivity and wood quality of eucalypt plantations. In: IUFRO international conference on eucalypt productivity, Hobart, pp 45–61

DeBell DS, Cole TG, Whitesell CD (1997) Growth, development, and yield in pure and mixed stands of eucalyptus and albizia. For Sci 43:286–298

Dias AC, Arroja L, Capela I (2007) Carbon dioxide emissions from forest operations in Portuguese eucalypt and maritime pine stands. Scand J For Res 22:422–432

Dixon RK, Brown S, Houghton RA, Solomon AM, Trexler MC, Wisniewski J (1994) Carbon pools and flux of global forest ecosystems. Science 263:185–190

Dore S, Kolb TE, Montes-Helu M, Eckert SE, Sullivan BW, Hungate BA, Kave JP, Hart SC, Koch GW, Finkral A (2010) Carbon and water fluxes from ponderosa pine forests disturbed by wildfier and thinning. Ecol Appl 20:663–683

du Toit B, Smith CW, Little KM, Boreham G, Pallett RN (2010) Intensive, site-specific silviculture: manipulating ersource availability at establishment for improved stand productivity: a review of South African research. For Ecol Manag 259:1836–1845

Elliott S, Navakitbumrung P, Kuarak C, Zangkum S, Anusarnsunthorn V, Blakesley D (2003) Selecting framework tree species for restoring seasonally dry tropical forests in northern Thailand based on field performance. For Ecol Manag 184:177–191

Engel V, Jobbagy EG, Stieglitz M, Williams M, Jackson RB (2005) Hydrological consequences of eucalyptus afforestation in the Argentine pampas. Water Resour Res 41:14

Eriksson E (2006) Thinning operations and their impact on biomass production in stands of Norway spruce and Scots pine. Biomass Bioenergy 30:848–854

Erskine PD, Lamb D, Bristow M (2006) Tree species diversity and ecosystem function: can tropical multi-species plantations generate greater productivity? For Ecol Manag 233:205–210

Evans BEI, Ashley J, Marsden SJ (2005) Abundance, habitat use, and movements of blue-winged Macaws (*Primolius maracana*) and other parrots in and around an Atlantic Forest Reserve. Wilson Bull 117:154–164

FAO (2000) Global forest resources assessment 2000. Forestry and Agricultural Organization, Rome

FAO (2006a) Global forest resources assessment 2005: progress towards sustainable forest management. FAO Forestry Paper 147

FAO (2006b) Global planted forests thematic study: results and analysis. Working Paper FP 38E

Farley KA, Kelly EF, Hofstede RGM (2004) Soil organic carbon and water retention following conversion of grasslands to pine plantations in the Ecuadoran Andes. Ecosystems 7:729–739

Farley KA, Jobbagy EG, Jackson RB (2005) Effects of afforestation on water yield: a global synthesis with implications for policy. Glob Change Biol 11:1565–1576

Finkral AJ, Evans AM (2008) Effects of a thinning treatment on carbon stocks in a northern Arizona ponderosa pine forest. For Ecol Manag 255:2743–2750

Forrester DI, Bauhus J, Cowie AL, Vanclay JK (2005) Mixed-species plantations of Eucalyptus with nitrogen-fixing trees: a review. In: Workshop on improving productivity in mixed-species plantations, Ballina, pp 211–230

Forrester DI, Bauhus J, Cowie AL, Vanclay JK (2006) Mixed-species plantations of eucalyptus with nitrogen-fixing trees: a review. For Ecol Manag 253:211–230

Forrester DI, Medhurst JL, Wood M, Beadle CL, Valencia JC (2010) Growth and physiological responses to silviculture for producing solid-wood products from *Eucalyptus* plantations: an Australian perspective. For Ecol Manag 259:1819–1835

Garber SM, Maguire DA (2004) Stand productivity and development in two mixed-species spacing trials in the central Oregon cascades. For Sci 50:92–105

Garcia-Quijano JF, Peters J, Cockx L, van Wyk G, Rosanov A, Deckmyn G, Ceulemans R, Ward SM, Holden NM, Van Orshoven J, Muys B (2007) Carbon sequestration and environmental effects of afforestation with *Pinus radiata* D. Don in the Western Cape, South Africa. Clim Change 83:323–355

Garen EJ, Saltonstall K, Ashton MS, Slusser JL, Mathias S, Hall JS (2011) The tree planting and protecting culture of cattle ranchers and small-scale agriculturalists in rural Panama: opportunities for reforestation and land restoration. For Ecol Manag 261(10):1684–1695

Gibbons P, Lindenmayer DB, Fischer J, Manning AD, Weinberg A, Seddon J, Ryan P, Barrett G (2008) The future of scattered trees in agricultural landscapes. Conserv Biol 22:1309–1319

Gibbs HK, Brown S, Niles JO, Foley JA (2007) Monitoring and estimating tropical forest carbon stocks: making REDD a reality. Environ Res Lett 2:1–13

Gower ST, Vogt KA, Grier CC (1992) Carbon dynamics of rocky-mountain douglas-fir – influence of water and nutrient availability. Ecol Monogr 62:43–65

Greenberg R, Bichier P, Sterling J (1997) Bird populations in rustic and planted shade coffee plantations of Eastern Chiapas, Mexico. Biotropica 29:501–514

Guo LB, Gifford RM (2002) Soil carbon stocks and land use change: a meta analysis. Glob Change Biol 8:345–360

Haile SG, Nair PKR, Nair VD (2008) Carbon storage of different soil-size fractions in Florida silvopastoral systems. J Environ Qual 37:1789–1797

Hall JS, Love BE, Garen EJ, Slusser JL, Saltonstall K, Mathias S, van Breugel M, Ibarra D, Bork EW, Spaner D, Wishnie MH, Ashton MS (2010) Tree plantations on farms: evaluating growth and potential for success. For Ecol Manag. doi:10.1016/j.foreco.2010.09.042

Hargreaves KJ, Milne R, Cannell MGR (2003) Carbon balance of afforested peatland in Scotland. Forestry 76:299–317

Harvey CA, Tucker NIJ, Estrada A (2004) Live fences, isolated trees, and windbreaks: tools for conserving biodiversity in fragmented tropical landscapes. In: Schroth G, de Fonseca GAB, Harvey CA, Gascon C, Vasconcelos HL, Izac AMN (eds) Agroforestry and biodiversity conservation in tropical landscapes. Island Press, Washington

Harvey CA, Medina A, Sánchez D, Vílchez S, Hernández B, Saénz J, Maes JM, Casanoves F, Sinclair FL (2006) Patterns of animal diversity associated with different forms of tree cover retained in agricultural landscapes. Ecol Appl 16:1986–1999

Harvey CA, Komar O, Chazdon R, Ferguson BG, Finegan B, Griffith D, Martinez-Ramos M, Morales H, Nigh R, Soto-Pinto L, van Breugel M, Wishnie M (2008) Integrating agricultural landscapes with biodiversity conservation in the Mesoamerican hotspot: opportunities and an action Agenda. Conserv Biol 22:8–15

Hassler SK, Zimmermann B, Breugel MV, Hall JS, Elsenbeer H (2010) Recovery of saturated conductivity under secondary succession on former pasture in the humid tropics. For Ecol Manag. doi:10.1016/j.foreco.2010.06.031

Hassler SK, Zimmermann B, van Breugel M, Hall JS, Elsenbeer H (2011) Recovery of saturated hydraulic conductivity under secondary succession on former pasture in the humid tropics. For Ecol Manag 261(10): 1634–1642

Hawes J, Barlow J, Gardner TA, Peres CA (2008) The value of forest strips for understorey birds in an Amazonian plantation landscape. Biol Conserv 141:2262–2278

Hirano T, Segah H, Harada T, Limin S, June T, Hirata R, Osaki M (2007) Carbon dioxide balance of a tropical peat swamp forest in Kalimantan, Indonesia. Glob Change Biol 13:412–425

Holl KD, Aide TM (2010) When and where to actively restore ecosystems? For Ecol Manag 261:1675–1683

Hooper E, Condit R, Legendre P (2002) Responses of 20 native tree species to reforestation strategies for abandoned farmland in Panama. Ecol Appl 12:1626–1641

Hu YL, Zeng DH, Fan ZP, Chen GS, Zhao Q, Pepper D (2008) Changes in ecosystem carbon stocks following grassland afforestation of semiarid sandy soil in the southeastern Keerqin Sandy Lands, China. J Arid Environ 72:2193–2200

Ibanez R, Condit R, Angehr G, Aguilar S, Garcia T, Martinez R, Sanjur A, Stallard R, Wright SJ, Rand AS, Heckadon S (2002) An ecosystem report on the Panama canal: monitoring the status of the forest communities and the watershed. Environ Monit Assess 80:65–95

Ilstedt U, Malmer A, Verbeeten E, Murdiyarso D (2007) The effect of afforestation on water infiltration in the tropics: a systematic review and meta-analysis. For Ecol Manag 251:45–51

IPCC (2007) Pachauri RK, Reisinger A (eds) Climate change 2007: synthesis report. Contribution of working groups I, II and III to the fourth assessment report of the intergovernmental panel on climate change. IPCC, Geneva, 104 p

Irianto RSB, Barry K, Hidayati N, Ito S, Fiani A, Rimbawanto A, Mohammed C (2006) Incidence and spatial analysis of root rot of *Acacia mangium* in Indonesia. J Trop For Sci 18:157–165

Jactel H, Brockerhoff E, Duelli SP (2005) A test of the biodiversity-stability theory: Meta-analysis of tree species diversity effects on an insect pest infestation and re-examination of responsible factors. Ecol Studies 176:235–262

Jaenicke J, Rieley JO, Mott C, Kimman P, Siegert F (2008) Determination of the amount of carbon stored in Indonesian peatlands. Geoderma 147:151–158

Jandl R, Lindner M, Vesterdal L, Bauwens B, Baritz R, Hagedorn F, Johnson DW, Minkkinen K, Byrne KA (2007) How strongly can forest management influence soil carbon sequestration? Geoderma 137:253–268

Jobbagy EG, Jackson RB (2003) Patterns and mechanisms of soil acidification in the conversion of grasslands to forests. Biogeochemistry 64:205–229

Jobbagy EG, Jackson RB (2004) Groundwater use and salinization with grassland afforestation. Glob Change Biol 10:1299–1312

Jones ER, Wishnie MH, Deago J, Sautu A, Cerezo A (2004) Facilitating natural regeneration in *Saccharum spontaneum* (L.) grasslands within the Panama Canal watershed: effects of tree species and tree structure on vegetation recruitment patterns. For Ecol Manag 191:171–183

Jones HS, Garrett LG, Beets PN, Kimberley MO, Oliver GR (2008) Impacts of harvest residue management on soil carbon stocks in a plantation forest. Soil Sci Soc Am J 72:1621–1627

Kelty MJ (2005) The role of species mixtures in plantation forestry. In: Workshop on improving productivity in mixed-species plantations, Ballina, pp 195–204

Kelty MJ (2006) The role of species mixtures in plantation forestry. For Ecol manag 233:195–204

Kim C, Son Y, Lee WK, Jeong J, Noh NJ (2009) Influences of forest tending works on carbon distribution and cycling in a *Pinus densiflora* S. et Z. stand in Korea. For Ecol Manag 257:1420–1426

Kirby KR, Potvin C (2007) Variation in carbon storage tree species: Implications for the management of a small-scale carbon sink project. For Ecol Manag 246:208–221

Kraenzel M, Castillo A, Moore T, Potvin C (2003) Carbon storage of harvest-age teak (*Tecona grandis*) plantations in Panama. For Ecol Manag 173:213–255

Laclau P (2003a) Biomass and carbon sequestration of ponderosa pine plantations and native cypress forests in northwest Patagonia. For Ecol Manag 180:317–333

Laclau P (2003b) Root biomass and carbon storage of ponderosa pine in a northwest Patagonia plantation. For Ecol Manag 173:353–360

Lamb D, Erskine PD, Parrotta JA (2005) Restoration of degraded tropical forest landscapes. Science 310:1628–1632

Licata JA, Gyenge JE, Fernandez ME, Schlichter TA, Bond BJ (2008) Increased water use by ponderosa pine plantations in northwestern Patagonia, Argentina compared with native forest vegetation. For Ecol Manag 255:753–764

Luxmoore RJ, Tharp ML, Post WM (2008) Simulated biomass and soil carbon of loblolly pine and cottonwood plantations across a thermal gradient in southeastern United States. For Ecol Manag 254:291–299

Mackensen J, Folster H (2000) Cost-analysis for a sustainable nutrient management of fast growing-tree plantations in East-Kalimantan, Indonesia. For Ecol Manag 131:239–253

Maeto K, Noerdjito W, Belokobylskij S, Fukuyama K (2009) Recovery of species diversity and composition of braconid parasitic wasps after reforestation of degraded grasslands in lowland East Kalimantan. J Insect Conserv 13:245–257

Malhi Y, Baldocchi DD, Jarvis PG (1999) The carbon balance of tropical, temperate and boreal forests. Plant Cell Environ 22:715–740

Malmer A, Murdiyarso D, Bruijnzeel LA, Ilstedt U (2010) Carbon sequestration in tropical forests and water: a critical look at the basis for commonly used generalisations. Glob Change Biol 16:599–604

Markewitz D (2006) Fossil fuel carbon emissions from silviculture: impacts on net carbon sequestration in forests. For Ecol Manag 236:153–161

Mather AS, Needle CL (1998) The forest transition: a theoretical basis. Area 30:117–124

Mendham DS, O'Connell AM, Grove TS, Rance SJ (2003) Residue management effects on soil carbon and nutrient contents and growth of second rotation eucalypts. For Ecol Manag 181:357–372

Montagnini F, Nair PKR (2004) Carbon sequestration: an underexploited environmental benefit of agroforestry systems. In: 1st World congress of agroforestry, Orlando, pp 281–295

Montagnini F, Porras C (1998) Evaluating the role of plantations as carbon sinks: an example of an integrative approach from the humid tropics. Environ Manage 22:459–470

Munoz F, Rubilar R, Espinosa A, Cancino J, Toro J, Herrera A (2008) The effect of pruning and thinning on above ground aerial biomass of *Eucalyptus nitens* (Deane & Maiden) Maiden. For Ecol Manag 255:365–373

Murgueitio E, Calle Z, Uribe F, Calle A, Solorio B (2010) Native trees and shrubs for the productive rehabilitation of tropical cattle ranching lands. For Ecol Manag 261:1654–1663

Nair PKR, Kumar BM, Nair VD (2009) Agroforestry as a strategy for carbon sequestration. Journal of Plant Nutrition and Soil Science-Zeitschrift Fur Pflanzenernahrung Und Bodenkunde 172:10–23

Neumann-Cosel L, Zimmermann B, Hall JS, van Breugel M, Elsenbeer H (2010) Soil carbon dynamics under young tropical secondary forests on former pastures- a case study from Panama. For Ecol Manag. doi:10.1016/j.foreco.2010.07.023

Neumann-Cosel L, Zimmermann B, Hall JS, van Breugel M, Elsenbeer H (2011) Soil carbon dynamics under young tropical secondary forests on former pastures—A case study from Panama. For Ecol Manag 261(10):1625–1633

Nichols JD, Bristow M, Vanclay JK (2006) Mixed species plantations: prospects and challenges. For Ecol Manag 233:383–390

Nilsen P, Strand LT (2008) Thinning intensity effects on carbon and nitrogen stores and fluxes in a Norway spruce (*Picea abies* (L.) Karst.) stand after 33 years. For Ecol Manag 256:201–208

Nosetto MD, Jobbagy EG, Paruelo JM (2006) Carbon sequestration in semi-arid rangelands: comparison of Pinus ponderosa plantations and grazing exclusion in NW Patagonia. J Arid Environ 67:142–156

Otsamo A, Adjers G, HadiTjuk S, Kuusipalo J, Vuokko R (1997) Evaluationof reforestation potential of 83 tree species plante don Imperata cylindrica domi-nated grassland. New For 14:127–143

Park A, van Breugel M, Ashton MS, Wishnie MH, Mariscal E, Deago J, Ibarra D, Cedeno N, Hall JS (2010) Local and regional environmental variation influences the growth of tropical trees in selection trials in the Republic of Panama. For Ecol Manag 260:12–21

Parrotta JA (1990) The role of plantation forests in rehabilitating degraded tropical ecosystems. In: Symp on the application of ecological principles to sustainable land-use systems at the 5th international congress of ecology, Yokohama, pp 115–133

Parrotta JA (1991) Secondary forest regeneration on degraded tropical lands – the role of plantations as foster ecosystems. In: Lieth H, Lohmann M (eds) Symposium on restoration of tropical forest ecosystems, Bonn, pp 63–73

Parrotta JA (1999) Productivity, nutrient cycling, and succession in single- and mixed-species plantations of casuarina equisetifolia, eucalyptus robusta, and leucaena leucocephala in puerto Rica. For Ecol manag 124:45–77

Paul KI, Polglase PJ, Nyakuengama JG, Khanna PK (2002) Change in soil carbon following afforestation. For Ecol Manag 168:241–257

Pautasso M, Holdenrieder O, Stenlid J (2005) Susceptibility to fungal pathogens of forests differing in tree diversity. In: Scherer-Lorenzen M, Körner C, Schulze ED (eds) Forest diversity and function: temperate and boreal systems, vol 176, Ecological Studies. Springer, Berlin/Heidelberg, pp 263–289

Piotto D, Craven D, Montagnini F, Alice F (2010) Silvicultural and economic aspects of pure and mixed native tree species plantations on degraded pasturelands in humid Costa Rica. New For 39:369–385

Piotto D, Montagnini F, Thomas W, Ashton M, Oliver C (2009) Forest recovery after swidden cultivation across a 40-year chronosequence in the Atlantic forest of southern Bahia, Brazil. Plant Eco 205:261–272

Piotto D, Viquez E, Montagnini F, Kanninen M (2004) Pure and mixed forest plantations with native species of the dry tropics of Costa Rica: a comparison of growth and productivity. For Ecol Manag 190:359–372

Plantinga AJ, Wu JJ (2003) Co-benefits from carbon sequestration in forests: evaluating reductions in agricultural externalities from an afforestation policy in Wisconsin. Land Econ 79:74–85

Raudsepp-Hearne C, Peterson GD, Bennett EM (2010) Ecosystem service bundles for analyzing tradeoffs in diverse landscapes. Proc Natl Acad Sci USA 107:5242–5247

Redondo-Brenes A (2007) Growth, carbon sequestration, and management of native tree plantations in humid regions of Costa Rica. New For 34:253–268

Redondo-Brenes A, Montagnini F (2006) Growth, productivity, aboveground biomass, and carbon sequestration of pure and mixed native tree plantations in the Caribbean lowlands of Costa Rica. For Ecol Manag 232:168–178

Rodrigues RR, Lima RAF, Gandolfi S, Nave AG (2009) On the restoration of high diversity forests: 30 years of experience in the Brazilian Atlantic Forest. Biol Conserv 142:1242–1251

Rodrigues RR, Gandolfi S, Nave AG, Aronson J, Barreto TE, Vidal CY, Brancalion PHS (2010) Large-scale ecological restoration of high diversity tropical forests in SE Brazil. For Ecol Manag. doi:10.1016/j.foreco.2010.07.005

Rudel TK, Bates D, Machinguiashi R (2002) A tropical forest transition? Agricultural change, out-migration, and secondary forests in the Ecuadorian Amazon. Ann Assoc Am Geogr 92:87–102

Rudel TK, Coomes OT, Moran E, Achard F, Angelsen A, Xu JC, Lambin E (2005) Forest transitions: towards a global understanding of land use change. Glob Environ Change-Hum Policy Dimens 15:23–31

Rydin H, Jeglum JK (2006) The biology of peatlands. Oxford University Press, Oxford

Saharjo BH, Watanabe H (2000) Estimation of litter fall and seed production of Acacia mangium in a forest plantation in South Sumatra, Indonesia. For Ecol Manag 130:265–268

Sampson DA, Waring RH, Maier CA, Gough CM, Ducey MJ, Johnsen KH (2006) Fertilization effects on forest carbon storage and exchange, and net primary production: a new hybrid process model for stand management. For Ecol Manag 221:91–109

Sarkhot DV, Comerford NB, Jokela EJ, Reeves JB, Harris WG (2007) Aggregation and aggregate carbon in a forested southeastern coastal plain spodosol. Soil Sci Soc Am J 71:1779–1787

Sartori F, Markewitz D, Borders BE (2007) Soil carbon storage and nitrogen and phosphorous availability in loblolly pine plantations over 4 to 16 years of herbicide and fertilizer treatments. Biogeochemistry 84:13–30

Sayer MAS, Goelz JCG, Chambers JL, Tang Z, Dean TJ, Haywood JD, Leduc DJ (2001) Long-term trends in loblolly pine productivity and stand characteristics in response to thinning and fertilization in the West Gulf region. In: 11th Biennial southern sivlicultural research conference on long-term production dynamics of lablolly pine stands in the Southern United States, Knoxville, pp 71–96

Schoeneberger MM (2005) Agroforestry: working trees for sequestering carbon on agricultural lands. In: 9th North American agroforestry conference, Rochester, pp 27–37

Schroth G, Harvey CA, Vincent G (2004) Complex agroforests: their structure, diversity, and potential role in landscape conservation. In: Schroth G, de Fonseca GAB, Harvey CA, Gascon C, Vasconcelos HL, Izac AMN (eds) Agroforestry and biodiversity conservation in tropical landscapes. Island Press, Washington, pp 227–260

Scott DF, Lesch W (1997) Streamflow responses to afforestation with *Eucalyptus grandis* and *Pinus patula* and to felling in the Mokobulaan experimental catchments, South Africa. J Hydrol 199:360–377

Scott DF, Bruijnzeel LA, Mackensen J (2005) The hydrological and soil impacts of forestation in the tropics. In: Bonell M, Bruijnzeel LA (eds) Forest–water–people in the humid tropics. Cambridge University Press, Cambridge, UK, pp 622–651

Selig MF, Seiler JR, Tyree MC (2008) Soil carbon and CO_2 efflux as influenced by the thinning of loblolly pine (*Pinus taeda* L.) plantations on the piedmont of Virginia. For Sci 54:58–66

Shan JP, Morris LA, Hendrick RL (2001) The effects of management on soil and plant carbon sequestration in slash pine plantations. J Appl Ecol 38:932–941

Sharrow SH, Ismail S (2004) Carbon and nitrogen storage in agroforests, tree plantations, and pastures in western Oregon, USA. Agrofor Syst 60:123–130

Shono K, Davies SJ, Chua YK (2007) Performance of 45 native tree species on degraded lands in Singapore. J Trop For Sci 19:25–34

Siddique I, Engel VL, Parrotta JA, Lamb D, Nardoto GB, Ometto JPHB, Martinelli LA, Schmidt S (2008) Dominance of legume trees alters nutrient relations in mixed species forest restoration plantings within seven years. Biogeochemistry 88:89–101

Sloan S (2008) Reforestation amidst deforestation: simultaneity and succession. Glob Environ Change-Hum Policy Dimens 18:425–441

Stallard RF, Ogden FL, Elsenbeer H, Hall JS (2010) The Panama Canal watershed experiment: Agua Salud Project. Water Resour Impact 12:17–20

Stallard RF, Ogden FL, Elsenbeer H, Hall JS (2010) The Panama Canal Watershed Experiment: Agua Salud Project. Water Resources Research 12(4):17–20.

Stape JL, Binkley D, Ryan MG (2008) Production and carbon allocation in a clonal eucalyptus plantation with water and nutrient manipulations. For Ecol Manag 255:920–930

Streed E, Nichols JD, Gallatin K (2006) A financial analysis of small-scale tropical reforestation with native species in Costa Rica. J For 104:276–282

Takimoto A, Nair PKR, Nair VD (2008) Carbon stock and sequestration potential of traditional and improved agroforestry systems in the West African Sahel. Agric Ecosyst Environ 125:159–166

Thomas SC, Malezewski G, Saprunoff M (2007) Assessing the potential of native tree species for carbon sequestration forestry in Northeast China. J Environ Manage 85:663–671

Tucker NIJ, Murphy TM (1997) The effects of ecological rehabilitation on vegetation recruitment: some observations from the wet tropics of North Queensland. For Ecol Manag 99:133–152

UNFCCC (2009) Clean development mechanism project website. http://cdm.unfccc.int/Projects/projsearch.html

van Bael SA, Bichier P, Ochoa I, Greenberg R (2007) Bird diversity in cacao farms and forest fragments of western Panama. Biodivers Conserv 16:2245–2256

van Breugel M, Hall JJ, Craven DJ, Gregoire TG, Park A, Dent DH, Wishnie MH, Mariscal E, Deago J, Ibarra D, Cedeno N, Ashton MS (2010) Early growth and survival of 49 tropical tree species across sitesdiffering soil fertility and rainfall in Panama. For Ecol Manag. doi:10.1016/j.foreco.2010.08.019

van Breugel M, Hall JS, Craven DJ, Gregoire TG, Park A, Dent DH, Wishnie MH, Mariscal E, Deago J, Ibarra D, Cedeño N, Ashton MS (2011) Early growth and survival of 49 tropical tree species across sites differing in soil fertility and rainfall in Panama. For Ecol Manag 261(10):1580–1589

van der Kamp J, Yassir I, Buurman P (2009) Soil carbon changes upon secondary succession in *Imperata* grasslands (East Kalimantan, Indonesia). Geoderma 149:76–83

Wang YH, Yu PT, Xiong W, Shen ZX, Guo MC, Shi ZJ, Du A, Wang LM (2008) Water-yield reduction after afforestation and related processes in the semiarid Liupan Mountains, Northwest China. J Am Water Resour Assoc 44:1086–1097

Wilk J, Hughes DA (2002) Simulating the impacts of land-use and climate change on water resource availability for a large south Indian catchment. Hydrol Sci J 47:19–30

Wishnie MH, Dent DH, Mariscal E, Deago J, Cedeno N, Ibarra D, Condit R, Ashton PMS (2007) Initial performance and reforestation potential of 24 tropical tree species planted across a precipitation gradient in the Republic of Panama. For Ecol Manag 243:39–49

Zamora CO, Montagnini F (2007) Seed rain and seed dispersal agents in pure and mixed plantations of native trees and abandoned pastures at La Selva Biological Station, Costa Rica. Restor Ecol 15:453–461

Zimmermann B, Papritz AJ, Elsenbeer H (2010) Asymmetric response to disturbance and recovery: changes of soil permeability under forest-pasture-forest transitions. Geoderma. doi:10.1016/j.geoderma.2010.07.013

Role of Forest Products in the Global Carbon Cycle: From the Forest to Final Disposal

12

Christopher Larson, Jeffrey Chatellier, Reid Lifset, and Thomas Graedel

Executive Summary

This chapter reviews the role of the production, use and end-of-life management of harvested wood products (HWPs) in the global carbon cycle. Harvested wood products can be long term reservoirs of carbon; however, solid wood products, paper, and paperboard manufacturing require large energy and heat inputs, and end-of-life pathways can further or hinder carbon sequestration, depending on management.

What We Know About Harvested Wood Products and the Carbon Cycle

- The global stock of carbon within forest products is estimated between 4,100 teragrams (Tg) carbon and 20,000 Tg carbon, with net sink rates estimated between 26 and 139 Tg carbon per year. The methods and assumptions used to estimate the role of HWPs in the global carbon cycle vary, resulting in a wide range of findings. Even assuming the high end of the estimates, these findings suggest that forest products are still are a minor component of the global carbon budget.
- Manufacturing processes operate on a mix of fossil energy and biomass energy, a by-product derived from wood waste. Emission reductions are achieved when energy generated from biomass displaces fossil fuel emissions.
- Newer wood products such as oriented strand board, laminated veneer lumber and I-joists use 80–216% of the energy needed to produce solid sawn lumber. It is unclear whether the lower density of newer wood product materials, given their increased strength and greater utilization of wood resources, offsets the energy intensity per unit of the newer materials.
- Paper products contain significantly more embodied fossil fuel (carbon) energy than solid wood products: 0.3–0.6 megagram carbon (MgC) in fossil energy used/MgC for virgin paper products vs. 0.07 MgC in fossil energy used /MgC for solid wood products in Finland in 1995. However, approximately 50% of U.S. paper production is manufactured using recycled paper as a feedstock. Recycled feedstock may reduce or increase GHG emissions relative to virgin pulping depending on the pulping process and energy sources.
- Global transport of wood and paper products is estimated to account for approximately 27% of total fossil carbon emitted within the manufacturing and distribution process.
- Several researchers assert that substitution of wood for other construction materials (e.g., steel and concrete) produces net GHG emissions

C. Larson
New Island Capital, 505 Sansone St, San Francisco, CA 94111

J. Chatellier
Forest Carbon Ltd, Bali, Indonesia

R. Lifset (✉) • T. Graedel
Yale School of Forestry & Environmental Studies, 360 Prospect St, New Haven, CT, USA
e-mail: reid.lifset@yale.edu

M.S. Ashton et al. (eds.), *Managing Forest Carbon in a Changing Climate*,
DOI 10.1007/978-94-007-2232-3_12, © Springer Science+Business Media B.V. 2012

reductions. These substitution effects may be up to 11 times larger than the total amount of carbon sequestered in harvested wood products annually. Quantification of substitution effects relies on many assumptions about particular counterfactual scenarios, most importantly linkages between increased/decreased forest products consumption and total extent of forestland.
- The end-of-life pathways of HWPs can augment GHG emissions reductions. Once discarded, HWPs can be burned for energy production, recycled or reused, or put in landfills, where the carbon can remain indefinitely due to anaerobic conditions. However, HWPs discarded in landfills create methane, a greenhouse gas that is 24 times more potent than CO_2, thus potentially offsetting gains from carbon storage.
- Inclusion of end-of-life pathways in HWP carbon stock calculation models is crucial, as failure to do leads to estimates with a high degree of error.

1 Introduction

Although many policy makers recognize the role forests play in carbon sequestration and climate change mitigation, to date there is no accepted methodology for quantifying and incorporating harvested wood products into global carbon budgets and carbon markets. While there is ample discussion surrounding sustainable forest management and the long-term sequestration of carbon in standing forests, the discussion rarely considers the life-cycle of wood and or the linkages between forest management and end markets for wood products.

Understanding the role of harvested wood products (HWPs) in the global carbon cycle is essential if appropriate policy concerning the treatment of HWPs as a carbon stock is to be implemented on a national or even international level under multi-lateral agreements in a post-Kyoto protocol regime (Rueter 2008). Studies that quantify current global stocks of HWPs vary greatly, as calculation methods are dependent on critical assumptions regarding product life, decay rates, and system boundaries (Pingoud et al. 2003; Green et al. 2006). A lack of data on the usage and disposal of HWPs adds to the difficulty of quantifying this global carbon stock (Kuchli 2008). Opinion on system boundaries is divided across the literature. The topic of landfills is a major part of this debate as models that include "end-of-life" within their system boundaries are intrinsically tied to assumptions made regarding the level of methane (CH_4) capture from landfills. The composition of materials in landfills has a significant impact on the magnitude of CH_4 generation, while the landfill design greatly influences the ability to capture landfill gases or convert methane passively through oxidation.

This chapter reviews the role of the forest products industry and harvested wood products within the context of global carbon stocks and flows, starting at the beginning of the HWP life cycle with production, turning then to the estimation of product life span and carbon stocks in products, and concluding with issues related to end-of-life management.

The chapter reviews the direct and indirect effects on greenhouse gas (GHG) emissions of production within the wood products and paper and paperboard sectors. It demonstrates the complexities of including wood products in use in such analyses by examining recent research in life-cycle assessment, manufacturing and use trend data, and literature on the impacts of materials substitution. The literature on the carbon stock of HWPs is reviewed and currently accepted research on the topic of product life spans and HWPs in landfills is summarized. The end-of-life pathways of HWPs and their carbon implications are then examined.

The chapter concludes by considering areas of further research, such as incorporation of the role of forest management in carbon sequestration via HWPs and further study of carbon benefits currently claimed by proponents of wood product substitution for more energy-intensive raw materials.

1.1 Overview and Framework for Assessing Role of HWPs in Carbon Cycle

From the perspective of global carbon stocks and flows, forest products are a heterogeneous

Table 12.1 A framework for evaluating the carbon profile of the forest products industry

Carbon Flux	Activity or Source
Direct emissions	Manufacturing
Indirect emissions	Transport
	Purchased power
	Landfill CH_4 emissions
Sequestration	Forests
	Products in use
	Products in landfills
Avoided emissions	Combined heat and power applications
	Product recycling
	Substitution effects

Adapted from Miner (2008)

Table 12.2 Emissions and sequestration estimates for the global forest products value chain

Value chain component	Est. Net emissions, Tg CO_2-eq. year^{-1}	Uncertainty[a]
Direct emissions: manufacturing	262	±20%
Indirect emissions: purchased power	193	±25%
Indirect emissions: transport	70	±50%
Indirect emissions: landfill-derived methane	250	−50% to +100%
Net forest sequestration	−60	±200%
Sequestration in forest products	−540	±50%
Avoided emissions: biomass fuels	−175	±200%
Avoided emissions: combined heat and power	−95	±200%
Avoided emissions: recycling	−150	±200%
Product substitution effects	Unknown	N/A

Modified after NCASI (2007)

[a]Uncertainty is based on professional judgment as presented in NCASI (2007)

group that consists of very short-lived products (e.g., newsprint) to very long-lived products (e.g., furniture, housing stock). Heterogeneity increases further due to different manufacturing processes, energy requirements for production, sources of energy within manufacturing, consumption patterns, end-of-life considerations, and substitution effects.

Miner (2008) presents a useful framework for evaluating the carbon profile of the forest products industry (Table 12.1). In general, direct and indirect emissions are quantified as positive GHG contributions, while sequestration and avoided emissions are quantified as negative GHG emissions—where negative emissions, i.e. reductions, are desired—relative to a business-as-usual scenario (Miner 2008). The net GHG profile is difficult to quantify because data for several of these processes are imprecise or unavailable. This is particularly true for sequestration and emissions within solid waste disposal sites as well as for substitution effects.

In light of the broad divestment of industrial timberland in the United States (Brown 1999) largely to timberland investment management owners, it is reasonable to ask whether forest carbon sequestration should be part of the carbon profile of the forest products industry. As many timberland buyers increasingly seek to manage, quantify, and monetize carbon sequestration benefits (Lippke and Perez-Garcia 2008) alongside traditional timberland management strategies, it is becoming increasingly important to apply a life-cycle framework to the industry, including the sequestration potential of forestlands as a source of raw material inputs. Similarly, it is also important to quantify products-in-use and in landfill sequestration since these are also key components of the life-cycle of forest carbon in use.

The National Council for Air and Stream Improvement, a U.S.-based industry-sponsored research group (NCASI 2007), presents estimates of net emissions of the forest products and forest carbon sequestration (Table 12.2). It is important to note that these figures do not take into account any product substitution effects, but do incorporate a large figure for forest carbon sequestration that may not be linked to the production of forest products. The largest areas of uncertainty in these estimates relate to transportation-related emissions, forest carbon sequestration, and methane emissions from landfilled forest products.

Forest products and the forest products industry are unique within the realm of carbon stocks and flows. First, industrial production of forest products typically uses a high proportion of its own feedstock (of biomass derived from production byproducts) as its energy source. Nevertheless, the vast majority of direct fossil CO_2 emissions are still generated in the production phase. Second, once in use, most forest products do not generate CO_2 emissions; upon disposal, they can generate varying degrees of CO_2 and methane (CH_4) depending on decomposition rates.

Forest products are often considered to be less energy- and emissions-intensive substitutes for other building materials, particularly concrete, steel, and aluminum (Wilson 2005; Upton et al. 2008). These substitution effects may play a much greater role in global CO_2 reduction schemes than improvements within the forest products manufacturing process itself (Kauppi and Sedjo 2001; Miner 2008). However, while some researchers (e.g., Burschel et al. 1993) regard product substitution as important, they point out that changes in forest management are even more significant. Denman et al. (2007) note that terrestrial ecosystems, and forests in particular, sequester amounts equal to approximately 25% of total anthropogenic emissions. Thus, the impacts of industry on forestland extent, stocking rates, and land-use conversion must be included in a comprehensive analysis of the carbon footprint of the industry.

Long-lived wood products-in-use constitute a carbon sink (Skog 2008), as do some wood products within solid waste disposal sites (Skog and Nicholson 1998; Micales and Skog 1997). NCASI (2007) estimates that within the United States the total gross emissions through the forest products value chain in 2005 were 212 Tg[1] carbon dioxide equivalents (CO_2e)[2] per year, while the forest carbon pool in products (in-use and landfills) grew by 108.5 Tg CO_2e per year. In 2005 in the U.S., landfilled wood products constituted 3% of total carbon stocks within the forest sector, but accounted for 27% of carbon sequestration (defined as flux in total carbon stocks), which is estimated to average 162 Tg carbon per year (Woodbury et al. 2007). The global stock of carbon within forest products is estimated between 4,100 Tg carbon (Han et al. 2007) and 20,000 Tg carbon (Sampson et al. 1993; IPCC 1996), with net sink rates estimated between 26 Tg carbon per year (IPCC 1996) to 139 Tg carbon per year (Winjum et al. 1998). The Intergovernmental Panel on Climate Change (IPCC 2007) estimates the total global standing forest carbon stock to be 3,590,000 Tg carbon for vegetation only and 11,460,000 Tg carbon for forest biomass and soils . Others estimate that the total carbon stock of the terrestrial biosphere is 24, 770,000 Tg C, including non-forest stocks (Fischlin et al. 2007). Even assuming the high estimate of 20,000 Tg carbon for forest products, this suggests that they still are a minor component of the global carbon budget.

However, Woodbury et al. (2007) assert that the forest products sector (including forest growth) provided net carbon sequestration equal to 10% of total U.S. CO_2 emissions in 2005. While forests accounted for 63% of net sequestration, changes in products-in-use and landfilled forest products accounted for 37% of net sequestration, implying that in 2005 the production and disposal of forest products was responsible for sequestering 3.7% of total U.S. CO_2 emissions. NCASI (2007) estimates that in 2005, 52% of gross emissions from the forest products industry was offset by carbon sequestration in products-in-use and products in landfill. It further estimates that annual forest growth on all private lands offset an additional 61% of gross emissions from the forest products industry. USEPA (2008) reports that total U.S. CO_2 emissions in 2005 were 7,130 Tg CO_2e, which, using figures from NCASI (2007), suggests that forest harvesting, clearance and disturbances (e.g. fire) represent only 1.8% of total U.S. emissions, and that forest products represent only 1.5% of total U.S. emissions.

A potentially broader set of carbon implications arises from the use of wood for energy or as a substitute for more carbon-intensive materials. These substitution effects have been explored extensively through comparisons of steel/aluminum/

[1]Teragrams – 1 Tg is equal to 1,000,000 Metric tons. All tons in this chapter are metric unless otherwise indicated.

[2]Carbon dioxide equivalent (CO_2e) is a measure for describing the climate-forcing strength of a quantity of greenhouse gases using the functionally equivalent amount of carbon dioxide (CO_2) as the reference.

Table 12.3 Production and production concentration of industrial roundwood, paper and paperboard, and wood fuel

	Industrial roundwood, 1,000 m^3 $year^{-1}$	Industrial roundwood, % of global production	Paper and paperboard, 1,000 ton $year^{-1}$	Paper and paperboard, % of global production	Woodfuel, 1,000 m^3 $year^{-1}$	Woodfuel, % of global production
Production, Top 10 Countries	1,175,185	71%	263,350	74%	1,061,620	60%
Production, Top 25 Countries	1,445,594	88%	331,510	94%	1,385,578	78%
Production, Total Global	1,645,681	100%	354,490	100%	1,771,978	100%
Top 10 Counties in Production	USA, Canada, Russia, Brazil, China, Sweden, Finland, Germany, Indonesia, France		USA, China, Japan, Canada, Germany, Finland, Sweden, South Korea, France, Italy		India, China, Brazil, Ethiopia, Indonesia, Dem. Rep. of Congo, Nigeria, Russia, USA, Mexico	

Derived from FAO (2004)

concrete vs. wood housing designs (Marceau and VanGeem 2002, 2008; Wilson 2005; Perez-Garcia et al. 2005; NCASI 2007) and use of biomass fuels (Sedjo 2008). There is less literature, however, on the effects of wood demand on maintaining tracts of forestland (Ince 1995). Together, these indirect effects may play a much larger role in GHG reduction than direct effects within the forest products sector.

1.2 Definition of Harvested Wood Products (HWPs) and Related Terms

Harvested wood products (HWPs) can be defined as wood-based materials that, following harvest, are transformed into commodities such as furniture, plywood, paper and paper-like products (Green et al. 2006). The term HWP is further simplified by the International Panel on Climate Change (IPCC) defining it as all wood material (including bark) that is transported off harvest sites. It does not include woody biomass, commonly referred to as slash or residual material, left at harvest sites (Pingoud et al. 2003).

Sawtimber refers to trees or logs large enough to be sawn into lumber. Once harvested, tree boles (i.e., the main stem of the tree) intended for human utilization are termed *roundwood*. The UN Food and Agriculture Organization (FAO) more formally defines roundwood as wood in its natural state after it has been harvested, including logs that have undergone minimal transformation and may be without bark, rounded, split, or roughly squared. Roundwood is used as either *woodfuel* or industrial roundwood, which is used to produce HWPs. Woodfuel, destined for heating, cooking and energy production, includes solids (fuelwood and charcoal), liquids (black liquor,[3] methanol, and pyrolitic oil) and gases from the gasification of these substances. *Fuelwood* is wood in the rough such as branches, twigs, logs, chips, sawdust and pellets, used for energy generation. Woodfuel is not analyzed here. While industrial roundwood and paper/paperboard production are concentrated in a few industrialized countries, fuelwood is less concentrated, and more prominent in lesser-developed countries (Table 12.3).

HWPs are categorized into two groups: *solid wood products* (SWPs) and *paper products*. Solid wood products consist of sawn wood and wood-based panels, typically measured in cubic meters. Paper products are defined as paper and paperboard (thicker, stronger and more rigid grades of paper) which are measured in dry tons (Green et al. 2006). In many cases, HWPs are further transformed into different product classes and categories throughout their lifecycle due to recycling (Pingoud et al. 2003).

[3] Black liquor is liquid residual from soda or sulfate pulping.

Table 12.4 Roundwood, pulpwood, woodfuel production and production ranking for selected countries and the global HWP industry

All values		Selected countries								
1,000 m³ year⁻¹	Global total	US	Finland	Canada	Russia	Japan	Brazil	China	Zambia	Mexico
Industrial roundwood	1,645,681	414,702	49,281	196,667	134,000	15,615	110,470	95,061	834	6,913
% as Pulpwood	*32%*	*41%*	*51%*	*14%*	*40%*	*23%*	*43%*	*7%*	*0%*	*14%*
Woodfuel	1,771,978	43,608	4,519	2,901	48,000	114	136,637	191,044	7,219	38,269
Total roundwood	**3,417,660**	**458,310**	**53,800**	**199,568**	**182,000**	**15,729**	**247,107**	**286,105**	**8,053**	**45,182**
% of Roundwood as fuelwood	*52%*	*10%*	*8%*	*1%*	*26%*	*1%*	*55%*	*67%*	*90%*	*85%*
Global rank of industrial roundwood production		1	7	2	3	18	4	5	78	33

Modified from data in FAO (2004)

2 The Global Forest Products Industry

For the purposes of this review, we make a distinction between the forestland harvests for land clearing/forest conversion versus the continuous production of forest products such as sawtimber, paper/pulp, biomass, and other forest products. Land clearing (deforestation) is considered a primary driver of anthropogenic CO_2 emissions, accounting for between 17% and 20% of total global CO_2 contributions between 1990 and 2002 (WRI 2006; Watson et al. 2000). Drivers of deforestation include a variety of sources ranging from fuelwood consumption, illegal logging, and expansion of agricultural land (Stern 2006).

The total global annual volume of harvested wood products in 2006 was 3.42 billion cubic meters (m^3) according to the U.N. Food and Agriculture Organization (FAO 2007). About 1.65 billion m^3 was industrial roundwood, while 1.77 billion m^3 was fuelwood. Others, however, suggest that harvest was slightly lower (approximately 3 billion m^3 per year), with approximately 1.8 billion m^3 as industrial roundwood, and 1.2 billion m^3 as fuelwood (Nabuurs et al. 2007). These figures may differ on the total fuelwood harvest, since economic data typically do not include fuelwood. In 2006, developed countries accounted for 70% of total global industrial roundwood consumption (USA, Canada, European Union, Japan) (FAO 2007). However, the largest producers, in order, were the USA, Canada, Russia, Brazil, and China.

The percentage of roundwood used for woodfuel varies greatly by country. Developed countries typically report low percentages used for fuelwood, while lesser-developed countries generally report a higher proportion of roundwood as fuelwood (Table 12.4).

3 Direct Effects of Forest Product Harvest, Manufacturing, and Distribution

3.1 Wood Harvesting

Wood products result from harvesting trees from natural forests and plantations. The harvesting process generates significant amounts of by-product, such as branches, leaves and other unmerchantable biomass, which are often either burned or left in the forest to decompose. The proportion of merchantable to unmerchantable biomass varies by forest type, species, and age at harvest. Representative values cited in the literature for North American forests suggest that 20–40% of tree biomass remains in the forest after harvest (Côté et al. 2002; Finkral and Evans 2008).

The variability reflects the diversity of commercially harvested species and forest types, as well as economic factors and harvest technologies.

Sustainable management and the use of a formal management plan should be requirements for any forest to be included as a carbon sink under national and international GHG accords. As of 2007, approximately 90% of developed country forests were harvested under sustained yield objectives within a management plan, while only 6% of developing country forests were similarly managed (Nabuurs et al. 2007).

3.2 Wood Products Manufacturing

Solid wood products manufacturing uses a majority of all global industrial roundwood volume. Its direct manufacturing emissions are a small fraction of total industry emissions, unlike paper and pulp manufacturing, which create much higher direct emissions. Globally, in 2004, 68% of all industrial roundwood volume went to the manufacturing of a variety of solid wood products (FAO 2007). Using 2004 FAO data, NCASI (2007) estimates that the global solid wood products industry emits 25 million tons (Mt) of fossil CO_2 per year. Major categories include solid sawn lumber (softwoods and hardwoods), structural panels (plywood, oriented strand board [OSB]), non-structural panels (e.g., particleboard), engineered wood products (laminated veneer lumber, I-joists, glulam), and miscellaneous uses (telephone poles, railroad tracks). Many of these products, with the exception of packaging/pallets, are manufactured for durable purposes. Since durable goods usually have product lives of 40–80 years, solid wood products have the potential to sequester carbon for significant periods. Furthermore, much of the solid wood stream is then deposited in a solid waste disposal site, where it may be sequestered near-permanently (Skog and Nicholson 1998).

McKeever (2002) estimates that in 1998 the United States, which leads the world in consumption of solid wood products, consumed 0.23 billion m^3 of solid wood products in the following proportions: solid sawn lumber (62%), structural panels (18%), nonstructural panels (12%), engineered wood products (1%) and miscellaneous (8%).[4] Researchers focused on the carbon sequestration potential of the solid wood products sector have offered several conclusions related to product mix and manufacturing:

1. Since 1970, the rate of resource utilization (the percentage of roundwood that ends up in final product form) of the U.S. solid wood products industry has increased significantly, despite a recognized reduction in size and quality of roundwood inputs. Yields from raw materials have increased, and inputs of petroleum-based additives in engineered and panel products have decreased.
2. Since 1970, the product mix within the solid wood products industry has shifted from solid sawn lumber and plywood to a mixture of engineered wood products and OSB. This is likely due to changes in quality of roundwood inputs and demand for uniform, high-performance engineered products (Meil et al. 2007).
3. The industry produces a substantial amount of its energy needs through biomass electricity and heat production, which are often adjacent to manufacturing facilities.

3.3 Carbon Management Implications of Trends in Solid Wood Product Manufacturing

Within the wood products sector, there is a strong trend toward engineered products such as glue-laminated lumber, I-joists, and non-plywood structural panels such as oriented strand board. Proponents recommend these products for their load-bearing strength and uniformity relative to solid sawn wood (Meil et al. 2007). They can also be manufactured from small-diameter roundwood and/or scraps from other processes. Because these products have been allowed under the two major international building codes (IBC/IRC), are favored by builders for their uniformity and strength, and allow for greater economic utilization of harvested fiber, it is unsurprising that this

[4] Data include domestic and imported products.

Table 12.5 Carbon dioxide (CO_2) emissions in the cradle-to-gate life cycle of a wood building product from the generation of the forest through product manufacturing, 2004

	Pacific northwest production				Southeast production				
Product	Glulam	Lumber	LVL	Plywood	Glulam	Lumber	LVL	Plywood	OSB
CO_2 emissions (biomass), kg/m^3	230	160	141	146	231	248	196	229	378
CO_2 emissions (fossil), kg/m^3	126	92	87	56	199	62	170	128	294
CO_2 emissions, total, kg/m^3	356	252	228	202	430	310	366	357	672
CO_2 emissions (biomass), kg/m^3	65%	63%	62%	72%	54%	80%	54%	64%	56%
CO_2 emissions (fossil), kg/m^3	35%	37%	38%	28%	46%	20%	46%	36%	44%
Total energy, MJ/m^3	5,367	3,705	4,684	3,638	6,244	3,492	6,156	5,649	11,145
Product yield, log to product		53%		51%		41%		50%	71%
Product yield, other wood inputs to product	82%		N/A		82%		N/A		
Description of other wood inputs	Dry, planed lumber		Veneer		Dry, planed lumber		Veneer		

Derived from Puettmann and Wilson (2005)
Note: I- joists are made of OSB and LVL, and could not be included in this table.
Glulam glue laminated timber beams, *LVL* laminated veneer lumber, *OSB* oriented strand board

is the fastest growing sub-segment of the solid wood products industry (Meil et al. 2007). By volume, these products made up about 1% of total U.S. roundwood consumption as of 1998. Sales growth of engineered wood products increased 30.2% over 2000–2004 (McKeever 2002). In contrast, the American Plywood Association projects that solid sawn lumber consumption will drop below 4 billion cubic feet (ft^3) in 2012, implying no growth in volume between 1998 and 2012.

Because solid wood manufacturing encompasses a mix of solid sawn wood and engineered wood products, it is worthwhile to examine the carbon footprint of each major segment. A series of studies by Wilson (2005) and Kline (2005) conducted as part of the CORRIM research program[5] provides carbon and energy consumption data for the production of various solid wood products (Table 12.5). CO_2 emissions by product type range from 202 to 672 kg CO_2/m^3, with U.S. Southern OSB production resulting in the highest emissions by volume[6] (Puettmann and Wilson 2005). Variability within a product type arises from differing regional energy sources and year of analysis. Solid sawn lumber production is not substantially lower in CO_2 emission/volume than the engineered wood products. However, Pacific Northwest plywood generated 24% lower CO_2 emissions than solid sawn wood.

Most engineered wood products contain (by mass) 5–15% in additives such as petroleum-based adhesives, waxes, and resins. These are created using more energy-intensive manufacturing processes. Because these products are stronger, less wood fiber is required within the construction process relative to solid-sawn lumber. For example, I-joists use approximately 62–65% of the wood fiber of a solid joist, but

[5]CORRIM is a research consortium focused on the environmental impact of the production, use, and disposal of wood and other bio-based materials. The Consortium includes US and Canadian research institution members and a number of contributing companies, associations and agencies related to the forest products industry.

[6]Meil et al. (2007) assert that this is largely due to the use of regenerative thermal oxidizer (RTO) units which are a critical element of air emissions control in OSB manufacturing.

their production is more energy-intensive. As a result, substitution of I-joists for solid-sawn lumber provides negligible opportunities for CO_2 emissions reduction (Perez-Garcia et al. 2005). Moreover, substitution of OSB for plywood reduces total carbon emissions only by 3–4%.

Resource utilization studies conducted in the U.S. in 1976 and again several decades later (Wernick et al. 1997; Meil et al. 2007) document increased utilization of by-products while providing interesting data on product yields from raw materials. In 1970, the softwood lumber industry had a 35% utilization rate by weight (e.g., conversion of raw logs into the primary product). This rose to 45% in 2000. Efficiency gains for softwood plywood showed a 7% improvement. However, a much greater proportion of plywood byproducts were used as raw materials for other products, such as nonstructural panels, rather than being burned or landfilled. In addition, over the same period, adhesive and resin content in plywood was reduced by 17% (Meil et al. 2007). The authors therefore calculate a reduction of 62.7 kg of fossil-derived CO_2/m^3 of softwood lumber produced in 2000 relative to 1970.

Biomass has been an important, carbon-neutral energy source for the forest products industry. Biomass is considered by Watson et al. (2000) and others to be a "carbon-neutral energy source" because it does not generate fossil carbon emissions. Within the forestry sector, forest regeneration is thought to offset carbon from energy production from biomass. Other researchers argue that such accounting masks important complexities (e.g., Luo et al. 2009; Walker et al. 2010). Regardless, within the IPCC framework, changes in forest regeneration are reported separately as land-use change (Watson et al. 2000; IPCC 2007), which makes it difficult for forest products industry bookkeeping to include the life cycle of their products for the purposes of calculating carbon stocks.

Over the past 30 years, the industry has improved utilization efficiency for materials by creating value from products once burned for energy, and by burning for energy products that were previously typically burned solely for disposal purposes (Wilson 2005). Historically, the industry burned bark and other "wet" residues in uncontrolled outdoor burners variously termed "teepee" or "beehive" burners, with significant particulate emissions and zero energy recovery. Only sawdust and planer shavings were converted into energy due to the cost and conversion efficiency of boilers. Today, in developed nations, it is more common for all residue, including bark, mill-ends, sawdust and shavings to be burned for the cogeneration of heat and electric power. These outputs are used to drive manufacturing processes within modern solid wood product mills. But they sometimes remain uncounted in carbon budgeting.

Two studies from the 1970s indicate that historically energy recovery was low. Grantham and Howard (1980) indicate that in 1970 25% of residual byproducts were used as fuel, and another 37% transferred to other facilities as raw materials. Corder et al. (1972) claim that 26% (for lumber) and 24% (for plywood) of byproducts were used for fuel in 1967. Between 1970 and 2000, bark and wet residues began to be used as fuel for combined heat and power applications at manufacturing sites.

3.4 Pulp and Paper Manufacturing

In 2004, the pulp and paper industry consumed approximately 32% of all industrial roundwood produced globally (FAO 2007). NCASI (2007) estimates that pulp and paper manufacturing processes globally emit 195–205 Mt of fossil CO_2 per year (compared to 25 Mt CO_2 per year for solid wood products). The pulp and paper industry generally produces products that are shorter-lived than the solid wood products segment, ranging from various grades of newsprint and paper to paperboard. The production of paper products from virgin fiber is considerably more energy-intensive than all solid wood products, since wood fiber must be converted (chemically or mechanically) from a mixture of cellulose, hemicellulose, and lignins into a cellulose-dominated pulp for papermaking. In 1998, the paper manufacturing industry ranked as the United States' fourth largest emitter of greenhouse gases,

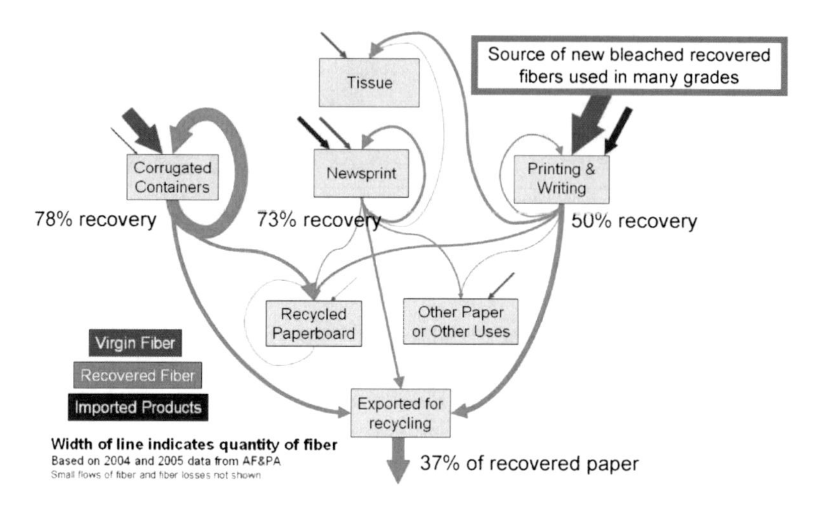

Fig. 12.1 The complex supply web of the forest products industry. Thickness of lines signify relative magnitude of flows. (*Source*: From Miner 2008. Reprinted with permission)

following petroleum, basic chemicals, and metals (EIA 2006). Using 1991 data, Subak and Craighill (1999) estimate that the paper and pulp industry directly and indirectly accounted for 1.3% of total global fossil carbon emissions in 1993.

Industry segments vary in production volumes, carbon intensity, manufacturing processes, and estimated service life. In 2006, the United States produced 41.8 million short tons of paper and 50.4 million short tons[7] of paperboard products. In contrast to flat or slowly growing markets for solid wood products (McKeever 2002), the U.S. paper and paperboard markets in total have been declining since 1999 (Irland 2008) likely the result of a transition away from newsprint consumption. Furthermore, production has dropped faster than consumption as significant industry segments have moved offshore (e.g., China now dominates global packaging markets) (FAO 2007).

International trade in pulp, paper, and paperboard products is considerably more developed than trade in raw sawtimber and solid wood products. A different set of nations is dominant within global production of paper and paperboard products (Table 12.3). Additionally, recycled fiber streams play a much greater role in paper manufacturing relative to solid wood products manufacturing (Falk and McKeever 2004).

Paper industry inputs vary by product type, and include (i) industrial roundwood, (ii) chips as a co-product of solid wood product manufacturing and (iii) recycled fiber. Certain products require more virgin fiber for tensile strength, while other products can be produced with predominantly recycled fiber. A NCASI (2007) report states: "Of the 352 million tons of paper and paperboard produced globally in 2005, 162 million tons, or 46%, was recovered rather than being disposed" (FAO 2006a). Miner (2008) also documented a complex fiber supply web within the industry (Fig. 12.1). Of the 100 million tons of paper consumed annually within the U.S, approximately 53.4 million was recovered for recycling in 2008. A 2008 press report from Forestweb (Irland 2008) indicates that paper manufactured from 100% recycled pulp results in 1,791 kg/ton of CO_2 emissions, vs. 4,245 kg/ton of CO_2 emissions from paper manufactured from virgin pulp. However, there is some debate over the role of recycled fiber in reducing GHG emissions within the industry. The de-inking and recycling process is energy-intensive, and typically involves 100% purchased power (vs. in-house biofuel-derived power in virgin pulp manufacturing). Some researchers suggest that the climate benefits of recycled material arise from the avoided CH_4

[7]One short ton = 2,000 pounds (lb) ≈ 907 kilograms (kg, SI) = .907 metric tons.

Table 12.6 Energy inputs and ratio of embodied** carbon in raw material vs. final product under a variety of pulping processes in Finland, 1995

	Total production, Gg year^{-1}	Direct fuels, MWh/Mg	Heat, MWh/Mg	Electricity, MWh/Mg	C in raw material/ C in final product
Mechanical					
GWP, B	801		0	1.55	1.2
GWP, NB	1,167		0	2.1	1.23
TMP, NB	923		−0.75	2.4	1.2
TMP, B	801		−1.17	3.37	1.24
CTMP	105		0.56	1.65	1.25
SCP	509		1.06	0.4	1.45
Chemical					
HSUP, B	2,174	0.39	3.07	0.69	2.46
SSUP, NB	680	0.52	2.77	0.57	2.56
SSUP, B	2,928	0.52	3.33	0.75	2.71
Recycled					
REC, NB	180	0	0	0.1	1.07
REC, B	272	0.25	0.17	0.4	1.17
Total	10,540				

From Pingoud and Lehtila (2002). Reprinted with permission

* In Finland, 51% of produced chemical pulp was dried in 1995 (Carlson and Heikkinen 1998). This is included in the energy demand figures

** Embodied energy, also known as embedded energy, refers to the energy consumed in the prior steps in the product chain

Abbreviations used: *GWP* ground wood pulp, *TMP* thermo-mechanical pulp, *CTMP* chemi-thermo-mechanical pulp, *SCP* semi-chemical pulp, *HSUP* hardwood sulphate pulp, *SSUP* softwood sulphate pulp, *REC* recycled pulp, *B* bleached, *NB* unbleached, *MWh* megawatt hours. One megawatt-hour (MWh) ≈ 3.6 × 10^9 joules (J, SI) ≈ 3.412 × 10^6 British Thermal Units (BTU).

emissions from decomposing paper within landfills (Subak and Craighill 1999; NCASI 2007).

Using data from Finland's forest products industry, Pingoud and Lehtilä (2002) estimate that in 1995 across pulping processes and fiber sources, the proportion of fossil-based carbon emissions per wood-based carbon in end products (Mg carbon/Mg carbon) is 0.07 for sawn wood and 0.3–0.6 for paper in the manufacturing stage, suggesting that paper is 428–857% more fossil carbon intensive than sawn wood by mass. They also found that direct fuel, heat, and electricity demands for the production of 11 grades of pulp in Finland in 1995 can dramatically vary (Pingoud and Lehtilä 2002) (Table 12.6).

Chemical pulping uses either a kraft (sulfate) process or sulfite process to dissolve lignins, which are burned with other derivatives to recover pulping chemicals and to provide process heat (Côté et al. 2002). This process leaves cellulose fibers largely intact for high-quality papermaking. Mechanical pulping uses fiber more efficiently, yielding a lesser amount for biofuel as a process energy, and increasing the need for purchased electricity (Pingoud and Lehtilä 2002). Chemical processes result in 50–55% loss of fiber by weight, while recovery of recycled paper results in a 16–18% loss of fiber by weight. Fiber that does not end up in the final product is generally burned in the production process or landfilled (Côté et al. 2002). Industry-wide in the U.S., 56% of all energy needs are met with biofuel co-products (Davidsdottir and Ruth 2004). Farahani et al. (2004) have highlighted a new technology, black liquor gasification-combined cycle (BLGCC), which has the potential, under certain conditions, to fully offset energy usage within the chemical pulping process. In this case, using less recycled feedstock actually improves the GHG emissions profile by providing greater opportunities to use biomass and black liquor as energy feedstock.

In general, mechanical pulping is less energy-intensive, although, as noted, it also uses a greater proportion of purchased electricity in its manufacture. Given the reputation for energy and process efficiency of the Nordic paper and pulp industry (Subak and Craighill 1999), the figures in Table 12.6 may not be globally representative, yet are among the few data points available on this topic.

Similar to trends within the solid wood products industry, the paper and pulp sector has experienced process, energy efficiency, and resource utilization improvements since 1970. IEA (2003) documents a 0.8% decrease in energy intensity of OECD-country paper and pulp making processes from 1968 to 1990. Nevertheless, as of 2002 in the U.S., the paper and pulp industry remains the second highest manufacturing sector on an energy intensity basis (with petroleum/coal as the highest) (Davidsdottir and Ruth 2004). This estimate does not take into account the relatively high proportion of energy derived from biomass fuels within the forest products industry, approximately 40% in the United States in 1998 (EIA 2008).

3.5 Carbon Implications of Transport and International Trade of Forest Products

Transportation of forest products, both as raw industrial roundwood and as consumer products, has been recognized as a significant potential source of fossil carbon emissions (Pingoud and Lehtilä 2002; NCASI 2007). Research indicates that some forest products can travel large distances prior to and following manufacture, via overland freight or cargo ship. Globally, NCASI (2007) estimates that product transport results in fossil carbon emissions of approximately 70 million tons CO_2 per year, or approximately 27% of total fossil carbon emitted within manufacturing and distribution processes. Pingoud and Lehtilä (2002) examined transportation related emissions in Finland, documenting a wide range of transportation modes and distances. Their research concluded that transportation from harvest site to mill, and from mill to consumer, accounted for 22% and 20% respectively of total fossil carbon emitted within manufacturing and distribution processes in 1995.

3.6 Indirect Effects of the Forest Products Industry on Carbon Emissions

As noted above, the forest products industry's contribution to total global GHG emissions is minor, despite its high energy intensity (NCASI 2007), partly due to its significant use of biomass fuels to power manufacturing processes, and its long-lived products, which sequester carbon in products-in-use and landfilled products. Beyond purchased power, transportation, and landfill methane emissions related to forest products, the forest products industry offers products that may be less fossil carbon-intensive than substitute materials such as concrete, aluminum, and steel. To the extent that increased use of forest products results in an expansion of timberlands operated on a sustained-yield basis, substitution effects may have a greater impact on net carbon sequestration beyond a comparison of the embodied energy within various substitutable building materials.

Gustavsson et al. (2006) describe four GHG emissions-related aspects to materials substitution: (i) emissions from fossil fuel use over the life cycle of the product (e.g., production, transportation, end use and waste management); (ii) replacement of fossil fuels with biomass energy within the production phase; (iii) carbon stock changes in forests, products-in-use and landfilled materials; and (iv) GHG emissions from industrial process reactions in such areas as cement and steel production. While it is impossible to accurately quantify all actual and counterfactual outcomes within this framework, Kauppi and Sedjo (2001) indicate that the range of possible substitution effects may be up to 11 times larger than the total amount of carbon sequestered in forest products annually. This suggests that minor changes in consumer preference for materials can have a big impact on the overall GHG emissions profile of the construction sector.

In many applications, forest products are substitutable with rival products, typically plastics, metals or concrete (Upton et al. 2008). Researchers have compared the carbon footprint of forest products relative to some of these materials, and concluded that increased use of forest products within the construction sector would result in decreased GHG emissions (Wilson 2005; Upton et al. 2008). Currently, in the U.S., wood framing techniques are used in approximately 90% of new housing starts (Upton et al. 2008). This percentage is much lower in other regions of the world, particularly outside of North America and Northern Europe (Gustavsson et al. 2006).

In the lifetime of a house, there are two primary sources of carbon emissions: the construction of the structure, and the energy requirements to heat and cool the structure over its lifetime. It is difficult to compare wood vs. other building materials because alternative materials have different thermal characteristics. For example, the thermal mass associated with concrete buildings may reduce heating and cooling costs, thereby lowering carbon emissions during building operation, suggesting that the advantages of wood could be overestimated (Nishioka et al. 2000; Upton et al. 2008).

Upton et al. (2008) project that wood-framed single-family houses require 15–16% less total energy for nonheating and cooling purposes and emit 20–50% less fossil CO_2 to build than non-wood houses made of steel framing products. This conclusion relies on several key assumptions about the ratio of embodied energy in housing relative to energy expended to heat and cool the house over its lifetime, as well as assumptions regarding the fate of forests used or not used for the production of industrial roundwood (Upton et al. 2008). Wilson (2005) found that the wood-framed house had a global warming potential index (a measure of total GHG emissions, not energy usage, as in Upton et al. (2008)) 26% and 31% lower, respectively, than model steel and concrete house designs. These figures represent only the embodied energy within the production of the house, not its operation. These figures are supported by Gustavsson and Sathre (2006), who conducted a sensitivity analysis around uncertainties and variability within the production of both concrete and wood. Using plausible inputs, wood building materials had lower embodied energy costs relative to concrete in all cases analyzed.

Perez-Garcia et al. (2005) characterize the substitution effects throughout the value chain from forest to landfilled product. The product life span of wood used in housing and therefore the carbon that is sequestered is unchanged, regardless of the length of the forest rotation. What changes is the carbon stock in the forest. Furthermore, with substitution effects, the use of wood products offsets concrete or metal construction, providing a greater benefit than either the forest carbon pool or the forest product carbon pool. In short, intensive forest practices create a "positive carbon leakage" through greater use of wood products in the market place.

Several studies examining substitution posit that greater use of forest products will result in greater retention of working forestlands, or conversely, that less use of forest products will hasten conversion of working forestlands to other land uses (Wilson 2005; Perez-Garcia et al. 2005; Upton et al. 2008). Regardless of the validity of this assumption, it is important to recognize that each author implicitly or explicitly recognizes that carbon fluxes within forestlands are several orders of magnitude greater than any identified substitution effect. Thus, it is worth examining how and whether the forest products industry has any effect on the extent and condition of forestlands relative to other factors.

4 Estimate of Carbon in HWPS

Global estimations of yearly HWP production are derived from statistics collected by FAO on the production of roundwood. In the United States, the USDA Forest Service keeps statistics on roundwood harvests and HWP production based on data collected from government agencies and industry. The FAO reports that, globally, 1.65 billion m^3 of roundwood is extracted annually for HWP production (FAO 2007). In 2002, the United States produced approximately 425 million m^3, or 25%, of global roundwood intended for HWPs (Howard 2006). If this global roundwood production figure were converted to carbon, it would be very large. However, production losses occur as roundwood is processed into different products, and assumptions on the magnitude of these losses greatly influence the final calculations.

Table 12.7 Global productiovn of HWPs in 2000 according to FAOSTAT 2002

	Billion m³/year	Pg C/ year
Primary products		
Roundwood	3.1	0.71
Wood fuel	1.5	0.37
Industrial roundwood	1.6	0.34
Pulpwood (Round & Split)	0.48	0.11
Sawlogs + Veneer Logs	0.95	0.20
Other Indust Roundwd	0.15	0.03
Semi-finished products		
Sawnwood	0.42	0.09
Panels + Fibreboard	0.22	
	billion tons/year	
Paper + Paperboard	0.32	0.15

Source: Pingoud et al. (2003)

The associated carbon fluxes have been estimated by assuming that the approximate dry weight of coniferous wood is 0.4 tons/m³ and non-coniferous is 0.5 tons/m³ and that the carbon fraction in biomass is 0.5. In addition, the estimated charcoal production was 0.04 billion tons/year (metric tons per year). The production of wood residues was 0.06 billion m³/year and chips and particles 0.16 billion m³/year, these being mainly by-products of wood processing

First, it is assumed that roughly 50% of harvested roundwood logs is lost as residues (Gardner et al. 2004), which brings the total to 825 million m³. Data from 2004 show that the paper products industry consumed 32% of the total roundwood production, which would account for 264 million m³, while solid wood products accounted for 561 million m³. However, these figures are further reduced when losses from final product finishing are taken into account. Skog and Nicholson (2000) assumed an 8% loss for solid wood products, and 5% for paper products, during finishing. This would mean that 516 million m³ of solid wood products and 251 m³ of paper products comprise the total annual global production of HWPs. Using the same assumptions on production losses, United States yearly production of HWPs would amount to 133 million m³ of solid wood products (SWP) and 65 million m³ of paper products (see Table 12.7).

In 1996, the world's forests produced 3.4 billion m³ of harvested roundwood. About 1.9 billion m³ (56%) of this harvest was fuelwood; the remainder (1.5 billion m³) was industrial roundwood (e.g., sawlogs and pulpwood). The industrial roundwood corresponds to a harvesting flux of about 0.3 Gt C year^{-1}(FAO 2004).

Estimates of the total carbon sequestered in HWPs globally vary widely from 4,200 Tg C (IPCC 2000) to 25,000 Tg C (Matthews et al. 1996). In another study, Harmon et al. (1990) suggest that global C stocks in long-lived products lie in the range 2–8 Pg C.

Similarly, estimates of the net annual sink from HWPs ranges from 26 to 139 Tg C/year in these same reports (IPCC 2000; Matthews et al. 1996). This compares to the 38,000 Tg CO_2e in estimated worldwide emissions in 2004 (IPCC 2007), which equates to 139,300 Tg C, thus the total amount of carbon sequestered annually in HWPs is small. There are several reasons to explain the wide range in these figures on HWP annual sink. First, estimates will vary based on the assumptions made about average production losses and wood densities. The choice of wood density can have considerable impact on the results (Stern 2008). Secondly, as described below, HWP stock estimates frequently do not distinguish between HWPs in use versus those in landfills (Pingoud et al. 2003). A standard methodology for converting HWP mass into carbon equivalents is needed to compare data reported from different countries along with better estimates for country-specific trends in landfill waste.

5 Calculating Useful Lifetimes of HWPs

The figures above give us a rough estimate of the potential yearly input to the global carbon stock of HWPs. However, since these calculations fail to recognize the finite life of HWPs, these rough estimates are inflated. Lifespans of HWPs vary significantly by product type and must be accounted for accordingly. The carbon embodied in short-lived products can be released quickly back into the atmosphere after rapid decomposition, while long-lived products can store carbon for many years. Some wood or paper items such as antiquities and historic buildings are expected to have very long lives (in excess of 100 years)

(Skog et al. 1998). However the majority of paper products have a high rate of retirement, lasting only weeks (Marland and Marland 2003).

The lifespan attributed to products has a major impact on the outcome of estimates on the stock of HWPs. Although it is critical to determine the lifespans of various HWPs, it is difficult due to a lack of data on product use and disposal (Stern 2008). In response, some believe that HWP lifespans should not be viewed as empirical but as parameter values used in models (Pingoud et al. 2003). Data on HWP use suggest that rate of retirement of HWPs from end uses is more or less constant for a period, then accelerates for a while near the median life, and finally slows down after the median life (Skog and Nicholson 2000). As a result, the average lifespan of HWPs is much shorter than some models would suggest (Pingoud et al. 2003). Because of the difficulty of determining lifespans, it is common to see conflicting values for the lifespan of the same products in different studies. For example, a review of studies has shown that the estimates of average lifespan of pallets range from 2 to 20 years (Pingoud et al. 2003).

Data on average lifespan can be used to model how HWPs are discarded and ultimately oxidized. Much of the literature uses the term decay to describe both the mathematical characterization of retirement of products from use as well as the biophysical decomposition of products (Dias et al. 2009). The use of the term decay in this way reflects the fact that researchers analyzing HWP lifespans frequently quantify both the length of time that a product is used and the time during which HWP generates carbon emissions during waste management using the same model. The decay parameters for products in use are nonetheless different from those out of use (in landfills, for example) where decay of HWPs may be halted almost completely. Most studies, however, do not separately model the decay of HWPs that are out of use (Pingoud et al. 2003). Instead these studies model the retirement of HWPs and assume that decomposition occurs at different rates as a function of the product's retirement function.

The type of decay model used has a significant impact on estimation of the HWP carbon stock as it determines the timing of carbon releases through oxidation during decomposition. Numerous methods for modeling the carbon release of HWPs exist. (Dias et al. 2009).

One method of modeling HWP oxidation is to assign an exponential decay rate to a product. This is often done by assigning each type of HWP a carbon half-life which represents the time in which half of the carbon embodied in the end-use product is no longer present and has been emitted back into the atmosphere. This exponential decay model assumes that 90% of the carbon in HWPs is released in 3.3 times the assigned half life. Under this model, carbon release begins immediately once a product is in use and occurs at a greater rate earlier on in the life of the product and slows as the product progresses through and end of life. Another approach assumes that products of this type all have the same age, which is set to the product's average lifespan. In the model, 100% of the carbon remains embodied in the HWP until it is discarded, at which time all the carbon in the HWP is then released into the atmosphere. A third method follows a linear model in which a percentage of the initial amount of carbon in the HWP is released each year. The year in which all the carbon has been released is the maximum lifespan of the HWP type. Half of the time needed to reach the maximum lifespan is the product's average lifespan. The emissions profile of these models can be linear, exponential or equal (Fig. 12.2) (Skog and Nicholson 2000).

The different methods in modeling carbon release from HWPs clearly show how assumptions concerning product lifespan can significantly alter estimates. The most rudimentary model is that which assumes products of a certain type have an equal age and release 100% of their carbon at the time of retirement. This model does not account for carbon that is released into the atmosphere from products that are discarded before reaching their average lifespan. This method may mask carbon emissions that are occurring from HWP end-of-life processes and may inflate estimates of the annual increase in the HWP carbon stock. The linear and exponential decay functions both have carbon emissions occurring from the start of a

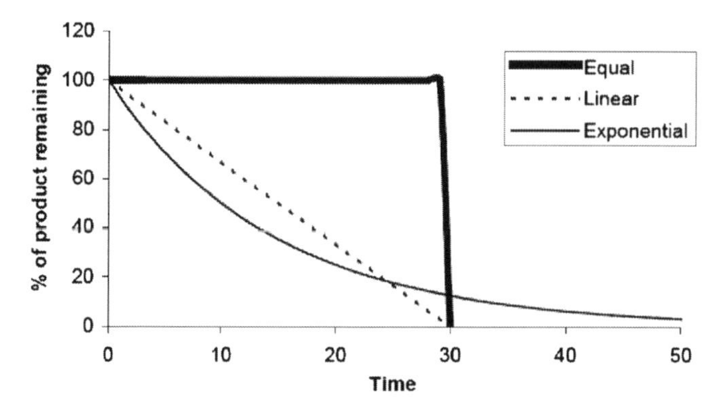

Fig. 12.2 A graphical representation of how carbon release is modeled using different methods of incorporating HWP product life into stock calculations (*Source*: Pingoud et al. 2003. Reprinted with permission)

product's life, which accounts for products that are discarded much earlier than those reaching the average lifespan. The exponential function creates a scenario where carbon emissions occur much faster in the beginning and slow as a product gets closer to reaching its average lifespan. HWP retirement most likely follows this decay function more closely as HWP retirement accelerates before reaching median life and finally slows down after the median life (Skog and Nicholson 2000). It must be noted that these decay functions do not effectively model conditions in landfills or bioenergy facilities.[8] Thus, they should only be used to model the rate of HWP retirement from use, which could then be incorporated into a larger model that more accurately portrays carbon emissions from HWP once they are discarded.

Determining accurate HWP lifespan values in order to create models that simulate real life conditions is difficult given a lack of data. There is room for vast improvement in reporting methods. Better data on product life for industrial uses of HWPs such as pallets may become available in the future as companies begin to label them with bar codes containing a pallet's age. Carbon markets may also encourage companies to keep better data on product life as they may in the future be able to sell temporary carbon credits based on their HWP stock. Efforts to develop global reporting standards and data sets on product life spans are underway (Murakami et al. 2010; Oguchi et al. 2010) Inclusion of HWPs in climate mitigation policy will require increased reporting which will lead to better data, allowing for more accurate product lifespans (Kuchli 2008).

6 End-of-life Pathways for HWPS

Harvested wood products can take several different pathways when they are discarded (CEPI 2007). Recent research has expanded the system boundaries of analysis to account for the different end-of-life pathways which can postpone carbon release of HWPs, store carbon indefinitely, displace fossil fuels, or even produce emissions at a significant level. HWPs can be recycled, burned (with or without energy recovery), composted, or disposed of in a dump or landfill (Fig. 12.3). Each of these pathways has different implications for carbon emissions. Calculations that do not account for these pathways are not accurately capturing the carbon effects. This is especially true in regards to the production of CH_4 resulting from the landfilling of HWPs. Research that includes end-of-life pathways has shown that from 2000 to 2005 the

[8]For example, carbon emissions from incineration of HWPs is immediate whereas, as described below, the release from landfilled products follows a longer and more complicated path. Composting of HWPs present an intermediate case.

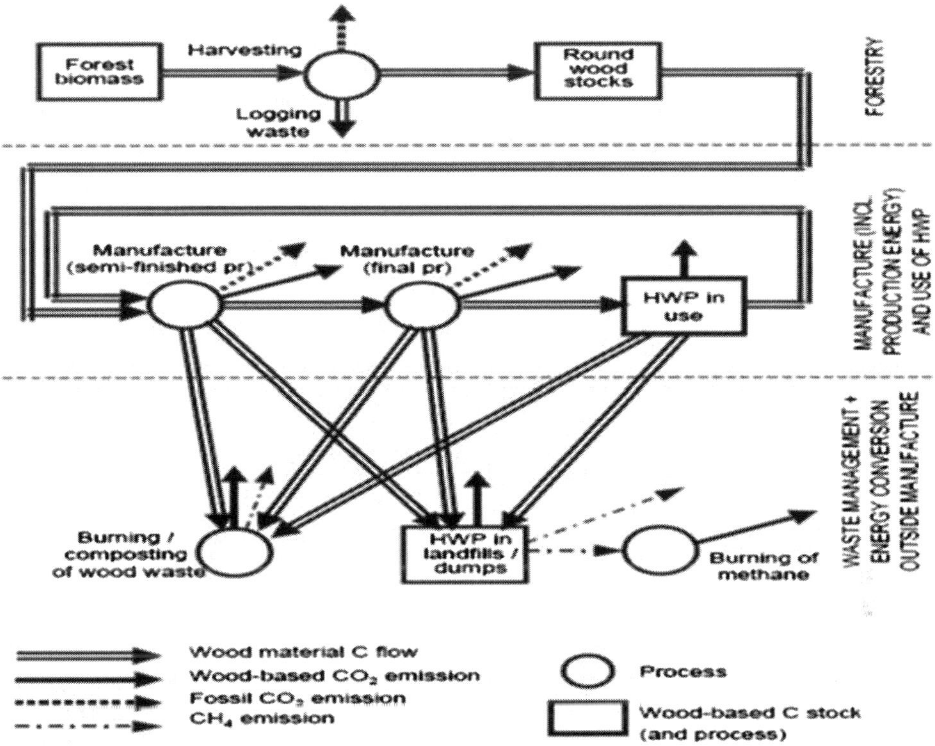

Fig. 12.3 Schematic representation of a lifecycle of HWP (*Source*: Pingoud et al. 2003)

global HWP stock had an average net increase of 147 Tg C/year, which is equivalent to 540 Tg CO_2/year (Miner 2008). These findings are at the higher end of the range compared to earlier studies due to the study's assumptions on landfills.

6.1 Burning HWPs

HWPs have the potential to be burned as a fuel. Short-lived wood products follow this pathway more often than long-lived products. Skog and Nicholson (2000) estimate that in 1993 in the United States, over 24% of paper and paperboard waste (after recycling) was burned. Although burning discarded wood or paper for energy is a carbon-emitting activity, it may result in lower net emissions if it has displaced more carbon-intensive fuel types (i.e., substitution effect). Using discarded HWPs for energy also reduces the amount that is put in landfills thus reducing the production of potent CH_4 gas. In order to evaluate whether burning HWPs for energy is superior to burning an alternative energy, a comparison of the two fuel chains must use a consistent methodology and a consistent definition of system boundaries.

6.2 Recycling

Recycling programs prolong the lifespan of carbon in HWPs, which keeps carbon stored in the product chain and extends carbon sequestration benefits. Recycling processes typically transform HWPs into products of lower wood content. This process can be repeated until the HWP is used to create bioenergy or otherwise disposed. This is known as a cascade effect (Kuchli 2008). HWP

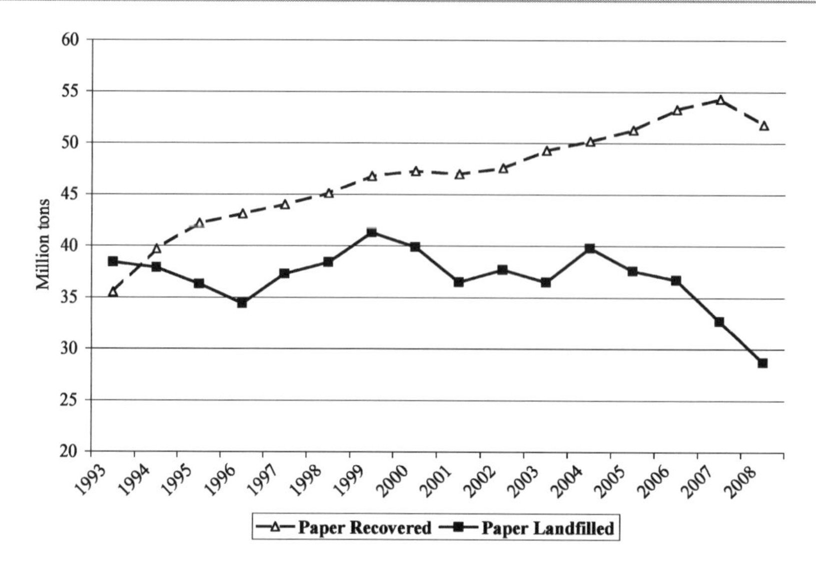

Fig. 12.4 Paper recovery vs. landfilling in the U.S. 1993–2007 (*Source*: Created with data from the American Forest & Paper Association, http://www.paperrecycles.org/stat_pages/recovery_vs_landfill.html)

recycling can thus reduce the rate of landfilling. This in turn reduces the amount of CH_4 produced by HWPs in landfills (CEPI 2007). This is particularly true for paper products, as these materials produce higher levels of CH_4 than landfilling of solid wood products (Skog et al. 2004).

As HWPs cascade into products of lower wood densities, however, their viability to be recycled is reduced. Once paper has reached a very low grade, such as tissue, it can no longer be recycled. Not surprisingly, in part because of the cascading effect and the downcycling of HWPs, low-grade paper products typically constitute a third of municipal solid waste (MSW) in landfills (Pingoud et al. 2003; EPA 2008).

The type of HWP plays a major role in whether or not it will be recycled. At the moment, recycling is only seen as a viable option for paper products.[9] The EPA reported that in 2007, 83 million tons (U.S.) of waste paper and paperboard were generated, of which 45 million tons (U.S.) (or 54%) were recovered through recycling (EPA 2008). In contrast, the recycling rate for HWPs used in construction is significantly lower. In 2007, the United States recycled only 1.3 million tons of durable wood products from the nearly 14 million tons generated (9%) (EPA 2008). This huge disparity in recycling rates is due to the nature of the products themselves. Newspaper is easily sorted and collected, while wood from construction demolition is very difficult to separate and re-use. Notably, data from the National Council for Air and Stream Improvement (NCASI) shows that while paper recovery is rising rapidly, the amount of paper products in landfills has decreased only nominally (Fig. 12.4) (Miner 2008). Still, reductions in the amount of HWP landfilled are expected to occur over time as HWP recycling processes modernize and become fueled by residue losses from the recycling process.

6.3 Landfills

Landfills have been criticized for their negative environmental impacts since the beginning of the environmental movement. Today, however, there are those in the scientific community who suggest that landfills could potentially act as a carbon *sink* for HWPs because HWP decomposition

[9]There is some niche market recycling of solid wood products—lumber from old buildings, etc. and some efforts to reclaim hardwood from pallets. (Technically, this is reuse and not recycling.)

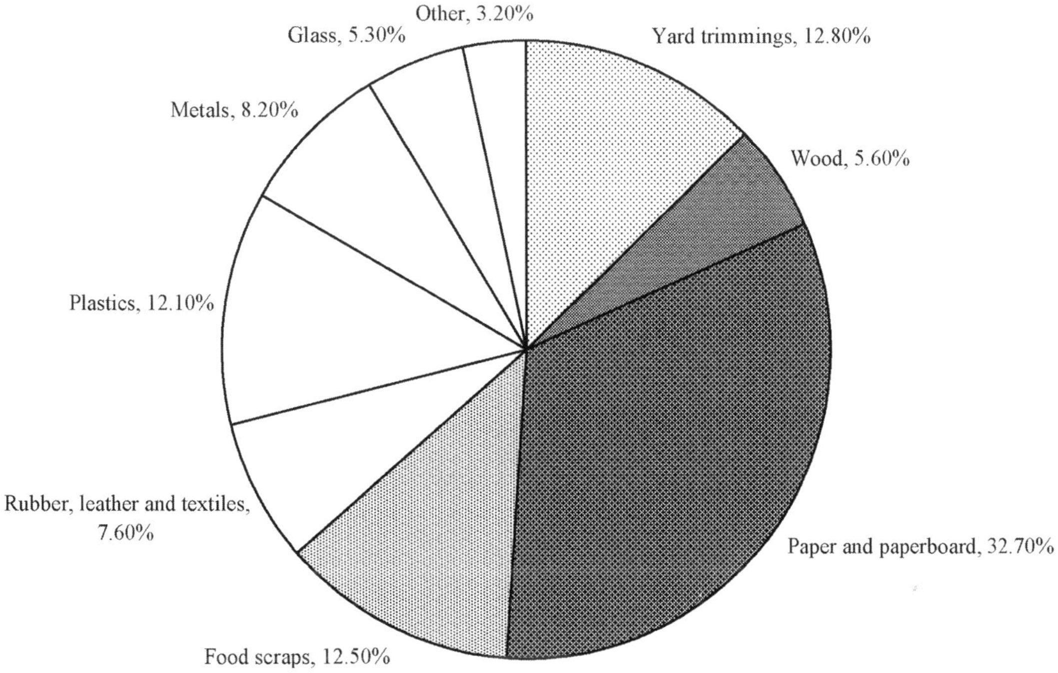

Fig. 12.5 Composition of waste materials destined for final disposal as Solid Waste in the United States, 2007 (*Source*. Data from EPA 2008)

can be very slow. Modern landfills are typically engineered to minimize the infiltration of water and the absence of moisture impedes biodegradation. Studies have shown that most wood products, when disposed of in a modern landfill, will experience a very slow decay (Bogner et al. 1993; Ximenes et al. 2008). This finding can have significant implications for calculating the stock of carbon in HWPs because it is estimated that biomass materials, such as paper, food, and wood, constitute about 63% of the municipal solid waste (MSW) in the U.S. (Fig. 12.5) (EPA 2008). The high proportion of HWPs in landfills further supports the case to expand the boundaries of analysis to include HWP end-of-life pathways. If CH_4 is captured and used for energy, carbon emission reductions can occur as carbon remains locked in HWPs at the same time that energy generated from landfill gases can displace fossil fuel emissions from traditional energy sources. Despite the attractiveness of using landfill gases for fuel, recent estimates indicate that only around 5 Tg C is captured worldwide, versus 15–20 Tg C of annual emissions from landfills (Spokas et al. 2006; Willumsen 2004). The large discrepancy between landfill gas (LFG) production and capture is best understood by analyzing how landfills work as well as current disposal practices. This may also help to forecast the likely impact from policies under debate to encourage or permit increased landfilling of HWPs.

6.4 Landfill Science

In a landfill, solid waste is buried. While this allows some biodegradable fractions of the waste to decompose via a complex series of microbial and abiotic reactions, the anaerobic conditions prevent a significant amount of decomposition. CH_4, or methane, is formed by methanogenic microorganisms under anoxic conditions, either

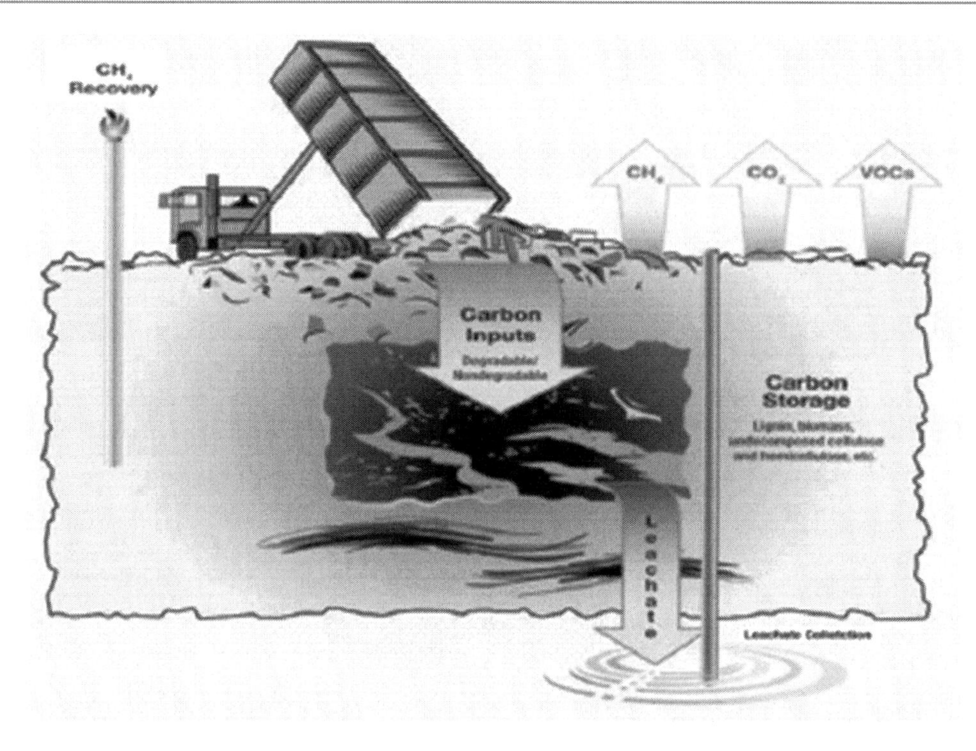

Fig. 12.6 Landfill carbon mass balance (*Source*: EPA 2006)

through the direct cleavage of acetate into CH_4 and carbon dioxide or the reduction of CO_2 with hydrogen (Fig. 12.6) (Spokas et al. 2006).

Since new layers of waste cannot be instantly covered, the waste is exposed to oxygen which allows white-rot fungus to decay wood. This type of decay, however, is limited because the available oxygen is rapidly consumed by the fungus, leaving only anaerobic bacteria. While anaerobic bacteria can break down hemicellulose and cellulose, these organisms cannot reach these materials if they are enclosed in lignin (Skog and Nicholson 2000). As a result, solid wood placed in landfills experiences low rates of decay. In newsprint, however, lignin content is only 20–27% and chemically-pulped paper has virtually no lignin, which results in a greater likelihood of decay than solid wood products, despite anaerobic conditions. Still, both wood and paper products experience low decay rates; in general, less than 50% of the carbon in these products is estimated to be ultimately converted to CO_2 or CH_4 (Table 12.8) (Skog and Nicholson 2000).

Table 12.8 Estimated maximum proportions of wood and paper that are converted to CO_2 or CH_4 in landfills

Product type	Maximum carbon converted (%)
Solid wood	3
Newsprint	16
Coated paper	18
Boxboard	32
Office paper	38

Source: Skog and Nicholson (2000)

Emissions created from anaerobic conditions are referred to generally as "landfill gas" (LFG) and encompass multiple gases, predominantly CO_2 and CH_4. According to Skog and Nicholson (2000), the proportion of carbon that is emitted as CO_2 and CH_4 in the gaseous product of MSW in landfills is skewed towards CH_4 at a rate of 1.5:1. Other studies suggest that the proportional difference between the two is not as great and that 1:1 should be used for commercial purposes (Johannessen 1999; Themelis and Ulloa 2007).

It is also important to note that emissions of various greenhouse gases occur on different temporal scales. On the one hand, CO_2 is released quickly as decomposition occurs while oxygen is still present in the system. Studies estimate that half of the total CO_2, is emitted in the first 3 years while the rest is emitted continually over time (Skog and Nicholson 2000). Methane, on the other hand, is released very slowly over time once all the oxygen is depleted, with half the total CH_4 emitted in approximately 20 years (Micales and Skog 1997). Moreover, Skog and Nicholson claim that 10% of the CH_4 is converted to CO_2 by microorganisms as it moves out of the landfill, which makes the landfill cover a de facto converter. According to Johansson, the conversion capacity for a landfill top cover varies depending on soil texture, moisture content, and the amount of organic matter available in the soil. Covers with porous soils and organic matter have achieved complete oxidation of methane (Johannessen 1999).

While LFG generation poses a problem in terms of carbon emissions, high LFG generation levels are desirable for operators of LFG recovery systems, particularly since such systems are capital intensive and often financed by energy sales. Although theoretically, 1 ton of biodegradable carbon can produce 1,800 m^3 of LFG, in practice, this number is much lower because of uneven and incomplete biodegradation. As a result, 200 m^3 is generally accepted as the maximum volume of LFG produced from 1 ton of land filled MSW (Johannessen 1999). Several factors influence the rate of capture to total volume of LFG generated. These include LFG losses to the atmosphere through the surface or through lateral gas migration; pre-closure loss due to decomposition of organic material under aerobic conditions; aerobic decomposition of the near-surface layer (e.g., air intrusion due to gas extraction); and washout of organic carbon via leachate (Johannessen 1999). All of these can reduce the potential LFG capture rate, and often tip the balance of whether landfills reduce emissions from carbon storage or serve as large sources of carbon emissions.

As of the beginning of this century there are more than 350 landfills in the United States with gas recovery plants, and more than 1,100 worldwide (Spokas et al. 2006). These landfills are very diverse with respect to the amounts of material placed in the landfill, the type of material, degradation rate, and LFG capture system. Moreover, within individual landfills, decomposition rates can vary even in adjacent areas of a landfill (Micales and Skog 1997). This variation makes it difficult to assign an average capture rate to all landfills (CEPI 2007). As one example, the EPA's Waste Reduction Model (WARM) uses a default value of 75% LFG capture rate. Compared to other reports, this figure is higher than average and likely varies greatly from region to region within the United States (Themelis and Ulloa 2007). Other studies are more conservative and claim that normal recovery rates are thought to range from 40% to 50% by volume (Johannessen 1999). In this case, even landfills with advanced cover systems are thought to recover just slightly over 60% of the LFG generated. However, a more recent study in France found LFG recovery rates ranged from 41% to 94% of the theoretical CH_4 production and were highly dependent on the engineered cover design (ADEME 2008). It further suggested that average LFG recovery rates could exceed 90% by excluding the poorest performing cover design from the study.

LFG generation and capture rates vary across a temporal scale. This has led the French environment agency (ADEME) to create different default values to account for landfill design and stage of operation with values ranging from 35% to 90% recovery (Spokas et al. 2006). The literature on this subject clearly shows that there is a high level of uncertainty when it comes to calculating emissions from landfills. However, industry experts believe that methane emissions from wood products in landfills will become a smaller part of the total carbon foot print from HWPs as technology improves and more LFG is captured (Miner 2008).

In 2007, 3.7 billion m^3 of methane was captured from landfills in the United States, of which 70% was used to generate thermal or electrical energy (Themelis and Ulloa 2007). The rest of the captured methane was flared since it was thought to have no economic value. Flaring of LFG and using it in energy production reduces the methane content to carbon dioxide and water (Johannessen 1999). Despite the fact that flaring reduces the potency of the methane, it still produces high levels of CO_2 emissions. It must also be noted that there are nearly 1,400 landfills in the United States (EPA 2008) that do not capture and flare any biogas. It is likely that HWPs in these sites are generating high levels of CH_4 emissions.

Including end-of-life conditions in HWP carbon stock models is critical due to the large potential emissions from landfills. Carbon released during end-of-life processes does not follow the simple decay functions most often used to model HWP retirement and discard. As described above, landfills may have varying conditions which will have a large impact on HWP carbon stocks. How these landfills are incorporated into HWP carbon stock accounting is key. In the United States, for example, only about 20% of 1,754 landfills are currently capturing LFG (EPA 2008). This figure raises serious doubts on the default LFG capture rate of 75% used by the EPA in the WARM model. Unfortunately, unrealistic default LFG capture rates have the potential to lead to significant miscalculation of the role of HWPs not only on a country basis but globally. Policies that promote or permit landfilling of HWPs could be aligned with policies that require high percentage LFG capture rates to ensure net emission reductions.

7 Management and Policy Implications

As policymakers focus on the role of forests and HWPs in mitigating climate change, additional research is needed to fully understand the relationships among climate policy, the forest products industry, consumers, and forests. Management and policy implications are summarized below.

7.1 Management Implications

- *Forestlands*. The potential to sequester carbon in forests is much larger than the potential to sequester carbon in forest products. Minor changes in forest extent have much greater impacts on GHG emissions than the forest products industry. Some researchers (Kauppi and Sedjo 2001; NCASI 2007) refer to the beneficial role that the forest products industry plays in maintaining sustained-yield forestland.
- The production and use of HWPs may postpone carbon emissions as carbon is stored in HWPs for a period after the initial harvest of roundwood. If the production of HWP exceeds the rate of retirement, then the amount of carbon bound in the HWP stock increases.
- Current methods for estimating carbon in HWPs are highly variable. A lack of data on product use makes it difficult to model HWP stocks; even assumptions on average wood density can significantly alter estimates of the conversion of HWP mass into carbon.
- *Substitution*. Each major building materials industry (wood, steel, and concrete) has published studies suggesting that their products are superior from the perspective of climate change mitigation. Given that climate considerations are currently an externality, more research is needed to understand what factors drive materials selection and whether a carbon price signal is sufficient to overcome these factors.
- *Landfills*. From the perspective of greenhouse gas emissions landfills with effective landfill gas collection could potentially be an acceptable final destination for discarded HWPs since HWPs have shown to have very low rates of decay in landfills. The production of LFG also fits into the "cascaded use of HWPs" framework because it can be converted into energy, displacing fossil fuels and further reducing global emissions. At the same time, it must be demonstrated that unintended consequences such as increased emissions elsewhere in the wood product life cycle are not triggered by strategies intended to enhance the HWP carbon stock at end of life.

7.2 Policy Implications

- The rise of biomass energy use in the forest products industry, as well as increasing utilization of wood products, has been driven by several factors. These include the competitive nature of the industry and the need to lower costs while seeking new sources of revenues, particularly for by-products and co-products that had historically not generated an economic return to the industry. Under certain economic conditions, however, forest products manufacturers may be inclined to alter manufacturing processes, which could result in incremental emitting activities under certain scenarios, particularly if it lowers costs for a profit maximizing entity.

- To date, policymakers have not fully considered the role harvested wood products can play in climate change mitigation and have not linked forest management practices to the full life cycle of harvested wood products. Incentives could be considered to support the use of recycled materials, to encourage such activities as product substitutions, industrial energy efficiency, and to encourage biomass fuel sources.
- There are many factors that will favor or disfavor wood as a construction material or energy source. These include relative price, technology, economic growth, policy, market efficiency, socioeconomic factors, and quality and quantity of energy and materials (Gustavsson et al. 2006). Recognizing that wood products are still largely a cyclical industry driven by global GDP, policies could begin to introduce longer- term, secular demand for wood products that encourage investment in wood that is both economically and environmentally sound.
- Recycling should be promoted heavily in policy intended to enhance the HWP carbon stock since recycling postpones carbon emissions of even short-lived HWPs. Recycling also fits very well in the "cascaded use of HWPs" concept where HWP are transformed multiple times within a tight recycling chain and finally converted into bioenergy.

References

ADEME, 2008. State-of-the-practices and implementation recommendations for non hazardous waste management using bioreactors landfills. French Environment and Energy Management Agency (Agence de l'Environnement et de la Maîtrise de l'Energie). http://www2.ademe.fr/servlet/getDoc?cid=96&m=3&id=51261&p2=17618&ref=17618

Bogner JE, Spokas K (1993) Landfill CH_4: rates, fates and global carbon cycle. Chemosphere 26:369–386

Brown R (1999) Timberland shedding trend: why are paper firms selling their land? Northern Logger Timber Processor 48:2

Burschel P, Kursten E, Larson BC, Weber M (1993) Present role of German forests and forestry in the national carbon budget and options to its increase. Water Air Soil Pollut 70:325–340

Carlson E, Heikkinen P (1998) Energy Consumption by the Finnish Pulp and Paper Industry and Comparison of Electricity Prices in Certain European Countries (Report in Finnish, Abstract in English), The Finnish Pulp and Paper Research institute, Sustainable Paper Report 21. 81 p

CEPI (2007) Framework for the development of carbon footprints of paper and board products

Corder SE, Scroggins TL, Meade WE, Everson GD (1972) Wood and bark residues in Oregon: trends in their use, research paper 11. Oregon State University, Forest Research Laboratory

Cote WA, Young RJ, Risse KB, Costanza AF, Tonelli JP, Lenocker C (2002) A carbon balance method for paper and wood products. Environ Pollut 116:S1–S6

Davidsdottir B, Ruth M (2004) Capital vintage and climate change policies: the case of US pulp and paper. Environ Sci Policy 7:221–233

Denman KL, Brasseur G, Chidthaisong A, Ciais P, Cox P, Dickinson RE, Haugustaine D, Heinze C, Holland E, Jacob D, Lohmann U, Ramachandran S, da Silva Dias PL, Wofsy SC, Zhang X (2007) Couplings between changes in the climate system and biogeochemistry. In: Solomon S, Qin D, Manning M, Chen Z, Marquis M, Avery KB, Tignor M, Miller HL (eds) Climate change 2007: the physical science basis, contribution of working group I to the fourth asssessment report of the intergovernmental panel on climate change. Cambridge University Press, Cambridge, pp 499–587

Dias AC, Louro, Margarida, Arroja L, Capela I (2009) Comparison of methods for estimating carbon in harvested wood products. Biomass and Bioenergy 33:213–222

Energy Information Administration (2008) Forest products industry analysis brief. http://www.eia.doe.gov/emeu/mecs/iab98/forest/. Accessed 30 Dec 08

EPA (2006) Solid waste management and greenhouse gases: a life-cycle assessment of emissions and sinks, 3rd edn. United States Environmental Protection Agency, Washington, DC

EPA (2008) Municipal solid waste in the United States: 2007 facts and figures. U.S. Environmental Protection Agency, Office of Solid Waste, EPA530-R-08-010

Falk RH, McKeever DB (2004) Recovering wood for reuse and recycling: a United States perspective. In: Gallis C (ed) Proceedings of management of recov-

ered wood recycling, bioenergy and other options. University Studio Press, Thessaloniki, pp 29–40
FAO (2004) FAOSTAT. United Nations Food and Agriculture Organization. http://faostat.fao.org/site/628/default.aspx
FAO (2007) State of the world's forests. United Nations Food and Agriculture Organization, Rome
Farahani S, Worrell E, Bryntse G (2004) CO_2-free paper? Resour Conserv Recycl 42:317–336
Finkral AJ, Evans AM (2008) Effects of a thinning treatment on carbon stocks in a northern Arizona ponderosa pine forest. For Ecol Manag 255:2743–2750
Fischlin A, Midgley GF, Price JT, Leemans R, Gopal B, Turley C, Rounsevell MDA, Dube OP, Tarazona J, Velichko AA (2007) Ecosystems, their properties, goods, and services. In: Parry ML, Canziani OF, Palutikof JP, van der Linden PJ, Hanson CE (eds) Climate change 2007: impacts, adaptation and vulnerability. Contribution of working group II to the fourth assessment report of the intergovernmental panel on climate change. Cambridge University Press, Cambridge, UK, pp 211–272
Gardner D, Cowie A, Ximenes F (2004) A new concept for determining the long-term storage of carbon in wood products. Cooperative Research Centre for Greenhouse Accounting
Grantham JB, Howard JO (1980) In: Sarkanen KV, Tillman DA (eds) Logging residues as an energy source in biomass conversion, vol 2. Academic, New York
Green C, Avitabile V, Farrell EP, Byrne KA (2006) Reporting harvested wood products in national greenhouse gas inventories: Implications for Ireland. Biomass and Bioenergy 30:105–114
Gustavsson L, Sathre R (2006) Variability in energy and carbon dioxide balances of wood and concrete building materials. Build Environ 41:940–951
Gustavsson L, Madlener R, Hoen HF, Jungmeier G, Karjalainen T, KlÖhn S, Mahapatra K, Pohjola J, Solberg B, Spelter H (2006) The role of wood material for greenhouse gas mitigation. Mitig Adapt Strateg Glob Change 11:1097–1127
Han FXX, Lindner JS, Wang CJ (2007) Making carbon sequestration a paying proposition. Naturwissenschaften 94:170–182
Howard J (2006) Estimation of U.S. timber harvest using roundwood equivalents In: McRoberts RE, Reams GA, Van Duesen PC, McWilliams WH (eds) Proceedings of the sixth annual forest inventory and analysis symposium; 2004 September 21–24; Denver, CO. Gen Tech Rep WO-70 Washington, DC: U.S. Department of Agriculture Forest Service 126 p
Ince PJ (1995) What Won't Get Harvested, Where and When: The effects of increased paper recycling on timber harvest. PSWP working paper #3. New Haven: Yale University School of Forestry & Environmental Studies.
International Energy Agency (IEA) (2003) Key world energy statistics 2003. International Energy Agency, Paris
IPCC (1996) Climate change 1995: impacts, adaptations and mitigation of climate change. Cambridge University Press, Cambridge, UK
IPCC (2000) Robert T. Watson, Ian R. Noble, Bert Bolin, Ravindranath NH, David J. Verardo, David J. Dokken (eds) Cambridge University Press, UK. pp 375
IPCC (2007) Forestry. In: Climate Change 2007: Mitigation. Contribution of Working Group III to the Fourth Assessment Report of the Intergovernmental Panel on Climate Change. Cambridge University Press, Cambridge/New York
Irland L (2008) Professor of Forest Finance, Yale School of Forestry and Environmental Studies, personal communication
Johannessen LM (1999) Guidance note on recuperation of landfill gas from municipal solid waste landfills. World Bank Urban Development Division
Kauppi P, Sedjo R (2001) Technological and economic potential of options to enhance, maintain, and manage biological carbon reservoirs and geo-engineering. In: Climate change mitigation, IPCC working group panel III report, pp 301–345
Kline DE (2005) Gate-to-gate life-cycle inventory of oriented strandboard production. Wood Fiber Sci 37:74–84
Kuchli C (2008) Chair's conclusions and recommendations In: Hetsch S (ed) Harvested wood products in the context of climate change policies, United Nations Palais des Nations, Geneva, Switzerland
Lippke B, Perez-Garcia J (2008) Will either cap and trade or a carbon emissions tax be effective in monetizing carbon as an ecosystem service? For Ecol Manag 256:2160–2165
Luo L, Evd V, Huppes G, Udo de Haes HA (2009) Allocation issues in LCA methodology: a case study of corn stover-based fuel ethanol. Int J LCA 14:529–539
Marceau ML, VanGeem MG (2002) Life cycle assessment of an insulating concrete form house compared to a wood frame house. PCA R&D SN2571. Skokie, Illinois, Portland Cement Association
Marceau ML, VanGeem MG (2008) Comparison of the life cycle assessments of an insulating concrete form house and a wood frame house. PCA R&D SN3041. Skokie, Illinois, Portland Cement Association
Marland E, Marland G (2003) The treatment of long lived, carbon containing products in inventories of carbon dioxide emissions to the atmosphere. Environ Sci Policy 6:139–152
Matthews RW, Nabuurs GJ, Alexyeyev V, Birdsey RA, Fischlin A, Maclaren JP, Marland G, Price DT (1996) Evaluating the role of forest management and forest products in the carbon cycle. In: Apps MJ, Price DT (eds) Forest ecosystems, forest management and the global carbon cycle. NATO ASI Series Vol. 40, pp. 293–301. Springer-Verlag, Berlin Heidelberg
McKeever DB (2002) Domestic market activity in solid wood products in the United States, 1950–1998. U.S. Department of Agriculture, Forest Service, Pacific Northwest Research Station
Meil J, Wilson J, O'Connor J, Dangerfield J (2007) An assessment of wood product processing technology

advancements between the CORRIM I and II studies. For Prod J 57:83–89
Micales J, Skog K (1997) The decomposition of forest products in landfills. Int Biodeterior Biodegrad 39:145–158
Miner R (2008) The carbon footprint of forest products. In: GAA environmental workshop, Geneva, 17–19 June 2008
Murakami S, Oguchi M, Tasaki T, Daigo I, Hashimoto S (2010) Lifespan of commodities, part I: The creation of a database and its review. J Ind Ecol 14(4):598–612
Nabuurs GJ, Pussinen A, van Brusselen J, Schelhaas MJ (2007) Future harvesting pressure on European forests. Eur J For Res 126:391–400
National Council for Air and Stream Improvement (2007) The greenhouse gas and carbon profile of the global forest products industry. NCASI Spec Rep 07–02:1–32
Nishioka Y, Yanagisawa Y, Spengler JD (2000) Saving energy versus saving materials: Life-cycle inventory analysis of housing in a cold-climate region of Japan. J Ind Ecol 4(1):119–136
Oguchi M, Murakami S, Tasaki T, Daigo I, Hashimoto S (2010) Lifespan of commodities, part II: Methodologies for estimating lifespan distribution of commodities. J Ind Ecol 14(4):613–626
Perez-Garcia J, Lippke B, Comnick J, Manriquez C (2005) An assessment of carbon pools, storage, and wood products market substitution using life-cycle analysis results. Wood Fiber Sci 37:140–148
Pingoud K, Lehtila A (2002) Fossil carbon emissions associated with carbon flows of wood products. Mitig Adap Strateg Glob Change 7:63–83
Pingoud K, Soimakallio S, Perala AL, Pussinen A (2003) Greenhouse gas impacts of harvested wood products: evaluation and development of methods. Espoo 2003. VTT Technical Research Center of Finland, Research Notes 2189, 120 p. + app. 16 p
Puettmann ME, Wilson JB (2005) Life-cycle analysis of wood products: cradle-to-gate LCI of residential wood building materials. Wood Fiber Sci 37:18–29
Rueter S (2008) Model for estimating carbon storage effects in wood products in Germany. In: Hetsch S (ed) Harvested wood products in the context of climate change policies, United Nations Palais des Nations, Geneva, Switzerland
Sampson RN, Apps M, Brown S, Cole CV, Downing J, Heath LS, Ojima DS, Smith TM, Solomon AM, Wisniewski J (1993) Workshop summary statement—terrestrial biospheric carbon fluxes—quantification of sinks and sources of CO_2. Water Air Soil Pollut 70:3–15
Sedjo RA (2008) Biofuels: think outside the cornfield. Science 320:1419–1420
Skog K (2008) Sequestration of carbon in harvested wood products for the United States. For Prod J 58:56–72
Skog K, Nicholson GA (2000) Carbon sequestration in wood and paper products. In: Joyce LA, Birdsey R, technical editors 2000. The impact of climate change on America's forests: a technical document supporting the 2000 USDA Forest Service RPA Assessment. Gen. Tech. Rep. RMRS-GTR-59. Fort Collins, CO: U.S. Department of Agriculture, Forest Service, Rocky Mountain Research Station, p 79–88
Skog KE, Nicholson GA (1998) Carbon cycling through wood products: the role of wood and paper products in carbon sequestration. For Prod J 48:75–83
Skog KE, Pingoud K, Smith JE (2004) A method countries can use to estimate changes in carbon stored in harvested wood products and the uncertainty of such estimates. Environ Manag 33:S65–S73
Spokas K, Bogner J, Chanton JP, Morcet M, Aran C, Graff C, Golvan YML, Hebe I (2006) Methane mass balance at three landfill sites: what is the efficiency of capture by gas collection systems? Waste Manag 26:516–525
Stern N (2006) Stern review on the economics of climate change. UK Office of Climate Change. http://www.hm-treasury.gov.uk/stern_review_final_report.htm
Stern T (2008) Contribution of the Austrian forest sector to climate change policies – industrial marketing implications. In: Hetsch S (ed) Harvested wood products in the context of climate change policies, United Nations Palais des Nations, Geneva, Switzerland
Subak S, Craighill A (1999) The contribution of the paper cycle to global warming. Mitig Adap Strateg Glob Change 4:113–136
Themelis NJ, Ulloa PA (2007) Methane generation in landfills. Renewable Energy 32:1243–1257
United States Environmental Protection Agency (2008) Inventory of U.S. greenhouse gas emissions and sinks: 1990–2006. USEPA #430-R-08-005
Upton B, Miner R, Spinney M, Heath LS (2008) The greenhouse gas and energy impacts of using wood instead of alternatives in residential construction in the United States. Biomass Bioenergy 32:1–10
Walker T, Cardellichio P, Colnes A, Gunn J, Kittler B, Perschel B, Recchia C, Saah D (2010) Biomass sustainability and carbon policy study. NCI-2010-03. Manomet Center for Conservation Sciences, Brunswick
Watson R, Noble IR, Bolin B, Ravindrath NH, Verardo DJ, Dokken D (2000) IPCC Special Report on Land Use, Land-Use Change and Forestry. IPCC, Geneva
Wernick IK, Waggoner PE, Ausubel JH (1997) Searching for leverage to conserve forests: the industrial ecology of wood products in the United States. J Ind Ecol 1(3):125–145
Willumsen H (2004) Number and types of landfill gas plants worldwide. Presentation to World Bank, November 18. http://deponigas.dk/uploads/media/Number_and_Types_of_LFG_Plants_Worldwide.pdf. Accessed 13 Oct 2011
Wilson J (2005) Documenting the environmental performance of wood building materials. Wood Fiber Sci 37:1–2

Winjum JK, Brown S, Schlamadinger B (1998) Forest harvests and wood products: Sources and sinks of atmospheric carbon dioxide. For Sci 44(2): 272–284

Woodbury PB, Smith JE, Heath LS (2007) Carbon sequestration in the US forest sector from 1990 to 2010. For Ecol Manag 241:14–27

World Resources Institute (2006) Climate analysis indicators tool (CAIT) on-line database version 3.0. World Resources Institute, Washington, DC. Available at. http://cait.wri.org

Ximenes FA, Gardner WD, Cowie AL (2008) The decomposition of wood products in landfills in Sydney, Australia. Waste Manag 28:2344–2354

Part IV

Socioeconomic and Policy Considerations for Carbon Management in Forests

Section Summary

While the biophysical characteristics of forests covered in the earlier parts of this book define the boundaries within which forest management can occur, actual management practices are driven by economic, policy and other cultural values. The purpose of the following chapters is to explore the economic and policy drivers that affect the opportunities for managing forests with carbon in mind. In the first three chapters, the economic pressures and incentives facing land managers are described: first, in tropical developing countries experiencing rapid rates of deforestation as land is converted to more remunerative agricultural uses; second, in tropical countries retaining large areas of relatively intact forests as a result of physical or market isolation; and third in the U.S., where the economics of developing land for buildings far outweighs the incentives for maintaining land as farms and forests. Finding ways to use policy to help overcome these incentives for land managers to convert forests to more lucrative uses of the land is the focus of the last two papers. The factors to be considered when deciding between use of the carbon markets (through offset projects) or direct public funding of forest conservation are described at both the global level as part of the REDD+ negotiations and at the federal level in the U.S. building on the experience in the voluntary carbon markets. While increasing numbers of people agree that forests and other land use issues have to be a significant part of the global response to climate change, the ways in which this goal will be achieved is still open to considerable debate.

Contributors toward organizing and editing this section were: *Bradford Gentry, Deborah Spalding, Mary L. Tyrrell and Lauren Goers*

Large and Intact Forests: Drivers and Inhibitors of Deforestation and Forest Degradation

13

Benjamin Blom, Ian Cummins, and Mark S. Ashton

Executive Summary

We examine the political, economic, geographic, and biophysical reasons for the presence of the remaining large and intact forests of the world. We discuss why these forests remain relatively undisturbed, and analyze current drivers of deforestation and degradation. Such forests are primarily in boreal (northern Eurasia and Canada) and tropical (South America and Central Africa) regions and are the main focus of this paper because of their disproportionate role in sequestering carbon and in influencing regional and potentially global climate. We conclude with recommendations for policymakers to help incorporate these forests and their carbon stocks into initiatives designed to mitigate the damaging effects of global climate change.

What We Know About Large Forests and Trends in Deforestation:

- Although clearing of forestland continues at high rates in many parts of the world, large tracts of continuous, intact forest still cover roughly a quarter of originally forested biomes. These forests are unique in that they represent stable, yet vulnerable, carbon stocks. Because of both their extent and the large amount of carbon stored within these forests, their protection must be a significant part of any global policy initiatives to combat climate change.
- Presently, the vast majority of the world's remaining tracts of intact forest are concentrated within continental interior boreal and tropical wet and semi-evergreen forest biomes. Within the boreal biome, these forests cover the northern and largely inaccessible regions of Canada, Alaska, and Russia. Within the tropics, vast wet and semi-evergreen forests are found within the Amazon Basin of South America and the Congo Basin of Central Africa.
- In terms of carbon storage, three of the countries with the largest area of remaining intact forestland (Brazil, Russia and the Democratic Republic of Congo) hold an estimated 384 billion tons of carbon dioxide equivalents in above and below ground biomass, both dead and living biomass. For comparison, global emissions from energy consumption were estimated at 29 billion tons of carbon dioxide in 2006.
- The significant amount of carbon stored within these intact forests contain higher densities of carbon in soils and living biomass than degraded or secondary forests because of proportionately higher numbers of slower-growing trees with denser wood

B. Blom
Bureau of Land Management, USDA, Washington, DC, USA

I. Cummins
Forest Carbon Ltd, Bali, Indonesia

M.S. Ashton (✉)
Yale School of Forestry & Environmental Studies, 360 Prospect St, Marsh Hall, New Haven, CT, USA
e-mail: mark.ashton@yale.edu

M.S. Ashton et al. (eds.), *Managing Forest Carbon in a Changing Climate*, DOI 10.1007/978-94-007-2232-3_13, © Springer Science+Business Media B.V. 2012

- In addition to the important role these forests play in the global carbon cycle, their protection from land conversion yields highly significant co-benefits. Evidence suggests that intact forests have significant cooling effects on both regional and global climates through the accumulation of clouds from forest evapotranspiration, which also recycles water and contributes to the region's precipitation.
- The low fertility and high vulnerability of the soils in interior regions of the tropics has slowed the development of permanent agriculture in these areas. Human communities that reside in these regions typically have low population densities and rely on hunting and migratory cultivation.
- Colonial history played a role in low human population densities of the large forest interiors of South America and Central Africa by being primarily resource-driven, resulting in less permanent European settlements outside of administrative extractive hubs. In addition, after settlers were established (largely on the coast), European diseases decimated native populations even in areas largely untouched by European settlers.
- The geography of remoteness is of critical importance in explaining why intact forests exist where they do, namely, in continental interiors. Much of the world's population is concentrated within 100 km of coasts, with population density decreasing as one moves to the interior. In addition, large mountain ranges (e.g. the Andes) and rugged topography (e.g. New Guinea) serve as barriers.
- A shared trait among the world's large and intact forests is a lack of foreign investment; however, globalization of markets and export products/crops has facilitated forest exploitation and land conversion of intact forests in recent years.
- Lack of government presence has resulted in poor infrastructure development, few government services, and an inability to integrate these regions into larger market and governmental/organizational structures.
- Some countries, in order to facilitate rural in-migration to the forest frontier, in part to secure sovereignty where there are adjacent country claims and in part as a "poverty release valve," have provided agricultural subsidies, free land, and seeds to colonial settlers.
- At the country scale, forest loss often follows a Kuznets curve, whereby deforestation rates are initially static, increase during industrialization when populations are growing, and finally stabilize into an equilibrium state. However, growing economies, increasing affluence, extreme levels of poverty, and rapid decreases in prosperity during periods of economic crisis can alter this trend and lead to unanticipated deforestation and forest degradation.
- Deforestation of large sections of the central Amazon Basin is directly attributable to governmental stimulus plans, road building programs, and subsidies for livestock production.
- The construction of roads linking both core forests and frontier forests to population centers and export markets is tied to increasing rates of deforestation. While such public highways have caused localized deforestation, the lack of parallel access outside of these roads leaves large tracts of forest intact. Unofficial roads, however, form extensive, dense networks to support transportation of the resources being harvested or extracted and can exacerbate deforestation.
- A lack of governance, coupled with the presence of infrastructure, is often a precondition for widespread illegal operations that promote deforestation (e.g. logging, illicit drug trade). However, a lack of governance with no infrastructure inhibits illegal operations that promote deforestation.

1 Introduction

Each year, approximately 13 million hectares of tropical forest, equivalent to the land area of Greece, are felled, burned, and converted to an alternative land use (FAO 2005). When such land is converted, carbon stored within above-ground biomass, downed woody debris, and soil is released from forests to the atmosphere as carbon dioxide and methane gas. Land use change is currently responsible for approximately 17% of global

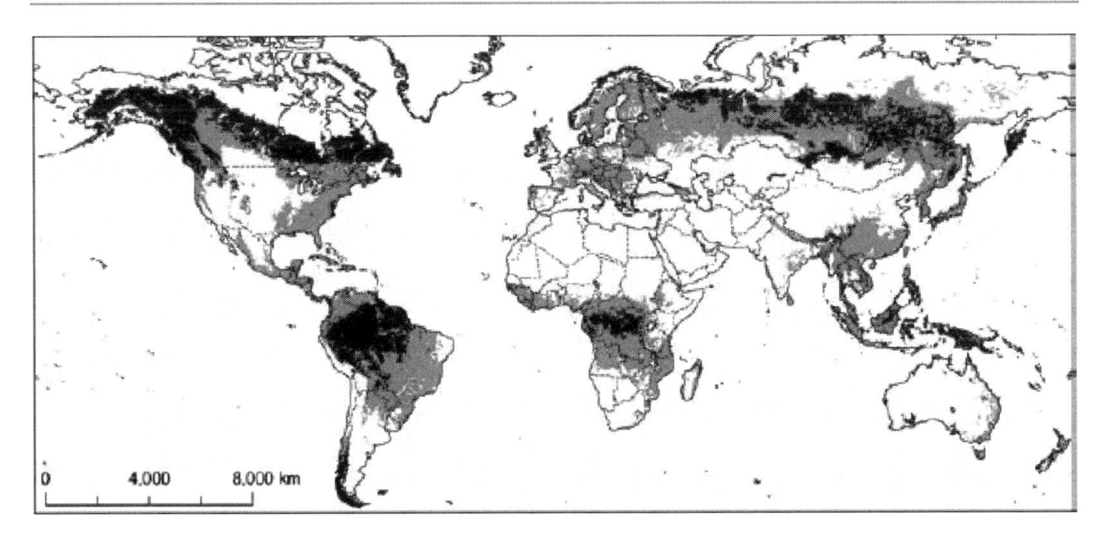

Fig. 13.1 Large and intact forests are highlighted in black. Other forested areas are highlighted in gray (From Potapov et al. 2008. Reprinted with permission)

greenhouse gas emissions, or 0.9 million tons of carbon dioxide equivalents per year (Defries et al. 2002; Pachauri and Reisinger 2007).

Although clearing of forestland continues at high rates in many parts of the world, large tracts of continuous, intact forest still cover roughly a quarter of originally forested biomes (Potapov et al. 2008). These forests are unique in that they represent stable, yet vulnerable, carbon sinks. Because of both the aerial extent and significant amount of carbon stored within the world's remaining large and intact forests, their protection must be part of any global policy initiatives to combat climate change. We examine the world's remaining large and intact forests and discuss the political, economic, geographic and biophysical reasons why these forests remain relatively undisturbed. We will analyze current drivers of deforestation and degradation and conclude with several recommendations for policymakers to help incorporate these forests and their carbon stocks into initiatives designed to mitigate the damaging effects of global climate change.

1.1 Defining Large and Intact Forests

Large and intact forests are defined as unbroken expanses of forest with negligible levels of human-induced degradation, resource exploitation, and fragmentation. As part of this definition, there is a continuous spatial threshold that these forests must meet (~100,000 km^2). Intact forests are functioning ecosystems characterized by full species assemblages, naturally occurring disturbance regimes, and unaltered hydrological patterns. They are distinct from exploited and/or degraded forests, which tend to occur as patches within a mosaic of developed and agricultural areas (Chomitz 2006). It is important to distinguish these degraded-mosaic forests from intact forests. While secondary, mosaic, and degraded forests play important roles for biodiversity, social values, carbon sequestration, and climate change amelioration, they are functionally distinct from large and intact forests in ecological terms, disturbance regimes, and management objectives. Within large and intact forests, deforestation primarily occurs along agricultural frontiers and generally does not occur from within the core interior forested areas (Chomitz 2006). Because large intact forests have high area-to-perimeter ratios, they have fewer access points than fragmented, mosaic forests, which help to shield the interior areas, at least in part, from deforestation. Presently, the vast majority of the world's remaining tracts of continuous, intact forest are concentrated within continental interior boreal and tropical wet and semi-evergreen forest biomes (Fig. 13.1).

Table 13.1 Forest and carbon data for countries and regions with large intact forests

	Forest area	Primary/intact forest area	Carbon in biomass	Carbon biomass density	Deforestation rate (2000–2005)	Loss of primary forest (2000–2005)
Amazon Basin/ Guayanas	(1,000 ha)	(1,000 ha)	Million tons CO_2 equivalents	(ton/ha)	(1,000 ha/yr)	(1,000 ha)
Bolivia	58,740	29,360	19,436	331	−270	−135.2
Brazil	477,698	415,890	181,059	379	−3,103	−3,466
Columbia	60,728	53,062	29,588	487	−47	−56.16
Guayana	15,104	9,314	6,320	418	0	0
Peru	68,742	61,065	Unknown	Unknown	−94	−224.6
Suriname	14,776	14,214	20,890	1,414	0	0
Venezuela	47,713	Unknown	Unknown	Unknown	−288	Unknown
Total Amazon/ Guayana Nations	743,501	582,905	257,293		−3,802	−3881.96
Congo Basin						
Cameroon	21,245	Unknown	6,980	329	−220	Unknown
CAR	22,755	Unknown	10,280	452	−30	Unknown
Congo	22,471	7,464	19,014	846	−17	−5.647
DRC	133,610	Unknown	85,045	637	−319	Unknown
Equatorial Guinea	1,632	Unknown	423	259	−15	Unknown
Gabon	21,775	Unknown	13,370	614	−10	Unknown
Total Congo Basin Nations	223,488	7,464	135,112	3136.188798	−611	−5.647
New Guinea						
Papua New Guinea	29,437	25,211	Unknown	Unknown	−139	−250.2
Total New Guinea	29,437	25,211	Unknown	Unknown	−139	−250.2
Boreal						
Russian Federation	808,790	255,470	118,211	146	−96	−532.2
Canada	310,134	165,424	Unknown	Unknown	0	0
Total Boreal Nations	1,118,924	420,894	118,211	146	−96	−532.2

Data compiled from FAO (FAO 2005). Cells shaded in grey contain incomplete data

Within the boreal biome, these forests cover the northern and largely inaccessible regions of Canada, Alaska, and Russia. They are characterized by short, cool summers followed by long, cold winters and are often dominated by single-stand coniferous forests with limited plant species diversity (Wieder and Vitt 2006). Within the tropics, vast wet and semi-evergreen forests are found within the Amazon Basin of South America and the Congo Basin of Central Africa. The Amazon Basin (5.5 million km^2) and the adjacent forests of the Guyana Shield contain the largest intact and contiguous tropical forest in the world. This forest is shared by nine nations of South America (Bolivia, Brazil, Columbia, Ecuador, French Guiana, Guyana, Peru, Suriname and Venezuela), although the majority of this forest lies within the borders of Brazil (Table 13.1) (Encyclopedia Britannica 2009).

The Congo Basin in Africa (3.5 million km^2) contains the world's second largest contiguous tropical forest and is largely located within the borders of six nations (Cameroon, Central African Republic, Congo, the Democratic Republic of the Congo, Equatorial Guinea, and Gabon) (CARPE 2001). Similar to the Amazon Basin, a single country (the Democratic Republic of the Congo (DRC)) contains a majority of the region's forested area (Table 13.1). Smaller, but significant, intact tropical forests are found on the islands of Sumatra, Borneo, and New Guinea, the highlands of mainland Southeast Asia, and the Atlantic coast of Central America. This chapter will focus on the boreal forests of Canada and Russia (including temperate forests bordering these boreal forests), the Amazon and Congo Basins, and New Guinea.[1] It will not consider forests in temperate regions, with the exception of temperate forests bordering boreal forests in Canada and Russia, since most forest cover in temperate parts of the world is patchy and dominated by secondary growth (Fig. 13.1) (Potapov et al. 2008).

A significant proportion of large and intact forests are within the borders of a small number of countries. For example, Brazil, Canada, and Russia contain 63.8% of the area of the world's remaining large and intact forests within their borders (Potapov et al. 2008) (Fig. 13.1).

2 Why Is Protecting Large and Intact Forests So Important?

Large and intact forests are extremely important for the multitude of ecosystem services they provide. Despite their global importance, however, only 18% of these forests had been designated as Protected Areas as of 2008 (Potapov et al. 2008).[2] Unfortunately, even with this designation, protection is minimal. While these forests have remained largely intact, they are often in areas under increasing pressure from land use conversion, road building, and timber extraction. As one example, recent history has seen rapid, large-scale deforestation in Borneo due to illegal logging and industrial-scale land conversion for agriculture (Curran et al. 2004). With increasing rates of deforestation, and its impact on global greenhouse gas emissions, there is broad consensus that continued illegal logging and aggressive industrial land conversion practices must be addressed immediately, either through market-based incentives such as carbon credits, regulatory structures to improve governance, or a combination of both (Zhang et al. 2006; Betts et al. 2008; Buchanan et al. 2008; Nepstad et al. 2008).

2.1 Carbon Sequestration and Storage[3]

Carbon markets may provide effective financial incentives to deter land conversion and illegal logging in large and intact forests simply due to the sheer amount of carbon stored in these forested areas. In terms of carbon storage, three of the four countries with the largest area of remaining intact forestland (Brazil, Russia, and the Democratic Republic of Congo) hold an estimated 384 billion tons of carbon dioxide equivalents in above and below ground biomass, including dead and living biomass (FAO 2005) (Table 13.1). In comparison, global emissions from energy consumption were estimated at 29 billion tons of carbon dioxide in 2006 (EIA 2006). The significant amount of carbon stored within these countries is due to the fact that primary or intact forests contain higher densities of carbon in soils and living biomass than degraded or secondary forests because of proportionately higher numbers of slower-growing trees with denser wood (Olson et al. 1985).[4]

[1]Because of the difficulty in segregating countrywide data in Indonesia from data specific to New Guinea (Irian Jaya), discussions regarding New Guinea will be focused on the Papua New Guinea half of the island.

[2]A Protected Area is defined as any land that is given protected status as an extractive reserve, national park, indigenous reserve or wildlife reserve.

[3]For more information regarding carbon sequestration and storage, see the science chapters of this book.

[4]While the total amount of carbon stored is highest in primary or intact forests, rates of carbon sequestration are highest in fast-growing secondary forest. This is why large and intact forests are considered carbon "reservoirs" as opposed to the carbon "sinks" of growing secondary forests.

2.2 Co-benefits of Protecting Large and Intact Forests

In addition to the important role these forests play in the global carbon cycle, their protection from land conversion yields highly significant co-benefits as well. First, large intact forests have been shown to play a role in regional climate regulation (Hoffman et al. 2003; Spracklen et al. 2008). In the boreal region, intact forests have a significant cooling effect on both regional and global climates through the accumulation of clouds from boreal forest evapo-transpiration (Spracklen et al. 2008). The cooling effect that large forests exert via evapo-transpiration has also been demonstrated in the tropics, particularly in the Amazon. A large portion of the precipitation in interior and continental regions of the Amazon Basin is derived from evapo-transpiration that is released over the course of a day (Makarieva and Gorshkov 2007). When there is deforestation of forest frontiers or edges, interior regions of wet tropical forests often cannot sustain their current forest type due to changes in precipitation patterns (Makarieva and Gorshkov 2007). When large swaths of previously intact tropical forests are cleared, evapotranspiration occurs much more rapidly, leading to disrupted precipitation patterns downwind of the deforestation (Roy et al. 2005). In one study, a model of precipitation in the Congo Basin suggested that rainfall could be reduced by 10% in certain regions as a result of deforestation (Roy et al. 2005).

Second, when changes to precipitation patterns occur in tropical forests, they can lead to altered fire regimes, which can impact the resilience of remaining forests. Many countries in the tropics with significant rates of deforestation and land conversion now experience much more frequent and severe fires (Siegert et al. 2001; Hoffman et al. 2003). These resulting fires can exacerbate deforestation and degradation rates in remaining forests, which in turn can have a large impact on global carbon emissions (Hoffman et al. 2003). This effect was seen in the 1997 fires on the island of Borneo, which released an estimated range of 8–25 billion tons of CO_2 equivalent into the atmosphere, equal to 13–40% of the mean annual global emissions from fossil fuels (Page 2002).

Third, there is ample evidence that forest fragmentation and degradation have significant effects on both floral and faunal species composition within a given region (Curran and Leighton 2000; Hoffman et al. 2003; Roy et al. 2005). Certain changes in plant species composition can compromise the resilience of an entire ecosystem and reduce its ability to withstand disturbance. Many plant species rely on large expanses of forest for their regeneration and cannot effectively reproduce in mosaic or fragmented forests (Curran and Leighton 2000). In addition to protecting plant biodiversity, these forests also provide some of the only remaining suitable habitat for wildlife in their respective regions (Joppa et al. 2008).

3 Common Features of the World's Remaining Large and Intact Forests

Today's large and intact forests share a number of common traits that have historically hindered deforestation. Many of these forests also, not surprisingly, share similar risks of potential deforestation. There are, however, regional variations which are important to keep in mind. For example, while industrial-scale agriculture and regional infrastructure play a strong role in deforestation in the Amazon, they are not considered significant threats to the forests of the Congo Basin. Reasons for regional differences are complex and often due to both local and international factors. (Table 13.2).

3.1 Why Have These Forests Remained Intact?

Deforestation rates in these forests are often much lower than other regions where land use conversion continues at a rapid pace. For example, in the Congo Basin nations of the Central African Republic, Congo, Democratic Republic

Table 13.2 A comparison among large intact forest regions of key factors facilitating persistence, and current drivers of deforestation and degradation

	Amazon Basin/Guyanas	Congo Basin	Boreal forests	New Guinea
Key historical factors allowing forest persistence				
Biophysical limitations				
Soil infertility	X	X	X	
Climatic barriers to agriculture			XX	
Low population density	X	X	XX	
Biogeographical isolation				
Inaccessibility to markets	XX	XX	XX	X
Governmental factors				
Lack of governmental capacity	X	XX		X
Lack of infrastructure	XX	XX	X	X
Low levels of foreign investment	X	XX	X	X
Current drivers of deforestation				
Poverty				
Subsistence extraction and agriculture	X	X		
Governance				
Land tenure insecurity	X	X		X
Poorly designed concession systems				
Corruption	X			X
Illegal resource extraction				
Infrastructure expansion	XX	X		
International trade and investment				
Poorly managed timber extraction				
Foreign investment	X	X		X
Current drivers of degradation				
Poverty				
Subsistence extraction and agriculture	X	XX		X
Governance				
Land tenure insecurity	X	XX		X
Poorly designed concession systems	X	X	X	X
Corruption	X	X	X	
Illegal resource extraction	X	X	X	X
Infrastructure expansion	XX	XX		
International trade and investment				
Poorly managed timber extraction	X	X	X	X
Foreign investment	X	X		X

A single X denotes regionally important factors and XX denotes highly important factors

of the Congo, and Gabon, average annual deforestation rates between 2000 and 2005 were only 0.13%, 0.076%, 0.24% and 0.046% respectively of their total forest area per year (Table 13.1) (FAO 2005). In contrast, the deforestation rate in Indonesia and Cambodia was 2.0% per year from 2000 to 2005 (FAO 2005). The remaining large and intact forests of the world persist to this day because of biophysical, biogeographical, demographic, governmental and economic factors that have allowed these forests to remain relatively undisturbed (Table 13.2), while primary forests in other parts of the world have gradually decreased in size and extent. An historical understanding of why deforestation rates in these areas have remained low will shed some light on the risks these forests may face as conditions change.

3.2 Biophysical Limitations

3.2.1 Tropics

The geography of human settlement is neither random nor uniform. Although the wet tropical rainforests of the world support an ecosystem of tremendous biodiversity, they are typically an inhospitable place for humans to live. The term "Counterfeit Paradise" has been coined to describe this paradox between biological richness and the physical impoverishment of many tropical forest dwellers (Meggers 1995). The majority of soils within the Congo Basin and Upper Amazon are classified as oxisols by the U.S. Department of Agriculture (Natural Resources Conservation Service 2005). These soils are characterized by extremely low levels of fertility, small nutrient reserves, low cation exchange capacities, and shallow organic layers. Because the nutrients in oxisols are rapidly leached by rainfall and because tropical forests receive extremely high amounts of precipitation, these forests must undertake rapid decomposition and nutrient cycling to prevent nutrient depletion (Markewitz et al. 2004). As a result, when they are converted to agriculture, these soils are typically only productive for a few years (Montagnini and Jordan 2005).

The low fertility and high vulnerability of the soils in interior regions of the tropics have prevented the development of permanent agriculture and led to cultures with low population densities which rely on hunting and migratory cultivation. These shifting cultivation/swidden cultures often do not put too much pressure on forest resources, which has helped to preserve large and intact forests in many of the areas they inhabit (Dove 1983).

While many of the interior regions of tropical areas have low soil fertility, other tropical areas can be highly suitable for agriculture. The volcanic, highly fertile soils of Java and the Great Lakes Region of Africa support some of the highest rural population densities in the world despite being located in areas that are classified as tropical rainforest (Natural Resources Conservation Service/USDA 2000). As a result of soil fertility and the ability to support large populations, the forests in these regions were largely converted to alternative land uses centuries ago. The relatively fertile highlands of New Guinea, which is one of the focal areas of this chapter, are an exception. This is mostly due to the fact that the soils of New Guinea are typically inceptisols, which although suitable for agriculture, are highly erodable on steep slopes, making agriculture logistically difficult (Natural Resources Conservation Service 2005).

Local climate may also play a role in deterring widespread agriculture within tropical basins. In the Brazilian Amazon, low levels of precipitation were shown to be the most important determining factor influencing the deforestation of land for agriculture and pasture. In fact, precipitation levels were found to be more important than access, soil fertility, and land protection status (Chomitz and Thomas 2003).

3.2.2 Boreal

Agriculture within the boreal ecosystems is inhibited by both poor soil quality and a climate that is unsuitable for most agriculture. Winters in the boreal zone are both long and extremely cold. Spring cold snaps and short growing seasons make agriculture in boreal forest regions unprofitable and unlikely to provide sufficient nourishment of large human settlements, particularly given seasonal risk (Wieder and Vitt 2006). Moreover, many boreal soils are classified as spodosols or gellisols. Spodosols tend to be acidic, have poor drainage, and low fertility while gellisols typically contain permafrost within 2 m of the soil surface (Natural Resources Conservation Service 2005). This makes it nearly impossible to undertake successful agricultural activities.

3.3 Population Density

3.3.1 Population Patterns

One of the more obvious shared traits among large and intact forests is that they are found where human populations are low. While low population densities are largely a result of the biophysical limitations of these regions, they are also due to biogeographical isolation and historical factors. The Amazon Basin, Congo Basin and boreal forests of North America and Eurasia all have population densities of less than ten people per km^2 (Natural Resources Conservation Service/ USDA 2000). Within these regions, rural population density is often much lower. For example, the population density in rural areas of the Peruvian Amazon was calculated to be about 1.6 people per km^2, in an area the size of roughly 715,000 km^2 (Instituto Nacional de Estadisticas y Informatica 2007). In many of our focal region nations, populations are highly urbanized and only a relatively small proportion of their populations live in rural areas. Notable exceptions to this are the DRC and Papua New Guinea, which both have largely rural populations (Table 13.3).

3.3.2 Colonial History in the Amazon Basin

In the Upper Amazon and Guyana Shield of South America, colonial history has played a large role in the low population densities of the interior portions of these countries. In tropical South America, for example, colonization was much more resource-driven, resulting in less permanent European settlements outside of administrative extractive hubs. In addition, after settlers were established (largely on the coast), European diseases decimated native populations even in areas largely untouched by European settlers (Diamond 1997). This assertion is supported by ongoing archeological research which indicates that pre-colonial indigenous populations within the Amazonian basin were significantly larger and more urbanized than those encountered after colonists arrived (Mann 2000). In contrast, the colonization of North America fits the "deep settler" model, in which Europe sent large numbers of immigrant families to settle permanently in the New World (Wolfe 1999). This led to increased fragmentation of forested landscapes from the onset of colonization.

As a result of its particular colonial legacy and the spatial-demographic patterns that resulted, population densities within the Amazon Basin in countries of Upper Amazonia (Peru, Bolivia, Ecuador, Columbia) and the Guyana Shield (Suriname, Guyana) have been much lower than those within the coastal and the Andean regions of South America. The 2005 national census found 75% of the Peruvian population to be urban dwelling, with the majority concentrated in coastal cities such as Lima, Trujillo, and Chiclayo (Instituto Nacional de Estadisticas yInformatica 2007). In many ways, the legacy of colonization is still seen in the population dynamics of the Amazon Basin today.

3.3.3 Population Growth

Despite having historically small populations, some regions with large and intact forests are experiencing rapid population growth. In Africa, the Democratic Republic of the Congo, Congo, and Gabon have estimated population growth rates of 3.0%, 2.6% and 2.2% respectively (Table 13.3). Papua New Guinea, another nation containing significant large and intact forests, has a population growth rate of 2.2% (Table 13.3). This suggests that while low population densities have historically inhibited deforestation and forest degradation in these regions, population may soon become a major deforestation and degradation driver. Meanwhile, Russia, on the other hand, is undergoing a 0.5% per year population decline (FAO 2005; Kaufmann et al. 2008).

Historically, larger populations have had higher rates of deforestation and forest degradation due to the need to support more people. Today, however, with increasingly globalized markets, local population growth may not play as significant a role in deforestation and forest degradation as one might imagine. As societies and economic trade become more global, populations growing in one region of the world can have large impacts on deforestation and forest degradation in another part of the world. One example is in the Russian Far East, an area of extremely low population density and growth, whose forests are rapidly being degraded as a result of China's economic and demographic expansion (World Wildlife Fund Forest Programme 2007).

3.4 Biogeographical Isolation

The geography of remoteness is of critical importance in explaining why intact forests exist where

Table 13.3 Data for some current drivers of deforestation and forest degradation in the large and intact forests

	2004 per capita GDP[c] ($US)	2004 Population density[c] (Pop/km²)	Population growth rate[c] (Annual %)	2004 Rural population[c] (% of Population)	Growing stock removed[c] (% harvested)	Road density[b] (km road/km² area)	Political stability[a] (Range from −2.5 to 2.5)	Control of corruption[a] (Range from −2.5 to 2.5)
Amazon basin/Guayana shield								
Bolivia	1,036	8.3	1.9	36.1	0	0.07	−0.99	−0.49
Brazil	3,675	21.1	1.2	16.4	0.4	0.1	−0.22	−0.24
Columbia	2,069	43.6	1.6	23.1	nd	0.4	−1.65	−0.28
Guayana	962	3.9	0.4	62	nd	0.03	−0.32	−0.64
Peru	2,207	21.5	1.5	25.8	nd	0.03	−0.83	−0.38
Suriname	2,388	2.8	1.1	23.4	0	0.1	0.23	−0.26
Venezuela	4,575	29.6	1.8	12.1	nd	0.04	−1.23	−1.04
Avg Amazon/Guayana Nations	2,416	18.7	1.4	28.4	0.1	0.11	−0.72	−0.48
Congo basin								
Cameroon	651	35.2	1.9	47.9	1.5	0.03	−0.39	−0.93
CAR	232	6.3	1.7	56.8	0.1	0.06	−1.78	−0.9
Congo	956	11.3	2.6	46.1	0.1	0.04	−0.83	−1.04
DRC	89	24.2	3	67.7	0.3	0.03	−2.26	−1.27
Equatorial Guinea	3,989	18	2.4	51	0.9	0.06	−0.16	−1.37
Gabon	3,859	5.3	2.2	15.6	0.1	0.07	0.2	−0.85
Avg Congo basin nations	1,629	17	2.3	47.5	0.5	0.05	−0.87	−1.06
New Guinea								
Papua New Guinea	622	12.4	2.2	86.8	0.8	0.03	−0.76	−1.05
Avg New Guinea	622	12.4	2.2	86.8	0.8	0.03	−0.76	−1.05
Boreal								
Russian Fed.	2,302	8.5	−0.4	26.7	0.2	0.6	−0.75	−0.92
Canada	24,712	3.5	0.9	19.2	0.7	1.1	1.02	2.09
Avg Boreal nations	13,507	6	0.25	23.0	0.45	0.8	0.135	0.59

Sources: [a]Kaufmann et al. (2008), [b]Central Intelligence Agency (2009), [c]FAO (2005)

they do and are not found around the periphery of New York or Shanghai. Much of the world's population is concentrated within 100 km of a coast, with population density decreasing as one moves to the interior (Small and Nicholls 2003). The Upper Amazonian regions of Peru, Colombia, Ecuador and Brazil are roughly 3,000 km from the Brazilian city of São Paolo. Although some of these areas are less than 300 km from the coast, the Andes Mountains, which span the length of the South American continent, create an effective natural barrier isolating large parts of the Upper Amazon from urban centers along coastal and intermountain population centers throughout the Andean region. This isolation has prevented the integration of the interior regions of the Amazon into regional and global markets, kept population densities low, and minimized rates of deforestation and forest degradation (Nepstad et al. 2008). Similarly, the large and intact forests of boreal Russia and North America, the Congo Basin, and New Guinea are also largely found in interior regions that have low accessibility to coastal regions (Fig. 13.1).

Geographical isolation also restricts the connection of these areas to natural resource markets. Many studies have examined the impact of distance to market on rates of deforestation and forest degradation. These studies have almost uniformly found that areas with longer travel times to market tend to have low rates of deforestation and forest exploitation (Chomitz and Gray 1996; Chomitz 2006). This subject is investigated in greater depth in the "Current Drivers" section in the discussion on roads and access.

3.5 Lack of Governance

One legacy of geographical isolation and low population densities is political isolation from centralized national governments. The lack of government presence within core forest regions has resulted in a lack of infrastructure development and government services, and an inability to integrate these regions into larger market and governmental structures. This has helped to maintain low levels of deforestation and forest degradation (Kaimowitz 1997). In the Amazon Basin nations, distrust between the largely indigenous inhabitants and representatives of the national governments, who tend to be of European descent, has resulted in very low governmental capacity and integration in the Amazon region. In the Congo Basin generally, and within the DRC in particular, armed conflict, ethnic tensions, and governmental instability have generally prevented large scale forest degradation and exploitation by discouraging investment of capital (Glew and Hudson 2007). Large sections of the upper Amazon within Colombia and neighboring Venezuela are violent and largely ungoverned due to the presence of the FARC, paramilitary groups, and large-scale cocaine trafficking. Isolated parts of the Peruvian Amazon have little government presence and are controlled by drug traffickers as well as remnants of the Shining Path guerilla group. While there are no studies linking war and conflict directly to lower deforestation rates, it is likely that their presence inhibits investment in roads, health care, and resource extraction, thus keeping overall land use conversion rates at low levels.

3.6 Low Levels of Foreign Investment

Another shared trait among many of the world's large and intact forests is a lack of foreign investment. Foreign investment can be a highly significant driver of deforestation and forest degradation, particularly through infrastructure development and natural resource extraction (Chomitz and Gray 1996; Carr et al. 2005). In many cases, foreign investment can be a catalyst for resource exploitation by giving projects sufficient capital to overcome high initial costs of resource extraction, turning an unprofitable endeavor into one that is economically viable.

A lack of project financing is often cited as a key constraint on logging expansion, particularly in areas such as the Congo Basin (Perez et al. 2006). There are many reasons why foreign investors may be less inclined to invest in resource extraction in certain forested regions. In the Congo Basin, it is likely the result of the high risks posed by violent armed conflicts and blatant corruption (Perez et al. 2006; Glew and Hudson

2007). By contrast, in the Amazon Basin, it is more likely driven by a lack of pre-existing infrastructure in the region (due to limited governmental capacity) and the absence of technology to make agricultural operations profitable. In recent years, however, with new technologies for soy cultivation in the southern Amazon, foreign investment, and consequently deforestation, in the region have accelerated (Wilcox 2008).

4 What Currently Drives Deforestation and Degradation of Large and Intact Forests?

There are a number of signs that, despite the lack of historical deforestation and degradation, many regions with large and intact forests are at risk in the near future. For example, some researchers believe that forests within the Congo Basin will fragment into three distinct and diminished forest blocks based on models predicting future population growth, road densities, and logging concessions (Zhang et al. 2006). Two of the blocks will be east of the Congo River in the Democratic Republic of Congo, with small patches remaining around the edges of the basin. Nepstad et al. (2008) have predicted that by 2050 the Amazon rainforest could be reduced to 51% of its initial extent due to a positive feedback mechanism from fires, land use conversion, and climate change. There are several interdependent factors driving the conversion and degradation of large and intact forests, including conflict, infrastructure expansion, unclear land tenure, poor governance, and global commodity flows. Each of these drivers will be discussed separately, however it is important to note that many of these factors are related and work in tandem to drive deforestation rates.

4.1 Poverty, Affluence, and The Kuznets Curve

At the country scale, forest loss often follows a Kuznets curve, whereby deforestation rates are initially static, increase during industrialization when populations are growing, and finally stabilize into an equilibrium state (Ehrhardt-Martinez et al. 2002). Users of this model often draw three conclusions from this trend.

First, they conclude that growing economies are most likely to exhibit rapid deforestation. This is fairly self-explanatory, as growing economies tend to increase their use of internal natural resources both to fund domestic economic growth and to participate in export markets.

Second, they conclude that increasing affluence during economic development accelerates the rates of deforestation and degradation. Studies have shown positive correlations between rising incomes, increasing agricultural exports, and forest degradation (Barbier et al. 2005; Carr et al. 2005). When agricultural operations are largely for domestic consumption, however, they do not tend to have the same impact on deforestation. For example, within the tropics, traditional shifting agriculture is responsible for only 6% of observed land use change, and only 26% of tropical deforestation is the result of small scale agriculture as a whole (Barbier et al. 2005; Martin 2008). Although the Kuznets curve suggests that deforestation and degradation increase with affluence during national development, it also suggests that once above a certain threshold, increasing affluence has a reverse effect (Ehrhardt-Martinez et al. 2002). This is likely due to the fact that as economies develop, their economic base becomes more diversified, with less reliance on natural resource commodities. At the same time, rising affluence tends to drive increased urbanization, which shifts populations from a decentralized agrarian base to more centralized, denser urban areas where residents do not engage in subsistence farming.

Third, users of the Kuznets model often conclude that the poorest members of society in developing nations are often not the major drivers of deforestation and degradation (Carr et al. 2005). Research has shown that within Latin America and Southeast Asia, poverty has had very little impact on increased deforestation (Chomitz 2006). In fact, studies within the Peruvian Amazon have shown that poverty actively constrains deforestation because labor and equipment inputs are prohibitively expensive (Zwane 2007). This trend is particularly evident where poor populations lack access to credit. In these cases, there are so few market and labor incentives that cultivation rarely grows beyond subsistence levels.

Nevertheless, some national governments continue to claim that subsistence forest inhabitants are driving deforestation and degradation. More often, however, it has been driven by government policies that welcome large, industrial scale conversion of land to agriculture, often for the benefit of multinational entities who pay hefty prices to governments for local access (Siegert et al. 2001; Doolittle 2007). The idea that poor, rural subsistence farmers are the chief cause of the deforestation and degradation of large and intact forests continues to be disputed. Indigenous inhabitants of large and intact forests have developed systems of resource extraction that usually, if allowed to continue undisturbed, have small impacts on the forests they inhabit (Dove 1983; Dugan 2007). In much of the world, deforestation and natural resource extraction are increasingly controlled by external actors who have few ties to the forests they impact (Lambin and Geist 2003).

One exception is in the Congo Basin. Here, extreme levels of poverty have led to unsustainable extraction of wood for fuel, a strong bushmeat trade, and the expansion of subsistence agriculture into the frontiers of intact forest. These activities have in fact been significant drivers of deforestation and forest degradation in the region (Iloweka 2004). In the DRC, many rural populations surrounding the city centers have come to rely on the collection of fuel wood for their livelihoods (Iloweka 2004). Sunderlin et al. (2000) examined the impact that Cameroon's economic downturn during the 1980s and 1990s had on deforestation in Cameroon's Congo Basin. As incomes decreased in the crisis, local landholders were forced to clear land to feed themselves, resulting in greatly increased rates of deforestation. This phenomenon of an economic crisis driving increased deforestation was also observed in Indonesia following the Asian financial crisis in the late 1990s. Small scale farmers significantly expanded their rubber holdings and other tree crops during the crisis, with the aim of increasing future income security (Sunderlin et al. 2001). Thus, despite the fact that in general forest loss and degradation follow a traditional Kuznets curve during economic development, extreme levels of poverty and rapid decreases in prosperity from periods of economic crisis can alter the trend and lead to unanticipated deforestation and forest degradation.

4.2 Governance

There is wide variation in the quality of governance among the regions discussed in this chapter (Table 13.3). The DRC lies at one end of the governance spectrum with poor governance while countries such as Canada lie at the other end. Good governance greatly increases the likelihood that countries will manage resources sustainably and take steps to control deforestation and forest degradation. In this section we will discuss the implications for forests of a lack of governance, as well as how poorly designed governmental policies can drive deforestation and forest degradation.

4.2.1 Problems Related to Lack of Governance and Land Tenure

The Democratic Republic of the Congo (DRC) and the Republic of the Congo are good examples of what happens to deforestation and forest degradation when there is a lack of national governance. As a result of violent conflicts in these two countries and in the neighboring country of Rwanda, there has emerged a large refugee population in the two Congo nations. At the same time, the lack of national governance has allowed many forested regions to remain controlled by rebel groups. The large refugee population in the region has led to illegal and unsustainable resource exploitation and is both the result of, and the cause of, continued armed conflict in the region (Glew and Hudson 2007). The most common forms of illegal natural resource extraction resulting from the presence of refugees are hunting for bushmeat, followed by illegal logging.[5]

[5]Although bushmeat hunting may not directly or initially impact the forest cover and carbon storage capacity of a forest, it has been shown to impact the floristic and faunal composition of tropical forests (Nunez-Iturri and Howe 2007).

It is important to note, however, that political instability has both positive and negative feedbacks on deforestation. On the one hand, the illegal harvest of forest products is often used to fund continued armed conflict in the region, thus perpetuating the cycle of lack of governance, increased numbers of refugees, and increased forest degradation (Glew and Hudson 2007). On the other hand, a lack of governance often means that national governments and foreign corporations are unable or unwilling to invest in infrastructure and resource extraction. For this reason, the DRC has a negligible deforestation rate, despite ranking as a bottom tier country in terms of corruption, government performance, and human livelihoods (Tables 13.1 and 13.3) (FAO 2005; Central Intelligence Agency 2009). In other words, the DRC may simply be too poorly governed to have a high net rate of deforestation.

In other regions, however, a lack of governance can be a key driver of deforestation and forest degradation. In the Amazon Basin, little government presence, alongside the presence of illicit actors, can increase localized deforestation rates and lead to the unsustainable extraction of timber. Increased rates of forest conversion within coca producing areas of Colombia and Peru have been directly linked to the traffic of cocaine in areas under the control of drug-related enterprises. The U.S. State Department has estimated that some 2.3 million hectares within the Peruvian Amazon Basin, accounting for 25% of deforestation, is directly the result of coca cultivation for cocaine (Beers 2002).

A lack of governance is often a precondition for widespread illegal logging. Illegal logging has been shown as a primary driver of forest degradation in the Russian Far East, parts of the Congo and Amazon Basin, and Southeast Asia (Auzel et al. 2004; Curran et al. 2004; World Wildlife Fund Forest Programme 2007). Often, the presence of illegal logging leads to significant resource loss, which may reinforce cycles of poverty and forest degradation if it is permitted to continue in an uncontrolled fashion (Auzel et al. 2004).

Landholders who have secure land tenure and confidence in the permanence of their residence are more likely to make long-term investments in their land (Chomitz 2006). When populations have tenuous land rights and risk being legally (or forcibly) removed from their land, they have little incentive to practice sustainable land management activities. Moreover, when landholders lack assurances that land will be protected from appropriation, they often practice unsustainable resource extraction that both degrades previously intact forests and contributes to continued poverty long term.

4.2.2 Problems Related to Poorly Designed Policies and Land Tenure Regimes

Beyond a simple lack of governance, poorly planned government policy can have a major influence on deforestation and forest degradation rates. Policies that create incentives to clear forests and build roads for industrial land-based operations are major drivers of deforestation and forest degradation in all of the regions covered in this chapter (Carr et al. 2005). For example, the deforestation of large sections of the central Brazilian Amazon is directly attributable to governmental stimulus plans, road building programs, and subsidies for livestock production (Fearnside 2007). In order to facilitate rural in-migration to the forest frontier, the Peruvian and Brazilian governments have provided agricultural subsidies, free land, and seeds to colonial settlers (Alvarez and Naughton-Treves 2003; Fearnside and De Alencastro Graça 2006). Within the Colombian Amazon, vague and un-enforced land tenure laws in the 1970s helped to promote deforestation (Armenteras et al. 2006). In Peru, since access to frontier land is free, colonists may gain legal title to the land once it has been deforested and put to agricultural use (Imbernon 1999). This has created a dynamic whereby agricultural settlers are encouraged to clear land in order to gain legal title. By contrast, comprehensive land tenure laws have been shown to incentivize good behavior. Research in the Honduran Miskito region found that properly demarcating land use tenure and assigning clear communal land rights lowered rates of agricultural expansion (Hayes 2007).

Concession policies are a corollary to land tenure issues and often drive deforestation and forest degradation in the tropics. Often, concessions are awarded for finite periods of time that are too short to make sustainable forest management a viable enterprise. Thus, concession holders often engage in short term resource extraction practices (Barr 2001). One way that better governance could improve natural resource management and decrease rates of deforestation and forest degradation would be to reform concession policies to encourage responsible forest management. Unfortunately, concession systems are extremely profitable for governments, which makes timber concession reform a highly contentious issue (Barbier et al. 2005). As a result, in most regions, concessions systems are designed to maximize short-term governmental profit at the expense of sustainability and the local inhabitants.

4.3 Roads, Infrastructure Expansion and Regional Market Integration

The construction of roads linking both core forests and frontier forests to population centers and export markets is invariably tied to increasing rates of deforestation. Econometric models have found that, within the Amazon Basin, roads directly cause local deforestation (Pfaff et al. 2007). Because roads decrease the transportation costs of labor inputs, equipment, and products, they greatly increase the economic feasibility of agriculture and extractive activities within affected areas. In the Congo Basin, roads provide accessibility to previously intact forested areas, allowing bushmeat extraction, illegal logging, and small land clearings (Makana and Thomas 2006; Perez et al. 2006). Fearnside (2007) also found that because roads greatly increase land values, they can lead to both violent confrontation and to increased rates of land use conversion by colonizers seeking to exert de facto ownership of their land.

Roads in the Amazon have typically been constructed to facilitate one or more of the following: natural resource extraction, extension of government control and services, and expansion of agricultural frontiers (Fearnside 2007). Within Amazonian Brazil and Peru, the construction of roads linking core forests and frontier forests to coastal population centers has historically been part of a concerted effort to populate and consolidate government control within the Amazon Basin (Alvarez and Naughton-Treves 2003; Fearnside 2007). Current road building and other infrastructural projects within the Amazon Basin are aimed at regional economic integration and the transportation of agricultural goods to export markets (Perz et al. 2008). The paving of the trans-oceanic highway is expected to link ports along Peru's Pacific coast to the Atlantic coast of Brazil and to facilitate export activities of participating countries and global markets.

It is unclear what effects these projects will have on deforestation rates long term. While roads have been associated with accelerated rates of deforestation, they are also seen as an essential component of economic and social development. In order to minimize deforestation, illegal land clearing, violence, and the displacement of indigenous groups along new road networks, there must be clear governance structures, enforceable land use tenure and zoning laws, and the strategic positioning of indigenous and natural reserves (Fearnside and De Alencastro Graça 2006).

4.3.1 Official vs. Unofficial Roads

Recent literature has focused on the proliferation of privately funded, unofficial road networks (Perz et al. 2005; Perz et al. 2008). Official roads tend to stretch for hundreds of kilometers and connect interior cities to population centers outside of the forest. They also tend to be financed with public funding and through international lending channels. On the other hand, the building of unofficial roads in the Amazon and Congo Basin tends to be driven by industrial scale resource extraction projects and typically does not serve population centers. The unofficial roads tend to be constructed by private interests to suit their particular needs. Perz et al. (2008) found that while public highways have caused localized deforestation, the lack of parallel access points generally leaves large tracts of forest intact. Unofficial roads, however, form extensive, dense networks to support transportation of the resources being harvested or extracted (Pfaff et al. 2007). In addition, when large roads are paved, they often stimulate the creation of extensive unofficial interior road systems that lead to deforestation and forest fragmentation (Perz et al. 2008).

4.4 International Trade and Investment

4.4.1 Global Trade

The globalization of international trade has been occurring at an unprecedented rate over the last 25 years. In many regions discussed in this chapter, particularly in the Amazon and in New Guinea, the most significant impact from globalization is an expanding agricultural sector that drives deforestation and forest degradation on the frontiers of large and intact forests. Another effect of globalization is the increased demand for timber products from these regions. Total international trade in wood and paper products has increased in value from just over 50 billion USD in 1983 to over 250 billion USD in 2005 (ACPWP 2007). Increased demand for forest products has led to widespread forest degradation, which is often exacerbated by over-harvesting and poor logging practices (FAO 2007).

Globalization of commodity markets, including timber, changes market dynamics that were once driven by local supply and demand. As a result, market forces in one part of the world can lead to forest degradation pressure in far removed regions, including those with large and intact forests. For example, domestic logging bans in China have led to an exponential increase in demand on Southeast Asia's timber producers to supply raw materials for China's rapidly expanding production of processed wood products (Lang and Wan Chan 2006). The impact of increasing Chinese wood and pulp demand has also been felt in the forests of the Russian Far East (FAO 2007). In this region, forest degradation has been particularly rapid as a result of poorly managed logging operations (World Wildlife Fund Forest Programme 2007). Often, the globalization of timber markets tends to favor large-scale industrial, export-oriented operations. This not only tends to accelerate the rate of forest loss, but it also has a negative impact on the viability and sustainability of smaller operations (Mertz et al. 2005).

Despite rampant forest degradation as a result of unsustainable logging practices, not all timber extraction and international trade in wood products leads to forest degradation or deforestation of these regions. A study of logging throughout the Congo Basin showed that major differences exist between timber concessions based on concession period, size, age, capital source and market focus (Perez et al. 2005). Large older concessions, particularly those granted to large established foreign entities, tend to utilize formal management plans with a longer term focus. They also tend to harvest trees in a slower, more deliberate fashion than locally financed and local market-focused concessions (Perez et al. 2005). While some of this may be due to governance and land tenure issues, it may also be due to the financial flexibility of concessionaires. Large multinational institutions may have greater flexibility in their capital structure and have greater access to working capital than smaller concessionaires who may be pressured to over-harvest to meet current cash flow needs (Perez et al. 2005). Forest degradation can also be partially mitigated through the use of reduced impact logging (RIL) techniques, which minimize unnecessary disturbance from harvesting operations. RIL practices have been shown to have fewer carbon losses from logging activities than conventional harvesting operations in the tropics, although the benefits of RIL are mostly seen in large scale operations and may not be as significant for small scale harvests (Feldpausch et al. 2005) (see Chap. 9 for further details on RIL).

4.4.2 Foreign Investment

Despite historically low levels of foreign investment in large and intact forests, New Guinea, the Amazon, and the Congo Basin have seen recent increases in foreign investment interests. Foreign investment can be a significant driver of deforestation and degradation if it provides sufficient capital to make certain land conversion projects viable that had previously been uneconomic. Oftentimes, foreign aid packages require economic liberalization policies that lead to increased resource extraction by multinational companies in the particular region. All of the regions discussed in this chapter have received funds from the International Monetary Fund since 1984 (IMF 2009). In many cases, these loans have provisions for structural adjustment policies, which require the recipient country to allow increased private investment from foreign companies. In Cameroon, structural adjustment policies implemented by international donors following an economic crisis in the 1990s have led to drastically increased foreign investment in the country's natural resources, which has accelerated deforestation and forest degradation (Kaimowitz et al. 1998). In the Congo, financial reforms pushed by international donors and development agencies have largely taken resource management control away from local entities and given multinational corporations greater control over these industries (Kuditshini 2008). Structural adjustment policies implemented in Indonesia following the Asian financial crisis also increased rates of deforestation and forest degradation (Dauvergne 2001).

The ways in which these structural adjustment policies impact rates of deforestation and forest degradation, however, are highly complex and interactive (Kaimowitz et al. 1998). Sometimes the presence of large multinational corporations can increase forest governance as these interests seek to protect their own investments through local regulation and oversight. Many multinational organizations have greater transparency in their operations and have active stakeholders who insist that management follow some degree of sustainable practices. This can have a positive effect on logging operations. Still, generally speaking, increased foreign investment leads to increased incremental demand for wood resources, so while governance may be improved, overall deforestation continues simply due to increases in absolute demand for forest products.

5 Conclusions and Policy Recommendations

Many of the world's large and intact forests have to date avoided significant anthropogenic disturbance due to a number of common factors. These include geographical isolation, low population densities, biophysical constraints on agriculture, a generalized lack of government presence, and low levels of foreign and domestic investment. It would be a mistake to assume, however, that these forests are not at risk. In the last two decades alone, large sections of formerly intact forests on the Indonesian islands of Borneo and Sumatra have been cleared. Although there is widespread agreement that curbing deforestation and forest degradation in the tropics is critical for many reasons, including the significant carbon releases from these activities, there are few mechanisms to change land management behavior. This in large part is due to the complex mix of deforestation and land degradation drivers. The players, markets, and governance mechanisms are both local and global. Incentives therefore must accommodate the needs of local households while recognizing the roles played by international corporations, banks, and national governments. They must recognize that forest products are a function of both local and global supply and demand forces. As a result, incentives to curb deforestation must be holistic, flexible, and reflect the myriad conditions at both a local and a multinational level.

Given the significant role played by deforestation and forest degradation in widespread global carbon emissions, and the need to reduce these emissions in the face of global warming, countries must make a joint and comprehensive effort to slow rates of deforestation. A primary focus of this effort should be on the world's remaining large and intact forests, particularly in the tropics. Many forest policymakers point out that the developed countries not only contribute the greatest amount of global greenhouse gas emissions, but they are often the key sources of timber and agricultural demand from these sensitive forests. As a result, it has been suggested that developed nations must help to underwrite incentives to compensate developing countries for the opportunity cost of not deforesting, including using carbon market incentives. Emerging mechanisms, including markets for Reducing Emissions from Deforestation and Degradation (REDD) and REDD+which goes beyond deforestation and forest degradation, and includes the role of conservation, sustainable management of forests and enhancement of forest carbon stocks, are seen as one example of market-based financial rewards for forest preservation and sustainable management. Without such incentives it may not be possible to stem the tide of forest loss in these regions. Policy recommendations for the protection of large and intact forests include the following:

- While countries with large and intact forests share many common variables, there is widespread disparity between governance structures and local drivers of deforestation. Thus, while international forest policies must share a common goal to help prevent continued forest loss, the policies enacted to implement these goals must reflect local conditions.
- In order to concentrate funds where they are needed most, a deforestation risk index should be established to rank and prioritize the disbursement of REDD/REDD + funds.
- Insecure land tenure is often a driver of deforestation and forest degradation. Countries receiving REDD/ REDD + funds should be required to have strong, functioning land tenure laws. These rights must extend not only to individuals and forest concessionaires but also to communally governed land.
- Resource extraction in large and intact forests does not always lead to widespread degradation or deforestation. Avoided deforestation should not preclude reasonable use, including sustainable forestry, hunting, or the use of non-timber forest products.
- Widespread tropical deforestation often occurs in countries that have some degree of infrastructure, an expanding agricultural sector, and an export-oriented economy. Improving existing governance is as important as establishing oversight in countries where there has been little to no governance.
- Road access to core areas of interior intact forest is likely to increase significantly in the near term. International and regional lending institutions should require an integrated forest management plan to limit deforestation and degradation along proposed and existing road networks.
- Management plans should be tailored to the physical, social, and economic realities of the site and should be shaped by the requirements of the funding agency, local governments, and civil society.
- Global mechanisms should be implemented to ensure that forest products in international trade come from sustainable management practices. This may include the use of certification schemes or other designations that identify wood from responsibly managed land.

References

ACPWP (2007) Global wood and wood products flow: trends and perspectives. Food and Agriculture Organization of the United Nations, Shanghai

Alvarez N, Naughton-Treves L (2003) Linking national agrarian policy to deforestation in the Peruvian Amazon. Ambio 32:269–274

Armenteras D, Rudas G, Rodriguez N, Sua S, Romero M (2006) Patterns and causes of deforestation in the Colombian Amazon. Ecol Indic 6:353–368

Auzel P, Feteke F, Fomete T, Nguiffo S, Djeukam R (2004) Social and environmental costs of illegal logging in a forest management unit in Eastern Cameroon. J Sustain For 19:153–180

Barbier E, Damania R, Leonard D (2005) Corruption, trade and resource conversion. J Environ Econ Manag 50:276–279

Barr C (2001) Timber concession reform: questioning the "sustainable logging" paradigm. In: Colfer CP, Pradnja Resosudarmo IA (eds) Which way forward? people, forests & policymaking in Indonesia. Resources for the Future, Washington, DC

Beers R (2002) Narco pollution: illicit drug trading in the Andes. Foreign Press Center Briefing, Washington, DC, 28 Jan 2002, US Department of State

Betts RA, Gornall J, Hughes J, Kaye N, McNeall D, Wiltshire A (2008) Forests and emissions: a contribution to the Eliasch review. Prepared for the Office of Climate Change by the Met Office Hadley Center, Exeter

Buchanan GM, Butchart SHM, Dutson G, Pilgrim JD, Steininger MK, Bishop KD, Mayaux P (2008) Using remote sensing to inform conservation status assessment: estimates of recent deforestation rates on New Britain and the impacts upon endemic birds. Biol Conserv 141:56–66

CARPE (2001) Taking action to manage and conserve forest resources in the Congo Basin: results and lessons learned from the first phase (1996–2000). Congo Basin Information Series, Central African Regional Program for the Environment, USAID, Washington, DC

Carr DL, Suter L, Barbieri A (2005) Population dynamics and tropical deforestation: state of the debate and conceptual challenges. Popul Environ 27:89–113

Central Intelligence Agency (2009) The world factbook. https://www.cia.gov/library/publications/the-world-factbook/

Chomitz K (2006) At Loggerheads? Agricultural expansion, poverty reduction, and environment in the tropical forests. The World Bank, Washington, DC

Chomitz KM, Gray DA (1996) Roads, land use, and deforestation: a spatial model applied to Belize. World Bank Econ Rev 10:487–512

Chomitz KM, Thomas TS (2003) Determinants of land use in Amazonia: a fine-scale spatial analysis. Am J Agric Econ 85:1016–1028

Curran LM, Leighton M (2000) Vertebrate responses to spatiotemporal variation in seed production of mast-fruiting Dipterocarpaceae. Ecol Monogr 70:101–128

Curran LM, Trigg SN, McDonald AK, Astiani D, Hardiono YM, Siregar P, Caniago I, Kasischke E (2004) Lowland forest loss in protected areas of Indonesian Borneo. Science 303:1000–1003

Dauvergne P (2001) The Asian financial crisis and forestry reforms. In: Loggers and degredation in the Asia-Pacific: corporations and environmental management. Cambridge University Press, Cambridge, UK

Defries RS, Houghton RA, Hansen MC, Field CB, Skole D, Townshend J (2002) Carbon emissions from tropical deforestation and regrowth based on satellite observations for the 1980s and 1990s. Proc Natl Acad Sci USA 99:14256–14261

Diamond J (1997) Guns, germs, and steel; the fates of human societies. W. W. Norton, New York

Doolittle AA (2007) Native land tenure, conservation, and development in a pseudo-democracy: Sabah, Malaysia. J Peasant Stud 34:474–497

Dove M (1983) Theories of swidden agriculture and the political economy of ignorance. Agrofor Syst 1:85–99

Dugan P (2007) Small-scale forest operations: examples from Asia and the Pacific. Unsasylva 58:60–63

Ehrhardt-Martinez K, Crenshaw EM, Jenkins JC (2002) Deforestation and the environmental Kuznets curve: a cross-national investigation of intervening mechanisms. Soc Sci Q 83:226–243

EIA (2006) Emissions from the consumption of energy. In: International energy annual 2006. Energy Information Administration, US Department of Energy

Encyclopedia Britannica (2009) The Encyclopedia Britannica Online, https:\\www.britannica.com

FAO (2005) Global forest resources assessment 2005. Food and Agriculture Organization of the United Nations, Rome

FAO (2007) Global wood and wood products flow: trends and perspectives. advisory committee on paper and wood products. Food and Agriculture Organization of the United Nations, Shanghai

Fearnside PM (2007) Brazil's Cuiabá- Santarém (BR-163) highway: the environmental cost of paving a soybean corridor through the Amazon. Environ Manage 39:601–614

Fearnside PM, De Alencastro Graça PML (2006) BR-319: Brazil's Manaus-Porto Velho highway and the potential impact of linking the arc of deforestation to central Amazonia. Environ Manage 38:705–716

Feldpausch TR, Jirka S, Passos CAM, Jasper F, Riha SJ (2005) When big trees fall: damage and carbon export by reduced impact logging in southern Amazonia. For Ecol Manag 219:199–215

Glew L, Hudson MD (2007) Gorillas in the midst: the impact of armed conflict on the conservation of protected areas in sub-Saharan Africa. Oryx 41:140–150

Hayes TM (2007) Does tenure matter? a comparative analysis of agricultural expansion in the Mosquitia Forest Corridor. Hum Ecol 35:733–747

Hoffman WA, Schroeder W, Jackson RB (2003) Regional feedbacks among fire, climate, and tropical deforestation. J Geophys Res D Atmos 108:ACL 4-1–ACL 4-11

Iloweka EM (2004) The deforestation of rural areas in the lower Congo province. Environ Monit Assess 99:245–250

Imbernon J (1999) A comparison of the driving forces behind deforestation in the Peruvian and the Brazilian Amazon. Ambio 28:509–513

IMF (2009) Country information. International Monetary Fund. http://www.imf.org/external/country/index.htm

Instituto Nacional de Estadisticas y Informatica (2007) XI Censo de Poblacion y VI de Vivienda 2007

Joppa LN, Loarie SR, Pimm SL (2008) On the protection of "protected areas". Proc Natl Acad Sci USA 105:6673–6678

Kaimowitz D (1997) Factors determining low deforestation: the Bolivian Amazon. Ambio 26:537–540

Kaimowitz D, Erwidodo, Ndoye O, Pacheco P, Sunderlin WD (1998) Considering the impact of structural adjustment policies on forests in Bolivia, Cameroon and Indonesia. Unsasylva 49:57–64

Kaufmann D, Kraay A, Mastruzzi M (2008) The world governance indicators project. In: Group TWB (ed) Governance matters 2008. The World Bank Group, Washington

Kuditshini JT (2008) Global governance and local government in the Congo: the role of the IMF, World Bank, the multinationals and the political elites. Int Rev Adm Sci 74:195–216

Lambin EF, Geist HJ (2003) The land managers who have lost control of their land use: implications for sustainability. Trop Ecol 44:15–24

Lang G, Wan Chan CH (2006) China's impact on forests in Southeast Asia. J Contemp Asia 36:167–194

Makana JR, Thomas SC (2006) Impacts of selective logging and agricultural clearing on forest structure, floristic composition and diversity, and timber tree regeneration in the Ituri Forest, Democratic Republic of Congo. Biodivers Conserv 15:1375–1397

Makarieva AM, Gorshkov VG (2007) Biotic pump of atmospheric moisture as driver of the hydrological cycle on land. Hydrol Earth Syst Sci 11:1013–1033

Mann T (2000) Earthmovers of the Amazon. Science 287:786–788

Markewitz D, Davidson E, Moutinho P, Nepstad D (2004) Nutrient loss and redistribution after forest clearing on a highly weathered soil in Amazonia. Ecol Appl 14(4 suppl):S177–S199

Martin RM (2008) Deforestation, land-use change and REDD. Unasylva 230(59):3–11

Meggers BJ (1995) Amazonia: man and culture in a counterfeit paradise. Smithsonian Institution Press, Washington, DC

Mertz O, Wadley RL, Christensen AE (2005) Local land use strategies in a globalizing world: Subsistence farming, cash crops and income diversification. Agric Syst 85 (3 Spec. Iss.)

Montagnini F, Jordan CF (2005) Tropical forest ecology: the basis for conservation and management. Springer, New York

Natural Resources Conservation Service (2005) Global soil regions. In: Service NRC (ed) Global soil regions map. USDA, Washington, DC

Natural Resources Conservation Service/USDA (2000) Global population density map – 1994. US Department of Agriculture

Nepstad DC, Stickler CM, Soares-Filho B, Merry FD (2008) Interactions among Amazon land use, forests and climate: prospects for a near-term forest tipping point. Philos Trans R Soc B Biol Sci 363: 1737–1746

Nunez-Iturri G, Howe HF (2007) Bushmeat and the fate of trees with seeds dispersed by large primates in a lowland rain forest in Western Amazonia. Biotropica 39:348–354

Olson JS, Watts JA, Allsion LJ (1985) Major world ecosystem complexes ranked by carbon in live vegetation: a database. ORNL/CDIAC-134, NDP-017. Carbon Dioxide Information Analysis Center, U.S. Department of Energy, Oak Ridge National Laboratory, Oak Ridge (Revised 2001)

Pachauri RK, Reisinger A (eds) (2007) Contribution of working groups I, II and III to the fourth assessment report of the intergovernmental panel on climate change. IPPC, Geneva, 104 p

Page S (2002) The amount of carbon released from peat and forest fires in Indonesia during 1997. Nature 420:61–65

Perez MR, De Blas DE, Nasi R, Sayer JA, Sassen M, Angoue C, Gami N, Ndoye O, Ngono G, Nguinguiri JC, Nzala D, Toirambe B, Yalibanda Y (2005) Logging in the Congo Basin: a multi-country characterization of timber companies. For Ecol Manag 214:221–236

Perez MR, De Blas DE, Nasi R, Sayer JA, Karsenty A, Sassen M, Angoue C, Gami N, Ndoye O, Ngono G, Nguinguiri JC, Nzala D, Toiramber B, Yalibanda Y (2006) Socioeconomic constraints, environmental impacts and drivers of change in the Congo Basin as perceived by logging companies. Environ Conserv 33:316–324

Perz SG, Aramburú C, Bremner J (2005) Population, land use and deforestation in the Pan Amazon Basin: a comparison of Brazil, Bolivia, Colombia, Ecuador, Perú and Venezuela. Environ Dev Sustain 7:23–49

Perz S, Brilhante S, Brown F, Caldas M, Ikeda S, Mendoza E, Overdevest C, Reis V, Reyes JF, Rojas D, Schmink M, Souza C, Walker R (2008) Road building, land use and climate change: prospects for environmental governance in the Amazon. Philos Trans R Soc A Math Phys Eng Sci 363:1889–1895

Pfaff A, Robalino J, Walker R, Aldrich S, Caldas M, Reis E, Perz S, Bohrer C, Arima E, Laurance W, Kirby K (2007) Road investments, spatial spillovers, and deforestation in the Brazilian Amazon. J Reg Sci 47:109–123

Potapov P, Yaroshenko A, Turubanova S, Dubinin M, Laestadius L, Thies C, Aksenov D, Egorov A, Yesipova Y, Glushkov I, Karpachevskiy M, Kostikova A, Manisha A, Tsybikova E, Zhuravleva I (2008) Mapping the world's intact forest landscapes by remote sensing. Ecol Soc 13:51

Roy SB, Walsh PD, Lichstein JW (2005) Can logging in equatorial Africa affect adjacent parks?. Ecol Soc 10: Article 6

Siegert F, Ruecker G, Hinrichs A, Hoffman AA (2001) Increased damage from fires in logged forests during droughts caused by El Nino. Nature 414:437–440

Small C, Nicholls R (2003) A global analysis of human settlement in coastal zones. J Coast Res 19:584–599

Spracklen DV, Bonn B, Carslaw KS (2008) Boreal forests, aerosols and the impacts on clouds and climate. Philos Trans R Soc A Math Phys Eng Sci 366:4613–4626

Sunderlin WD, Ndoye O, Bikie H, Laporte N, Mertens B, Pokam J (2000) Economic crisis, small-scale agriculture, and forest cover change in southern Cameroon. Environ Conserv 27:284–290

Sunderlin WD, Angelsen A, Daju Pradnja Resosudarmo AD (2001) Economic crisis, small farmer well-being, and forest cover change in Indonesia. World Dev 29:767–782

Wieder RK, Vitt DH (eds) (2006) Boreal peatland ecosystems. Springer, New York

Wilcox RW (2008) Ranching modernization in tropical Brazil: foreign investment and environment in Mato Grosso, 1900–1950. Agric Hist 82:366–392

Wolfe P (1999) Settler colonialism and the transformation of anthropology: the politics and poetics of an ethnographic event (writing past imperialism). Continuum International Publishing Group, London

World Wildlife Fund Forest Programme (2007) The Russian-Chinese timber trade: export, supply chains, consumption, and illegal logging. World Wildlife Fund

Zhang Q, Justice CO, Jiang M, Brunner J, Wilkie DS (2006) A GIS-based assessment on the vulnerability and future extent of the tropical forests of the Congo Basin. Environ Monit Assess 114:107–121

Zwane AP (2007) Does poverty constrain deforestation? econometric evidence from Peru. J Dev Econ 84:330–349

Economic Drivers of Tropical Deforestation for Agriculture

14

Lauren Goers, Janet Lawson, and Eva Garen

Executive Summary

Land use change from deforestation in the tropics is a major source of greenhouse gas (GHG) emissions. In order to develop policies that address this significant portion of emissions that contribute to global climate change, it is essential to understand the primary factors driving deforestation in the tropics. This chapter examines the socioeconomic, institutional and economic drivers of tropical deforestation for agriculture in order to gain a better understanding of how incentives to store and sequester carbon in forests may or may not impact deforestation rates. While the circumstances that drive deforestation must be examined within the particular context of each locality and depend upon a variety of factors that include social, political and geographical considerations, there are some general lessons that can be learned from our review of the literature. Government-driven development efforts such as infrastructure development in forested areas, for example, are correlated with deforestation throughout the tropical region. Institutional factors, such as land tenure laws that incentivize forest clearing or macroeconomic policies that provide agricultural subsidies, also influence deforestation rates in a number of tropical countries. In most regions, the factors driving deforestation are complex and interrelated and have significant implications for global climate negotiations where the international community seeks to negotiate a mechanism to reduce emissions from deforestation and forest degradation (REDD+). Future REDD+ policies and programs must incentivize countries to address the underlying factors that drive deforestation, many of which are related to agricultural production and will require national governments to implement profound economic and institutional reforms.

What We Know About Drivers of Tropical Deforestation:

- Population growth and poverty have often been overstated as drivers of deforestation.
- Roads are strongly correlated with deforestation.
- Fluctuating commodity prices for crops directly affect household decision-making to deforest for agriculture or to maintain the forest.
- Economic agricultural policies at the national level – including subsidies and access to credit – play a key role in influencing deforestation.
- Agricultural technologies that increase yields are capital intensive, and allow farmers to employ less labor but may exert stronger pressure on individuals to deforest.

L. Goers (✉)
World Resources Institute, 10 G Street NE, Washington, DC 20002, USA
e-mail: lgoers@wri.org

J. Lawson
USAID, U.S. Department of State, Washington, DC 20585, USA

E. Garen
Yale School of Forestry & Environmental Studies, 360 Prospect St, New Haven, CT 06511, USA

M.S. Ashton et al. (eds.), *Managing Forest Carbon in a Changing Climate*,
DOI 10.1007/978-94-007-2232-3_14, © Springer Science+Business Media B.V. 2012

- Interaction between drivers of deforestation at different scales suggests that no single policy can be effective in slowing or halting deforestation.
- Regional models of deforestation must account for heterogeneity across landscapes and the complexity of interacting drivers.

What We Do Not Know About Drivers of Tropical Deforestation

- Significant drivers of deforestation are usually shaped by complex historical circumstance and are affected by local political, socioeconomic, cultural, and biophysical factors that make it difficult to generalize.
- It is unknown how these drivers will continue to shift over time since demographic trends, institutional factors, and economic policies are constantly changing.
- In order to have a successful REDD+ mechanism, it will be essential to address many of these underlying drivers of deforestation in tropical regions and to understand context specific circumstance of each country and region.

1 Introduction

Fossil fuel combustion is frequently cited as the primary driver of human-induced climate change (Barker et al. 2007; Betts et al. 2008). Although the extraction and use of fossil fuel are indeed large contributors to greenhouse gas (GHG) emissions, the impacts of land cover change and deforestation, particularly in the tropics, also account for a significant percentage of annual GHG emissions. According to the Intergovernmental Panel on Climate Change (IPCC), land cover change and deforestation account for an estimated 17.4% of GHG emissions (Barker et al. 2007).

While fossil fuel emissions are generated primarily by activities occurring in developed nations with high levels of industrialization, consumption, and vehicle use, emissions from land use change and forestry largely stem from cutting down tropical forests for agriculture and other purposes in developing countries, such as Brazil and Indonesia (FAO 2010). Altering the forest and land management practices that lead to deforestation in these regions is considered to be an important component of efforts to reduce global GHG emissions that may be quicker and less expensive to implement than restructuring the economies and infrastructure of developed countries. This approach also is thought to have other benefits beyond climate change amelioration, including generating funding for capacity-building and technology transfer to developing countries to help implement changes in forest management practices.

This chapter reviews the current research on the drivers of deforestation in the tropics with a focus on land clearing for agricultural purposes. It considers the impact of socioeconomic, institutional and economic factors on drivers of deforestation for agriculture, particularly in key countries with large emissions from deforestation, such as Brazil and Indonesia. It concludes by outlining several issues that policy-makers and land managers must consider when developing incentives to prevent GHG emissions from forest conversion and degradation.

Currently, emissions from forest land use conversion and change activities are estimated to produce 17.4% of greenhouse gas emissions (Fig. 14.1), largely from deforestation. For the purposes of this chapter, deforestation is defined as the conversion of forest to another land cover when tree canopy falls below a certain established minimum threshold (Lepers et al. 2005). The Food and Agriculture Organization (FAO) uses a tree canopy cover of 10% to classify areas as forested (FAO 2010).

Land clearing of tropical forests for agricultural purposes comprises a significant portion of the total GHG emissions and is a primary driver of tropical deforestation (Angelsen 1995; Angelsen and Kaimowitz 2001; Achard et al. 2002). This trend in land clearing is attributable, in part, to the fact that nearly 700 million people live near tropical forests and depend on forest land or resources for food, fuel and a source of income (Chomitz et al. 2007). Forests are converted into different kinds of agricultural systems, including swidden agriculture practiced by indigenous groups, subsistence farming by

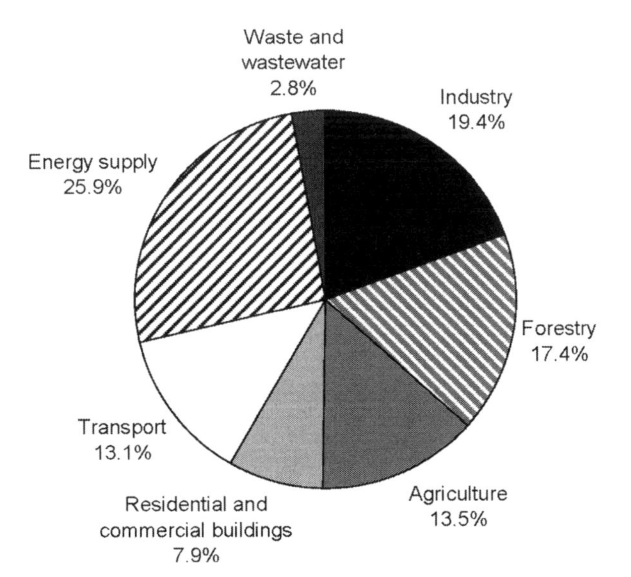

Fig. 14.1 Greenhouse gas emissions by sector. Forestry is defined by forest clearance primarily for plantations and agriculture. Forestry defined here is not the professional activity but all activities that impact forest degradation – clearance and or conversion to agriculture, pasture or from logging (*Source*: Betts et al. (2008). Reprinted with permission)

smallholders, industrial plantations for commercial use, and pasture land for cattle, all of which contribute to the rapid loss of tropical forests worldwide (Barbier and Burgess 2001).

1.1 Deforestation Trends

To understand the complex relationship between deforestation and agriculture, and to better implement carbon policies aimed at reduced deforestation, it is important to identify where and at what rate deforestation is occurring. During the 1980s, the FAO estimated that nearly 15.4 million hectares of tropical forests were cleared each year (Angelsen and Kaimowitz 1999). Subsequent studies have shown a slight decrease in overall forest loss in the 1990s, but changing definitions of forest could account for some of that loss (Angelsen and Kaimowitz 1999).

Worldwide figures showing the scale of forest loss are important tools for understanding the magnitude of the problem. However, examining regional variations in forest loss is important to understand the underlying drivers of deforestation in different regions of the world. The world's three major tropical regions differ in amount and rate of forest loss (Table 14.1). According to Achard et al. (2002), Southeast Asia has the highest rate of tropical forest conversion for the period spanning 1990–1997. Although deforestation rates in Africa and Latin America are lower within that time frame, the total area of forest converted is similar in Latin America and Southeast Asia.

Degraded forest lands, defined as forests where changes have negatively altered the structure or function of the site (including the capacity to sequester carbon), show a similar trend (FAO 2005) – the change in area of degraded forest is highest in Southeast Asia, followed by Latin America and Africa.

Tropical deforestation for agricultural purposes has significant implications for local, regional and global climate trends. As noted above, forest impacts (mostly the conversion of forests to other land uses) contributes 17.4% of global GHG emissions. Coupled with emissions from agriculture at 13.5%, total land use activities generate nearly one third of global emissions. Since tropical forests account for approximately 37% of the world's forested area, they are also a

Table 14.1 Humid tropical forest and annual changes 1990–1997 (millions of hectares)

	Latin America	Africa	Southeast Asia	Global
Total study area	1155	337	446	1937
Forest cover in 1990	669 ± 57	198 ± 13	283 ± 31	1150 ± 54
Forest cover in 1997	653 ± 56	193 ± 13	270 ± 30	1116 ± 53
Annual deforested area	2.5 ± 1.4	0.85 ± 0.30	2.5 ± 0.8	5.8 ± 1.4
Rate	0.38%	0.43%	0.91%	0.52%
Annual regrowth area	0.28 ± 0.22	0.14 ± 0.11	0.53 ± 0.25	1.0 ± 0.32
Rate	0.04%	0.07%	0.19%	0.08%
Annual net cover change	−2.2 ± 1.2	−0.71 ± 0.31	−2.0 ± 0.8	−4.9 ± 1.3
Rate	0.33%	0.36%	0.71%	0.43%
Annual degraded area	0.83 ± 0.67	0.39 ± 0.19	1.1 ± 0.44	2.3 ± 0.71
Rate	0.13%	0.21%	0.42%	0.20%

Source: Achard et al. (2002) Reprinted with permission from AAAS

critical carbon sink (Betts et al. 2008). Continued deforestation of tropical forested ecosystems has the potential to release vast amounts of stored carbon, which would have significant consequences for global climate.

1.2 Conversion to Different Types of Agricultural Land Uses

Landowners convert forested land for a variety of different agricultural purposes. Their decision often is based on a combination of site characteristics and economic, political, and social drivers. Sixty-nine percent, or 3,488 million hectares, of the 5,023 million hectares designated worldwide as agricultural land are used for pasture or forage crops (Smith et al. 2007; Lambin et al. 2003). Lands with marginal productivity typically do not generate significant return on the investment of capital and labor for growing crops and, therefore, are converted to less intensive agriculture, including pasture for cattle (Lambin et al. 2003). In contrast, intensive agriculture is often placed on higher quality, more productive lands. While intensive agriculture supports increased food production, it often also has higher input requirements per unit of area, relying upon mechanization, fertilizers and agrochemicals. Agroforestry systems are mixed systems that can combine trees, shrubs, crops, grasses, and animals and may have high carbon sequestration and storage potential compared to other productive land use options (Ilany and Lawson 2009). Fallow lands are agricultural lands that have been idle for one or more growing season.

While land conversion itself is a significant source of carbon emissions, carbon may be sequestered once agricultural systems are implemented. Carbon sequestration rates will differ, however, depending on the type of agriculture and the productivity of the site. In a study comparing the potential of different land use systems to sequester carbon in eastern Panama, for example, managed forests were found to store an average of 335 Mg C per ha, traditional agroforestry systems stored an average of 145 Mg C per ha, and pastures stored an average of 46 Mg C per ha (Kirby and Potvin 2007). Another study in Saskatchewan, Canada, compared the median ecosystem carbon density for forests, pastures, and cultivated fields (Fitzsimmons et al. 2004). Forest ecosystems contained a median of 158 Mg C per ha, while pastures contained 63 Mg C per ha and cultivated fields contained 81 Mg C per ha (Fitzsimmons et al. 2004). The level of carbon sequestration within a forestry or agricultural system varies between sites in relation to different biophysical characteristics and climatic variations, as well as different land use and management techniques.

There is clear regional variability in the types of agricultural management implemented on deforested land. While in Latin America the primary driver of deforestation is the establishment of pastures for livestock, forest conversion in Africa

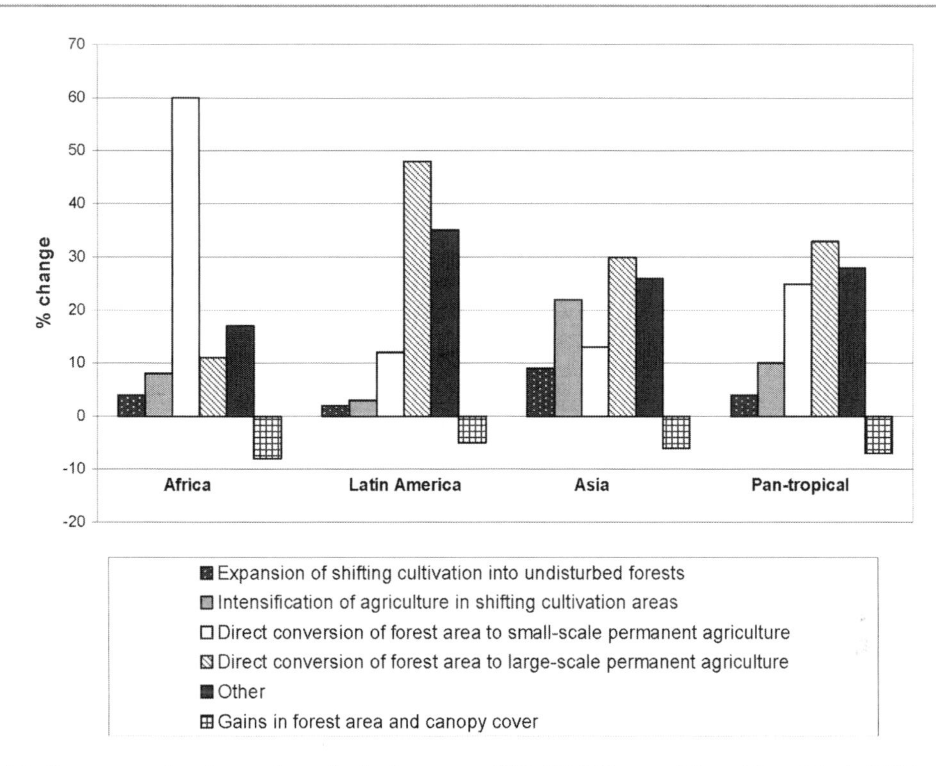

Fig. 14.2 Percentage of total area change by land use type, 1980–2000 (*Source*: Adapted from Martin 2008)

is driven by the establishment of small farm croplands (Lambin et al. 2003). In Asia, forest loss is attributable both to widespread logging and the establishment of permanent agricultural crops, including crops that supply increasing demands for biofuel feedstocks, such as oil palm (Kummer and Turner 1994; Lambin et al. 2003; Koh 2007; Butler and Laurance 2008; Gibs et al. 2010).

Shifts in land use between forest and agricultural systems often are dynamic processes. In the Brazilian Amazon, for example, after initially deforesting the land, landholders often establish annual crops for an average of 2 years and then shift to establish pasture or perennial crops or leave the land fallow (Vosti et al. 2001). It is estimated that the conversion of Brazilian tropical rainforest to arable land releases 703–767 Mg CO_2 equivalent per hectare (Reijnders and Huijbregts 2008). The remote sensing data in Fig. 14.2 compares changes in land use between 1980 and 2000 in Africa, Asia, and Latin America (Martin 2008). With the differences in land use and forest conversion to agriculture in these regions, it appears that a variety of interrelated factors drive regional differences. These factors are explored in the following section.

2 Drivers of Tropical Deforestation

Although there is general acceptance that land conversion for agricultural purposes accounts for a significant amount of tropical deforestation, the factors that drive the conversion of forests for agriculture are less clear. Academic debate has ranged from simple, single driver hypotheses, such as population growth or poverty, as the primary causes of land use conversion to more complex models that list combinations of market-based explanations and other socio-political factors (Geist and Lambin 2002, 2003). Econometric models and empirical studies are often used to explain the combination of factors that drive deforestation in an effort to design better policies that will slow forest loss while addressing the

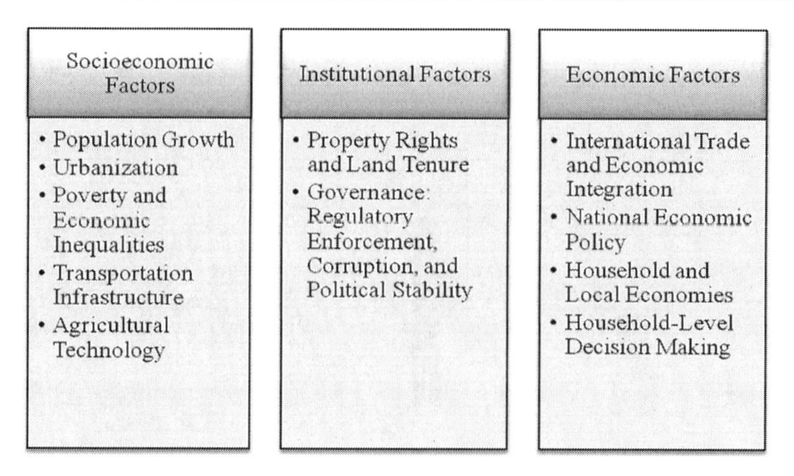

Fig. 14.3 Drivers of deforestation for agriculture

underlying causes of encroachment into forest areas. A review of the literature (i.e. Allen and Barnes 1985; Angelsen and Kaimowitz 2001; Barbier and Burgess 2001; Lambin et al. 2001; Geist and Lambin 2001, 2002; Achard et al. 2002; Fearnside 2005) indicates that there are three major categories of deforestation drivers in the tropics: socioeconomic, institutional, and economic factors (Fig. 14.3).

2.1 Socioeconomic Factors

2.1.1 Population Growth

Population growth is frequently cited as a major driver of deforestation for agriculture in the developing world (Lambin et al. 2001; Allen and Barnes 1985). However, the argument that population growth adequately explains deforestation rates is not as robust as previously thought. For example, researchers frequently attribute tropical deforestation to increasing populations of shifting agriculturalists, despite the fact that recent FAO data estimates that shifting cultivators account for only 5% of pan-tropical forest conversion (Chomitz et al. 2007). Some models indicate that there is a correlation between population growth and clearing of forest land at the national level, but analyses reveal that populations that move into forested areas and subsequently clear land are driven to do so by a host of other factors that include access to infrastructure, high quality soils, off-farm employment opportunities and distance to markets (Angelsen and Kaimowitz 1999). Population growth as an independent factor to explain deforestation in many regions around the globe fails to account for the complex cultural and political context driving population growth in these regions. The causes of tropical deforestation for agriculture cannot be understood without an accurate accounting of the historical relationships between people and the environment (Fairhead and Leach 2008).

More recent work focuses less on the impacts of overall population growth and instead seeks to characterize deforestation trends as they relate to different population types. For example, Jorgenson and Burns (2007) focus on patterns in urban and rural population growth, migration patterns, and economic development to draw contrasts between the location of population growth and the impacts on forest cover (Jorgenson and Burns 2007). Their results indicate that while rural population growth does drive deforestation, urban population increases actually have a slowing effect on forest conversion for agriculture as subsistence farmers migrate to urban centers for work. Other work on population and deforestation suggests that the location of population growth is significant; the first people entering a frontier area have much more impact on deforestation in an area than population growth or migration in an already populated area (Pfaff 1996). These findings may be significant for forest policy, as they indicate the importance of spatial heterogeneity of population density in addressing deforestation rates.

2.1.2 Urbanization

Urbanization and the movement of human populations also are important factors to consider when examining the relationship between human populations and deforestation. While urban areas tend to be more compact and require less land, changing urban diets and consumption patterns ultimately lead to a greater strain on rural natural resources. Additionally, land use change from urban areas frequently expands into nearby agricultural land, thus pushing agricultural pressures into forested areas (Lambin et al. 2003). Overall, the impacts of urbanization on land use change in forests need to be studied more closely at the local level. Urbanization trends lead to complex and nonlinear feedback mechanisms that include rural encroachment, the migration of landless workers from urban centers back to rural areas, or abandonment of agricultural lands that leads to secondary growth (Jorgenson and Burns 2007).

Case studies of the impacts of population on forest cover reveal the complexities associated with determining what drives deforestation in an area. For example, population growth on the Indonesia island of Java led the Indonesian government and the World Bank to sponsor a transmigration program that transplanted Javanese urban dwellers to the more remote, largely forested islands of Kalimantan (Borneo) and Irian Jaya (West Papua). These government policies to reduce population density in urban areas had important implications for deforestation in Indonesia during the program in the late 1970s and early 1980s (Fearnside 1997). The lack of traditional agricultural knowledge on the part of the migrants and the influx of spontaneous migrants who were not part of the government-sponsored program led to increased deforestation for agricultural purposes. Subsequent government-sanctioned migration programs in Indonesia have increased migration to the outer islands as a means of subsidizing labor for timber plantations, primarily oil palm. The impacts of transmigration policy on forests in Indonesia is estimated to range from 2 to 5 ha per family for the early programs that encouraged subsistence farming to nearly 20 ha per family for industrial plantation farming (Fearnside 1997).

Similarly, in Brazil, along with many other Latin American countries, government sponsored settlement projects often have provided urban poor plots of land in the Amazon to farm that cannot sustain agriculture for any extended period of time, leading to their abandonment and often to be purchased by richer landowners for cattle ranching (Laurance et al. 2001; Fearnside 2005).

2.1.3 Poverty

Like population, the poverty hypothesis has traditionally been cited by scholars as a key reason that deforestation for agriculture occurs in developing countries. The logic is that poorer farmers have more of an incentive to deforest in the short term rather than waiting for longer term potential profits from other land uses (Lambin et al. 2001; Angelsen and Kaimowitz 1999). However, this view attributes much of the deforestation that is occurring for agriculture purposes in tropical countries to poor smallholders rather than to larger industrial plantations, government-sponsored concessions or other macro-scale land uses and policies (Dove 1987, 1993; Angelsen 1995; Fearnside 2005).

An alternate view of the poverty hypothesis contends that smallholders do clear some of the forest for subsistence purposes, but they lack the capital, labor, and access to credit that is required to invest in large-scale forest clearing (Angelsen and Kaimowitz 1999). This conclusion supports the finding of Chomitz et al. (2007) that conversion of forest to large-scale agriculture accounts for approximately 45% of land clearing in Asia and 30% in Latin America, whereas shifting cultivation by smallholders only accounts for approximately 5% of forest clearing. The situation in tropical Africa does differ somewhat, as over half of land use change is attributed to forest clearing for permanent, small-scale agricultural endeavors (Chomitz et al. 2007), although it is necessary to further examine the factors that drive this trend in each particular context. While the reasons for these regional differences are complex, one contributing factor is the high global demand for the timber species found in the tropical forests of Asia as compared to Latin America and Africa (Geist and Lambin 2001, 2002; Chomitz et al. 2007).

2.1.4 Economic Inequalities

Due to economic inequalities on a local and regional scale, access to economic opportunities, technology, and land differs across households and regions, which impacts deforestation trends. During the 1970s, for example, subsidized credit for machinery and chemical inputs for soybean production in Brazil was given primarily to large-scale land owners (Kaimowitz and Smith 2001). Not surprisingly, the high commodity price of soy and subsidized credit led to increases in land prices. Facing high land costs, expensive machinery, and chemical inputs for producing mechanized soy, small farmers could not compete, which resulted in land consolidation by large operators (Kaimowitz and Smith 2001). Estimates indicate that in Brazil, the expansion and mechanization of soybean production lead to the displacement of 11 farm workers for every worker employed (Altieri and Bravo 2006). With a total of almost three million people displaced by soybean production in the Brazilian states of Parana and Rio Grande do Sul in the 1970s, many of these displaced individuals moved to the Amazon and subsequently cleared forest for agriculture (Altieri and Bravo 2006). In this particular context, the disproportionate access to economic opportunities, technology, and land at the expense of small landowners exacerbated income inequalities and further increased deforestation trends.

2.1.5 Transportation

In most tropical forest regions, roads are frequently shown to be highly correlated with an increase in deforestation, including roads constructed for agricultural purposes (Angelsen and Kaimowitz 1999; Laurance et al. 2001). Increased infrastructure allows for greater access to interior forests and to end markets for products. While there is a general consensus in the literature that increased access will lead to less forest, roads are both direct facilitators of deforestation activities as well as by-products of other economic activities that may already be causing deforestation (Lambin et al. 2003; Angelsen and Kaimowitz 1999). In some cases, such as in central Africa, roads built for logging concessions typically lead to an influx of new residents who may clear the forest for agricultural purposes (Burgess 1993). While roads are considered to be a primary driver of deforestation in most tropical areas, there are some noteworthy exceptions. For example, areas with low population density or pressure from growth, such as West Kalimantan, Indonesia, do not show a strong correlation between the presence of paved roads and deforestation pressure (Curran et al. 2004). It is not known, however, whether this trend is a short term observation or one that will continue long term.

The role that roads have in the landscape varies by geography and other factors. In the case of West Kalimantan, the high value of dipterocarp timber species and the power of the timber industry in the region have a much stronger impact on deforestation than the presence of either roads or people. Additionally, roads can promote connectivity between rural areas and nearby towns, thereby providing individuals with jobs that might reduce their need to clear forestland for income (Chomitz et al. 2007). Thus, although roads are a primary driver of deforestation in most parts of the tropics, their local impact can vary.

2.1.6 Technology

Depending on the local economy, technologies that increase agricultural productivity have generally been associated with both forest loss and avoided deforestation. While several hypotheses have been developed to explore the causal links between technology and deforestation, two in particular stand out. First, the Borlaug hypothesis asserts that new higher-yielding technologies can increase agricultural production and profitability, thereby reducing deforestation pressures (Angelsen and Kaimowitz 2001). Although this hypothesis might prove true for global food production, it has been shown that commodity prices have a greater impact on agricultural expansion than technological change at the local and regional levels, and particularly on forest frontiers. Second, the economic development hypothesis proposes that increased agricultural productivity due to technology will enhance overall economic development, thus decreasing poverty and pressures on forests (Angelsen and Kaimowitz 2001).

While these two hypotheses indicate that technology advancements in agriculture reduce deforestation pressures, in reality the impacts of agricultural technology on deforestation depend on a myriad of factors, including farmer characteristics, the scale of adoption, how the technology impacts labor and migration, and the profitability of agriculture on the forest frontier (Lambin et al. 2003; Angelsen and Kaimowitz 2001). Technologies that allow farmers to save capital and to create jobs while also driving increased productivity will be most successful at diminishing pressures on forests. However, the mechanization of agricultural production can lead to land degradation due to soil erosion, compaction, and loss of fertility, thus increasing pressure on forests for agricultural land conversion. The industrialization of agriculture can also lead to land consolidation and loss of rural employment, leading to the displacement of small-scale farmers and farm workers to marginal lands or the forest frontier (Lambin et al. 2003). In the Brazilian Amazon, for example, mechanized agriculture increased by more than 3.6 million hectares between 2001 and 2004, mainly for soybean plantations. As a result, cattle ranchers have been displaced and are placing increasing pressures on the forest frontier (Azevedo-Ramos 2007). Therefore, the complexity of factors affecting technological innovation and adoption, as well as the diversity of consequences resulting from such innovation, can lead to either an increase or a decrease in the rate of forest loss.

2.2 Institutional Factors

2.2.1 Land Tenure

Property and land tenure rights are another important driver of deforestation for agricultural purposes. There is a large literature on this subject of which only some that is most relevant is reported here (Dove 1987; Godoy et al. 1998; Angelsen and Kaimowitz 1999; Geist and Lambin 2001). Many countries with high rates of deforestation and agricultural production are still developing economically and may have weak institutional governance and forest law enforcement (Binswanger et al. 1995; Sponsel et al. 1996). In these countries, forest clearing can be the primary mechanism for claiming property rights (Dove 1987; Godoy et al. 1998; Angelsen and Kaimowitz 1999). For example Land tenure laws in the Brazilian Amazon incentivize deforestation by granting title to settlers who "improve" the land by clearing forests (Mendelsohn 1994). While Panamanian landholders are prohibited to cut down the forest for development purposes, such as the construction of hotels and resorts, they can obtain permits to clear forests for agricultural purposes (Nelson et al. 2001).

Studies that correlate land tenure security to deforestation have found that, in some instances, even secure tenure is not enough to stop forest clearing (Angelsen and Kaimowitz 1999; Geist and Lambin 2001), especially when governments have established incentives to clear the forest. In order for the landowner to see forest preservation as a viable management option, the financial benefits of keeping the forest intact must outweigh the net present value of clearing the forest for agricultural production. The relationship between land tenure and forest clearing ultimately will depend on factors such as enforcement and governance. Regional level studies in Latin America, for example, have shown that stronger land tenure support by the state is correlated with slowed deforestation (Godoy et al. 1998; Angelsen and Kaimowitz 1999; Geist and Lambin 2001).

2.2.2 Institutions and Governance

Institutional factors, such as governance and political instability, contribute to deforestation in a variety of contexts. The structure of property rights, environmental laws, and decision making systems are all important aspects of government that influence which groups are granted forest concessions or are allowed to extract natural resources. Governments also have an enforcement responsibility. Due to corruption and lack of regulatory enforcement, however, many countries with significant tropical forest resources do not monitor and prevent deforestation in areas where it is illegal (Lambin et al. 2003). Protected areas in some countries oftentimes are subject to illegal logging simply due to lack of enforcement. For example, researchers found that in Gunung Palung National Park in West Kalimantan, Indonesia, approximately 38% of the lowland forests were illegally deforested in a 14 year span (Curran et al. 2004).

Over the past several decades, developing nations have increasingly adopted decentralization policies as a strategy to improve governance, local empowerment, and natural resource management (Tacconi 2007). A study commissioned by the World Bank, for example, found that over 80% of developing countries with populations greater than five million were attempting to decentralize their governance structures (Silver 2003). Donor agencies and development organizations, such as the World Bank, The U.S. Agency for International Development (USAID), the International Monetary Fund, espouse decentralization as a means to increasing accountability, transparency, and democracy in developing countries (McCarthy 2004).

The popularity of decentralization policies among key donors agencies and academic theorists has resulted in attempts by many developing countries with significant forest resources to transfer power over forest resources from central to local governments. While, in theory, local control over resources leads to improved resource governance, in practice, decentralization has led to power struggles over resources and confusion over delegation of powers (Ribot et al. 2006; Thorburn 2002). Rhetoric surrounding the decentralization of natural resource management supports the idea that local government control will lead to a scaling up of community-based natural resource management and more sustainable forestry practices in countries like Indonesia, yet the impacts of decentralization on Indonesia's forests reveal a significantly different outcome across much of the country (McCarthy 2004). Once decentralization was put into place and local districts were allowed to grant small forest concessions, the result in some areas was a rapid harvest of remaining lowland forest (Curran et al. 2004).

2.3 Economic Factors

2.3.1 International Trade and Economic Integration

International trade, as well as the push for economic liberalization and integration, also has shaped land use trends related to agriculture. Economic liberalization policies, such as the institutionalization of free trade and the removal of tariffs and trade barriers, have typically encouraged incremental land conversion for agricultural purposes. These policies can change capital flows and investments in a region, leading to land use changes that may include deforestation (Lambin et al. 2003). As governments continue to remove barriers to trade and focus on export markets, individuals become increasingly driven by market price fluctuations. Consequently, conversion of land to agriculture becomes more closely correlated to global commodity markets. Governments also have been influenced by the International Monetary Fund (IMF) to institute structural adjustment programs that can change agricultural practices by removing price supports, subsidies, and barriers to trade (Roebeling and Ruben 2001). This trend may or may not increase pressures to drive land conversion for agriculture, depending on current commodity prices and economic cycles.

The net impact of economic liberalization, however, is not clear. On the one hand, economic liberalization can increase investment in industrial agriculture, leading to higher levels of deforestation and land degradation. On the other hand, economic liberalization can increase productivity and drive the implementation of more environmentally-sustainable agricultural technologies. With the right incentives, it may also encourage participation in alternative markets that support improved environmental practices through eco-labeling and green certification systems (Lambin et al. 2003).

2.3.2 National Economic Policy

National economic policies are largely driven by the need for economic growth and national security and tend not to be designed to consider resulting impacts on the forest (Naughton-Treves 2004). Depending on the region, economic policies driving deforestation for agriculture include credit policies, subsidies for agricultural inputs and outputs, taxation schemes, and agricultural price supports (Naughton-Treves 2004; Martin 2008). Currency devaluation has also been correlated with deforestation for agriculture because it encourages individuals to increase agricultural production in order to compensate for economic insecurity (Mertens et al. 2000; Richards 2000). Not surprisingly, when dollar-denominated, global commodity prices are high and the cost of local farm inputs are steady or decreasing, deforestation generally increases (Chomitz et al. 2007).

While commodity prices are most directly affected by subsidies, currency devaluation, exchange rates, and international trade, farm input prices vary most significantly in response to credit access and subsidies (Chomitz et al. 2007). In an economic simulation for Costa Rica, a 20% increase in input price subsidies resulted in a 2% decline in forested area (Roebeling and Ruben 2001). Similarly, a 20% increase in the availability of formal credit also led to a 2% decrease in forestland (Roebeling and Ruben 2001). Government subsidies and access to credit for farm equipment can lead to mechanization and intensification of agricultural production, lowering overall costs and further driving land conversion for agriculture (Azevedo-Ramos 2007). Consequently, national economic policies can create unintended and perverse incentives to deforest land for agriculture.

The interplay between international commodity markets and national economic policies can result in deforestation for agricultural uses. In the Brazilian Amazon, for example, a combination of government incentives for forest conversion coinciding with an increase in beef prices led to the conversion of millions of hectares to low-productivity pasture lands (Chomitz et al. 2007; Azevedo-Ramos 2007). A similar process is currently underway in the Amazon in context of producing biofuels feedsocks, such as oil palm, sugar cane, and soybeans (Laurance and Fearnside citations). In Cameroon, when cocoa and coffee prices began to decline in 1985 and the country entered an economic crisis, the government increased subsidies for agricultural inputs, leading to the expansion of agricultural cultivation into forested lands (Mertens et al. 2000).

2.3.3 Household and Local Economies

At the household level, land use decisions are directly linked to local market access and fluctuations in on-farm and off-farm wages. Access to local markets is generally constrained by insufficient roads and transportation infrastructure. When greater market access and economic opportunities emerge, individuals will often respond by increasing production of valuable commodities and expanding agricultural operations (Lambin et al. 2003). In Cameroon, for example, the villages with the greatest increase in access to local markets through improved food distribution networks also were found to have the highest rates of forest loss (Mertens et al. 2000).

In terms of labor markets, decreases in on-farm wage rates have been linked to agricultural conversion, while increases in off-farm wages and employment have been associated with decreased deforestation rates (Barbier and Burgess 2001). In Puerto Rico, when coffee prices dropped and city wages increased, there was migration to the cities, leading to decreased deforestation and forest regeneration (Chomitz et al. 2007). Remittances from family members abroad can also serve to reduce deforestation as these households feel less economic pressure to expand croplands (Lambin et al. 2003). Thus, in some cases improved market access and decreased on-farm wage rates can encourage households to make decisions to deforest for agricultural expansion, while in other areas improved off-farm wages and opportunities can lead to decreased rates of deforestation and even forest regeneration.

2.3.4 Culture and Household-Level Decision Making

Individuals make daily land use decisions based on cultural preferences, available information, and cultural and economic expectations (Lambin et al. 2003). The aggregation of these individual decisions can translate into extensive deforestation and land use change. Properly organized and with proper incentives, they can also lead to conservation and avoided deforestation. Influenced by the political economy, biophysical characteristics of the land, and culture of the region, individuals will make rational decisions as to what type of land use they choose to implement, varying from swidden agriculture, diversified production systems, and agro-silvopastoral systems to intensive monoculture plantations and pasture (Lambin et al. 2003; Bebbington 1996). This process is important to consider when designing carbon storage and sequestration incentives, particularly since activities related to carbon storage and sequestration will be one of many land use choices available to landowners.

While subsistence agriculture and agroforestry systems have been associated with lower rates of deforestation, the establishment of pastures can contribute to higher deforestation rates (Lambin et al. 2003). In the case of the Atlantic Forest, which extends into Argentina, Paraguay, and Brazil, only 7.5% of the primary vegetation is still intact due to land use change (Myers et al. 2000). In the province of Misiones, Argentina, high rates of deforestation of the Atlantic Forest are the result of national agricultural and economic policies, as well as the increased use of mechanized agricultural production methods. These trends not only have resulted in loss of forests, but also have led to the establishment of increased monoculture agricultural and forestry plantations (Carrere 2005; Lawson 2009). The regional political and economic context, combined with cultural preferences (see Moran 1993; Kaimowitz and Angelsen 1998), affect the decision to adopt a particular agricultural system and its management techniques, which can have positive or negative effects on forestland acreage.

3 The Role of Climate Policy in Reducing Tropical Deforestation

One potential mechanism for addressing tropical deforestation has emerged through the international climate negotiations under the United Nations Framework Convention on Climate Change (UNFCCC). Policy incentives to reduce deforestation and forest degradation, or REDD, are being considered as part of a new climate agreement. Major progress was made in Cancun in November, 2010, and will continue to be negotiated at the next meeting to be held in South Africa in 2011. "REDD+" goes beyond deforestation and forest degradation, and includes the role of conservation, sustainable management of forests and enhancement of forest carbon stocks.

There are many issues that must be taken into account when designing policies to protect forests either using a fund or carbon markets. Since national, regional and local-level government entities would ultimately administer domestic REDD+ programs, implementation challenges in developing countries must be taken into account when allocating funds for REDD+. For governments that have weak regulatory enforcement structures, it is difficult to monitor and enforce behavior that maintains the carbon stock of standing forests. Similarly, for governments where corruption is an issue, it may be difficult to ensure that REDD+ funding and profits are equitably distributed to individuals who are reducing deforestation on their lands or increasing carbon sequestration via sustainable land use practices.

Addressing land tenure issues and economic inequalities are important factors when establishing institutional capacity for REDD+. For farmers who do not have formal title to their land, there must be other incentive structures established to promote forest and agricultural management for carbon storage and sequestration. It is unclear today whether farmers will have access to REDD+ funding if they lack ownership of the land. One solution might be to promote existing cooperatives and farmers' associations to channel REDD+ funds to smallholders who keep their land forested or who establish agroforestry and silvopastoral systems to increase carbon storage and sequestration. Cooperatives, farmers' associations, and extension agencies also could serve as a mechanism to provide training on REDD+ and to assist smallholders in obtaining payments to support reduced deforestation and other sustainable land use practices. However, clear land tenure does not always lead to clear ownership of carbon credits from trees and forests. The development of effective laws and institutions that clarify land tenure and rights to receive benefits from the sale of carbon credits at the local, regional and national levels are essential to promote reduced deforestation and emissions from land use change and allow for equitable access to revenue generate by REDD+.

Since REDD+ policies and programs ultimately will be administered by national governments, assessments and reforms of contradictory government-led policies and programs that lead to widespread deforestation in tropical countries also must take place before REDD+ can be a successful strategy. National governments cannot simultaneously promote forests conservation and restoration policies and programs (i.e., REDD+) while at the same time provide incentives for agricultural expansion into forested areas (either directly or indirectly) via subsidies and laws that support and promote these practices. These contradictory practices expand beyond the agricultural frontier, as many national governments promote mining in tropical forests as a primary economic development strategy and illegal logging, which oftentimes is connected to government officials, is rampant.

4 Conclusions and Policy Recommendations

There is no single model to explain economic drivers of tropical deforestation for agriculture across all regions and scenarios. The circumstances that drive deforestation are locally based and depend upon a variety of factors that include social, political, historical, and geographical considerations. A comprehensive look at these drivers requires a multi-scale analysis that addresses how these factors interact. For example, at a local scale, population pressure and poverty can be shown to lead to deforestation, but these explanations are limited in their ability to describe the scale of deforestation that many tropical countries have experienced in recent years. Policies to address the drivers of deforestation must, therefore, be multidimensional and historically-grounded, and must examine the underlying causes of socioeconomic factors along with larger macroeconomic policies and institutional arrangements that may affect local level land use decisions.

As REDD+ negotiations continue to consider the various ways carbon financing can be used to help preserve carbon stored in standing tropical forests and to promote sustainable land use and management practices that increase carbon storage and sequestration, it is important to consider what are generally accepted economic drivers of tropical deforestation, alongside what is less well understood:

- Significant drivers of deforestation are frequently context-specific and are affected by local political, socioeconomic, cultural, and biophysical factors that are shaped by complex historical circumstance.
- The role of population growth and poverty in driving deforestation have often been overstated for certain regions.
- Transportation infrastructure is strongly correlated with deforestation. Therefore, supporting national policies that reduce development pressure on forests or require improved land use planning could be an effective method for reducing deforestation along roads.
- Fluctuating commodity prices for agricultural crops, timber, and livestock can directly affect household decision-making to deforest for agriculture or to maintain the forest.
- Economic policies at the national level – including subsidies and access to credit – can play a key role in influencing deforestation for agriculture.

- Agricultural technologies that improve productivity, save capital, and create jobs may not necessarily increase deforestation pressure. Agricultural technologies that increase yields, are capital intensive, and allow farmers to employ less labor in fact may exert stronger pressure on individuals to deforest.
- The complex interaction between drivers of deforestation at different scales suggests that no single policy can be effective in slowing or halting deforestation, even with a REDD+ scheme.
- Regional models of deforestation drivers must account for heterogeneity across landscapes and regions, as well as the complexity of interacting drivers.
- It is unknown how these drivers will continue to shift over time since demographic trends, institutional factors, and economic policies are constantly changing.
- In order to have a successful REDD+ mechanism, it will be essential to address many of these underlying drivers of deforestation in tropical regions. REDD+ should provide incentives or contain eligibility criteria for countries seeking REDD+ money to start undertaking some of these broader economic and governance reforms.
- National governments that adopt REDD+ must reconcile their support of conflicting economic growth policies and agendas that promote deforestation (i.e., subsidies and incentives that promote agricultural expansion into forests (either directly or indirectly) and the expansion of mining activities).
- Secure land tenure does not necessarily insure equitable distribution of REDD+ benefits to landholders.

References

Achard F, Eva HD, Stibig HJ, Mayaux P, Gallego J, Richards T, Malingreau JP (2002) Determination of deforestation rates of the world's humid tropical forests. Science 297:999–1002

Allen JC, Barnes DF (1985) The causes of deforestation in developing countries. Ann Assoc Am Geogr 75:163–184

Altieri M, Bravo E (2006) The ecological and social tragedy of crop-based biofuel production in the Americas. http://www.globalbioenergy.org/uploads/media

Angelsen A (1995) Shifting cultivation and deforestation – a study from Indonesia. World Dev 23:1713–1729

Angelsen A, Kaimowitz D (1999) Rethinking the causes of deforestation: lessons from economic models. World Bank Res Obs 14:73–98

Angelsen A, Kaimowitz D (2001) Introduction: the role of agricultural technologies in tropical deforestation. In: Angelsen A, Kaimowitz D (eds) Agricultural technologies and tropical deforestation. CABI Publishing, New York, pp 1–17

Azevedo-Ramos C (2007) Sustainable development and challenging deforestation in the Brazilian Amazon: the good, the bad and the ugly. In: Our common ground: innovations in land use decision-making. Vancouver. http://www.fao.org/docrep/011/i0440e/i0440e03.htm

Barbier EB, Burgess JC (2001) The economics of tropical deforestation. J Econ Surv 15:413–433

Barker T, Bashmakov I, Bernstein L, Bogner JE, Bosch PR, Dave R, Davidson OR, Fisher BS, Gupta S, Halsnæs K, Heij GJ, Ribeiro SK, Kobayashi S, Levine MD, Martino DL, Masera O, Metz B, Meyer LA, Nabuurs G-J, Najam A, Nakicenovic N, Rogner H-H, Roy J, Sathaye J, Schock R, Shukla P, Sims REH, Smith P, Tirpak DA, Urge-Vorsatz D, Zhou D (2007) Technical summary. In: Metz B, Davidson OR, Bosch PR, Dave R, Meyer LA (eds) Climate change 2007: mitigation. contribution of working group III to the fourth assessment report of the intergovernmental panel on climate change. Cambridge University Press, Cambridge/New York

Bebbington A (1996) Movements, modernizations, and markets: indigenous organizations and agrarian strategies in Ecuador. In: Peet R, Watts M (eds) Liberation ecologies: environment, development, social movements. Routledge, New York

Betts R, Gornall J, Hughes J, Kaye N, McNeall D, Wiltshire A (2008) Forest and emissions: contribution to the Eliasch review. http://www.occ.gov.uk/activities/eliasch.htm. Accessed 25 Feb 2009

Binswanger HP, Deininger K, Feder G (1995) Chapter 42 Power, distortions, revolt and reform in agricultural land relations. Handb Dev Econ 3:2659–2772

Burgess JC (1993) Timber production, timber trade and tropical deforestation. Ambio 22:136–143

Butler RA, Laurance WF (2008) New strategies for conserving tropical forests. Trends Ecol Evol 23:469–472

Carrere R (2005) Misiones: La Selva Quiroga Convertida en Pinos para Celulosa. Available: http://www.guayubira.org.uy/celulosa/informeMisiones.html

Chomitz K, Buys P, De Luca G, Thomas T, Wertz-Kanounnikoff S (2007) At loggerheads? agricultural expansion, poverty reduction, and environment in the tropical forests. World Bank, Washington, DC

Curran LM, Trigg SN, McDonald AK, Astiani D, Hardiono YM, Siregar P, Caniago I, Kasischke E

(2004) Lowland forest loss in protected areas of Indonesian Borneo. Science 303:1000–1003
Dove MR (1987) The perception of peasant land rights in Indonesian development: causes and implications. In: Raintree JB (ed) Land, trees and tenure. ICRAF, Nairobi, pp 265–271
Dove MR (1993) A revisionist view of tropical deforestation and development. Environ Conserv 20:17–24
Fairhead J, Leach M (2008) False forest history, complicit social analysis: rethinking some West African environmental narratives. In: Environmental anthropology: a historical reader. Blackwell Publishing, Malden, pp 102–117
FAO (2005) Global forest resources assessment: progress towards sustainable forest management. Food and Agriculture Organization, Rome FAO Forestry Paper 147
Fearnside PM (1997) Transmigration in Indonesia: lessons from its environmental and social impacts. Environ Manage 21:553–570
Fearnside PM (2005) Deforestation in Brazilian Amazonia: history, rates, and consequences. Conserv Biol 19:680–688
Fitzsimmons MJ, Pennock DJ, Thorpe J (2004) Effects of deforestation on ecosystem carbon densities in central Saskatchewan, Canada. For Ecol Manag 188:349–361
Food and Agricultural Organization (2010) Global forest resources assessment 2010: progress towards sustainable forest management
Geist HJ, Lambin EF (2001) What drives tropical deforestetation? a meta analysis of proximate and underlying cause of feforestation based on subnational case study evidence. LUCC Report Series No. 4. CIACO
Geist HJ, Lambin EF (2002) Proximate causes and underlying driving forces of tropical deforestation. Bioscience 52:143–150
Geist H, Lambin E (2003) Is poverty the cause of tropical deforestation? Int For Rev 5:64–67
Gibs HK, Ruesch AS, Achard F, Clayton MK, Holmgren P, Ramankutty N, Foley JA (2010) Tropical forests were the primary sources of new agricultural land in the 1980s and 1990s. Proc Natl Acad Sci USA 107:16732–16737
Godoy R, Jacobson M, DeCastro J, Aliaga V, Romero J, Davis A (1998) The role of tenure security and private time preference in neotropical deforestation. Land Econ 74:162–170
Ilany T, Lawson J (2009) The future of small yerba mate farmers in Argentina: an opportunity for agroforestry. Tropical Resources Institute, Yale School of Forestry & Environmental Studies, New Haven, May 2009
Jorgenson AK, Burns TJ (2007) Effects of rural and urban population dynamics and national development on deforestation in less-developed countries, 1990–2000. Sociol Inq 77:460–482
Kaimowitz D, Angelsen A (1998) Economic models of tropical deforestation: a review. CIFOR, Bogor, p 139
Kaimowitz D, Smith J (2001) Soybean technology and the loss of natural vegetation in Brazil and Bolivia. In: Angelsen A, Kaimowitz D (eds) Agricultural technologies and tropical deforestation. CABI Publishing, New York, pp 195–213
Kirby KR, Potvin C (2007) Variation in carbon storage among tree species: implications for the management of a small-scale carbon sink project. For Ecol Manag 246:208–221
Koh LP (2007) Potential habitat and biodiversity losses from intensified biodiesel feedstock production. Conserv Biol 321:1373–1375
Kummer DM, Turner BL II (1994) The human causes of deforestation in Southeast Asia. Bioscience 44:323–328
Lambin EF, Turner BL, Geist HJ, Agbola SB, Angelsen A, Bruce JW, Coomes OT, Dirzo R, Fischer G, Folke C, George PS, Homewood K, Imbernon J, Leemans R, Li XB, Moran EF, Mortimore M, Ramakrishnan PS, Richards JF, Skanes H, Steffen W, Stone GD, Svedin U, Veldkamp TA, Vogel C, Xu JC (2001) The causes of land-use and land-cover change: moving beyond the myths. Glob Environ Change-Hum Policy Dimens 11:261–269
Lambin EF, Geist HJ, Lepers E (2003) Dynamics of land-use and land-cover change in tropical regions. Ann Rev Environ Resour 28:205–241
Laurance WF, Cochrane MA, Bergen S, Fearnside PM, Dalamonica P, Barber C, D'Angelo S, Fernandes T (2001) The future of the Brazillian Amazon. Science 291:438–439
Lawson J (2009) Cultivating green gold: the political ecology of land use changes for small yerba mate farmers in Misiones, Argentina. MESc thesis, Yale School of Forestry & Environmental Studies, New Haven
Lepers E, Lambin EF, Janetos AC, DeFries R, Achard F, Ramankutty N, Scholes RJ (2005) A synthesis of information on rapid land-cover change for the period 1981–2000. Bioscience 55:115–124
Martin RM (2008) Deforestation, land-use change and REDD. http://www.fao.org/docrep/011/i0440e/i0440e02.htm#box
McCarthy J (2004) Changing to gray: decentralization and the emergence of volatile socio-legal configurations in central Kalimantan, Indonesia. World Dev 32:1199–1223
Mendelsohn R (1994) Property-rights and tropical deforestation. Oxford Econ Pap-New Ser 46:750–756
Mertens B, Sunderlin WD, Ndoye O, Lambin EF (2000) Impact of macroeconomic change on deforestation in South Cameroon: integration of household survey and remotely-sensed data. World Dev 28:983–999
Moran EF (1993) Deforestation and land use in the Brazilian Amazon. Hum Ecol 21:1–21
Myers N, Mittermeier R, Mittermeier C, da Fonseca G, Kent J (2000) Biodiversity hotspots for conservation priorities. Nature 403:853–858
Naughton-Treves L (2004) Deforestation and carbon emissions at tropical frontiers: a case study from the Peruvian Amazon. World Dev 32:173–190

Nelson GC, Harris V, Stone SW (2001) Deforestation, land use, and property rights: empirical evidnce from Darien, Panama. Land Econ 77:187–205

Pfaff A (1996) What drives deforestation in the Brazilian Amazon? evidence from satellite and socioeconomic data. The World Bank, Policy Research Department Environment, Infrastructure, and Agriculture Division, Washington DC

Reijnders L, Huijbregts MAJ (2008) Biogenic greenhouse gas emissions linked to the life cycles of biodiesel derived from European rapeseed and Brazilian soybeans. J Clean Prod 16:1943–1948

Ribot JC, Agrawal A, Larson AM (2006) Recentralizing while decentralizing: how national governments reappropriate forest resources. World Dev 34:1864–1886

Richards M (2000) Can sustainable tropical forestry be made profitable? the potential and limitations of innovative incentive mechanisms. World Dev 28:1001–1016

Roebeling P, Ruben R (2001) Technological progress versus economic policy as tools to control deforestation: the atlantic zone of Costa Rica. In: Angelsen A, Kaimowitz D (eds) Agricultural technologies and tropical deforestation. CABI Publishing, New York, pp 135–152

Silver C (2003) Do the donors have it right? decentralization and changing local governance in Indonesia. Ann Reg Sci 37:421–434

Smith P, Martino D, Cai Z, Gwary D, Janzen H, Kumar P, McCarl B, Ogle S, O'Mara F, Rice C, Scholes B, Sirotenko O (2007) Agriculture. In: Metz B, Davidson OR, Bosch PR, Dave R, Meyer LA (eds) Climate change 2007: mitigation. Contribution of working group III to the fourth assessment report of the intergovernmental panel on climate change. Cambridge University Press, Cambridge/New York

Sponsel LE, Headland TN, Bailey RC (1996) Tropical deforestation: the human dimension. Columbia Universty Press, New York, 343 p

Tacconi L (2007) Decentralization, forests and livelihoods: theory and narrative. Glob Environ Change-Hum Policy Dimens 17:338–348

Thorburn C (2002) Regime change: prospects for community-based resource management in post-new order Indonesia. Soc Nat Resour 15:617–628

Vosti SA, Carpentier CL, Witcover J, Valentim JF (2001) Intensified small-scale livestock systems in the Western Brazilian Amazon. In: Angelsen A, Kaimowitz D (eds) Agricultural technologies and tropical deforestation. CABI Publishing, New York, pp 113–133

The Economic Drivers of Forest Land Use and the Role of Markets in the United States

15

Lisa Henke, Caitlin O'Brady, Deborah Spalding, and Mary L. Tyrrell

Executive Summary

While forests in the U.S. have been both a net source and sink for CO_2 at different times throughout history, today they are a weak net carbon sink, largely as a result of changes in land use patterns over time. The capacity of forests to continue to serve as a carbon sink makes them potentially valuable as mitigation tools to offset the damaging effects of greenhouse gas emissions. However, policymakers must recognize that urbanization and development in the U.S. will continually pressure forests, leading to reduced forest cover and fragmented landscapes. From a purely economic standpoint, development is often the highest and best use of land, particularly if financial returns are the primary driver in land use decision-making. Finding the right balance between competing land uses has become an area of focus for economists and policymakers. As policymakers promote carbon strategies for U.S. forests, they should consider what is generally accepted in terms of the economic drivers of land use, and what is less well understood, as outlined below:

L. Henke
The Nature Conservancy, 1917 1st Avenue Seattle, WA, USA

C. O'Brady
Ecotrust, 721 Northwest 9th Avenue #200, Portland, OR, USA

D. Spalding (✉) • M.L. Tyrrell
Working Lands Investment Partners, LLC, New Haven, CT 06510, USA
e-mail: deporah.spalding@yale.edu; Mary.tyrrell@yale.edu

What Is Known About the Economic Drivers of Land Use Change in the U.S.

- Land use change can have a significant impact on carbon storage. While we have seen little net loss of U.S. forestlands in recent decades, increasing pressure to convert forests to other uses has raised concerns about the reduced potential to store carbon in the U.S. land base.
- Residential and commercial development often represent the "highest and best use" for a parcel of land, resulting in the permanent conversion of forestlands, with negative results on U.S. carbon stocks.
- Subsidies and other government programs alter the balance between forestry, agriculture, and development, including which land use is most profitable at any point in time. Adding forest carbon into the mix of values a landowner can derive from the land may make forests more economically viable.

What We Do Not Know About the Economic Drivers of Land Use Change in the U.S.

- The economic viability of forest carbon projects is still unproven. While models have been developed to predict landowner behaviors when carbon is introduced at various prices, these models have not been widely tested. Additionally, price and project risks continue to challenge the economic attractiveness of potential carbon projects.
- Information on land use changes across the country is incomplete. While general trends in

M.S. Ashton et al. (eds.), *Managing Forest Carbon in a Changing Climate*,
DOI 10.1007/978-94-007-2232-3_15, © Springer Science+Business Media B.V. 2012

land use change can be determined from national inventory, satellite and remote sensing data, local data is not consistently available at a scale useful to land use planning. Analysis must include not only site specific data, but also local rules and regulatory structures that impact behavior.

Some Factors to Consider When Formulating Carbon Policies for Forestry Projects

- Land use change is the primary driver of carbon sequestration trends in the U.S.
- Residential and commercial development is still considered by many to be the "highest and best use" for a parcel of land, and will continue to be a primary driver of land use decision-making.
- If federally mandated market mechanisms emerge in the US to manage carbon emissions, the use of carbon credits from forests may help to improve the economic viability of forested landscapes.
- Whether financially attractive forest carbon projects can be successfully executed is still unproven. Price and project risks continue to challenge the economic attractiveness of potential carbon projects.
- Information on land use changes across the country is incomplete.

1 Introduction

In the United States, land conversion from forests to other uses not only has a significant impact on the amount of carbon stored on the landscape, but it often leads to large carbon emissions during the conversion process. Yet, clearing forestland for development is often a superior economic choice due to higher financial returns versus keeping lands forested. In order to create effective carbon policies that help forests compete with other land uses while maintaining them as a large carbon sink, it is necessary to understand the forces driving landowner behavior and choices. This chapter explores the economic drivers of forestland conversion in the U.S. and the role carbon policy can play in helping forests become a more economically viable land use. It begins by examining the current status of forest as a land use within the historical context of large shifts between forest, agriculture, and development; the role of markets both traditional and emerging; and the contribution of US forests to global carbon stocks. It then explores the primary factors contributing to forestland conversion, including shifts in forest ownership and economic incentives to convert forests to other uses, as well as projected future forest cover and carbon stocks. It will consider the role of carbon-related market incentives and the degree to which they can serve as economic drivers of land use. It will conclude with policy recommendations for how to improve the economic viability of forests versus conversion to other land uses.

Today, forests cover 33% of the U.S. land base and are an important carbon sink within the nation's total carbon budget. According to the EPA's 2009 greenhouse gas survey, the United States currently emits 6,103 Tg of carbon dioxide (CO_2), of which 94% is from fossil fuel emissions, primarily electricity generation (USEPA 2009). Forests, including vegetation, soils, and harvested wood, are currently the largest carbon sink in the United States. Sequestering between 630 Tg (Xiao et al. 2011) and 790 Tg (Heath et al. 2011) of CO_2 per year, they play a key role in offsetting emissions from other sectors.

The amount of CO_2 sequestered in U.S. forests has grown very slowly, just under 0.5% annually since 1990. In fact, according to US Forest Service estimates, the extent of forest cover in the U.S. has not changed significantly since 1950, although estimates for the last two decades vary based on whether it is land use or land cover that is being measured. Inventory data show an increase in forest land use, whereas remote sensing data show a decrease in forest land cover, which will be discussed in more detail below.

1.1 History of Forest Cover in the United States

Both the extent of forestland in the U.S. and the rate of carbon sequestration in forests have varied

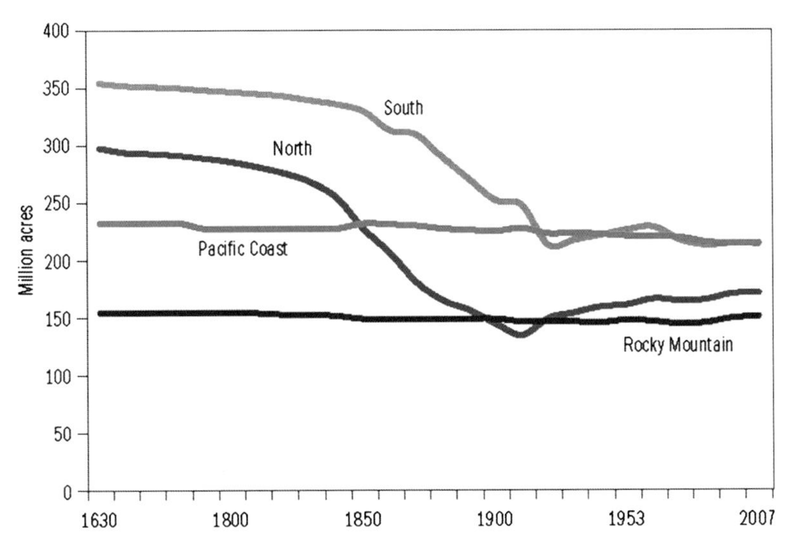

Fig. 15.1 Changes in forest area by region (1630–2007) (*Source*: Smith et al. 2009)

over time, reflecting centuries of changing land use by native populations and European settlers. As human values and resource needs have shifted, forest extent and growth patterns have changed. When Europeans began to settle in North America during the 1600s, forests covered approximately one million acres of what is today the United States (Clawson 1979). Although forest loss during the seventeenth and eighteenth centuries was fairly modest on a national basis, much of it was concentrated in the northeast and to a lesser extent in the southeast where early settlements expanded (Fig. 15.1).

By the early 1800s, however, the trend began to shift. As infrastructure building across the country led to an unprecedented demand for wood products, forests were cut for fuelwood and sawtimber, and as the population grew, more land was cleared for agriculture. This resulted in significant carbon emissions, which continued to increase until peaking at 2,931 Tg CO_2 annually (not including soil emissions) at the turn of the twentieth century (Birdsey et al. 2006). Although some portion of emissions was offset by sequestration in long -lived wood products, nevertheless, this dramatic change in forest cover had a significant impact on the U.S. carbon budget which is still apparent today.

A reversal of this trend occurred in the twentieth century as forests shifted from a net source to a net sink of carbon. This occurred largely because landscapes which were once heavily deforested began to regenerate back into forests (Smith et al. 2009). Regionally, however, the patterns of reforestation following land clearing have been quite diverse. In the South, a large percentage of former pasture and agricultural land has been converted into pine plantations (Sohngen and Brown 2006). The net impact of this transition (from pre-agricultural clearing to post-agricultural replanting) on forest carbon storage has generally been negative. Due to intensive management practices and, in some cases, the use of genetically altered seedlings, many of these plantations have higher rates of net primary production than naturally regenerating stands (Hicke et al. 2002). However, carbon emissions during the harvesting process as well as a tendency for intensively managed plantations to have lower biomass (or carbon per hectare) versus natural stands can more than offset incremental carbon sequestration from higher rates of net primary production, even when a portion of the harvested materials remains sequestered in wood products (Sohngen and Brown 2006). To some extent, this has been partially offset by forest encroachment on savannas, also in the southeast. On these sites, carbon storage rates are higher than historical levels (Rhemtulla et al. 2009).

In the northeast, land originally cleared for timber and agriculture during early colonial history has naturally regenerated back into forest as settlers abandoned farms for more fertile land in the midwest. Not only has reforestation of these lands caused carbon stocks to increase, but these forests continue to sequester incremental carbon as they mature and transition to hardwood- dominated stands.

In the western U.S., many of the old growth forests were heavily harvested in the twentieth century, but are now federally protected. As a result, there has subsequently been significant forest regeneration in these areas leading to an abundance of early/mid successional classes of softwoods across the landscape (Hicke et al. 2007). Because of the large extent of publicly owned forestland in the west that is protected from intensive timber harvesting, carbon losses and changes in forest cover in this region tend to be driven by natural disturbance such as fire and insect outbreaks, in contrast to other regions such as the north and southeast where carbon stocks remain driven by timber management and real estate development activities on private lands.

More recent trends in forest cover are unclear as forest inventory and remote sensing data tell different stories. Although forest inventory data show a seven million hectare increase in forest area between 1990 and 2002, and another four million increase by 2008 (Heath et al. 2011), the National Land Cover Data (NLCD) show a net forest cover loss of 4.7 million hectares between 1992 and 2001, with the greatest loss in the eastern US (Wickham et al. 2008). Proportionally more forest was lost in the "mid" class of fragmentation (patches of at least 60% forest), than in either interior or highly fragmented forest, which suggest a trend of increasing fragmentation. These differences in amount and direction of change between inventory and remote sensed data present challenges for carbon accounting as well as predicting future changes in carbon stocks.

A recent study using remote sensing technology of land cover in the eastern US showed a 4% decline in the total area of forest between 1973 and 2002, with the highest rates of loss in the northeastern highlands, central Appalachians, the Piedmont, and along the coastal plains (Drummond and Loveland 2010). The largest changes in forest cover were due to timber harvesting, reservoir construction, or land clearing for mining, which altogether led to a decrease of 3.2 million hectares of forest cover during the 30 year period. Another 1.9 million hectares were converted to development. The authors argue that intensive timber management practices can lead to a reduced carbon stock on the landscape over time as more and more land is in early successional stages (Drummond and Loveland 2010).

Longer term, carbon uptake rates for all U.S. forests will be a function of multiple factors, including soil fertility, stand age, natural and anthropogenic disturbance patterns, and longer term climate effects, including CO_2 fertilization, increased temperature, drought stress, and disturbances such as fire and insect outbreaks. Nevertheless, the most significant factor influencing how much carbon is being sequestered and stored in forests in the United States, as elsewhere, is the area of land that is in forest.

1.2 Current Forest Land Use

Forest land use is a broad term that generally denotes land intended to be kept as forest for specific purposes. These purposes can be subdivided into five main categories: public goods (federal, tribal, and state lands); commercial timber (industry, TIMO and some family lands); investment (TIMO, developer-owned and some family lands); conservation (NGO and easement lands); and lifestyle (family and individual owners).

Although there are no data on the exact proportion of land owned for each of these purposes, we can provide reasonable estimates from the data that are available. Public good lands are about 46% (44% federal, state, local; 2% tribal) (Butler and Leatherberry 2004; Gordon et al. 2003); commercial timber and investment combined are approximately 10% (Sample 2007; SFFI 2011); family lifestyle, 31% (Butler and

Leatherberry 2004); and conservation lands about 13%. These are based on actual data available for public, tribal, and family lands.

2 Economic Drivers of Land Use in the U.S.

The drivers of land use change often correlate with the highest economic benefits that can be derived from the land, although there are significant variations in this trend over time and across different regions of the United States. Forestland values are driven by a number of interdependent factors that relate to traditional commodity markets, real estate development trends, natural resource values such as recreation, mineral and water rights, and the existence of encumbrances including conservation easements, deed restrictions and rights of way. More recently, emerging markets for ecosystem services such as water quality, wetland functioning, habitat, and carbon sequestration have begun to emerge as mechanisms for assigning environmental values to lands beyond those recognized by traditional markets.

Empirical studies have shown that land "rents" are the key determinant of most private land use decisions (Ahn et al. 2002). The concept of land rent, whereby property owners make land use decisions based on the highest rate of return, has been used to explain landowner behavior since the nineteenth century (Alig et al. 2004). Today, the term "highest and best use" or HBU has been used in the real estate industry to describe the greatest value that can be derived from a property. Highest and best use is defined as that land use that is legally permissible, physically possible, financially feasible, and optimally productive. For many landowners, selling land to traditional real estate developers often represents its highest and best use. Indeed, corporate owners of timberland often separate out their HBU land from other land managed for timber in their financial reporting, or set up real estate subsidiaries, since land best suited for real estate development carries a higher value than timberland (Weyerhaeuser 2007).

The notion of economically rational behavior regarding land use decisions does not apply to all private landowners, however. Many smaller family or individual landowners are motivated primarily by lifestyle preferences (Butler et al. 2007), and although finances are important, decisions about their land are not solely based on which land use can provide the highest land "rent". Decisions to sell are more likely to be driven by financial necessity, such as the need to pay property taxes, a need to generate income or to offset unexpected large expenses, or disinterest on the part of heirs when land is transferred to the next generation (Stone and Tyrrell 2012).

2.1 Forest Ownership

Over the last decade, there has been a significant change in the composition of forest ownership in the U.S. This has not only driven changes in the way forests have been managed, but it has altered the landscape itself, which will undoubtedly impact future carbon sequestration. Of the 750 million forested acres in the U.S. today, 44% remains in public lands, primarily in the West (Fig. 15.2). The remaining 56% is privately held by families, corporations, conservation trusts, tribes, and financial investors.

Each type of forest owner has different, and often complex, reasons for maintaining land in a forested state. The likelihood of their converting forestland to other uses, such as residential or commercial development, is dependent on diverse factors closely related to their reasons for owning the land, as well as economic and social factors.

Perhaps the most significant shift in ownership patterns in recent years has been large scale land sales by vertically integrated forest products companies to financial investors, including timber investment management organizations (TIMOs), real estate investment trusts (REITs), pension funds, and endowments. The shift began in the mid 1970s when Congress passed the 1974 Employee Retirement Income Security Act (ERISA). ERISA was enacted to encourage pension plans to diversify their investment portfolios away from significant fixed income

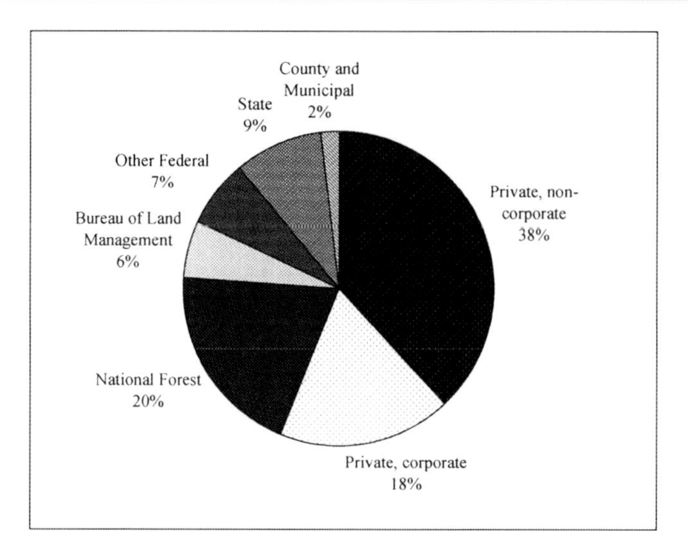

Fig. 15.2 Forestland owners in the United States, 2007 (*Source*: Smith et al. 2009; USDA Forest Service, Northern Research Station, Forest Inventory and Analysis, Family Forest Research Center)

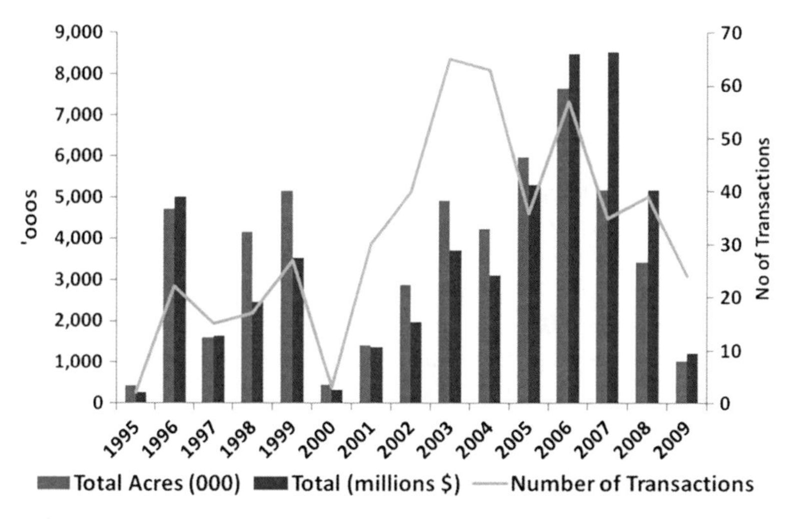

Fig. 15.3 Transactions in U.S. forestland 1995 – 2009 (*Source*: Data from www.RISI.com)

allocations and into other asset classes. Over time, these asset classes have expanded to include real estate and timber (JP Morgan 2009).

Historically, most forest products companies were vertically integrated. Timberland ownership was viewed as a way to secure guaranteed access to raw materials that could be processed at company-owned mills and converted into consumer products. However, as increased real estate demand drove up land prices, and as a growing global fiber supply market offered raw materials at cheaper prices than internal supply, forest products companies began to reconsider the value of holding their timberlands. During the 1990s, paper and forest products companies were increasingly pressured to divest their forestland by shareholders seeking avenues to unlock the value of land holdings, particularly those lands with attractive real estate development potential. As a result, forest products companies began to divest timberland holdings and/or convert to new corporate structures as a way to monetize these values (Fig. 15.3; Binkley 2007).

Due to favorable regulation that stimulated increasing investor interest in timberland, coupled with corporations eager to sell their landholdings, timber investment management organizations (TIMOs) began to emerge to capitalize on this growing investment asset class and to offer professional management services to investors seeking to implement timber strategies in their portfolios. Over the past few decades, traditional timber companies have sold an estimated 25 million acres of timberlands, largely to institutional investors (Stein 2005). Today, financial investors, including TIMOs, timber real estate investment trusts (timber REITs), pension funds, endowments, insurance companies, and investment management firms own approximately 40 million acres throughout the country.[1]

While companies such as Kimberly-Clark, Georgia Pacific, International Paper, and Temple Inland divested their timberland, other companies such as Plum Creek, Rayonier, Potlach, and Weyerhauser divested their forestry operations and converted their corporate structure into real estate investment trusts (REITs). REITs function much like high dividend paying stocks by offering a liquid, tradable vehicle offering high yields to investors. Not only are REITs required to pay out nearly all of their net profits as dividends (to qualify for REIT status), but these dividends are taxed at much lower capital gains rates than traditional dividends, which are subject to ordinary income taxes (Chun et al. 2005). This makes them a very attractive investment vehicle for investors looking for regular income.

The transition from corporate ownership of timberlands for operational needs to financial ownership of timberlands for portfolio returns is likely to have a significant long term impact on the way forestlands are bought, sold, and managed. In theory, ownership of timberlands by forest products companies represents a perpetual interest in the land, since its role in the corporation is to supply necessary raw materials for ongoing business operations. On the other hand, the underlying goal of a financial investor is investment gain over a reasonable time horizon, which for most investment funds is typically 10 years or less. Although a significant percentage of forest acreage transferred over the last 15 years still remains as working forest, there is a risk that financial owners will ultimately convert the lands for HBU values, primarily development, to meet high investor return requirements and as an exit strategy when the terms of the investment funds end and capital must be returned to the investors (Rinehart 1985; Stein 2006).

Despite continued forestland conversion by institutional investors, many TIMOs explicitly state a commitment to responsible forest stewardship. In fact, some claim that combining timber returns with conservation strategies can be as profitable as traditional management. This is one reason why certain TIMOs seek to sell working forest conservation easements on their properties. While working forest conservation easements preclude intensive management activities in favor of more sustainable harvest plans (which reduces long term timber returns), an easement sale to a conservation organization can result in significant cash flow. If this sale is executed early in the life of the property investment, the cash flow will return to investors more quickly, which, due to the time value of money, may yield attractive net returns over the life of the investment. (Stein 2006; Binkley et al. 2006).

The other significant trend to emerge in forestland ownership is the increasing number of family forest owners. Increasing affluence and a desire to own land in rural areas has stimulated sales of forestland to individuals for personal use. Many of these landowners own forested parcels for non-market values such as recreation, aesthetics, and a commitment to conservation (Butler et al. 2007), confounding traditional economic models of land use decisions, such as the "land rent" concept. There are a number of consequences to this trend. First, an increase in individual and family owners has resulted in decreasing parcel size as large tracts are carved into smaller units, particularly in rural areas. This fragmentation of the landscape could have a significant impact on the ecological

[1]Estimate based on REIT company websites, Yale Program on Private Forests Fact Sheet, "Institutional Timberland Investment," and research conducted by the Open Space Institute.

resiliency of these lands as well as their potential to sequester carbon. Second, few of these owners are active managers of their forestland and often do not harvest the land for timber. Most of these owners also lack long term management plans for their forests. Third, individual and family owners represent an aging demographic (34% are above 65 years (Butler and Leatherberry 2004)), which increases the risk that these lands will be fragmented further when they transition to the next generation.

2.2 Forest Conversion to Development

Population and income growth put pressure on natural resources and alter the way humans impact their landscapes. This pressure leads to conversion of forestlands to increasingly intensive uses, including housing, commercial development, and natural resource extraction, including energy and food production (Kline and Alig 2005).

The U.S. Census projects a 35% population increase from 2000 to 2025, including a 79% increase in developed areas, predominantly at the expense of forests, which are projected to decline by 26 million acres, or 3.5% of total forest area between 2000 and 2030. It is expected that this trend will be most prevalent in the southeast and Pacific northwest, where population growth projections are highest. In contrast, the Rocky Mountains and the Corn Belt are most likely to see a decrease in forest cover as a result of expansion of agriculture, pastureland, and rangeland (Alig et al. 2004).

The demand for land is highest near infrastructure, public and private services, and in places that are relatively affordable to develop in terms of cost and regulatory hurdles. Because agriculture and forestry uses are typically not as financially attractive as traditional real estate development, owners of forest and agricultural land will often sell when there is strong development demand (Kline et al. 2004; Zhang et al. 2005; Mundell et al. 2010). As rural lands are developed, and as infrastructure and impervious cover increase, natural ecosystem functions can be compromised. For this reason, federal, state and local governments often regulate certain types of land conversion in order to preserve the social and ecological benefits of open space.

Along with population increases, U.S. household incomes have been on the rise. Median household income climbed from $32,264 in 1994 to $50,233 in 2008, which has exceeded inflation rates during the period (U.S. Census 2009). Although populations have grown at the fastest rate in the southeast and west, the greatest increases in incomes have occurred in the southeast and midwest. Second home development, facilitated by higher disposable incomes, has been a driver of forest fragmentation in areas close to major metropolitan centers, such as the Catskills in New York (Tyrrell et al. 2005).

2.3 Change in Forest Area with Development

Although net forestland area has remained fairly constant over the past several decades, almost 50 million acres shifted in or out of forest cover between 1982 and 1997 (Alig et al. 2004). Consequently, some forests have undergone dramatic change. Much of the land that has been deforested has been comprised of largely mature forests with significant carbon pools and structural diversity. The forests regenerating in their place today are much younger, have lower biomass values, and, if regenerated as plantations, also have reduced structural diversity[2] (Alig and Plantinga 2004; Wimberly and Ohmann 2004).

2.4 Urban and Rural Sprawl

Urban sprawl (sometimes called "suburban sprawl") is the term used to describe the conversion of open lands to development in a sprawling pattern, typically radiating out from a metropolitan area. The first major trend in suburban sprawl

[2]The dynamics of carbon sequestration as these stands develop is discussed in more detail in Chap. 3 in this volume.

took place following World War II (Radeloff et al. 2005). Greater personal use of automobiles and road funding from the Federal Aid Highway Act of 1956 increased access to the fringes of metropolitan areas, which led to greater low-density development (Jeffords et al. 1999). *Rural sprawl*, on the other hand, refers to scattered houses and other structures built on landscapes in non-metropolitan areas. Typically, rural sprawl occurs in areas with high natural amenity values. Since the 1970s, this type of growth has been driven by a desire by homeowners to purchase primary or vacation homes in naturally beautiful areas with affordable transportation connections to nearby metropolitan areas (Radeloff et al. 2005; Ward et al. 2005). These two development trends differ in that urban sprawl impacts less total area than rural sprawl but has more intense effects, whereas rural sprawl can have a much larger spatial impact, though it may be less intense on a particular parcel of land (Radeloff et al. 2005). Carbon emissions from sprawling development occur immediately as forests are cleared for homesites, and at a higher rate over a longer period of time as a result of increased fossil fuel burning from personal vehicles traveling greater distances.

Lands that are close to dense population centers, and in proximity to affluent areas, natural amenities, and public services such as water and electricity, are more susceptible to conversion (Plantinga et al. 2001; Ahn et al. 2002). The financial incentive to sell forestland in these areas can be very strong. For example, in the southeast and the Pacific Northwest, forestland is worth 25–141 times less (respectively) than urban land values (Alig et al. 2004). As the relative profitability of selling rural land for development grows versus retaining open landscapes, landowners often accelerate sales to capture high valuations (Alig 2007).

3 Government Policies Affecting Rate of Forest Conversion to Development

Government policies can both mitigate and exacerbate development pressure on rural lands. Federally owned forests are currently well protected from development pressures. Typically, they are managed for a variety of uses, including timber extraction, recreation, watershed health, and wildlife habitat. Private forests are much more susceptible to conversion. Development rates on these lands are impacted by federal, state and local policies, including forest protection incentives, tax incentives, zoning, transportation funding, infrastructure projects, and mortgage incentives, among others.

Federal forestry assistance is provided through United States Department of Agriculture (USDA), either through the Farm Bill or through the United States Forest Service (USFS). As of 2008, assistance included technical and financial aid for forest management, forest protection, forest recovery and restoration, and economic assistance. The Forest Legacy program specifically authorizes the Forest Service to protect forestlands at risk of conversion to development or agriculture by purchasing the lands or funding a conservation easement on the property. Similarly, the Community Forest and Open Space Conservation program, established by the 2008 Farm Bill, provides funding to purchase titles to at-risk forestlands. State funding administered through state environmental agencies provides complementary funding for private landowners to keep their land forested and help support proper management of these properties. Additionally, various tax incentives are targeted at non-industrial private forest owners to reduce the cost burden of forest management activities (Riitters et al. 2002; Gorte 2007).

While these initiatives have helped to protect forestland and promote healthy forested landscapes, other policies have facilitated conversion of forestland to urban and rural sprawl- related uses, sometimes unintentionally. The most commonly cited federal drivers of sprawl include highway spending, water and sewer system requirements, and subsidies for suburban homeownership. Other government policies can contribute to sprawl as well, including economic development incentives such as income tax credits or local zoning regulations that lack provisions to protect open space. Overall, it would appear that there is a tension between government programs to combat sprawl, and subsidy and regulatory programs which encourage sprawl, either intentionally or unintentionally (Jeffords et al. 1999). Still, when the U.S. General Accounting Office (now the Government Accountability Office) reviewed the influence of these federal policies on sprawl, they concluded that their impact is "unclear" (Jeffords et al. 1999). In the end, job creation, affordability and public and private services are often a greater priority for governments than managing sprawl. And in the recent challenging economic environment, a number of programs that promote open space have seen their funding reduced or eliminated.

3.1 Fragmentation

Rural sprawl has a unique impact on forests by physically fragmenting, or breaking up, large contiguous patches of forestland with houses, roads, and other development. Even rural roads create forest edges and reduce the extent of interior forest. Currently 62% of forestland in the contiguous United States is located within 150 m of a forest edge (Riitters et al. 2002).

Increasing forest edge and reducing interior area can compromise ecological values, which may in turn influence land values. In fragmented forests, certain plants, animals and insects gain a competitive advantage from increased light or a change in nutrient cycling. For example, invasive plants and animals which can thrive along forest edges may outcompete native species, particularly if invasive plants are better able to weather roadside runoff laden with salts and other pollutants. Increased access to forests can lead to illicit dumping and campfires, which can create hazards for neighboring communities and increase the property's susceptibility to wildfire (Theobald and Romme 2007). As a landscape becomes fragmented, its resilience to large disturbances may be reduced. This may in turn threaten developments at the edge of forested landscapes. Additionally, ecosystem services such as water quality or stormwater catchment may be reduced by fragmentation. In these ways, fragmenting landscapes may have consequences that can ultimately impact the financial value of developed lands. Unfortunately, preventative measures require long term land management planning, which may be outside the time horizon of most homeowners and local zoning officials.

At the same time, residential development near working forests can pose difficulties for forestry operations. New homeowners who move close to the forest for its amenity values, such as scenic vistas and tranquility, may not want forest operations in the area. Frustrated with the noise and pollution associated with harvest activities, residential landowners may require forest managers to alter their management plans by altering transportation routes for harvested wood, dictating when harvesting activities can take place, and precluding certain ecologically beneficial management strategies, such as prescribed burning. Neighbors may even file expensive nuisance lawsuits against forest owners in order to restrict forestry activities (Alig and Plantinga 2004). This can impact both the productivity of the landscape as well as the cost structure for the forest manager. This may make forest management a less profitable activity in these areas, further reducing the economic viability of working forests and increasing the pressure to convert to other land uses.

3.2 Parcelization

Parcelization is the division of a large property into smaller land holdings more appropriately sized for development. Landowners are often driven to parcelize their land, particularly if property taxes reflect high land values from embedded subdivision rights. When carrying costs of land are high, the owner may feel pressure to sell one or more parcels to reduce the overall carrying cost. Parcelization can also compromise the economies of scale in forestry operations. Smaller parcels are often less profitable than larger parcels since the cost of equipment and other fixed cost components cannot be spread over a large acreage. Moreover, larger parcels may give the landowner flexibility in timing and type of harvests. If harvests are implemented in stages, this may provide a steadier, long term cash flow stream while creating age and structural diversity. This may not be feasible, however, with a smaller parcel. Insufficient scale may also make it difficult to implement viable projects for carbon sequestration, water quality, and wildlife habitat (Plantinga et al. 2001). As a result, even if parcelization results in smaller parcels that remain undeveloped, forestry and agricultural uses may be priced out of the land use options altogether (Alig and Plantinga 2004).

3.3 Managing Sprawl Through Land Conservation

Concern over the lack of open space and the loss of the ecological and social values it provides has increased as more forestland is fragmented or lost to development. The amount of forestland per capita in the U.S. has declined by almost half in the past 50 years, and by as much as two thirds in the Pacific Northwest, largely due to population growth (Kline et al. 2004). In fact, in April, 2007, forestland per capita fell below 1 ha for the first time in U.S. history (Woodall and Miles 2008). Historically, government policies to protect public lands through national and state forests and parks have been instrumental in protecting open space for the public good (Alig 2007). However, these policies have not helped to protect private lands. Since real estate markets have not been able to efficiently value the social benefits of private forest land, including aesthetics, recreation, water resources, and resilient ecosystems, they are often left out of land buyer decision-making processes. (Kline et al. 2004). For this reason, policies have emerged to encourage open space preservation, such as urban growth boundaries, open space set asides, and tax incentives for keeping forests intact (Plantinga et al. 2001).

However, such policies can have unintended consequences. In their 2003 study on the influence of public space on urban landscapes, Wu and Plantinga (2003) found two potential sprawl effects from designating public open space outside the city boundary. First, they showed that urban residential communities may expand to envelope the open space. This is particularly likely if the open space is sufficiently close to the city (ease of access) and if it provides a high level of amenities. A second possibility is that open space set asides create "leap frog development," bypassing the restricted open space to build further away from existing centers of development (Wu and Plantinga 2003). Ironically, their research suggests that while delineating public open space may protect some portion of land, it may inadvertently encourage outward sprawl that is greater than what would have occurred without the set asides.

4 Government Policies Affecting Forestry and Agricultural Land Uses

Markets for both forest and agricultural land are influenced by government policies, including agricultural subsidies, conservation programs, and timber harvesting regulations. These policies exist to promote economic livelihoods, but also to protect socially desirable quantities of certain land types and ecosystem processes. They can also alter land use decisions by influencing commodity prices and the supply of and demand for natural resources (Alig et al. 1998; Ahn et al. 2002). For example, research suggests that in some cases agricultural subsidies and guaranteed prices paid to farmers have promoted more forest conversion to agricultural land than would have taken place without such subsidies (Alig 2007). The impacts of government programs and policies on land use change are enormously complex and vary by program.

4.1 The U.S. Farm Bill

Competition between forestry, agriculture, and other land uses has been heavily driven by government subsidies since the New Deal in the 1930s. At that time, the U.S. government intervened in agricultural markets by providing subsidies to farmers to stabilize crop prices. The Agricultural Adjustment Act (AAA) was passed in 1933 to raise prices for basic agricultural commodities in order to increase the rent attained on agricultural lands (Skocpol and Finegold 1982). The AAA was the first of many federal agricultural subsidy programs. The U.S. Farm Bill has since replaced the AAA as the government's predominant food and agriculture policy tool. The practice of subsidizing agricultural products has increased over time; in fact, between 1997 and 2006, government payments accounted for 30% of net farm income (Jordan et al. 2007).

Farm subsidies that favor agriculture often inflate the agricultural land value relative to other uses. While agricultural subsidies may reduce the likelihood that agricultural land is converted to developed land uses, it may increase the likelihood that other rural lands, such as forests, will be developed. While the Farm Bill continues to support agricultural subsidies, recently, new provisions have been added to address the importance of maintaining working forests. For example, the 2002 Farm Bill authorized the Forest Land Enhancement Program, a multi-million dollar forestry program designed to assist non-industrial private forest landowners. The 2008 Farm Bill (Food, Conservation, and Energy Act of 2008, P.L. 110–246) has taken a step further and outlined a set of principles specifically aimed at forest stewardship. These include:

- Conserving and managing working forests for multiple values and uses;
- Protecting forests from threats, including "catastrophic wildfires, hurricanes, tornados, windstorms, snow or ice storms, flooding, drought, invasive species, insect or disease outbreak, or development," and restoring appropriate forest types in response to such threats;
- Enhancing public benefits from private forests, including air and water quality, soil conservation, biological diversity, carbon storage, forest products, forestry jobs, production of renewable energy, wildlife, wildlife corridors, and wildlife habitat, and recreation.

4.2 The Conservation Reserve Program

The U.S. Department of Agriculture's Conservation Reserve Program (CRP) subsidizes the maintenance of ecosystem functions by compensating farmers for converting erodible and ecologically sensitive agricultural land into alternative land uses such as forestry or grassland (Roberts and Lubowski 2007). Established in 1985, the program has preserved approximately 34 million acres for up to 15 years (Sullivan et al. 2004), of which approximately 7% was put into forest use (Plantinga et al. 2001).

The CRP is often held up as a model for how the government might create incentives for other ecological values, including carbon sequestration. Some researchers claim that from 1986 to 1991 the CRP effectively made land conservation a competitive choice from a market perspective (Plantinga et al. 2001). During this time, CRP administrators specified maximum allowable rental rates for retiring agricultural lands by region while allowing unlimited enrollment of lands. By using farmers' behavioral responses to the CRP as a proxy for likely actions under a carbon market, the authors (Plantinga et al. 2001) suggest that carbon management will require an annual return threshold of $40–$80 per acre (depending on the region) in order to stimulate conversion of agricultural lands into reforestation projects.

4.3 Timber Management Regulations

Pressure to protect forested areas for wildlife and other ecological services has led to government policies that restrict timber extraction on public lands. While this has ensured protection of habitat on public lands, it has often led to timber harvesting activities being diverted to unregulated private lands. For example, in the Pacific northwest, protection of the endangered northern spotted owl shifted timber harvests to other parts of the country, particularly the southeast. At the same time, reduced harvesting on public lands has led to significant build up of fuel loads in federal forests, hence increasing the risk of wildland fire and its associated greenhouse gas emissions. It has also led to more dense forests, which are at greater risk of disease and pest outbreak. Curtailments on harvesting on federal lands have simply shifted harvesting and its commensurate carbon emissions elsewhere while increasing the risk that existing carbon stocks on public land may be lost to large scale disturbance.

Recently, policies have been proposed to regulate timber harvests on private lands to protect ecosystem values. Research has shown that extended timber rotations provide greater stand structure and age class diversity, which is positive for wildlife and carbon management. Proponents of policies designed to lengthen timber rotations claim that, in the long term, forestry land rents should increase, since more mature wood commands a higher price in timber markets. This may create an effective incentive for landowners to keep lands in a forested state for a longer term (Alig et al. 1998).

4.4 Agricultural Biofuels Subsidies

Due to greater national interest in renewable energy sources, liquid biofuels (or simply "biofuels") have become an increasingly attractive land management option. As demand for biofuels has increased, this has put pressure on forestlands to convert to agricultural biofuel production (Murray 2009). Biofuels (e.g. biodiesel and ethanol) are fuels that are derived from organic matter. In the U.S., biofuels are primarily produced from feedstocks such as corn, soybeans and sorghum. In many countries around the world, including the U.S., use of biofuels has been mandated for transportation fuels. Increased demand for biofuels in recent decades has led to what a recent report from the International Institute for Sustainable Development called "frenzied" expansion in the industry (Koplow 2006). Because of major subsidies at all levels of the supply chain, many rural landowners have shifted into production of these lucrative crops. In 2007, 24% of corn harvested in the U.S. went to ethanol production (Howarth and Bringezu 2009). As biofuel production increases, so does demand for agricultural land, which shifts economic rents in favor of fields over forests. While it is unclear today how biofuel subsidies will impact land use decisions longer term, they have raised significant concern among stakeholder groups interested in preservation of forestland and agricultural land for food crops (Koplow 2006). Ironically, high demand for biofuels may have unintended carbon implications. Not only does land cleared for biofuel production lead to greenhouse gas emissions during the conversion process, but the increased use of nitrogen fertilizers prevalent on land managed for biofuels may also offset the carbon savings from substituting biofuels for fossil fuels (Howarth and Bringezu 2009).

Still, forestland owners may benefit from biofuels subsidies through incentives to encourage the use of wood biomass. This could serve to increase the economic returns from forestland as new markets emerge for wood fuels, particularly for thermal use in residential, commercial, and public structures. However, while increased wood fuel use may reduce carbon emissions from the energy industry, it may also compromise carbon sequestration rates on standing forests, particularly if forests are transitioned to shorter rotation plantations to serve the wood fuel market.

4.5 Carbon Storage

Over the past decade, a new potential economic driver has emerged: carbon sequestration. While forests were not included in initial agreements to mitigate global climate change, they have since been acknowledged as important carbon sinks by scientists, policymakers, and government officials. There is much interest in pushing for the inclusion of forests in the next iteration of a United Nations Framework Convention on Climate Change (UNFCCC), which is expected to replace the Kyoto Protocol when it expires in 2012. Meanwhile, the U.S. Congress continues to debate new federal climate change legislation, including the role the forests might play in future mitigation efforts.

It is much too early to know the impact that carbon related policies will have on forests and forestland conversion in the U.S. Among carbon market architects and participants, there is a high level of uncertainty over the economic viability of forest carbon as both a climate mitigation tool and as an economic incentive to preserve forestland. Pricing, project costs, risk management, and the balance between competing land uses all arise as key issues when exploring the potential for forest carbon sequestration as an economic and policy tool.

4.5.1 Pricing and Transaction Costs

Pricing trends for carbon credits around the world have been mixed. In recent years, carbon offset credits have averaged approximately $16/ton in the international regulatory market (Hamiton et al. 2010). Transactions in the U.S. voluntary market have been even more variable, ranging from US$0.20 to $111/ton in 2009 with an average of just over $4/ton (Hamilton et al. 2010). The lack of price stability, transparency, and clear rules for market participants have inhibited many potential carbon project developers.

At the same time, transaction costs for creating carbon offsets have remained high. According to economists van Kooten and Sohngen (2007), costs associated with forest carbon projects have ranged from $3 to $280 per ton of carbon. Important factors contributing to this disparity are the location of the project, the project type (reforestation, avoided deforestation or managed forest), whether opportunity costs have been included, and the methodology used to measure the carbon itself. Despite the wide variety of costs, and the high probability that they may exceed expected carbon revenues, efforts to promote forest carbon projects remain strong. Proponents often cite current high levels of expertise in forest management (particularly in developed countries) alongside an expectation that new technologies will emerge to lower the overall cost structure of forest carbon projects (Murray 2009).

4.6 Trending to the Future

Without major policy or market intervention, carbon stocks in US forests are projected to decline over the next century (Drummond and Loveland 2010; Bierwagen et al. 2010; USFS 2010). This is due to a variety of factors influencing land use change from forest to developed land use including population expansion out of urban and suburban areas, expansion of mining and reservoir construction, and demand for biomass, pulp and wood products influencing timber cutting cycles.

5 Conclusions and Policy Recommendations

Economic drivers have influenced land use in the United States since the first European settlers arrived. Carbon stocks have fluctuated as land has been forested, managed for agriculture, or converted to development. Whereas U.S. forests served as carbon sources during the 1800s, changes in land use over the past 100 years have transformed U.S. forests into a net carbon sink. Many landowners, scientists, and policy makers acknowledge the potential for these forests to capture greenhouse gas emissions and are seeking to determine the appropriate mixes of land use for carbon sequestration and other values, considering both economic and ecological factors. Current drivers of land use change include demographics, market demands, government policies, and owner preferences, and others are complex and ever changing. Whether or not carbon-related incentives will alter land use decision-making is still undetermined. While the introduction of carbon markets is often seen as a potential driver to help forests compete with other land uses, the economic viability of forest carbon projects remains undetermined.

As the debate continues over the role of forests in climate change mitigation, scientists and policy makers will need to consider several unanswered questions:

- What policies would give forests (as carbon sinks) a competitive economic advantage over other land uses?
- How might using forests to mitigate climate change impact ecosystem services and other valued land use options?
- What are the larger impacts of forest carbon projects on public welfare?
- How will we measure and model land use change? Are there tools that provide simple, accurate, accessible information for policy makers?

References

Ahn SE, Plantinga AJ, Alig RJ (2002) Determinants and projections of land use in the South Central United States. South J Appl For 26:78–84

Alig RJ (2007) A United States view on changes in land use and land values affecting sustainable forest management. J Sustain For 24:209–227

Alig RJ, Plantinga AJ (2004) Future forestland area: impacts from population growth and other factors that affect land values. J For 102:19–24

Alig RJ, Adams DM, McCarl BA (1998) Ecological and economic impacts of forest policies: interactions across forestry and agriculture. Ecol Econ 27:63–78

Alig RJ, Kline JD, Lichtenstein M (2004) Urbanization on the U.S. landscape: looking ahead in the 21st century. Landsc Urban Plan 69:219–234

Bierwagen BG, Theobald DM, Pyke CR, Choate A, Groth P, Thomas JV, Morefield P (2010) National housing and impervious surface scenarios for integrated climate impact assessments. Proceedings of the National Academy of Sciences of the United States of America 107(49):20887–20892

Binkley CS (2007) The rise and fall of the timber investment management organizations: ownership changes in US forestlands 2007. Pinchot Distinguished Lecture, Pinchot Institute for Conservation, Washington, DC

Binkley CS, Beebe SB, New DA, von Hagen B (2006) An ecosystem-based forestry investment strategy for the coastal temperate rainforests of North America. Ecotrust white paper, 18 p

Birdsey R, Pregitzer K, Lucier A (2006) Forest carbon management in the United States: 1600–2100. J Environ Qual 35:1461–1469

Butler BJ, Leatherberry EC (2004) America's family forest owners. J For 102:4–9

Butler BJ, Tyrrell M, Feinberg G, VanManen S, Wiseman L, Wallinger S (2007) Understanding and reaching family forest owners: lessons from social marketing research. J For 105(7):348–357

Chun C, Wilde M, Butler M (2005) Timber REITs: growing value. Deutsche Bank Report, 83 p

Clawson M (1979) Forests in the long sweep of American history. Science 204(4398):1168–1174

Drummond MA, Loveland TR (2010) Land-use pressure and a transition to forest-cover loss in the eastern United States. Bioscience 60(4):286–298

Gordon J, Berry J, Ferrucci M, Franklin J, Johnson KN, Mukumoto C, Patton D, Sessions J (2003) An assessment of Indian forests and forest management in the United States, by the second Indian forest management assessment team for the intertribal timber council. Interforest, LLC

Gorte RW (2007) Carbon sequestration in forests. Congressional Research Service

Hamilton K, Sjardin M, Peters-Stanley M, Marcello T, Ecosystem Marketplace, New Energy Finance, Bloomberg (2010) Building bridges: state of the voluntary carbon markets 2010. Ecosystem Marketplace, Washington, DC

Heath LS, Smith JE, Skog KE, Nowak DJ, Woodall CW (2011) Managed forest carbon estimates for the US greenhouse gas inventory, 1990–2008. J For 109(3): 167–173.

Hicke JA, Asner GP, Randerson JT, Tucker C, Los S, Birdsey R, Jenkins JC, Field C (2002) Trends in North American net primary productivity derived from satellite observations, 1982–1998. Glob Biogeochem Cycl 16(2):2-1–2-22

Hicke JA, Jenkins JC, Ojima DS, Ducey M (2007) Spatial patterns of forest characteristics in the western United States derived from inventories. Ecol Appl 17(8):2387–2402

Howarth RW, Bringezu S (2009) Biofuels: environmental consequences and implications of changing land use. In: Proceedings of the scientific committee on problems of the environment international biofuels project rapid assessment. Cornell University Press, Ithaca, New York

Jeffords JM, Levin C, DeGette DL, Baucus MS (1999) Community development: extent of federal influence on "urban sprawl" is unclear. United States General Accounting Office, Washington, DC

Jordan N, Boody G, Broussard W, Glover JD, Keeney D, McCown BH, McIsaac G, Muller M, Murray H, Neal J, Pansing C, Turner RE, Warner K, Wyse D (2007) Environment – sustainable development of the agricultural bio-economy. Science 316:1570–1571

JP Morgan Investment Analytics & Consulting (IAC) (2009) Investing in timberland: another means of diversification. JP Morgan: The Pensions Perspective 1Q09 Edition, 1–3

Kline JD, Alig RJ (2005) Forestland development and private forestry with examples from Oregon (USA). For Policy Econ 7:709–720

Kline JD, Alig RJ, Garber-Yonts B (2004) Forestland social values and open space preservation. J For 102:39–45

Koplow D (2006) Biofuels-at what cost? government support for ethanol and biodiesel in the United States. International Institute for Sustainable Development, Geneva

Mundell J, Taff SJ, Kilgore MA, Snyder SA (2010) Using real estate records to assess forestland parcelization and development: a Minnesota case study. Landscape and Urban Planning 94(2):71–76

Murray BC (2009) Economic drivers of forest land use choices: relevance for U.S. climate policy. Seminar, Yale University

Plantinga AJ, Alig R, Cheng HT (2001) The supply of land for conservation uses: evidence from the conservation reserve program. Resour Conserv Recycl 31:199–215

Radeloff VC, Hammer RB, Stewart SI (2005) Rural and suburban sprawl in the U.S. Midwest from 1940 to 2000 and its relation to forest fragmentation. Conserv Biol 19:793–805

Rhemtulla JM, Mladenoff DJ, Clayton MK (2009) Legacies of historical land use on regional forest composition and structure in Wisconsin, USA (mid-1800s–1930s – 2000s). Ecol Appl 19(4):1061–1078

Riitters KH, Wickham JD, O'Neill RV, Jones KB, Smith ER, Coulston JW, Wade TG, Smith JH (2002) Fragmentation of continental United States forests. Ecosystems 5:815–822

Rinehart JA (1985) Institutional investment in US timberlands. For Prod J 35(5):13–19

Roberts MJ, Lubowski RN (2007) Enduring impacts of land retirement policies: evidence from the conservation reserve program. Land Econ 83(4):516–538

Sample A (2007) Introduction to Binkley, C. The rise and fall of the timber investment management organizations: ownership changes in US forestlands. Pinchot Distinguished Lecture. Pinchot Institute for Conservation, Washington, DC

Skocpol T, Finegold K (1982) State capacity and economic intervention in the early new deal. Political Sci Q 97:255–278

Smith W, Miles P, Perry C, Pugh S (2009) Forest resources of the United States, 2007. US Department of Agriculture, Forest Service, Washington, DC

Sohngen B, Brown S (2006) The influence of conversion of forest types on carbon sequestration and other ecosystem services in the South Central United States. Ecol Econ 57:698–708

Stein PR (2005) Changes in timberland ownership. The Lyme Timber Company, Hanover

Stein P (2006) Future of the forest. Adirondack Explorer interview, 8–12

Stone RS, Tyrrell ML (2012) Forestland parcelization in the catskill/delaware watersheds of New York. Journal of Forestry. In press

Sullivan P, Hellerstein D, Hanson L, Johansson R, Koenig S, Lubowski R, McBride W, McGranahan D, Roberts M, Vogel S, Bucholtz S (2004) The conservation reserve program: economic implications for rural America. United States Department of Agriculture, Washington, DC

Sustaining Family Forests Initiative (SFFI) (2011) Tools for Engaging Landowners Effectively, Supplemental Income Owner Profile. Available online at http://www.engaginglandowners.org/profile/national/national/at3/1?selection=United%2BStates Accessed 19 May 2011

Theobald DM, Romme WH (2007) Expansion of the U.S. wildland-urban interface. Landsc Urban Plan 83: 340–354

Tyrrell ML, Hall MHP, Sampson RN (2005) Dynamic models of land use change in Northeastern USA: developing tools, techniques, and talents for effective conservation action. In: Laband, David N (ed) Emerging issues along Urban-rural interfaces: linking science and society. Conference Proceedings, Atlanta, 13–16 Mar 2005

United States Census Bureau (2009) Median household income tables. http://www.census.gov/hhes/www/income/reports.html. Accessed Sept 2009

United States Environmental Protection Agency (2009) Inventory of U.S. greenhouse gas emissions and sinks: 1990 – 2007. U.S. Environmental Protection Agency, Washington, DC

van Kooten GC, Sohngen B (2007) Economics of forest ecosystem carbon sinks: a review. University of Victoria, Department of Economics, Resource Economics and Policy Analysis Research Group, Columbus

Ward BC, Mladenoff DJ, Scheller RM (2005) Simulating landscape-level effects of constraints to public forest regeneration harvests due to adjacent residential development in northern Wisconsin. For Sci 51:616–632

Weyerhaeuser Corporation (2007) Second quarter earnings report. Available online at www.investor.weyerhaeuser.com

Wickham JD, Riitters KH, Wade TG, Homer C (2008) Temporal change in fragmentation of continental US forests. Landscape Ecol 23(8):891–898

Wimberly MC, Ohmann JL (2004) A multi-scale assessment of human and environmental constraints on forest land cover change on the Oregon (USA) coast range. Landsc Ecol 19:631–646

Woodall CW, Miles PD (2008) Reaching a forest land per capita milestone in the United States. Environmentalist 28:315–317

Wu JJ, Plantinga AJ (2003) The influence of public open space on urban spatial structure. J Environ Econ Manag 46:288–309

Xiao JF, Zhuang QL, Law BE, Baldocchi DD, Chen JQ, Richardson AD, Melillo JM, Davis KJ, Hollinger DY, Wharton S, Oren R, Noormets A, Fischer ML, Verma SB, Cook DR, Sun G, McNulty S, Wofsy SC, Bolstad PV, Burns SP, Curtis PS, Drake BG, Falk M, Foster DR, Gu LH, Hadley JL, Katulk GG, Litvak M, Ma SY, Martinz TA, Matamala R, Meyers TP, Monson RK, Munger JW, Oechel WC, Paw UKT, Schmid HP, Scott RL, Starr G, Suyker AE, Torn MS (2011) Assessing net ecosystem carbon exchange of U.S. terrestrial ecosystems by integrating eddy covariance flux measurements and satellite observations. Agricultural and Forest Meteorology 151(1):60–69. doi:10.1016/j.agrformet.2010.09.002

Zhang Y, Zhang D, Schelhas J (2005) Small-scale non-industrial private forest ownership in the United States: rationale and implications for forest management. Silva Fennica 39(3):443–454

United States Legislative Proposals on Forest Carbon

16

Jaime Carlson, Ramon Olivas, Bradford Gentry, and Anton Chiono

Executive Summary

This chapter provides an overview of the role of managing forests to store carbon in the efforts to adopt U.S. climate legislation at the national level (as of September 2010). While the U.S. has not ratified the Kyoto Protocol or adopted national climate legislation yet, considerable efforts have been underway to reduce emissions of greenhouse gasses at the regional (Northeastern U.S.), state (California), municipal, corporate, and individual levels. The issue of storage of carbon in forests and farmland has played a major role in U.S. emission reduction efforts, particularly in the voluntary carbon markets. As the demand for land-based carbon offsets has grown, so too has the demand for rules to define high quality, real offsets. The U.S. market has responded with a range of such rules, from those directly supported by governments, to those that are purely voluntary. Some of these rules cover how best to account for carbon in forest systems, such as: the types of forests/forestry operations covered; the pools of carbon in the forest that are included; the location of acceptable projects; and the "business as usual"/baseline emissions to be considered. Others go more directly to the quality of the offset produced, namely, whether the emission reductions are truly "additional" to those that would have happened anyway; how best to monitor and verify that the promised storage has occurred; how to protect against "leakage," i.e. that the emissions just move to another location; and how to ensure that the storage is permanent or how to protect against potential releases in the future. As federal efforts to adopt climate legislation intensify, these lessons learned from the voluntary carbon markets are being incorporated into the draft bills. It is clear that any U.S. federal climate legislation will include provisions to encourage the storage of carbon in forest and agricultural lands – both through the markets for carbon offsets, as well as direct public funding. The details of these programs, however, are likely to be delegated to the U.S. Department of Agriculture and other federal agencies to be worked out.

What We Know About Forest Carbon Policy

While it is extremely difficult to predict how U.S. federal climate policy will evolve, there are a few areas where the likely results seem clear:

- If and when the U.S. adopts federal climate policy, forests and other land uses are likely to play a major part in both the market and public funding approaches adopted.

J. Carlson
Secretary of Energy, U.S. Department of Energy, Washington, DC 20585, USA

R. Olivas
Emerging Energy and Environment LLC, Naucalpan de Juarez, Mexico

B. Gentry (✉)
Yale School of Forestry and Environmental Studies, New Haven, CT, USA
e-mail: bradford.gentry@yale.edu

A. Chiono
Pacific Forest Trust,

M.S. Ashton et al. (eds.), *Managing Forest Carbon in a Changing Climate*,
DOI 10.1007/978-94-007-2232-3_16, © Springer Science+Business Media B.V. 2012

- A wide range of land uses seems likely to be included, such as afforestation/ reforestation and managed forests, as well as soil carbon in farm and range lands. The inclusion of harvested wood products as approved project activities seems less likely at this time.
- Both domestic and international offsets from forest projects seem likely to be included. One open question is whether credits from international projects should be discounted compared to those from domestic projects.
- Substantial requirements will be imposed to help ensure that the offsets are "real." Finding the right balance between lower cost and higher accuracy will be difficult in the areas of monitoring and verification.
- While any policy will refer to the need to address leakage, few concrete measures to do so outside of project or entity boundaries seem likely to be required.
- Some combination of dedicating land to carbon storage for a lengthy period of time (through a conservation easement or contractual arrangement) and requiring that a portion of the credits be held for use as a buffer against unexpected changes seems likely. While there is some discussion of temporary credits, experience in the CDM market suggests that other ways should be used to address permanence issues.

What We Do Not Know About Forest Carbon Policy

The role of forests in likely future climate policy in the U.S. is a much larger set of questions, encompassing not only the unresolved scientific questions covered in other chapters in this volume, but also the constantly shifting efforts to build political coalitions in favor of federal legislation.

1 Introduction

1.1 The United States and Climate Change Policy

In the past decade, climate change has moved to the forefront of environmental concern in the United States. While 183 countries have ratified the Kyoto Protocol to the UN Framework Convention on Climate Change (UNFCCC 2009), the U.S. has not. Among the reasons given are concerns that the Kyoto Protocol does not set realistic goals and does not include emissions from rapidly growing developing counties (Barrett and Stavins 2003; Stavins 2005).

In the absence of action by the U.S. at the international and national levels, regional, state and municipal level climate initiatives have emerged, along with voluntary efforts. For example, at the local level, on February 16, 2005, the date the Kyoto Protocol became law in 141 countries, Seattle Mayor Greg Nickels launched the U.S. Mayors Climate Protection Agreement. The agreement represents a local effort to meet or beat Kyoto Protocol targets in communities across the U.S. By 2008, 916 cities and towns from 50 states, Washington D.C. and Puerto Rico had joined the Mayors Climate Agreement, representing more than 83 million citizens (U.S. Conference of Mayors 2009). At the state level, in September 2006, Governor Schwarzenegger of California signed the Global Warming Solutions Act, making California the first state to cap greenhouse gas (GHG) emissions in the U.S. (California AB32 2006). Similarly, under the Regional Greenhouse Gas Initiative (RGGI), ten Northeastern and Mid-Atlantic states have agreed to cap and reduce emissions from the power sector by 10% by 2018 (RGGI 2009). As of 2009, RGGI is the first mandatory, market-based effort in the United States.

Many argue that it is in the best interest of the U.S. to develop a national GHG program that will allow the U.S. to be part of any future global climate agreements, particularly after the Kyoto Protocol expires in 2012. The Obama administration and the Democratically controlled Congress have indicated support for this view. As a result, there has been a surge in efforts to design a national emissions cap and trade program, as part of the federal response to climate change. For example, a consortium of major corporations (e.g. Alcoa, BP, DuPont, GE, Pepsi, Shell) and leading environmental groups (e.g. World Resources Institute, Natural Resources Defense Council, Environmental Defense Fund, The Nature Conservancy) formed the United States

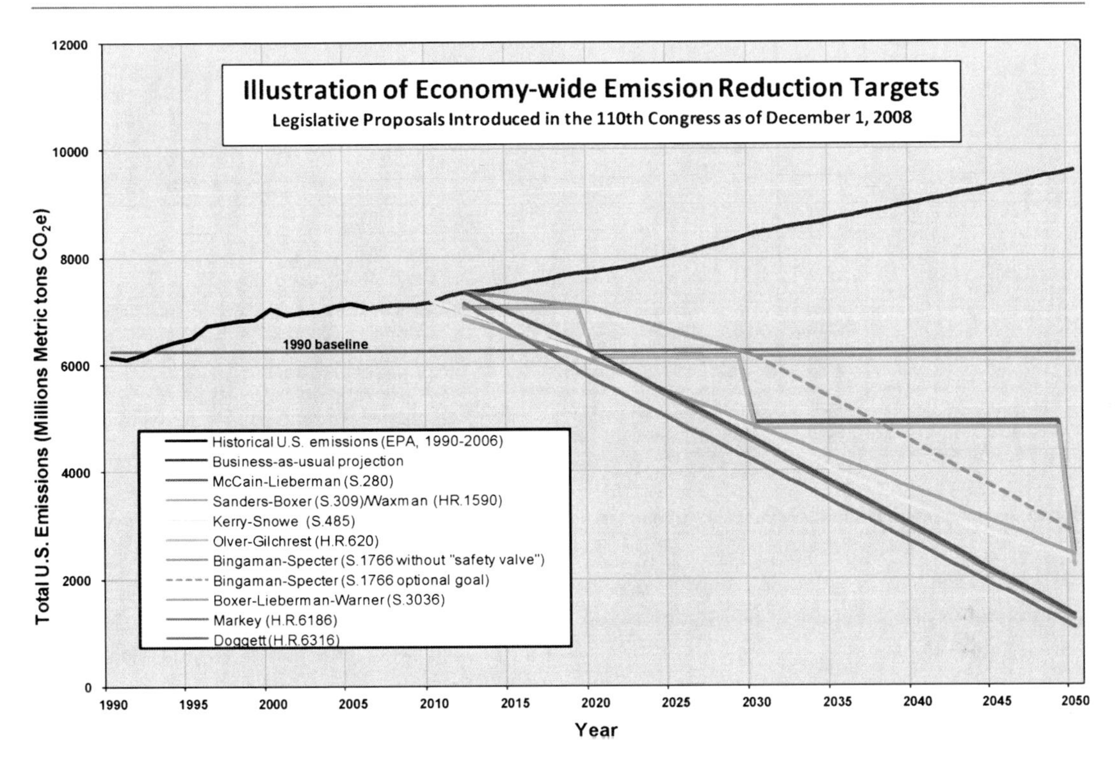

Fig. 16.1 Future emission reductions considered in the 110th Congress (*Source*: Pew Center on Global Climate Change 2008. Reprinted with permission)

Climate Action Partnership (USCAP) "to call on the federal government to quickly enact strong national legislation to require significant reductions of greenhouse gas emissions" (USCAP 2009a). The negotiations behind USCAP's Blueprint for Legislative Action have influenced the design of many recent Congressional proposals (USCAP 2009b).

As of December 2008, there were ten economy-wide cap-and-trade proposals under consideration in the 110th Congress (Pew 2008). Figure 16.1 shows the emission reduction goals of the ten proposals. In the 111th Congress, the House of Representatives passed the Waxman-Markey American Clean Energy and Security Act of 2009, which included an economy-wide cap-and-trade program (U.S. Congress 2009).,The two subsequent economy-wide cap-and-trade proposals in the Senate, the Kerry-Boxer Clean Energy Jobs and American Power Act (S.1733) and the Kerry-Lieberman American Power Act, drew in part from climate provisions in Waxman-Markey (U.S. Congress 2009, 2010). While Kerry-Boxer, Kerry-Lieberman, and Waxman-Markey each contained economy-wide cap-and-trade provisions, a pared, utility-only cap-and-trade approach appeared to have the greatest political viability as climate and energy negotiations entered the 2010 midterm election season and beyond.

1.2 The Voluntary Carbon Market in the United States

At the same time that the municipal, state, regional and national efforts to address climate change in the U.S. have expanded, so too has the work by corporations, academic institutions, individual U.S. citizens and others to reduce their carbon footprints. One part of these efforts is an active market for voluntary carbon offsets – for example where the owner of a car pays to have a farmer reduce GHG emissions from farm operations (Hamilton et al. 2009). Many corporate buyers in the voluntary market participate in order to better understand the transaction process

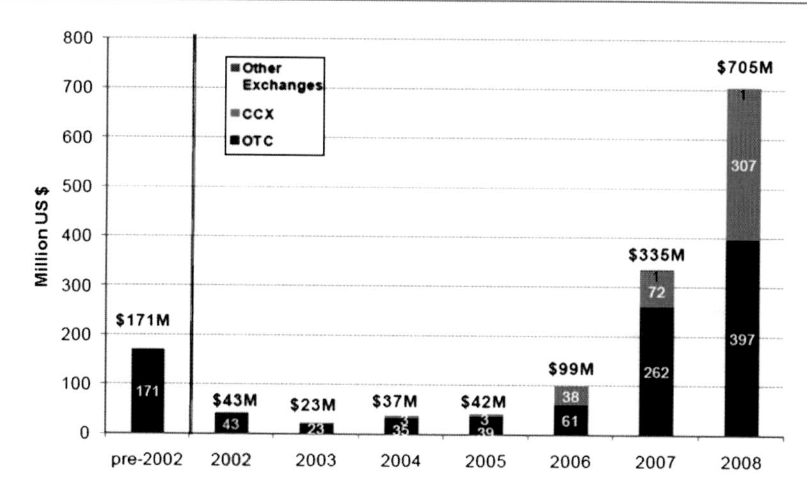

Fig. 16.2 Historic growth in the voluntary carbon market (*Source*: Hamilton 2009. Reprinted with permission)

in anticipation of a federal cap and trade system that includes offsets.

In an effort to lend some structure to the voluntary carbon markets, a number of different organizations have developed rules for ensuring that offsets actually lead to emission reductions. It is expected that the standards and guidelines ultimately included in a federal GHG regime will draw heavily from the experience and rules being used in the voluntary carbon markets.

While the first voluntary carbon transaction in the United States occurred in 1988 (16 years before the first one in the Kyoto Protocol), when carbon offsets were purchased from a forestry project (Hamilton et al. 2009), transactions in the over-the-counter (OTC) market did not gain significant momentum until the early 2000s (see Fig. 16.2). The launch of the Chicago Climate Exchange (CCX) in 2003 added further depth to the voluntary carbon markets. In 2005, voluntary markets scaled up as offsetting emissions entered the mainstream and there was an increase in transactions, as well as both praise and criticism.

Ecosystem Marketplace and New Carbon Finance tracked a total of 66 million tons of carbon dioxide equivalent ($MtCO_2e$) worth of offsets traded in the U.S. voluntary carbon market in 2007 and 123.4 $MtCO_2e$ in 2008 (Hamilton et al. 2009). Of the 123.4 $MtCO_2e$ traded in 2008, 54 $MtCO_2e$ (44%) involved exchanges on the OTC market and 69.2 $MtCO_2e$ (56%) were conducted on the CCX. The total value of the transactions for the year was U.S. $705 million (Hamilton et al. 2009).

1.3 Forests as Part of the U.S. Climate Change Strategy

Forests influence greenhouse gas concentrations because they are both a potential CO_2 sink (sequestering carbon) when they grow, as well as a potential source of CO_2 when they are disturbed. According to a 2007 report from the Intergovernmental Panel on Climate Change (IPCC), deforestation and subsequent land use change accounted for 17.4% of total anthropogenic global greenhouse gas emissions in 2004 (IPCC 2007). Any comprehensive climate change policy must address these emissions.

At the same time, forests have a significant potential to sequester carbon. Compared to alternatives such as industrial carbon capture and storage, forest offset projects are regarded as a less expensive means of carbon storage (Enkvist et al. 2007). Moreover, financial incentives for carbon sequestration in forests would help fund biodiversity conservation efforts in the U.S. and abroad. The "technology" or ability to sequester and store carbon in forests already exists, and keeping forests as forests (i.e. preventing deforestation) is the most straightforward way of

Table 16.1 Proves an overview of how the three U.S. legislative initiatives address forests as part of their climate mitigation strategies

Proposal/program	Rules for forest offsets	Public funding programs
RGGI	Detailed requirements for afforestation Offsets set forth in RGGI Model rule	None provided
Kerry-Lieberman	USDA will establish a GHG Reduction and Sequestration Advisory Committee to determine guidelines for forestry offsets. They will give "due consideration" to existing methodologies	From 2019 to 2034, 1.5–6.0% of allocations are divided equally between domestic natural resource adaptation and international adaptation and global security. Carbon Conservation Program created to reward conservation easements and other non-offset emission reduction/storage activities from forestry and agriculture, but no funding is allocated.
Waxman-Markey	USDA will establish a GHG Reduction and Sequestration Advisory Committee to determine guidelines for forestry offsets. They will give "due consideration" to existing methodologies.	EPA to sell a declining percentage (from 5% to 2%) of annual allowances to generate funds for incentivizing reduced deforestation in developing countries.

Source: Data from RGGI (2008a) and U.S. Congress (2010, 2009)

maintaining carbon stocks. Given its large land base, the U.S. is also well positioned to use its domestic forests to help meet any future national emission reduction targets. If carbon storage in farm and range lands is included, this may help acquire votes from senators in the key Midwestern states in favor of U.S. climate legislation.

For these reasons, many of the proposals for U.S. climate change legislation include incentives – either through the carbon markets or public funding – for activities that increase forest carbon sequestration and reduce emissions from deforestation and degradation. If such incentives are approved, they would go beyond the existing structures of the Kyoto Protocol and the Clean Development Mechanism (CDM), which have not yet included forests as a significant source of offset projects. While the rules under the Kyoto Protocol include forests as a verifiable GHG sink, tradable credits are only granted for afforestation/reforestation projects established after 1990 (European Commission 2009). Moreover, these credits under Kyoto are "temporary credits" and typically trade at a discounted price (Hamilton 2009).

Generally speaking, the prices paid for offsets in the U.S. voluntary carbon markets are lower than those paid in international compliance markets, such as in the EU's Emissions Trading System (Carbon Positive 2009). However, forestry credits have remained some of the highest priced offset credits in the U.S. For example, in 2006 and 2007, credits from afforestation/reforestation projects received the highest prices of any offset projects in the U.S. voluntary markets (Table 16.1) (Hamilton et al. 2007). Moreover, many (though not all) forest carbon projects have a higher value on the OTC market for the social and environmental cobenefits they offer. For companies buying voluntary credits for the sake of public relations, the tangible nature of conserved land and general understanding of trees in the carbon cycle adds to their appeal – so-called "charismatic carbon" (Hamilton et al. 2007; Conte and Kotchen 2009).

However, even in the U.S., the groups setting rules for and creating registries to track offsets have faced difficulties incorporating forest carbon projects into their frameworks. This is due, in part, to the variable nature of forest growth and the risks associated with the potential impact of natural disasters (e.g. fires or disease) on carbon stocks. Given the complexity and variability in forest systems, forest-based offset projects have raised many debates over how to account for and ensure the quality of the credited emission reductions over time.

In addition, forestry or land use offset projects can involve a number of different types, each posing a range of issues. For example, "biological carbon sequestration projects" made up 26% of the transactions in the voluntary carbon markets in 2007 (Hamilton et al. 2007). This included projects that involved storing carbon through a range of different activities:

- Afforestation/reforestation with native species: 42%
- Avoided deforestation: 28%
- Agricultural soil management: 16%
- Afforestation/reforestation in plantation monocultures: 13%
- Other biological sequestration (such as wetlands preservation): 0.1%

2 Overview of U.S. Legislative Initiatives and the Role of Forest Carbon

In order to illustrate the issues to be addressed as part of any U.S. national policy on forest carbon, this chapter focuses on three legislative initiatives and four sets of carbon market rules. Each is described briefly below (as of September 2010). We then dig more deeply into how each of these initiatives addresses key issues facing forest carbon offsets and public funding for forest carbon sequestration efforts.

2.1 Legislative Initiatives Covered

Three legislative initiatives are considered: one regional and two national.

2.1.1 Regional Greenhouse Gas Initiative (RGGI)

RGGI is a multi-state, mandatory cap and trade program to reduce CO_2 emissions from electricity generation in the northeastern U.S. (RGGI 2009). It was established in 2005 by the governors of seven states in the Northeast and Mid-Atlantic regions and has since expanded to include ten states.

RGGI began in 2009 as the first mandatory cap and trade program for GHGs in the U.S. Its objective is to reduce CO_2 emissions from the regulated energy sector by 10% from 2009 to 2018. It starts by setting a regional cap to stabilize emissions from 2009 to 2014 and then reducing the cap by 2.5% each year until 2018. RGGI's first 3-year compliance period started in January 2009. The program is expected to cap CO_2 emissions at 188 million short tons to the end of 2014. The first auction of RGGI emission allowances was held in September 2008. All the allowances were sold at a price above the auction reserve price, selling for $3.07 per ton (RGGI 2008b).

Offsets serve as a limited alternative compliance mechanism for regulated facilities under the RGGI program (RGGI 2008a, § XX-10). Five types of offsets are defined in the rule as qualifying for use in the program (see discussion below) (RGGI 2008a, § XX-10.3.a.1). While the amount of emissions that can be offset is limited, the use of offsets can be expanded if the price of emission allowances rises beyond $7 per ton. As such, it remains to be seen what the future role and size the offset market will be under the RGGI program.

2.1.2 Kerry-Lieberman American Power Act of 2010

Two economy-wide cap-and-trade bills were proposed before the Senate in the 111th Congress: the Kerry-Boxer Clean Energy Jobs and American Power Act (S.1733), and the Kerry-Lieberman American Power Act (U.S. Congress 2009, 2010). Despite a Republican boycott of the final committee vote on Kerry-Boxer, the bill was passed 11–1 by the Senate Committee on Environment and Public Works in November of 2009. Its initial passage distinguished it as the only economy-wide cap-and-trade proposal to be accepted in the 111th Congress by any of the six Senate committees with jurisdiction over climate and energy legislation. Despite clearing this initial hurdle, the decided lack of bi-partisan support for Kerry-Boxer gave rise to serious doubts as to its political viability in the broader Senate.

In an effort to craft a proposal capable of garnering the 60 votes necessary to overcome a Senate filibuster, Senators John F. Kerry, Joseph I. Lieberman, and Lindsey Graham launched a tri-partisan effort to draft a comprehensive climate and clean energy bill in early 2010. Despite Senator Graham's decision to withdraw from the effort in

April of 2010, Senators Kerry and Lieberman released their comprehensive climate and energy proposal, the Kerry-Lieberman American Power Act, the following month. The Kerry-Lieberman bill establishes a market-based cap-and-trade program for GHG emissions in the United States, and creates economy-wide emissions reductions targets of 20% by 2020, 42% by 2030, and 83% by 2050 (relative to 2005 levels). The cap covers emissions from U.S. electric power, petroleum fuels, distributors of natural gas, producers of fluorinated gases and other specified sources that together account for about 85% of U.S. GHG emissions (Pew 2010). The bill also includes incentives for agriculture and forestry programs that cut emissions, but may not qualify as offsets.

2.1.3 Waxman-Markey Clean Energy and Security Act of 2009

The American Clean Energy and Security Act of 2009 (ACES) is a climate change and energy bill presented by Chairman Henry Waxman of the Energy and Commerce Committee and Chairman Edward J. Markey of the Energy and Environment Subcommittee. Their "discussion draft" went to the U.S. House Committee on Energy and Commerce in March 2009 and a substantially revised version was passed by the full House in June 2009. The ACES aims to "create jobs, help end our dangerous dependence on foreign oil, and combat global warming." To meet these goals, the legislation has four titles:

- A clean energy title that promotes renewable sources of energy, carbon capture and sequestration technologies, low-carbon fuels, clean electric vehicles, the smart grid, and expanded electricity transmission;
- An energy efficiency title that aims to increase energy efficiency across all sectors of the economy, including buildings, appliances, transportation, and industry;
- A global warming title that places limits on emissions of GHGs; and
- A transitioning title aimed at protecting U.S. consumers and industry while promoting green jobs during the transition to a clean energy economy (H.R. 2454 2009).

While most legislative initiatives specify the types of forest systems covered and the funding methods proposed, few (other than RGGI) offer many details on forest carbon accounting or quality assurance issues. Rather, many refer to existing efforts in the voluntary carbon markets to define rules and establish registries for the types of carbon offsets that will be accepted for trading.

2.2 Rules for Offsets Sold in the Voluntary Markets

In addition to the three legislative initiatives described above, this chapter analyzes four sets of rules in the U.S. voluntary carbon markets: the Climate Action Reserve (CAR), the Voluntary Carbon Standard (VCS), the Chicago Climate Exchange (CCX), and the U.S. Department of Energy's Reporting Program 1605(b). These initiatives serve as reference points for future regulatory efforts given that they provide both applied market experiences, as well as examples of rules formulated in the complex political realities of the U.S.

The protocols for offset projects across these groups vary significantly. This is partially due to the complicated nature of accounting for forest projects to ensure that sequestration is real and based in environmental integrity. It also is indicative of the regional priorities in climate change policy. The diverse frameworks for registering forest projects attest to the complexity of designing a forest carbon accounting system.

2.2.1 Climate Action Reserve (CAR)

The CAR is a private non-profit organization originally formed by the State of California. It is the parent organization of the California Climate Action Registry, a body that registers and tracks voluntary greenhouse gas emission reduction projects. CAR's purpose is to establish regulatory-quality standards for the development, quantification and verification of GHG emissions reduction projects (CAR 2009a). For projects meeting its rules, carbon offset credits known as Climate Reserve Tonnes (CRTs) are issued for the emission reductions generated. Sales and ownership of CRTs are tracked over time in a publicly accessible registry system (CAR 2009a). The rules set by CAR are likely to have a major influence on defining the offsets that will qualify under any cap and trade program that the state of California may adopt under its climate legislation.

2.2.2 Voluntary Carbon Standard (VCS)

The Voluntary Carbon Standard was established by the World Economic Forum and the International Emissions Trading Association in 2005 (www.ieta.org). It is a global program working to provide a standard and a mechanism for approval of credible voluntary carbon offsets across multiple voluntary programs. It has established the voluntary carbon unit (VCU) as a means of providing tradable offset credits. The VCS focuses on a chain of ownership through its multiple registries and publically available central project database, striving to prevent voluntary offsets from being used twice. The VCS has approved the offset rules under the UNFCCC's Clean Development Mechanism and Joint Implementation Program, as well as the Climate Action Reserve Program, as meeting its rigorous registry criteria (VCS 2008).

2.2.3 Chicago Climate Exchange (CCX)

The Chicago Climate Exchange was launched in 2002 as a voluntary greenhouse gas (GHG) emission cap and trade system for North America (CCX 2009a; Kollmuss et al. 2008). Although participation in the CCX cap and trade program is voluntary, once entities elect to participate and commit to emission reduction targets, compliance is legally binding. Members can comply by cutting their emissions internally, trading emission allowances with other CCX members, or purchasing offsets generated under the CCX offset program. There are no limits on the use of offsets for compliance with parties' emission reduction commitments.

U.S. Department of Energy's Reporting Program [1605(b)]

Section 1605(b) of the Energy Policy Act of 1992 established a program on the Voluntary Reporting of Greenhouse Gases (USDOE 2007). Its purpose was to encourage corporations, government agencies, non-profit organizations, households, and other private and public entities to submit annual reports of their greenhouse gas emissions, emission reductions, and sequestration activities. Included are rules on reporting emissions and emission reductions from forest and other land-based activities (USDOE 2007).

Taken together, these three legislative initiatives and four sets of offset rules offer a range of options for including forest carbon in future U.S. legislation – from market-based approaches involving offsets for emission reduction projects, to public funding for forestry activities. The implications of the different approaches taken are explored below.

3 Market Approaches: Including Offsets from Forest Carbon Projects in U.S. Policy

This section presents a guide to the treatment of forest carbon projects under the legislative initiatives and carbon market rules covered in this analysis. Each of the following aspects is considered:

- *Forest carbon accounting*: Forest systems; carbon pools; project sites; and baselines
- *Quality assurance:* Additionality; monitoring and verification; leakage; and permanence

3.1 Forest Carbon Accounting

3.1.1 Scope of Forest Systems Allowed

As shown in Table 16.2, while most of the legislative initiatives allow carbon offset credits generated from afforestation and reforestation projects, there has been much debate over whether to include managed forests, conservation forests, harvested wood products and other forest systems as approved carbon sinks. Part of the concern over expanding offset eligibility is the ability to track carbon storage effectively. This is especially true in the case of harvested wood products, given the uncertainties associated with their end use once they leave the forest (i.e. incorporated into a solid wood product, burned, decayed).

Under RGGI, afforestation is the only approved forestry-related offset project type (RGGI 2008a, § XX-10.3.a.1). In contrast, the Waxman-Markey bill passed by the U.S. House of Representatives in June 2009 contains an extensive list of forestry projects as approved offset types and includes forest projects that are not commonly accepted as

Table 16.2 Forest systems allowed as offsets in US Legislation

Legislative proposal/program	Eligible domestic forest offset projects	Other eligible offset projects
RGGI	Afforestation	Landfill methane capture and combustion; sulfur hexafluoride (SF_6) capture and recycling; end-use fossil fuel (natural gas, propane, and heating oil) energy efficiency; methane (CH_4) capture.
Kerry-Lieberman	Afforestation/reforestation; forest management; land management (e.g. restoration, avoided conversion/ reduced deforestation; urban tree-planting); forest-based manufactured products.	Methane collection; fugitive emissions capture; Biochar production; Agriculture and rangeland sequestration (e.g. altered tillage, winter cover cropping, conversion of cropland to grassland, fertilizer reduction); Manure management and disposal; reduced agricultural production intensity. List subject to revision by USDA
Waxman-Markey	Afforestation/reforestation; conservation forestry; improved forest management; reduced deforestation; urban forestry; agroforestry; management of peatland; harvested wood products.	Agricultural, grassland, and rangeland sequestration (e.g. altered tillage practices, winter cover cropping, reduction in fertilizer, etc.); Manure management and disposal (e.g. waste aeration, biogas capture, substitute or commercial fertilizer).

Source: Data from RGGI (2008a) and U.S. Congress (2010, 2009)

forest offsets (H.R. 2454 2009, § 733). For example, urban forestry, harvested wood products, and peatland management are currently included as potential offset projects. While this list is subject to further revision, the bill requires the Secretary of Agriculture to publish within 1 year an official list of offset projects types that will be allowed under a federal system. Acknowledging that there is still some uncertainty as to what offset projects are truly verifiable, additional, and permanent, the bill requires this list of offset practices to be revised every 2 years by the Secretary.

Kerry-Lieberman similarly directs the Secretary of Agriculture to establish and maintain a list of eligible offset project types that includes a minimum initial list specified in the bill. As in Waxman-Markey, the Secretary may add projects to the list via regulation, and must periodically assess whether or not any project types cease to satisfy the purposes of the Act (i.e. no longer create emissions reductions that are additional, measurable, verifiable, and enforceable), and should be removed. As Table 16.2 demonstrates, the minimum initial list included in Kerry-Lieberman closely resembles that in Waxman-Markey, and is similarly expansive. In addition to publishing and maintaining the list of eligible project types, Kerry-Lieberman directs the Secretary to gather data on forest carbon stocks, fluxes and sequestration rates to assist landowners who are preparing to undertake an offset project. To assist in gathering this information, the bill includes provisions directing the EPA, in collaboration with Secretaries of Agriculture and the Interior, to establish a national-scale forest carbon accounting program to more effectively track forest carbon storage nationally (U.S. Congress 2010, § 807).

Table 16.3 Offset rules and scope of forest systems

Offset rules	Scope: forest systems
CAR	Reforestation, Improved Forest Management (IFM), and Avoided Conversion (AC). All must utilize natural forest management practices
VCS	Afforestation/reforestation (A/R); Improved forest management (IFM); Forest conservation (REDD)
CCX	Afforestation/reforestation; Improved forest management; Harevested wood products; Rangeland soil and carbon management
DOE 1605b	Afforestation/reforestation; Agroforestry; Forest conservation; Sustainably managed forests; Urban forestry; Short rotation biomass; Harvested wood products

Source: Data from RGGI (2008a), CAR (2010), VCS (2009), CCX (2009b), and USDOE (2007)

As Table 16.3 highlights, all of the market rules permit afforestation. Historically, reforestation and afforestation have been the favored

Table 16.4 Offset rules and carbon pools.

Offset rules	Carbon pools
RGGI	Above-ground living tree biomass; soil carbon; dead biomass (unless pool is at or near zero, in which case it is optional); above-ground non-tree biomass (optional).
CAR	Standing live trees; shrubs and herbaceous understory (optional for IFM and AC); standing dead biomass; lying dead wood (optional); litter (optional); soil (optional); wood products.
VCS	Above-ground tree biomass (non-tree excluded); Below-ground biomass (A/R required); deadwood (IFM required); harvested wood products (IFM/REDD required).
CCZ	Above-ground living trees; below-ground living biomass; Soil carbon
DOE1605b	(All optional). Above-ground living biomass; below-ground living biomass; standing and down dead trees; below-ground dead trees; litter; soil carbon; harvested wood products.

Source: Data from RGGI (2008a), CAR (2010), VCS (2008, 2009), CCX (2009b), and USDOE (2007)

forest carbon project types, partly due to the ease in calculating baselines and additionality. More recently, CAR, CCX and the DOE 1605(b) have also incorporated sustainably managed forests as approved forest offset systems.

The selection of forest system types permitted in an offset regime will prove to be an important decision. The portfolio of approved forest offset projects must ensure that offset supply will be sufficiently large to assure liquidity in the market (i.e. ease of buying and selling offset credits without causing a significant movement in price). At the same time, carbon storage and uptake in these forest systems must be efficiently and accurately quantified and verifiable. This becomes increasingly complicated with forest types such as managed and conservation forests where it is difficult to accurately quantify baselines or flows of carbon storage over time (e.g. change in carbon stores post-harvest) compared to afforestation and reforestation projects (that start with little or no stored carbon).

3.1.2 Carbon Pools

Carbon pools are the parts of a forest system in which carbon is stored. These pools may include above-ground biomass, below-ground biomass, soils and wood products, among others.

To date, all offset market rules include, and most require,[1] that above-ground biomass be used as an approved carbon pool (Table 16.4). However, there is still much debate around whether to account for carbon stored in below-ground biomass, soils, and harvested wood products in an offset regime.

The ultimate decision on what pools should be included will involve a balance of costs and benefits. For example, soils are known to be a significant carbon pool. However, soil carbon is highly variable depending on site conditions and land use history. This variability increases the amount of sampling required and, therefore, the cost to accurately estimate soil carbon stocks. If the quantification and verification of carbon pools is too expensive, it may not make sense to include these pools. However, if a higher price is obtained in the carbon markets for more accurate measurements of carbon pools, then it will make sense for carbon developers to incur these increased costs. As such, carbon pools that can be accurately quantified may eventually be included, regardless of monitoring costs.

In addition to the question of what carbon pools should be included in the rules for allowable offsets, a different question remains around what carbon pools should be measured in order to ensure the environmental integrity of forest carbon projects. As it currently stands, above-ground biomass is the only carbon pool that is approved and required in all registry systems. However, it is possible that in order to avoid deleterious ecosystem effects, other carbon pools should also be accounted for. For example, afforestation of inappropriate sites can result in an increase in above-ground carbon stores, but depletion of below-ground soil carbon for an overall net loss of carbon (Paul et al. 2002). Moreover, some afforestation or reforestation

[1] Note: Required in all except DOE 1605(b) in which all pools are optional.

Table 16.5 Descriptions of offsets from forests inside and outside the US by proposal/program

Proposal/program	Offsets from forests in the US	Offsets from forests outside the US
RGGI	Offsets are limited to 3.3% of a facility's emissions, but amount can be increased when allowance price exceeds $7. Afforestation only type of forest-based offset recognized.	Not accepted
Kerry-Lieberman	Two billion tons of GHG emissions per year can be offset. 75% of these offsets, or 1.5 billion tons, can come from domestic agricultural or forestry projects.	International forest offsets are allowed, but may only be used to offset 25%, or 500 million tons, of GHG emissions per year unless the domestic offset supply is insufficient. After 2018, international offsets are discounted by 20% (i.e. 1.25 international offset credits must be submitted for every 1 emission allowance). International offsets are limited to credits from sector-based projects, those issued by an international body, or from reduced deforestation
Waxman-Markey	Two billion tons of GHG emissions per year can be offset. Half of these offsets can come from domestic agricultural of forestry projects	Fifty percent of total offsets (1 billion tons) can come from international offsets. This may be extended to 1.5 billion tons of domestic offsets. Market is limited.

Source: Data from RGGI (2008a) and U.S. Congress (2010, 2009)

projects may result in water quantity and/or quality loss (Farley et al. 2005).

To protect against such negative ecological results, impacts on both above-ground biomass and soil carbon should be considered in afforestation projects. However, this is not currently required under any legislative or carbon market rules.

3.1.3 Spatial Scale

Another key decision in developing an offset market is determining from where offset credits can be sourced. For example, RGGI currently requires that offset credits originate from projects in one of the ten Northeastern or Mid-Atlantic states that has signed the RGGI protocol (Table 16.5) (RGGI 2008a). This rule has raised concerns over whether the RGGI offset market will be large enough to be liquid and efficient. The Massachusetts Department of Environmental Protection was sufficiently concerned about this that they decided to expand the offset project location rules to include international offset projects. They stated that insufficient offsets were available in the U.S. for facilities to achieve compliance (MADEP 2007).

Both Kerry-Lieberman and Waxman-Markey also favor domestic offsets by discounting international offsets by 20% (i.e. by requiring 1.25 international offset credits in lieu of an emission allowance), although this discount does not take effect until after 2018 (H.R. 2454 2009, § 722; U.S. Congress 2010, § 722). In both bills, offsets can be used by covered entities to satisfy a percentage of their compliance obligation, up to a total of approximately 2 billion tons of CO_2e per year (H.R. 2454 2009, § 722; U.S. Congress 2010, § 722). Under Waxman-Markey, up to 50% of these offsets (or 1 billion tons per year) may come from domestic forest and agricultural offsets or from international reduced deforestation projects. If supplies of U.S. offsets prove to be limited, the Secretary of Agriculture may permit an increase in the number of international offsets to up to 1.5 billion tons, but the overall 2 billion ton limit on offsets will still hold. Kerry-Lieberman favors domestic offsets to an even greater extent, and initially allows no more than 25% (or 500 million tons per year) of the annual offset supply to come from international offsets. However, if it appears that domestic supplies will be insufficient, the limit on allowable international offsets can be increased to up to 50% of the total 2 billion ton annual offset supply.

Table 16.6 Baseline requirements for the different offset rules

Offset rules	Baseline
RGGI	Base year approach; net increase in carbon relative to the base year (often the year prior to beginning of the offset project).
CAR	Common-practice baseline for IFM projects; Avoided Conversion project baseline characterized by stocking levels that would occur in the event of conversion.
VCS	Business as usual baseline. With IFM, baseline is "most likely land use in absence of project". Three means to establishing (REDD) baseline depending on type of REDD activity. CAR and Clean Development Mechanism (CDM) baselines accepted.
CCX	Base year approach; net increase in carbon relative to previous year.
DOE 1605b	Base year approach; net increase in carbon relative to previous year.

Source: Data from RGGI (2008a), CAR (2010), VCS (2009), CCX (2009b), and USDOE (2007)

3.1.4 Baselines

Baselines are a quantitative assessment of the likely amount of carbon stored (or emissions produced) if the offset project had never taken place – such as what would have happened as part of "business-as-usual" if the offset developer had not taken steps to increase carbon sequestration (Pfaff et al. 2000). Baselines are critical measurements, as most carbon markets only grant offset credits for the extra or "additional" carbon stored by the project.

As such, the methods for establishing baselines are an important policy choice, as they dictate what forest-based activities are incentivized and qualify for carbon offsets. Baselines may be calculated by extrapolating from recent regional trends, current growth rates, existing project emissions or other quantitative measures. The most common methods used for establishing forest carbon baselines are "business as usual," "base year" or "without-project." Table 16.6 summarizes how baselines are calculated under RGGI and the carbon market rules covered in this chapter.

The "business as usual" (BAU) scenario establishes a project baseline based on estimates of future emissions of a project, in the absence of carbon offset policy and without any commitments to carbon reduction (Pfaff et al. 2000). Essentially, the BAU baseline relies on projections of project-specific carbon sequestration and storage if the project proceeded untouched by carbon policy or offset credits. This approach is used by CAR (where it is referred to as "common practice") and VCS for afforestation, reforestation and forest management projects (CAR 2010 §6; VCS 2009, §3.1).

The base-year (BY) approach to establishing a baseline is similar to BAU in that the baseline is based on current project emissions and carbon storage in the absence of carbon offset policy. However, BY does not require developers to project future trends in the project's carbon sequestration and storage. Rather, it chooses a base-year (often the year prior to beginning the offset project) to serve as the baseline and from which all "additional" carbon is measured (RGGI 2008a, § XX-10.5.c.4). So if a forest owner was planning to leave his or her plantation to grow for the next 10 years without harvesting it, they would receive no offset credits for additional carbon sequestered under a BAU approach, unless they did something above and beyond normal operations. However, under the BY approach, the forest owner would be eligible for credits for all carbon sequestered by the plantation in the next 10 years that is above the initial base year, regardless of the forest owner's original intent. The BY approach is used by CCX, DOE 1605(b) and RGGI (CCX 2009c; USDOE 2007, § 2.3; RGGI 2008a, § XX-10.5.c.4).

The concern with using the business-as-usual or base-year baseline is that they tend to reward project developers that have not previously adopted carbon sequestering or storage practices (Fenderson et al. 2009). For example, land managers that have been clearcutting forests would have a lower baseline than those who had historically managed their forests according to an ecologically sensitive selective harvesting regime or with longer rotations. Despite the fact that the latter's project could sequester the same amount

Table 16.7 Additionality and offset rules

Offset rules	Additionality
RGGI	Must be actions beyond those required by regulations or law. No credits for electric generation within RGGI states. No funding from any system or customer benefit fund. No credits or allowances awarded under any other mandatory or voluntary GHG program
CAR	Any net increase in carbon stocks caused by the project activity relative to baseline. Baseline estimates must reflect the legal, physical, and economic factors that influence changes in carbon stocks on a project. "Anti-depletion" credits are given for carbon stocks exceeding a common-practice baseline at the time of project initiation.
VCS	Proved through regulatory, economic or technology factors. Project must not be mandated by law and must face a barrier (technology, investment or institutional) that demonstrates that it would not occur otherwise.
CCX	All changes in carbon store after base year are considered additional.
DOE1605b	Not specifically required. All changes in carbon store after base year are considered additional.

Source: Data from RGGI (2008a), CAR (2010), VCS (2009), CCX (2009b), and USDOE (2007)

or possibly more carbon over the lifetime of the project, they would essentially receive fewer offsets credits than the first land manager as their baseline began at a higher value. In essence, this creates a system that penalizes good actors that have already incorporated silviculture practices that increase carbon sequestration into their forest management regime.

"Business as usual" and base-year approaches differ from baselines established by a "without-project case" method. The "Without-project" case approach can either establish the baseline according to the carbon stored under the previous land use system (prior to the forest carbon project) or based on regional trends (from forest inventory data). Integrating a regional average data baseline such as the methodology used in the 1605(b) guidelines (USDOE 2007, § 2.3) establishes baselines based on general land use practices in the project region. This type of baseline works well for forest offset projects that occur in regions with low forest density and a high threat of agricultural conversion or sprawl. They also reduce the costs of calculating the baseline.

3.2 Quality Assurance

In addition to the basic rules on accounting for carbon stored in forests discussed above, any future U.S. policy allowing forest carbon offset projects will need to ensure that the quality of the offsets is high enough to justify their use to meet emission reduction requirements. In doing so, the policy will need to address the quality assurance issues of additionality, monitoring and verification, leakage, and permanence.

3.2.1 Additionality

Offsets credits are granted only when an offset project's activities (i.e. avoiding deforestation, lengthening rotations, reforesting previously cut sites, etc.) are considered 'additional' to those that would have occurred in any event (i.e. those reflected in the baseline scenario). Different approaches are used to demonstrate additionality across various rules for the carbon markets, such as direct measurement of the additional carbon sequestered, removal of barriers, performance beyond that required, and/or intent (Table 16.7).

3.2.2 Monitoring and Verification

While some offset regimes require that projects be monitored on an annual basis (CCX 2009c; USDOE 2007, 1605(b) § 1; CAR 2010, § 8), others only require periodic reviews on a 2, 5 or 10-year basis. For example, RGGI requires that overall carbon stocks be assessed in afforestation projects at least every 5 years (Table 16.8) (RGGI 2008a, § XX – 10.5.c.5).

Likewise, certain initiatives require direct sampling of carbon stocks (CAR 2010, Appendix A; RGGI 2008a, § XX-8), while the DOE's 1605(b) protocol estimates carbon stock based on

Table 16.8 Offset rules and verification

Offset rules	Monitoring and verification
RGGI	Validation through an accredited independent verifier
CAR	Third-party verification must be conducted within 30 months of project submission; on-site verification is required at least once every 6 years thereafter. Monitoring reports must be prepared for each 12-month reporting period, and include estimates of carbon stocks using approved inventory methodologies. Confidence deductions are applied according to inventory sampling error, which may not exceed 20%.
VCS	Project monitoring and ex-post calculation of net GHG emission reduction required. Project monitoring should also include monitoring of project implementation, land use change and carbon stocks. Ex ante accounting system, but when there is low precision then calculation should be revised based on ex post monitoring.
CCX	Validtin through a CCX accredited verifier. Small projects may use either direct measures or CCX-approved default tables.
DOE 1605b	Changes in carbon stocks are accounted for by periodic inventory and reporting. Default tables used for region, species, management intensity, productivity class. If negative balance (carbon stock losses), the losses are reported in ELA documents and the entity cannot register additional reductions. Monitoring over a 5 year period.

Source: Data from RGGI (2008a), CAR (2010), VCS (2009), CCX (2009b), and DOE (2007)

tables for region, forest type and age (USDOE 2007 § H). While the results from direct sampling are more robust than estimates, they often require 3rd party assistance and/or verification, and thus the costs are higher for landowners.

While the carbon calculation default tables in the DOE's voluntary 1605(b) reporting are a simple and inexpensive approach (USDOE 2007 § H), using them may raise concerns regarding accuracy and environmental integrity since the uncertainties surrounding any individual project can be high. In addition, this methodology may not be correct for calculating all forest carbon pools. For example, the DOE 1605(b) recommends something called the flow approach for estimating changes in soil carbon (USDOE 2007 § H). It also provides a detailed format for estimating carbon captured and stored in harvested timber products. As a result of concerns regarding harvested forest product quantification methodologies, neither RGGI nor CCX has moved to offer credits for wood products.

3.2.3 Leakage

Most forest carbon accounting regimes attempt to incorporate indirect impacts. This is to ensure that a forest sequestration project in one location does not result in increased logging and higher emissions in another region. Leakage is the unanticipated loss or gain in carbon benefits outside of the project's boundary as a result of the project activities. It is perhaps one of the most difficult items to measure, especially considering that it is often unintended and not under the control of the offset project developer.

Leakage can be divided into two types: activity shifting and market effects (Brown 2009). Activity shifting is primary leakage – it occurs when the activity causing the carbon loss in the project area is displaced outside the project boundary (e.g., preventing deforestation in the project area may send the deforestation elsewhere). One difficult question to address with primary leakage is how large the "carbon shed" of the offset project should be. If the area of project influence is of manageable scale, primary leakage could potentially be addressed by establishing leakage prevention activities (e.g. alternative community development strategies) or including a buffer pool (setting aside a percent of the credits generated to cover leakage) (Hamilton et al. 2009).

Secondary leakage can occur as a result of market effects. Market effect leakage occurs when project activities change the supply and demand equilibrium. For example, if an offset project reduces the supply of wood products, it may cause an increase in forest logging in other regions to meet demand. Secondary leakage is difficult to

Table 16.9 Offset rules and leakage

Offset rules	Leakage
RGGI	There are no guidelines for addressing leakage
CAR	Secondary effects (activity-shifting leakage) arising within entity boundaries assumed to be negligible due to certification or sustained yield plan requirements; equations are provided to estimate external leakage from reduced harvest on project areas (thus displacing harvest to other forests). Carbon credits are discounted according to external leakage estimates.
VCS	VCS provides a table of adjustments to be made to account for offsite leakage. Project developers must demonstrate there is no activity shifting or leakage within their operations – i.e. on lands outside the project, but within their management control.
CCX	Must verify that there is no internal leakage. There are no guidelines for addressing external leakage.
DOE 1605b	Small emitters must prove that reductions are not likely to cause increases elsewhere in entity (internal leakage). No requirements for external leakage.

Source: Data from RGGI (2008a), CAR (2010), VCS (2009), CCX (2009b), and USDOE (2007)

monitor as market transactions are not always transparent. Moreover, market effects may occur at a regional, national and/or international scale.

While most guidelines for offset regimes mention the importance of addressing leakage, none of the U.S. regimes covered in this chapter include concrete measures to address this concern (Table 16.9). The regulations set out by CAR and DOE 1605(b) ensure that there is no internal (project or entity) leakage (CAR 2010, § 6; USDOE 2007), but do not provide measures for monitoring external leakage. CAR and VCS provide tables/worksheets for calculating the probability that there is leakage from the activity of the forest offset project and associated adjustments for verifiable carbon stocks eligible for credits (CAR 2010, § 6; VCS 2009, § 5). Those projects with a higher probability of leakage may be awarded a discounted number of credits.

3.2.4 Permanence

Permanence is the main technical issue that differentiates forestry-based projects from many other emission-reducing projects (Richard et al. 2006). The concern revolves around the length of time for which carbon will remain stored in the forest and the possible loss of carbon stocks either naturally (e.g. decomposition of ephemeral tree tissues; respiration), on purpose (e.g. timber harvests) or as a result of natural disasters (Aukland and Costa 2002). For example, while CAR considers permanence of forest projects on a 100 year basis, CCX only requires forest carbon offsets to be secured for 15 years (Table 16.10) (CAR 2010, § 7; CCX 2009b).

Approaches proposed for addressing issues of permanence in forest offset projects include:

- Discounting the number of credits allowed from forest offset projects (so as to create a pool of unused credits to help cover any future increases in emissions) (VCS);
- Placement of a perpetual forest easement on the project site (RGGI, CAR); and/or
- Designing formal insurance contracts that provide buffer credits in the event of a loss (CAR, CCX, VCS).[2]

4 Public Funding for Forest Carbon

In addition to the carbon markets, various federal programs have the potential to incentivize forestry practices that increase carbon sequestration. One approach is to implement more carbon-sequestering forestry practices on federal lands. Another is to provide technical and

[2] In order to hedge the risk of carbon delivery failure (due to natural disaster, improper management, etc.), insurance contracts can be used to ensure there are "buffer" credits from forests that are sequestering carbon, but that are not currently accounted for in offset crediting. These buffer credits can either be from the same site as the offset credits being traded or potentially from a "buffer forest" pool.

Table 16.10 Offset rules and permanence

Offset rules	Permanence
RGGI	A legally binding premanent conservation easement is required
CAR	100-year period required. Developer signs a Project Implementation Agreement (contract) and must insure against reversal by contributing to a buffer pool of credits. For IFM projects, a perpetual conservation easement reduces a project's required buffer pool contribution. Easements are required for Avoided Conversion projects.
VCS	An accounting method must be employed that deals with non-permanence issues from project start. The VCS approach for addressing non-permanence is to require that projects maintain adequate buffer reserves of non-tradeable carbon credits to cover unforseen losses in carbon stocks. The buffer credits from all projects are held in a single pooled VCS buffer account.
CCX	Landowners must sign a contract with their aggregators attesting that the land will be maintained as forest for at least 15 years from the date of enrollment in CCX. All issuance of A/R projects shall require placement of 20% of earned Exchange Forestry Offests in a Forest Carbon Reserve Pool.
DOE 1605b	Permanence not seen as an issue because the periodic inventory and annual reports reflect changes in net carbon flows. If the effects of natiral disturbance can be separated fom other causes in carbon pools, the estimated changes should not be deducted from the annual estimate for the entity.

Source: Data from RGGI (2008a), CAR (2010), VCS (2009), CCX (2009b), and USDOE (2007)

Table 16.11 Sources of public funding for each program

Proposal/program	Public funding programs
RGGI	None provided
Kerry-Lieberman	From 2019 to 2034, allocates between 1.5% and 6.0% of allowance value to be divided equally between domestic natural resource adaptation, and international adaptation and global security. Establishes a Carbon Conservation Program to promote conservation easements and other practices that increase carbon storage; creates a program for reducing emissions from avoided deforestation in developing countries. Authorizes appropriations, but provides no allowance allocations for these two programs.
Waxman-Markey	EPA to sell a percentage of annual allowances and use funds to incentivize reduced deforestation in developing countries: • 2012–2025 – 5% p.a. • 2026–2030 – 3% p.a. • 2031–2050 – 2% p.a.

Source: Data from RGGI (2008a), U.S. Congress (2010, 2009)

financial assistance on forest management practices to private landowners. A third is to offer tax incentives to encourage carbon-sequestering forestry practices by private landowners.

In addition to these existing federal programs, the two climate bills propose to establish programs and use a portion of the proceeds from auctioning emission allowances to promote a range of activities related to forest carbon (Table 16.11).

For example, the Kerry-Lieberman bill allocates 1.5% of the total allowances issued in 2019 to be divided equally between domestic natural resource adaptation and international adaptation and global security. This allocation increases to 6.0% by 2030, where it remains until 2034 (U.S. Congress 2010, §781). For international adaptation, assistance generated from allowance allocations must be distributed to promote adaptation in the developing countries most vulnerable to climate change. While assistance may be distributed bilaterally, between 40% and 60% must be allotted to one or more multilateral funds or international institutions for disbursal. In the United States, 20% of the proceeds from domestic adaptation allocations are to be distributed to the Land and Water Conservation Fund, 8% to the Department of Interior for cooperative grant programs, and 8% to the Forest Service for adaptation activities in national forests, grasslands and state and private forests (U.S. Congress 2010, §6008). In addition to its adaptation provisions, the Kerry-Lieberman bill also establishes a Carbon Conservation Program to provide incentives for forest and agricultural landowners who undertake activities that reduce GHG emissions but do not qualify as offsets. Funding for this program is prioritized according to the quantity and duration of carbon storage provided by a project, and at least 30% of the funds disbursed under the program must be dedicated to funding conservation easements. While the bill creates an associated fund to carry out the Carbon Conservation Program, no allowances are allocated to this fund.

Similarly, the Waxman-Markey bill requires the investment of a percentage of the quarterly strategic auction proceeds in programs that will further reduce the costs of climate policy, spur the development of advanced low-carbon technologies, grow the U.S. economy, and address unavoidable impacts of climate change (H.R. 2454 2009, § 726). Included is funding for:

- U.S. farmers and forest landowners to reduce greenhouse gas emissions and increase carbon storage in agricultural soils and forests;
- Green jobs training and assistance for workers to transition into the new jobs of a low-carbon economy;
- Reduction of deforestation and deployment of clean technologies in developing countries; and
- Programs to increase resilience to climate change impacts in the United States and in developing countries.

The bill also allows the EPA Administrator to set aside an additional percentage of annual allowances to incentivize reduced deforestation in developing countries (Table 16.11).

5 Conclusions and Policy Implications

In addition to the forest carbon accounting and quality assurance factors outlined above, there are a number of overarching topics that will need to be addressed as part of the forest policy discussion.

5.1 Balancing Public Benefits Against Potential Detriments from Forest Carbon Projects

Proper management and/or conservation of forests represent an opportunity to sequester carbon dioxide and mitigate climate change. Moreover, these forest systems offer a multitude of other ecosystem services (e.g. water quality and quantity) (Graedel and van der Voet 2009). The idea of making payments for these multiple services (in addition to carbon) may serve to make conservation financially attractive for landowners.

However, the focus should not be purely on the public benefits provided by forest offset projects – attention should also be paid to the potential for deleterious ecological impacts from carbon-focused forestry activities. For example, while afforestation may increase the carbon stored in a piece of previously unforested land, it is important to consider whether it is ecologically beneficial for the land to support trees. Afforestation or reforestation activities that require soil drainage or conversion of wetlands, as well as those that add stress to water-scarce areas, could create more public detriment than benefits.

5.2 Accuracy Versus Simplicity in Measurement/Crediting

Accuracy in accounting for sequestered forest carbon varies according to scale: global, national, and project or site-based. The larger the area considered, the greater the uncertainties. National-level accounting is significantly more accurate than at the worldwide scale. It is believed that project-level accounting for sequestration and release of forest carbon can be achieved with 90–95% accuracy (Brown 2009). These measurements are critical to calculating the carbon additionality of forest systems and awarding offset credits.

The accuracy vs. simplicity issue is also posed when considering different methodologies, such as for calculating carbon storage. While most market rules recommend direct sampling by a 3rd party verifier, DOE's 1605(b) recommends the use of look-up tables of forest conditions for a region, ownership class, forest type and productivity as a simpler and less expensive way to estimate forest carbon content (USDOE 2007, § 1.I). While DOE notes that more elaborate models may be more accurate than look-up tables for specific activities or entities, it argues that they require more effort and significantly higher costs for not a lot of extra benefit (USDOE 2007, § 1.I.2.6.2).

Ultimately, tradeoffs between accuracy and cost will have to be made. One way to address these choices is to link accuracy and cost to the number of credits awarded, i.e. the more accurate your methodologies, the more credits you are issued (Brown 2009).

5.3 Incentives: How to Make a Difference in Land Managers' Decision Making

Other than specialist carbon developers, most land managers are not participating in the carbon markets. In part, this is because of the complexity of the various regimes, as well as the constant changes in rules and relatively low prices for land-based carbon compared to other land uses. The lack of standardized methodologies has limited the capacity of landowners to evaluate the feasibility of investments that utilize forest management as a tool to offset GHG emissions. Furthermore, the lack of publicly available, documented experience deters landowners from taking the risk of developing carbon offsets that might or might not find a market at a worthwhile carbon price.

Two of the major questions landowners should ask as the legislative debates move forward on incentives for managing forest land for carbon sequestration are the following:

- Does the legislation allow complementary funding for other environmental co-benefits, thereby increasing its attractiveness to land owners?
- How are timber management practices likely to be affected by each proposal? What are the major practical differences between them?

References

Aukland L, Costa P (2002) Review of methodologies relating to the issue of permanence for LULUCF projects. Winrock International

Barrett S, Stavins R (2003) Increasing participation and compliance in international climate change agreements. Int Environ Agreem Politics Law Econ 3:346–376

Brown S (2009) Design and implementation of forest-based Projects. Winrock International. Presentation at the Yale School of Forestry & Environmental Studies

California Global Warming Solutions Act (AB32) (2006) Health & SC § 38500–38598. http://www.arb.ca.gov/cc/docs/ab32text.pdf

Carbon Positive (2009) Carbon trading prices. www.carbonpositive.net

Chicago Climate Exchange (CCX) (2009a) Chicago climate exchange. http://www.chicagoclimatex.com/

Chicago Climate Exchange (CCX) (2009b) Protocol for CCX afforestation projects

Chicago Climate Exchange (CCX) (2009c) Protocol for sustainably managed forests

Climate Action Reserve (CAR) (2009a) Climate action reserve: about us. http://www.climateactionreserve.org/about-us/

Climate Action Reserve (CAR) (2010) Forest project protocol version 3.2, August 2010

Conte M, Kotchen M (2009) Explaining the price of voluntary carbon offsets. National Bureau of Economic Research Working Paper Series. http://www.nber.org/papers/w15294.pdf

Enkvist PA, Nauclér T, Rosaner J (2007) A cost curve for greenhouse gas reduction. The McKinsey Quarterly, February 2007

European Commission (2009) Europa Climate Change Unit AFOLU Data. http://afoludata.jrc.ec.europa.eu/index.php/public_area/reporting_requirements

Farley K, Jobbágy E, Jackson R (2005) Effects of afforestation on water yield: a global synthesis with implications for policy. Glob Change Biol 11:1565–1576

Fenderson J, Kline B, Love J, Simpson H (2009) Guiding principles for a practical and sustainable approach to forest carbon sequestration projects in the southern United States. Southern Group of State Foresters. http://southernforests.org/documents/SUM%20Carbon%20Paper%2015may2009%20d2.pdf

Graedel T, van der Voet E (eds) (2009) Linkages of sustainability. MIT Press, Cambridge

Hamilton I (2009) Carbon positive news, July 3 2009: temporary carbon credits revived in U.S. climate bill. http://www.carbonpositive.net/viewarticle.aspx?articleID=1593

Hamilton K, Sjardin M, Shapiro A, Marcello T, Xu Gordon (2007) Forging a frontier: state of the voluntary carbon markets 2008. A report by Ecosystem Marketplace and New Carbon Finance, May 2007. http://portal.conservation.org/portal/server.pt/gateway/

PTARGS_0_2_144221_0_0_18/09.%20State%20 of%20Voluntary%20Carbon%20Market.pdf
Hamilton K, Sjardin M, Shapiro A, Marcello T (2009) A Fortifying the foundation: state of the voluntary carbon markets 2009. A report by Ecosystem Marketplace and New Carbon Finance, May 2009. http://ecosystemmarketplace.com/documents/cms_documents/StateOfTheVoluntaryCarbonMarkets_2009.pdf
Intergovernmental Panel on Climate Change (IPCC) (2007) Climate change 2007: synthesis report. IPCC fourth assessment report. IPCC, Geneva
Kollmuss A, Lazarus M, Lee C, Polycarp C (2008) A review of offset programs: trading systems, funds, protocols, standards and retailers. Research report. Stockholm Environment Institute, Stockholm
Massachusetts Department of Environmental Protection (MA DEP) (2007) Notice of public commitment period. http://www.mass.gov/envir/mepa/notices/122407em/10.pdf
Paul K, Polglase P, Nyakuengama N, Khanna P (2002) Change in soil carbon following afforestation. For Ecol Manag 154:395–407
Pew Center on Global Climate Change (2008) Economy-wide cap-and-trade proposals in the 110th congress. http://www.pewclimate.org/federal/analysis/congress/110/cap-trade-bills
Pew Center on Global Climate Change (2010) Pew Center Summary of the American Power Act of 2010 (Kerry-Lieberman). http://www.pewclimate.org/federal/congress/111/kerry-lieberman-american-power-act
Pfaff A, Kerr S, Hughes R et al (2000) The Kyoto Protocol and payments for tropical forest: an interdisciplinary method for estimating carbon-offset supply and increasing the feasibility of a carbon market under the CDM. Ecol Econ 35:203–221
Regional Greenhouse Gas Initiative (RGGI) (2008a) Regional Greenhouse Gas initiative model rule: part XX CO2 budget trading program. http://www.rggi.org/docs/Model%20Rule%20Revised%2012.31.08.pdf. Accessed 31 Dec 2008
Regional Greenhouse Gas Initiative (RGGI) (2008b) RGGI states' first CO2 auction off to a strong start. http://www.rggi.org/docs/rggi_press_9_29_2008.pdf
Regional Greenhouse Gas Initiative (RGGI) (2009) The Regional Greenhouse Gas Initiative: CO2 budget trading program online. http://www.rggi.org
Richard K, Sampson N, Brown S (2006) Agricultural and forestlands: U.S. carbon policy strategies. Pew Center on Global Climate Change, Arlington
Stavins R, Richards K (2005) The cost of U.S. forest-based carbon sequestration. Pew Center on Global Climate Change, Arlington
U.S. Climate Action Partnership (USCAP) (2009a) The U.S. climate action partnership. www.us-cap.org
U.S. Climate Action Partnership (USCAP) (2009b) Blueprint for legislative action: consensus recommendations for U.S. climate protection legislation. http://www.us-cap.org/blueprint/
U.S. Conference of Mayors Climate Protection Center (2009) Mayors leading the way on climate protection. http://www.usmayors.org/climateprotection/revised/
U.S. Congress (2009). H.R. 2454: American clean energy and security act of 2009
U.S. Congress (2010) Kerry-Lieberman American power act of 2010
U.S. Department of Energy (USDOE) (2007) Technical guidelines: voluntary reporting of Greenhouse gases (1605(b)) program. Office of Policy and International Affairs: U.S. Department of Energy. January 2007
United Nations Framework Convention on Climate Change (UNFCCC) (2009) Kyoto protocol status of ratification. http://unfccc.int/files/kyoto_protocol/status_of_ratification/application/pdf/kp_ratification.pdf
Voluntary Carbon Standard (VCS) (2008) Tool for AFOLU methodological issues. www.v-c-s.org
Voluntary Carbon Standard (VCS) (2009) Improved forest management methodology: estimating Greenhouse Gas Emission reductions from planned degradation. Document Version: BSTP – 1:300709. July 2009. www.v-c-s.org

REDD+ Policy Options: Including Forests in an International Climate Change Agreement

17

Eliot Logan-Hines, Lauren Goers, Mark Evidente, and Benjamin Cashore

Executive Summary

We an overview of the role of tropical forests in the international efforts to negotiate a new global climate treaty. Under the existing treaty, the Kyoto Protocol and its "flexible mechanisms" – particularly the Clean Development Mechanism (CDM) – have succeeded in building a billion dollar market for emission reduction projects in developing countries. The role of forests and land use in those markets has been a major source of controversy, and debate however. Owing, in part, to concerns about a focus on forests taking pressure off of other industrial emissions, only afforestation and reforestation projects can be included under the CDM with deforestation and degradation efforts ineligible. However, increasing political support following the "2007 Bali Action Plan" for reducing emissions from deforestation and degradation (REDD), which was spearheaded by developing countries themselves, as well as scientific evidence about the current emissions from tropical forests surpassing those of the global transportation sector has led to a consensus that a post Kyoto architecture will expand to include forest activities excluded under the CDM. The positions taken by different countries on what is now referred to as "REDD +" are often explained by the condition, and state, of their own forests. The overarching issues to be decided in developing the framework of a REDD+ mechanism include: the scope of the forestry activities to be covered; the scale of accounting for forestry activities and the baseline for measuring reference emissions levels; the type of financing to be provided for REDD+ activities; how to address fundamental issues of capacity and governance; and the consideration of co-benefits. Many of these remain contentious so that the role of REDD+ in addressing tropical forest challenges remains uncertain.

Despite the many different interests of the countries seeking to take part in a REDD+ mechanism and their different positions, it is possible summarize to what is known and what is not known about the key components of a REDD+ mechanism and where the debate stands on these issues as of the fall of 2010.

What We Know About Forest Carbon Policy

- There will almost certainly be a number of interrelated approaches to REDD+ rather than one overarching approach.
- The scope of these efforts will include deforestation, degradation, enhancement of carbon

E. Logan-Hines
Fundación Runa, Quito, Ecuador, USA

L. Goers
World Resources Institute, 10 G-Street NE, Washington, DC 20002, USA

M. Evidente
TWO ECO, Manila, Philippines

B. Cashore (✉)
Yale School of Forestry & Environmental Studies, 360 Prospect Street, New Haven, CT, USA
e-mail: benjamin.cashore@yale.edu

M.S. Ashton et al. (eds.), *Managing Forest Carbon in a Changing Climate*, DOI 10.1007/978-94-007-2232-3_17, © Springer Science+Business Media B.V. 2012

stocks, as well as the 3Es: "efficiency, effectiveness and equity" (Angelsen 2009).
- Uncertainty about where international agreements are headed is leading to the development of REDD+ "reddiness" projects with which to prepare for, and "learn" about policy initiatives.
- Sub-national level accounting is likely to be allowed under the REDD mechanism as an interim measure while countries build technical capacity; however, there is consensus that a national level baseline must ultimately be reached. Therefore, the approach of scaling up from sub-national to national for countries that need time and investment to develop monitoring is likely.
- The success of a REDD+ mechanisms will require significant investments not just in technical capacity, but in governance reforms and institutional capacity-building.

What We Do Not Know About Forest Carbon Policy
- Which carbon pools will be included in a REDD(+) mechanism.
- The definition of national baselines. While the majority of the proposals argue for the use of historic baselines, many proposals include provisions for "national circumstances" or "development adjustment factors".
- How social safeguards or ecosystems and biodiversity standards might be incorporated into criteria or eligibility for REDD+ funding.

1 Introduction

The problem of global climate change is of increasing concern to the scientific and political communities. While activity in industrialized countries bears the major responsibility for these emissions, activities in developing countries, including land conversion for agriculture and ranching are also contributing to the problem. Because emissions from land use change make up a significant proportion of global emissions, efforts are underway to develop a new strategy for bringing developing country emissions from land use change into the new climate treaty. Known by the acronym REDD+, or Reducing Emissions from Deforestation and forest Degradation, it is an effort to generate resources for reducing emissions from forestland conversion in the tropics.

In this chapter, we review the history of forests in the climate negotiations, the key considerations in the negotiations among countries on the role of forests, and the major issues that will need to be worked out as part of a REDD mechanism. We close by summarizing what is known and not known about the potential framework for a REDD agreement based on the current status of the negotiations.

2 The Role of Forests in the Global Climate Negotiations

2.1 The Basic Structure of the Global Climate Treaty

The United Nations Framework Convention on Climate Change (UNFCCC) was agreed in 1992 as a means for addressing a changing climate brought about by increased concentrations of carbon dioxide (CO_2) and other greenhouse gases (GHG) in the atmosphere (UNFCCC 1992). The UNFCCC established core principles of how climate change should be addressed and called for cooperation between states in information-gathering, study, and planning. Each year, the Conference of the Parties (COP) to the Convention meets to assess progress in achieving the goals of the treaty.

A Secretariat to the UNFCCC was established to provide support to the COP and the other institutions involved in addressing climate change at the international level. In addition, the UNFCCC set up two subsidiary bodies, one to provide scientific and technical advice (SBSTA) and the other to work on implementation of the treaty (SBI). SBSTA's work includes advice on technical methodologies, such as accounting for carbon in forests. The SBI's efforts include reviewing the financial assistance given by industrialized (Annex I[1]) countries to developing (non-Annex I) countries, as well as assessing the national emissions inventories submitted by parties. In addition, periodic reviews of the science of climate change are conducted by the Intergovernmental Panel on Climate Change (IPCC), a joint project of the UN Environment Program (UNEP) and the World Meteorological Organization (WMO).

While the UNFCCC imposed general duties on all the Parties (some more than others), specific emission reduction commitments for industrialized countries and the methods for achieving them were agreed in 1997 with the adoption of the Kyoto Protocol. The Protocol entered into force in 2005 and expires in 2012. It establishes both collective and individual emission reduction commitments for industrialized (Annex I) countries. Annex I nations as a whole committed to reduce their GHG emissions to 5.2% below 1990 levels by 2012 (UNFCCC 1997). Countries' individual targets vary according to national circumstances. For example, the Protocol requires the European Union to limit emissions to 8% below 1990 levels, while Iceland and Australia were allowed to increase their emissions by a specified amount. Non-Annex I (developing) countries were not required to make binding commitments to reduce their emissions under the Protocol (UNFCCC 1997).

The Kyoto Protocol allows Annex I countries to meet their emission reduction commitments in two general ways: (1) through domestic action; or (2) by using one of several "flexible mechanisms." Measures to reduce domestic emissions can take many forms, such as carbon cap and trade regimes, taxes, regulatory limits, incentive programs or information requirements. The flexible mechanisms allow Annex I countries to pay other countries or organizations in other countries to reduce their emissions, rather than having to reduce domestic emissions even further. One mechanism allows Annex I countries that have reduced their domestic emissions to lower than required levels to sell some of their unused national rights to emit (Assigned Amount Units or AAUs) to other Annex I countries that are having trouble meeting their targets (so-called Emissions Trading) (UNFCCC 1997).

The other two flexible mechanisms under the Kyoto Protocol take place at the project level, rather than at the national level. Both allow Annex I governments or emitters, in effect, to help meet their emission reduction requirements by paying an emitter in another country to reduce its emissions instead. When the emission reduction project is located in another Annex I country, it is called a Joint Implementation (JI) project (UNFCCC 1997). When the project is in a developing or non-Annex I country, it is done under the Clean Development Mechanism (CDM) (UNFCCC 1997). Given that non-Annex I countries are not subject to emission reduction commitments under the Kyoto Protocol, protections were put in place to ensure that the emission reductions from CDM projects are real and deserving of credit against the commitments by Annex I countries. The CDM Executive Board (EB) was established under the UNFCCC to oversee the crediting process, from approving project methodologies to issuing tradable emission reduction credits (Certified Emission Reductions or CERs) (Paulsson 2009). A useful source of information on the extensive rules governing the CDM program is provided in the CDM Rulebook at http://cdmrulebook.org/.

While the CDM has faced its share of critics (for example, see Paulsson 2009), it has been remarkably successful in increasing the amount of private investment in emission reduction projects in developing countries. For example, since 2001 the total volume of credits under the CDM program rose from zero to a high of almost 550 $MtCO_2e$ in 2007 (Fig. 17.1) (Capoor and Ambrosi 2009).

[1]Annex I Parties to the United Nations Framework Convention on Climate Change (UNFCCC) include the industrialized countries that were members of the OECD (Organisation for Economic Co-operation and Development) in 1992, plus countries with economies in transition (the EIT Parties), including the Russian Federation, the Baltic States, and several Central and Eastern European States.

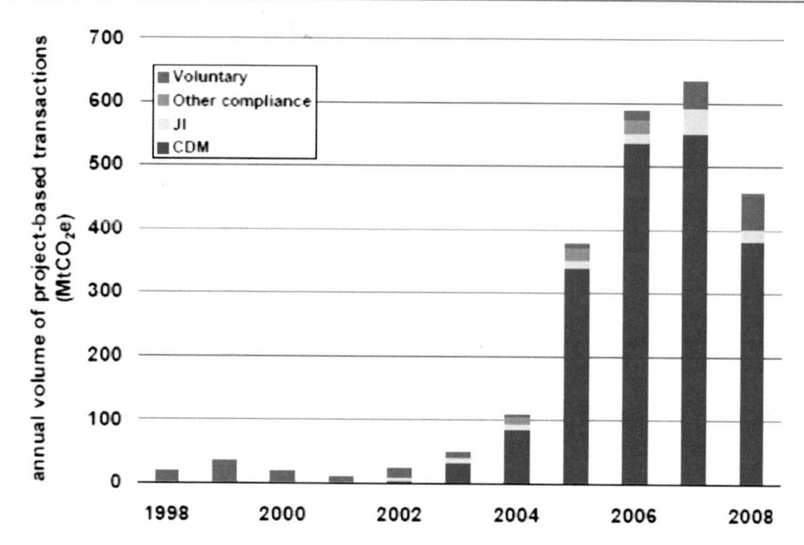

Fig. 17.1 Annual volumes ($MtCO_2e$) of project-based emission reductions transactions (*Source*: Capoor and Ambrosi 2009. Reprinted with permission)

In large part, this increase stems from the fact that European countries have allowed CDM credits to be recognized and traded as part of the EU's GHG Emissions Trading System.[2]

As the international focus shifts from the implementation of the Kyoto Protocol to what will take its place when it expires in 2012, questions about the roles of carbon markets and project based credits, as well as the common, but differentiated responsibilities of industrialized and developing countries continue to pose real issues for negotiators. The debates over the role of forests in the global response to climate change reflect the difficulty of negotiating a climate agreement that seeks to balance the historical responsibility of Annex I countries with the need for all parties to undertake mitigation actions.

2.2 REDD and REDD+

Reducing Emissions from Deforestation and Forest Degradation (REDD) is being developed to use market mechanisms and financial incentives in order to reduce the emissions of green house gases. Its original objective was to reduce green house gases but it can also deliver "co-benefits" such as biodiversity conservation and poverty alleviation which many interest groups and countries find desirable. REDD implies a distinction between deforestation and degradation. The process of identifying the two is what raises questions about how to measure each. Deforestation is defined as the permanent removal of forests and conversion of land to non-forest use. Forest degradation refers to declines in forest composition, structure and area that limit its production capacity. In the last few years countries and organizations have voiced support for moving beyond REDD. Hence REDD+ goes beyond deforestation and forest degradation, and includes the role of conservation, sustainable management of forests and enhancement of forest carbon stocks. REDD+ calls for activities with serious implications directed towards the local communities, indigenous people and forests which relate to reducing emissions from deforestation and forest degradation.

2.3 Forests, the UNFCCC and the Kyoto Protocol

Land use, land use change and forestry (LULUCF) issues have traditionally played

[2] For information on the EU ETS see: http://ec.europa.eu/environment/climat/emission/implementation_en.htm

second fiddle to energy issues in the global climate discussions (Schlamadinger et al. 2007). This is true for a variety of reasons, including that most emissions of GHGs come from the burning of fossil fuels; forests are complex and changing systems – both storing and emitting GHGs over time; and different countries have different opportunities to include forests and other land uses as part of their response to climate change (Boyd et al. 2008).

As a result, the current discussions over options for REDD and REDD+ in tropical forests build on a contentious history of decisions to limit the role of LULUCF in the global climate policy instruments. Article 4 of the UNFCCC starts with commitments by all parties to: inventory the storage of GHGs in sinks (such as forests); promote processes that reduce GHG emissions in the agriculture and forestry sectors; and promote the enhancement of sinks in forests and other ecosystems (UNFCCC 1992). At the first Conference of the Parties in 1995, a pilot program of "Activities Implemented Jointly" (AIJ) projects was launched to reduce emissions, including many in the forestry sector (Boyd et al. 2008).

However, by the time negotiations over the Kyoto Protocol began in earnest, LULUCF and the role of sinks more generally had become one of the most controversial issues facing the parties (Bettelheim and D'Origny 2002). In addition to the reasons noted above, there were concerns that allowing tradable credits from forestry projects would swamp the nascent carbon markets. The fear was that this would both delay action by industrialized countries to reduce their own emissions, as well as depress the market price for carbon credits thereby undermining the incentives for changing energy systems (Wainwright et al. 2008). Concerns were also expressed that the methodologies implemented might not be robust enough to ensure real reductions of carbon emissions and that the benefits of forest carbon projects might not accrue to the local communities living in the forests (Skutsch et al. 2007). As such, while the Kyoto Protocol allows Annex I countries to claim credit for the use of domestic sinks and expressly includes sinks in the JI Program, it is silent on the use of sinks under the CDM (UNFCCC 1997).

The continuing debates over the roles of sinks and emissions trading contributed to the failure of the parties to reach agreement during COP-6 in The Hague in 2000 (Bettelheim and D'Origny 2002). Soon thereafter, the Bush administration announced that it would not ratify the Kyoto Protocol. Meeting in Bonn later in 2001, the other parties agreed that while Annex I countries could use domestic sinks to help meet a portion of their emission reduction commitments, only afforestation and reforestation (A/R) projects (not avoided deforestation or forest management) would qualify for tradable credits under the CDM.

Further limits on the use of A/R projects were imposed at COP-9 in Milan in 2003. Most important was the decision to address permanence and leakage concerns by making credits from A/R projects temporary (Boyd et al. 2008). The decision to allow only temporary credits (tCERs or lCERs) from A/R projects meant that complicated, time-consuming rules had to be followed to generate a less valuable carbon commodity compared to the CERs from all other types of approved emission reduction projects.[3] The decision by the EU not to recognize forestry-based projects in its Emissions Trading System was another blow to the markets for A/R credits (EU Linking Directive 2004).

The result has been that land use, land use change and forestry remain sidelined in the global carbon markets (Schlamadinger et al. 2007). Of the 2,148 CDM projects registered by the Executive Board as of July 30, 2009, only six were A/R projects – less than 0.3% (UNFCCC 2009a), and agro-forestry projects made up less than 0.1% of the volume of CDM projects in 2008 (Capoor and Ambrosi 2009).

[3] For a detailed description of the rules governing A/R projects, see Pearson et al. 2006, *A Guidebook for the Formulation of Afforestation and Reforestation Projects under the Clean Development Mechanism.*

2.4 REDD and the 2007 Bali Action Plan

As recognition of the urgent need for major emission reduction has grown, forests and other land use issues are coming back into the mainstream of the global climate negotiations. Again, this is happening for a number of different reasons, including the fact that GHG emissions from deforestation are larger than those from the entire transportation sector (IPCC 2007); that there is the opportunity to bring developing countries with large forested areas more directly into the global climate negotiations; and that the cost of reducing emissions/storing carbon in forests or grasslands is lower compared to many other mitigation options (Boyd et al. 2008).

This new round of discussions around forests started in earnest at the 2005 Conference of the Parties. Two key members of the Coalition for Rainforest Nations, Costa Rica and Papua New Guinea, introduced a proposal on reducing emissions from deforestation (RED) (Wainwright et al. 2008). Their submission suggested that a mechanism for preserving tropical forests could be financed through the carbon markets and could also provide a sustainable way forward for developing countries to mitigate their emissions from forests. Brazil proposed a quite different approach in 2006. Instead of relying on the carbon markets and private investment, Brazil's position was that reducing emissions from tropical deforestation should be paid for by a public fund (from donations by industrialized countries) that is used to create positive incentives for Non-Annex I countries to reduce their own emissions, rather than offsetting emissions from Annex I countries (Wainwright et al. 2008).

While the initial proposals put forth by Costa Rica/Papua New Guinea and Brazil focused on deforestation, countries with significant forest degradation (such as in the Congo Basin) or those with little remaining forest cover but active reforestation programs (such as India), objected to proposals focusing solely on reducing emissions from deforestation. Their position was that degradation and forest management/conservation also needed to be part of the package (REDD or REDD+) (Potvin and Bovarnick 2008).

At the 2007 Conference of the Parties in Bali, a compromise was reached as part of the Bali Action Plan (UNFCCC 2007). Included is a commitment to include REDD as part of the national and international mitigation actions to be undertaken under the successor agreement to the Kyoto Procotol, including provisions addressing "issues relating to reducing emissions from deforestation and forest degradation in developing countries;" as well as "the role of conservation, sustainable management of forests and enhancement of forest carbon stocks in developing countries" (UNFCCC 2007).

2.5 Forests on the Road to Copenhagen 2009

Since the Bali Agreement, the negotiations under the UNFCCC have proceeded along two tracks:

- the Ad Hoc Working Group on the Kyoto Protocol (AWG-KP).
- the Ad Hoc Working Group on Long-Term Cooperative Action (AWG-LCA).

Both working groups have held several negotiating sessions between the annual COP meetings. These meetings are an attempt to work out major conceptual issues and make progress towards developing a text for the negotiations in Copenhagen in 2009.

The bulk of the negotiations on REDD took place through the AWG-LCA, which is tasked with leading a "comprehensive process to enable the full, effective and sustained implementation of the Convention through long-term cooperative action" by all parties, industrialized and developing, with the goal of signing a new climate agreement at COP-15 in Copenhagen (UNFCCC 2009b). The work of the AWG-KP, to the extent that it addresses the CDM and the role of forests in any future carbon trading regime, seems likely to raise issues similar to those in the REDD discussions – but the two groups are currently working separately.

As such, the rest of this paper deals primarily with the party submissions that have been made to the AWG-LCA over the past several years and their implications for the more detailed structure of the REDD mechanism. However, before reviewing the details of the REDD submissions, it is important to consider the complex dynamics of international governance and its implications for expanding the role of forests in a new climate agreement.

The most significant immediate outcome on REDD from Copenhagen seems to be pledges of billions of dollars towards REDD from a handfull of developed nations, much of which may be distributed in association with bilateral agreements with developing countries (e.g. NORAD funding, US debt-for-nature REDD funding). The emerging structure of these bilateral agreements will presumably set precedents for the eventual structure of the multilateral REDD mechanism. Norway continues to be the UN-REDD Program's first and largest donor. Since the Program was launched in 2008, Denmark and Spain committed as donor countries along with Norway to join the UN-REDD Program.

3 International Governance and Country Perspectives in the Climate Negotiations

The UN-REDD Program supports nationally-led REDD (+) processes and promotes the involvement of all stakeholders (e.g. indigenous peoples and other forest-dependent communities), in national and international REDD+ implementation. The Program also works to build awareness and agreement about REDD (+) mechanisms and its potential role in a future climate change agreement.

The UN-REDD Program is not the only program helping developing countries that wish to become involved with REDD (+) activities. Some of the other principle organizations and programs include Norway's International Climate and Forest Initiative, the Global Environment Facility, the World Bank's Forest Carbon Partnership Facility, and Australia's International Forest Carbon Initiative.

3.1 The Basic Themes of International Governance

The match between international 'willingness to pay' and national 'willingness to play' is essential for the success of REDD and REDD+. The need for balancing the different interests of many parties presents a major challenge for negotiating a climate change treaty. This has been nicely summarized by Angelsen (2009), who, with colleagues, sought to examine how participating countries would participate in REDD+ and what new institutions, processes, and projects are needed.

When a problem transcends an individual state's borders and affects enough people, international solutions are often developed to solve it. Some of these solutions occur between individual states in bilateral agreements or on a regional level through a variety of cooperative arrangements – such as the European Union or the North American Free Trade Agreement – or with the world community at large through international treaties such as the Convention on International Trade in Endangered Species or the Montreal Protocol on Substances that Deplete the Ozone Layer.

The system of international governance does not proceed from a single authority that possesses a mandate to govern and has the ability to enforce compliance. International law is based on voluntary agreements between states, of which treaties are the most concrete form. Obligations are articulated in these agreements, and states fulfill these obligations by implementing them within their territories or through their subjects (Brownlie 1998).

As voluntary agreements, international treaties must be the product of discussion and consensus. It is often the case that choices have to be made between a strict system with few members, or a less robust agreement with broader participation (Speth and Haas 2006). For instance, a state that agrees with the principles of a treaty but has few resources to enforce it effectively might be persuaded to participate if it is assured of the assistance of other states and that lapses on its part will not be punished. Moreover, as with any agreement, such a system can be affected by parties with strong interests in various outcomes, or by those that can use other resources to generate or hinder consensus. But in issues where the participation of many is as important as enforcement, a balance between the two must be struck, and the role of interest and power must be factored in the process of negotiation.

Because the scale of the problem of climate change is global, it is necessary to bring all states into a regulatory framework to address climate change. Operationalizing that framework, however, can be problematic. The UNFCCC itself has near-universal membership, reflecting broad agreement on its principles. But in working out specific obligations for different members under Kyoto, the interests of key emitting states were insufficiently addressed, reducing the overall effectiveness of the agreement. Some developed states felt that they were taking on the full burden of addressing climate change, while rapidly developing states were not bound to reduce their own emissions. On the other hand, many developing states emphasized their own need for economic development and that adopting emission restrictions would hamper their ability to secure the material well-being of their citizens (Hunter et al. 2007).

The UNFCCC is premised on the "common but differentiated responsibilities and respective capabilities" of states that are parties to the Convention, but it stresses that developed countries should be leaders in mitigating climate change and recognizes the vulnerability of many developing countries, particularly small island states and least developed countries (Stone 2004). While all parties should take "precautionary measures to anticipate, prevent or minimize the causes of climate change and mitigate its adverse effects" such measures should "be appropriate for the specific conditions of each party, and be integrated with national development programs, taking into account that economic development is essential for adopting measures to address climate change" (UNFCCC 1992) .

3.2 REDD (+) as a Mitigation Strategy

As the Ad-Hoc Working Group on Long-Term Cooperative Action (AWG-LCA) stated at its 5th session in Bonn in April 2009, "a REDD mechanism should be designed to accommodate differing national circumstances and respective capabilities within and between developing countries on issues relating to reducing emissions from deforestation and forest degradation, and the role of conservation, sustainable management of forests, and enhancement of forest carbon stocks" (AWG-LCA 2009). Reducing emissions from deforestation by safeguarding the world's remaining forests can be a critical step on that path.

REDD and REDD+ has thus received increased attention for its potential to address the concerns of both developed and developing countries. Developed countries that bear most of the responsibility for the current global emissions see REDD as a cost effective mitigation tool that will enable them to help meet their own emission reduction targets through the sale of carbon offsets or credits. On the other hand, many developing countries see the role of REDD and REDD+ as a way of meeting their own mitigation goals. While the Kyoto Protocol did not require non-Annex I countries to make commitments, as a part of the Bali Action Plan, developing countries are tasked with developing "nationally appropriate mitigation actions" (NAMAs) that allow them to contribute to climate change mitigation. Many developing countries see REDD as a NAMA that can be used to meet this goal of developing national mitigation strategies. Whether REDD is an offset mechanism for Annex I countries or contributes towards non-Annex I mitigation goals will be a contentious issue in negotiations. For example, both Brazil and Panama have emphasized that the REDD mechanism should not be a means for developed countries to meet their emission reduction commitments under Kyoto (Wainwright et al. 2008).

3.3 Country Perspectives

In order to understand the different country perspectives on REDD and REDD+, it is essential to consider the differences in national circumstances with regards to forest cover and historical rates of forest loss in developing countries with tropical forests. Da Fonseca et al. (2007) have categorized countries based on their

Table 17.1 Developing country circumstances classified by forest cover and deforestation rates

	Low forest cover (<50%)	High forest cover (>50%)
High deforestation rate (>0.22%/year)	*Quadrant I*	*Quadrant III*
	e.g., Guatemala, Thailand, Madagascar	e.g., Papua New Guinea, Brazil, Dem. Rep. of Congo
	High potential for RED credits	High potential for RED credits
	High potential for reforestation payments under CDM	Low potential for reforestation payments under CDM
	Number of countries:44	Number of countries:10
	Forest area: 28%	Forest area: 39%
	Forest carbon (total): 22%	Forest carbon (total): 48%
	Deforestation carbon (annual): 48%	Deforestation carbon (annual): 47%
Low deforestation rate (<0.22%/year)	*Quadrant II*	*Quadrant IV – HFLD countries*
	e.g., Dominican Republic, Angola, Vietnam	e.g., Suriname, Gabon, Belize
	Low potential for RED credits	Low potential for RED credits
	High potential for reforestation payments under CDM	Low potential for reforestation payments under CDM
	Number of countries: 15	Number of countries: 11
	Forest area: 20%	Forest area: 13%
	Forest carbon (total): 12%	Forest carbon (total): 18%
	Deforestation carbon (annual): 1%	Deforestation carbon (annual): 3%

Source: da Fonseca et al. (2007). Reprinted with permission

(1) remaining forest cover and (2) deforestation rate as a way of highlighting the fact that the state of the forests in a country plays a crucial role in that country's ability to benefit from – and therefore its views on – a REDD mechanism (Table 17.1). Specifically, these differential circumstances underpin debates such as the type of forestry activities to be included in the mechanism, the establishment of reference emissions levels for generating carbon credits, or the scale at which activities should be undertaken.

The primary focus of REDD in the negotiations since Bali has been on providing incentives for countries to reduce deforestation rates. The countries with the highest amount of emissions from forests are Brazil and Indonesia, and both countries have moderate to high rates of deforestation (The Nature Conservancy 2009). Therefore, the greatest potential activity for mitigating emissions from land use conversion is to reduce and eventually halt deforestation rates in these countries. In countries with significant deforestation, reducing the rate of that deforestation compared to an established baseline is the key way to generate credits or offsets for emissions reductions. However, countries with historically low deforestation rates, such as Guyana and Suriname, argue that countries that have left more of their forests standing should not be penalized for having lower rates of deforestation by having fewer REDD credits available to them. Explicitly emphasizing that focusing on reversing deforestation will leave out countries that have the highest percentage of remaining forest cover and lowest rates of deforestation, Suriname proposes that a REDD program must in fact focus on these countries and provide support for their economic development that doesn't involve cutting down their forests. Considering the situation of these countries would reduce the likelihood of leakage, and the importance as well of providing ex ante funding to avoid development pressures (SBSTA 2008).

There is another group of countries, such as India and many of the West African nations, that have little forest cover remaining but are eager to participate in REDD through afforestation, reforestation and sustainable forest management activities (REDD+). The REDD mechanism negotiated at Copenhagen must therefore balance the different national circumstances with regard to the state of forests, but also seek to develop a REDD architecture that ensures environmental integrity and ultimately reduces emissions from forests by stopping deforestation.

4 Key Considerations for a REDD Program and How They Are Addressed in the Major Proposals

A strategy to reduce global emissions from deforestation and forest degradation must consider the diversity of interests of the states involved, and will therefore necessarily delve into many complex issues. The overarching questions to be decided in developing the framework of a REDD (+) mechanism can be divided into several topics that have been a source of debate among both developed and developing country parties at the intersessional negotiations. The major questions can be categorized as follows:

1. The scope of forestry activities to be included in the REDD (+) mechanism;
2. The scale of accounting for forestry activities and the baseline for measuring reference emissions levels;
3. The type of financing to be provided for REDD (+) activities;
4. Capacity building and REDD (+) "readiness"; and
5. The role of co-benefits.

4.1 Scope of a REDD Program

4.1.1 Activity Scope

Over the course of the debate, there have been three ideas of what activities should be included in a program to reduce emissions from forests:

1. REDD proposals limit included activities to only those which result in reduction in deforestation rates;
2. REDD includes reducing emissions from deforestation and forest degradation; and
3. REDD+ further expands the scope of REDD by including activities such as sustainable forest management, conservation, and enhancement of carbon stocks of forests and plantations. (Parker et al. 2008).

Some countries, such as Australia and the United States, have articulated in their submissions to the AWG-LCA that the long term goal should be for countries to do full land-based accounting, or that developing a REDD strategy should ultimately form part of a sustainable land use management plan as part of a low carbon development strategy. Recent findings from climate models suggest that if policy is focused just on energy use, the land use implications – in terms of increased deforestation for biofuels – will be huge and negative (Wise et al. 2009).

Resolving the debate about the scope of a potential REDD mechanism was the key to adopting the decision to move forward on REDD as part of the Bali Agreement at COP13 (Potvin and Bovarnick 2008). The root of this debate is the difference between carbon sinks and sources across tropical countries. Deforestation and degradation in forests are a source of emissions, while the intact forest itself is a sink for carbon dioxide. Brazil wanted a RED mechanism that only focused on deforestation because of the fear that the inclusion of the carbon stocks of intact forests for conservation would dilute the funding sources needed for combating deforestation (Potvin and Bovarnick 2008). As a country that has little forest cover remaining and a relatively low rate of deforestation, India proposes a common methodology that assesses: (i) changes in carbon stocks and GHG emissions due to conservation and sustainable management of forests, and (ii) reductions in emissions from deforestation and degradation (SBSTA 2008). While some countries cite concerns over methodologies for going beyond monitoring deforestation, Canada asserts in its submission to COP 13 that, while methodological difficulties do exist in assessing and quantifying forest degradation, countries should not be excluded from incentives to reduce emissions from deforestation (SBSTA 2008).

The REDD architecture is still in its developing stages as the negotiations progress, but in reviewing the most recent Party submissions on REDD to the AWG-LCA, there is convergence around the inclusion of the "plus" activities in a REDD mechanism. However, there are still important technical and financing issues to consider within the scope of the discussion. For example, the feasibility of developing countries performing carbon accounting for activities beyond degradation may ultimately limit the initial scope of REDD in many developing countries. Additionally, the distribution and types of funding available could be contingent on the scope of the activity being implemented. Both India and Mexico have proposed financing mechanisms that are linked to the type of forestry activity being performed. Their proposals relate back to the discussion of the forest as source and sink: India proposes that market mechanisms may be suitable for activities such as deforestation (which is comparatively easier to measure), while fund-based financing may be necessary for activities that enhance carbon stocks (AWG-LCA 2009). It will also be important, at some point, to bring together the ideas on forests from the AWG-LCA and the AWG-KP. The lessons learned from the efforts to include A/R projects (a part of many REDD+ formulations) in the CDM should be useful for discussions of REDD+. In addition, protections against double-counting projects will have to be included to ensure that any particular project is not counted by both a funding country and a host country against each of their commitments.

4.1.2 Carbon Pools/Ecosystem Types Covered

Because articulating a framework for REDD (+) is still in early stages, most submissions to the AWG-LCA do not contain a great level of detail on issues such as the carbon pools that will be eligible for generating emissions reductions, i.e. above or below ground biomass, soils, or wood products (Parker et al. 2008). This is due in part to the difficulties associated with measuring carbon pools such as soil and below ground biomass (see Chap. 2, this volume). However, a few countries such as Australia and the United States propose the eventual inclusion of all forms of terrestrial carbon (grasslands, woodlands, peatlands, etc.), not just forests. They advocate that these other ecosystems be phased in as science develops methods to quantify their respective carbon benefits (Ashton 2008). While this perspective ties in with the goal of sustainable land use management, it is unclear whether or not any REDD agreement reached in Copenhagen will deal explicitly with the different carbon pools in forests. While carbon stored in above-ground biomass will certainly be included in any future REDD system, the inclusion of other carbon pools is much less certain.

4.2 Baselines and Accounting

The REDD mechanism also has a myriad of options for how emissions reductions can be measured. The reference level has to define the way in which emissions reductions or carbon stock enhancements will be compared to a chosen baseline, and the scale at which carbon accounting is done will impact whether REDD is implemented at a project/sub-national or a national scale (e.g. Griscom et al. 2009).

4.2.1 Reference Emission Levels

The establishment of a baseline scenario of deforestation/degradation over a defined scale in a business-as-usual scenario is the first step in accounting for REDD (+). This reference level greatly affects which countries will be able to generate emissions reductions and the amount of credits that will be available due to varying levels of deforestation. Because of the differences in national circumstances enumerated above, most party submissions state the need for flexibility in reference levels that allow countries with low rates of deforestation to participate in the generation of emission reductions. This flexibility is important because looking only at current deforestation rates across countries does not take into account either the historical or future drivers of deforestation that must be addressed in order to reduce emissions.

4.2.2 Historical Baselines

One potential reference scenario is measuring emissions reductions against a historical baseline. This method could be set by choosing a baseline year and comparing rates of deforestation. Many countries are proposing baseline years for their own national emissions reductions strategies, and baselines vary from 1990, 2000, or 2005 levels depending on the proposal (Olander et al. 2007). Alternately, historical reference scenarios could be developed by taking the average rates of deforestation over a defined period of time. For example, Brazil has already set its own goal of reducing its emissions against a baseline taken from its average area of deforestation from 1996 to 2005 (Carbon Positive 2009).

While historical baselines can be relatively simple to calculate, those countries with high, current deforestation rates will gain much more from emissions reductions under a REDD program. Malaysia's proposal voices concerns that using a historic baseline will create a perverse incentive to increase timber harvests in the years before the first commitment period (Parker et al. 2008). While other countries seek to avoid this problem by not taking into account the most recent time period in developing historical baselines, the majority of proposals for REDD are not focusing on historical baselines for reasons of both equity across countries and because many developing countries lack consistent historical data on deforestation rates.

4.2.3 Projected Baselines

The Centre for International Sustainable Development Law (CISDL) advocates the use of projected baselines outlined in their "Carbon Stock Approach" (Climate Focus 2007). They see the lack of ex ante funding (payments to countries up front for capacity building and strategy development), as a major roadblock in implementing REDD (+) activities in developing countries that have very little capital to invest in such projects. In this scenario, a projected baseline of emissions from deforestation would be created in order to set aside a certain stock of forest carbon that is expected in the future. Countries might achieve this through using current deforestation data, information about the country's development pathway, population growth, or other data on drivers of deforestation to create predictions or use econometric models to develop future emissions scenarios.

The major criticism of the projected baseline approach is the difficulty in accurately predicting future forest trends. Many critics see room for distortion and corruption. Colombia proposes that projected baselines could be based on either an extrapolation of past trends into the future, prevailing technology or practice, or logical arguments made by activity participants based on observed trends (Parker et al. 2008).

4.2.4 Historical Adjusted Baseline

At this point in the negotiations, many countries are suggesting an approach that takes into account both historical trends and national circumstances such as drivers of deforestation or development pathways that can have significant impacts on forest cover. While party submissions are vaguely worded, proposals from the United States, India, Papua New Guinea, Australia, and Norway all reference the need to develop baselines that take into account national circumstances and capacities (AWG-LCA 2009). Norway's proposal is the most detailed. It recommends establishing reference emission levels using a formula based on inputs such as historical emissions, forest cover, and measures of per capita GNP to factor into an adjusted baseline (AWG-LCA 2009). Setting these reference levels could be done by an expert body or technical panel that is in charge of overseeing a standardized process for setting baselines. While this type of proposal might be successful in taking into account national circumstances beyond simple measures of forest cover or deforestation rate, a methodology would still need to be developed and agreed upon by the interested parties under the Convention.

The discussion of reference levels underscores the fact that countries are likely to favor different methods of setting reference scenarios depending on their national circumstances with regards to forest area. It is unlikely that there will be a single solution for developing reference levels; developing national reference levels will likely fall to individual countries as they develop their REDD strategies, and a technical or scientific body of experts under the COP could be responsible for review and approval of country proposals.

4.3 Scale

The scale at which REDD (+) activities are implemented – project or national – is another key issue to be decided as part of the UNFCCC negotiations in Denmark. The scale of REDD poses risks to the environmental integrity of the mechanism because being able to implement REDD at a national scale requires measurement and monitoring capacity that many developing countries may lack.

4.3.1 National Level

From the perspective of ensuring the environmental integrity of REDD (+) projects, establishing a national level baseline is a key strategy that is acknowledged by most parties as the most effective way to prevent displacement of emissions within a country from the site or from a REDD project to another area (frequently called leakage). National-level baselines also empower host countries to pursue a broader set of policy tools and take ownership of their projects (Angelsen et al. 2008b). India proposes a national baseline to prevent double counting and national-level leakage (SBSTA 2008). They also propose that CDM A/R activities be debited from a national inventory for REDD accounting in order to address additionality concerns that forestry activity implemented under the CDM would be counted twice because of overlap between REDD and A/R mechanisms.

4.3.2 Sub-national/Project Level

The argument for a sub-national approach presented by the Latin American coalition and Malaysia is that it (1) is easier to monitor and verify, (2) encourages investment from the private sector, and (3) could provide more direct benefits to forest-dwelling people (Potvin and Bovarnick 2008; Cortez et al. 2010). They argue that relying only on national level baselines is problematic as many countries lack the capacity, governance, and control of territory to effectively implement a national baseline. Use of project level baselines means that many developing countries that do not have resources to create a national carbon accounting mechanism will nevertheless have flexibility to engage in REDD activities (Parker et al. 2008). Most countries supporting a sub-national approach see it as an interim measure, acknowledging the need to eventually work towards national scale accounting.

4.3.3 Global Level

The Centre for Social and Economic Research on the Global Environment (CSERGE), a research centre based out of the University of East Anglia, UK, proposes that credits should be generated relative to a global baseline as a way of eliminating international leakage (Strassburg et al. 2008). The Joint Research Centre, the European Commission's research organization, proposes an "Incentive Accounting" program where countries with emissions less than half of a global average baseline be rewarded for maintaining carbon stocks, whereas those with higher than the global average are rewarded for reducing emissions from forest conversion (Parker et al. 2008).

With regards to the scale of REDD (+) activities, there is convergence around the idea that national baselines are the ultimate goal, but that sub-national projects should be allowed as part of a readiness or scaling up phase in order to allow countries that will not be immediately ready to do national level monitoring time to improve technical and institutional capacity.

4.4 Financing

Generating financing for REDD (+) activities at an adequate and sustainable scale is crucial to its success. This is particularly true in order to create incentives and payment systems for government actions and specific projects to reduce emissions that overcome the drivers of deforestation and forest degradation. Recognizing the varied interests and institutional capacities of states, various proposals are being discussed, ranging from creating a market for the trading of forest carbon emission reduction credits, establishing a public fund from contributions by states and financial institutions to pay directly for such reductions, or combinations of the two approaches.

4.4.1 Market System

A market system to finance REDD (+) presupposes the existence of a working cap-and-trade market for carbon credits (such as the EU's Emissions Trading System). The system will need to cover large portions of the industrial sectors of developed states (and perhaps increasing numbers of developing states with the passage of time), for which the total amount of GHG emissions will be set (the "cap") and permits for those emissions will be sold and traded between emitters (the "trade"). Under the cap-and-trade system, REDD projects could be issued to offset credits once they have achieved emission reductions by protecting forests. These credits in turn can also be traded on the cap-and-trade market, and the proceeds from the sale of REDD credits would ensure continuing REDD activities.

As discussed above, many questions have been raised about the wisdom of including forest-based credits in the carbon markets. One major concern in the REDD discussions is that the price of REDD credits would be much cheaper than the cost of reducing the same amount of emissions from the energy sector, such that the economic incentives to change our use of fossil fuels would be undermined. The fear is that REDD credit prices will be so much lower than the cost of a clean energy project, for example, that it will bring large volumes of new credits into the market, thus easily meeting the demand for credits and driving the overall market price for credits lower.

As such, some proposals suggest that any tradable credits from REDD projects should not be traded at face value on any cap-and-trade market (Dutschke 2008). Greenpeace recommends that REDD credits should be sold with a surcharge to make them more expensive and that the proceeds from the surcharge could be used to fund other institutional or capacity-building activities, over and above specific measures against deforestation and degradation (Thies and Czebiniak 2008). Of course, proponents of market approaches argue that imposing any such additional costs on REDD projects will only limit their use, and hence, their effectiveness in mitigating or storing emissions.

While the price of REDD credits will directly benefit the entity operating the REDD project, many people believe that separate funding will also be necessary to address the broader social and political factors that contribute to deforestation, such as insecure land tenure and indigenous peoples' rights, enforcement and monitoring capabilities, economic and agricultural policy coherence, among others (Thies and Czebiniak 2008).

4.4.2 Fund-Based Mechanism

Another option for providing REDD (+) financing is for governments, financial institutions or private entities[4] to contribute to a fund, which can then be disbursed to support REDD projects. How those contributions are generated can take a variety of forms. There are proposals, for instance, that the Assigned Amount Units (AAUs) for allowable emissions from Annex I countries be auctioned and a part of the auction proceeds be used for REDD projects (AWG-LCA 2009). In

[4] Some carbon funds currently administered by the World Bank have voluntary private sector contributors, and there appears to be no hindrance to allowing similar arrangements for REDD funding. See its BioCarbon Fund, among others. The amount of this "charitable" or "learning" capital, however, is relatively limited. More extensive incentives for private investors to put money into REDD projects will have to be included if substantial amounts of private investment are to be expected.

other proposals, participating governments can impose taxes within their own states or make annual appropriations and remit these to a central REDD fund. In any case, parallel to such a central fund, it would still be possible that REDD activities in one country can be directly financed by foreign governments, corporations, or financial institutions to comply with their own emissions targets through bilateral agreements such as the Amazon Fund, a fund run through the Brazilian Development Bank that is currently funded by the government of Norway (BNDES 2009).

While most parties agree that some degree of public funding should be involved in a REDD mechanism, the major arguments against solely a fund-based mechanism is that it is unlikely to be able to generate funding at the required scale to effectively provide support for emissions reductions activities and building the capacity to monitor those activities, and that funding will not be continued long-term.

At present, some non-market funds are in place to help countries prepare for what is being termed REDD "readiness." The Forest Carbon Partnership Facility of the World Bank and the UN-REDD Programme – run as a collaboration between the Food and Agriculture Organization (FAO), United Nations Development Programme (UNDP) and United Nations Environment Programme (UNEP) – are working with developing countries to provide support for the development of REDD strategies. The donors for these programs thus far are predominantly national governments (for example the UN-REDD Programme is funded by a $52 million grant from Norway) (UNDP 2008).

4.4.3 Hybrid Funding Mechanism

Noting that both market and non-market systems have their limitations, there are various combinations being explored that make the most of the strengths of each system. Some proposals suggest that the type of financing should depend on the type of action being undertaken. For example, efforts to build capacity and improve forest governance could be separated from activities that directly result in emission reductions. Contributions from funds could be used to finance the governance activities, while market-linked or direct market financing could be used to finance the actual emissions reductions. The International Institute for Environment and Development (IIED) proposes a system in which governmental transfers would focus on improving institutions and governance – improving monitoring and law enforcement, land tenure reform and indigenous rights, agricultural and economic policies, among others – while carbon markets would direct resources to people and communities to provide the incentive and support to manage forests at the ground level (Viana 2009).

Another proposal that has gained significant support through the negotiation process leading up to Copenhagen is the phased approach enumerated by the Norwegian government. Recognizing that different countries are at different levels of institutional capacity to effectively utilize market-based financing, Norway proposes three phases of REDD that consist of: (1) a capacity building phase; (2) a scaling up phase to include government policies and measures addressing drivers of deforestation as well as demonstration activities for emissions reductions; and (3) full implementation (AWG-LCA 2009). In this scenario, the type of funding available would depend on the phase of REDD, with initial phases being supported by non-market funds for planning, and institution- and capacity-building activities at the national level. When institutions develop sufficient capacity for monitoring and demonstrating emissions reductions, countries could proceed to a full implementation phase in which they would access the carbon market (Angelsen et al. 2008a).

While hybrid systems attempt to address the deficiencies of both market and non-market financing schemes, they also create a complicated system that will require its own bureaucracy. Transaction costs will thus increase, and target communities and programs may actually receive fewer funds. IIED explored the strengths and weaknesses of market and non-market (government) strategies (Table 17.2) (Viana 2009).

Table 17.2 Strengths and weaknesses of government and market finance for REDD

Effectiveness	Efficiency	Equity	Urgency
Government			
+ Strong support of rain forest governments encourages sound policies	+ Lower international transaction costs	+ Facilitates international transfers between rich and poor countries	– Slow implementation of intergovernmental funding
– Limited effectiveness of government-based policies	– Higher domestic costs	– Favours middle-income countries	– Slow implementation of government programmes
+ Captures domestic leakage	+ Greater incentives for governmental policies	– Risk of domestic distribution inequities	
– Does not capture international leakage	– Greater risk of policy and governance failure		
– Limited attractiveness to private funders	+ Lower monitoring costs		
Market-based			
– Weak support to encourage sound policies by rain forest governments	– Higher international transaction costs for small projects	+ Increases funding from market to forest communities in poor countries	+ Quicker implementation of project-based activities
+ Greater effectiveness of field project-based activities	+ Lower bureaucracy and administrative costs	+ Does not favour middle-income countries	+ Quicker impacts on reduction of deforestation and degradation
– Does not capture domestic leakage	– Smaller incentives for governmental policies	+ Smaller risk of inequitable distribution of benefits to local communities	
+ Increases area of forests under protection with positive impacts on international forest leakage	+ Smaller risk of policy and governance failure	– Potential risk of inequitable distribution of benefits to local communities if project certification schemes are ineffective	
+ Greater attractiveness to private funders	– Greater monitoring costs		

Source: Viana (2009). Reprinted with permission

Market and non-market sources of financing have their strengths and weaknesses. An attempt to have the best of both will likely result in a hybrid system, and while such a system may be considerably more complex and more costly to operate, it may also be able to be more flexible in addressing the different needs of different states.

4.5 Capacity Building and Readiness

While scope, scale of activity, and sources of financing are all critical topics to be discussed when developing a REDD (+) mechanism, it is also essential to consider the obstacles that the limited capacity for forest and revenue management in many developing countries may present to the successful implementation of REDD. As illustrated in prior chapters of this book the drivers of deforestation in many developing countries are complex, operate across multiple spatial scales, and are not confined to the activities that directly impact forests such as logging or agricultural conversion. Additionally, forest governance, institutional capacity, and technical expertise must all be improved in order to achieve the long term goals of REDD.

There is increasing recognition that these issues of institutional capacity and national circumstance must be addressed in order for countries to more effectively engage the global REDD system. Thus, it is likely that any agreement on REDD will involve a complex system of distributing REDD benefits in order to address the differences among states in the hope of actually achieving reductions in emissions from deforestation and forest degradation. One approach with the potential to address these governance and capacity issues directly is the phased approach to REDD.

4.5.1 Phased Approach to REDD

Considering the need to address institutional obstacles alongside activities directly linked to emission reductions, Norway proposes a phased approach to REDD proceeding first from readiness planning, then to strategy implementation (where incentives are given based on proxies for emission reductions), and ultimately to a phase where incentives are given for actual emission reductions (AWG-LCA 2009). In each phase, obligations become more rigorous and defined.

In the first phase, funding is voluntary or bilateral in nature, depending on the country's commitment to REDD. Thus far the FCPF and UN-REDD Programme have operated in such a way that the work being done by developing countries to create REDD strategies could inform this initial phase of REDD through lessons learned and best practices in developing stakeholder buy-in and creating an implementation strategy for REDD. In the second phase, funding will be directed towards capacity building and institutional reform, financed through the auction of AAUs. Phase three will be financed through a market, with sales of forest-based credits from reduced emissions relative to an agreed baseline, dependent on an operational national GHG forest inventory. Norway's proposal also puts considerable emphasis on the role of forest stakeholders, the need to improve forest governance, respect for rights of indigenous peoples, and biodiversity conservation in the implementation of REDD (AWG-LCA 2009).

4.6 Co-benefits

Many policymakers see a REDD (+) mechanism as a method of incentivizing countries to reduce emissions while simultaneously preserving "co-benefits," such as protecting biodiversity in forests, preserving ecosystem services, and helping to improve the livelihoods of indigenous peoples and other communities that reside in forests (Angelsen et al. 2008a).

While many of the submissions to the AWG-LCA contain language about respecting the property rights of indigenous peoples and local communities, as well as biodiversity, there have been few attempts to flesh out what these social and environmental safeguards might look like. The idea of social safeguards is prevalent in the development community and it is possible that one idea for mainstreaming biodiversity concerns into REDD is to make it part of countries' eligibility for funding. Another issue concerning biodiversity is that of plantation forests. While most countries do not address this issue specifically, Bolivia's proposal seeks to ensure that REDD activities do not result in the clearing of natural forests to be replaced by plantations, which might generate carbon credits as a REDD+ project but would have negative impacts on forest biodiversity (AWG-LCA 2009).

Another serious concern in the NGO community and civil society in potential REDD countries is the impact that REDD could have on local communities and indigenous peoples who dwell in forests. One key way in which this concern can be addressed is by meaningful inclusion of all stakeholders, including local communities and indigenous peoples, in the design, development and implementation of national REDD strategies. The FCPF and UN-REDD processes thus far have required consultation and stakeholder plans to be submitted along with country proposals or strategies for creating a REDD plan. However, thus far these consultations have tended more towards education and awareness raising rather than engaging stakeholders in inclusive consultations that incorporate local perspectives into national REDD plans (Daviet et al. 2009).

Overall, the approach to considering the rights of indigenous peoples and local communities, as well as language for protection of biodiversity and ecosystem services, will need to be elaborated in order for REDD to proceed with adequate safeguards in place. The UNFCCC process may be able to address these issues by building off of the readiness processes that are currently in place. Full inclusion of all relevant stakeholders and recognition of indigenous rights, and safeguards to ensure that biodiversity is not adversely affected by REDD projects, are both essential to the integrity of the REDD mechanism.

5 Conclusions and Policy Implications

This review of the major substantive issues and the proposals to the AWG-LCA has identified two countervailing trends. On the one hand, forests provided a modicum of optimism for generating a global focus around the climate aspects of deforestation and degradation – a critically important question not only for climate concerns, but also for ecosystem structure and function, and the rights and responsibilities of forest dependent communities.

This potential was illustrated following general frustration globally at failure of COP 15 in Copenhagen in December 2009 to reach consensus on a post-2012 climate agreement where growing interests in forests was, for many, one of the few positive developments. Such interest eventually coalesced through the establishment of a "REDD+ Partnership formed by 58 Partner countries on 27 May 2010 in Oslo, Norway" which was designed to "complement and feed into the UNFCCC process."(Gluck et al. 2005). And most pointedly, it sought to coalesce efforts to enhance funding and provide coordination and knowledge sharing over the increasing array of REDD+ related projects, programs and activities.

On the other hand, this review has found that though there is convergence on the goals and objectives of the REDD+ concept many contentious issues remain that may require significant effort to resolve. As Gluck et al. (2005) nicely summarize, "Institutions are needed to manage the flow of information on changes in forest carbon stocks (or proxies of that), and the flow of funding from domestic and international sources." Many actors will be seeking REDD+ rents, and the successful implementation of REDD+ will hinge on good governance and domestically driven reforms.

Recognition of this requires much greater attention to the governance architecture that will be needed if REDD is to continue to make advances, which requires securing a much higher level of international financing commitments, and much more attention to promoting on the ground implementation (Gluck et al. 2005). This latter effort, Karnowski et al. (2011) argue, should begin by developing and nurturing "national, subnational and local capacities" to implement existing forest conservation and management requirements in ways that are consistent with the principles of "good forest governance", including transparency, equity, and inclusionary processes (Brown et al. 2008; Phelps et al. 2010; Sikor et al. 2010).

What is clear that the process of reaching an agreement on REDD through the UNFCCC negotiations will be complex and challenging. Such negotiations must be consistent with, and draw on, the strategic interests and norms that motivate countries and other relevant stakeholders. But most importantly, negotiations over approaches must begin, and end, with an honest assessment of their ability to address the problems for which they were created (Levin et al. 2008). Unlike many other policy problems the acute crisis facing global emissions and degradation means that compromise must be replaced with innovation, so that problems, rather than politics, drive outcomes.

The most significant immediate outcome on REDD from Copenhagen seems to be pledges of billions of dollars towards REDD from a handfull of developed nations, much of which may be distributed in association with bilateral agreements with developing countries (e.g. NORAD funding, US debt-for-nature REDD funding). The emerging structure of these bilateral agreements will, presumably, set precedents for the eventual structure of multilateral REDD mechanism.

State-level initiatives (i.e. Governor's Climate Task Force), representing international state-to-state agreements, may play a lead role in setting precendents for international REDD framework prior to national-to-national international agreements.

References

Angelsen A (ed) (2009) Realising REDD+: national strategy and ploicy options. CIFOR, Bogor, 362 p

Angelsen A, Brown S, Loisel C, Peskett L, Streck C, Zarin D (2008a) REDD: an options assessment report. Meridian Institute, Washington, DC

Angelsen A, Streck C, Peskett L, Brown J, Luttrell C (2008b) What is the right scale for REDD? the implications of national, subnational, and nested approaches. Center for International Forestry Research, Bogor

Ashton R (2008) How to include terrestrial carbon in developing nations in the overall climate change solution. The Terrestrial Carbon Group, Canberra

AWG-LCA (2009) Ideas and proposals on the elements contained in paragraph 1 of the Bali action plan. http://unfccc.int/resource/docs/2009/awglca6/eng/misc04p01.pdf

Bettelheim EC, D'Origny G (2002) Carbon sinks and emissions trading under the Kyoto protocol: a legal analysis. Philos Trans R Soc Lond Ser A Math Phys Eng Sci 360:1827–1851

Boyd E, Corbera E, Estrada M (2008) UNFCCC negotiations (pre Kyoto to COP-9): what the process says about the politics of CDM-sinks. Int Environ Agreem-Politics Law Econ 8:95–112

Brazilian Development Bank (BNDES) (2009) The Amazon fund. http://www.amazonfund.gov.br/. Accessed 20 Aug 2009

Brown D, Seymour F, Peskett L (2008) How do we achieve REDD co-benefits and avoid doing harm? In: Angelsen A (ed) Moving ahead with REDD. Centre for International Forestry Research, Bogor, pp 107–118

Brownlie I (1998) Principles of public international law. Oxford University, Oxford

Capoor K, Ambrosi P (2009) State and trends of the carbon market 2009. The World Bank, Washington, DC

Carbon Positive (2009) Brazil sets 10-year deforestation target. http://www.carbonpositive.net/viewarticle.aspx?articleID=1327. Accessed 25 Aug 2009

Climate Focus (2007) Briefing note: carbon stock approach to crediting avoided deforestation

Cortez R, Saines R, Griscom B, Martin M, De Deo D, Fishbein G, Kerkering J, Marsh J (2010) A nested approach to REDD+: structuring effective and transparent incentive mechanisms for REDD+ implementation at multiple scales. The Nature Conservancy/Baker and McKenzie, Arlington/Chicago, 46 pp

da Fonseca GAB, Rodriguez CM, Midgley G, Busch J, Hannah L, Mittermeier RA (2007) No forest left behind. PLoS Biol 5:1645–1646

Daviet F, Davis C, Goers L, Nakhooda S (2009) Ready or not? a review of the World Bank forest carbon partnership R-plans and the UN-REDD joint program documents. WRI Working Paper. World Resources Institute, Washington, DC

Dutschke M (2008) How do we match country needs with financing resources? In: Angelsen A (ed) Moving ahead with REDD. CIFOR, Bogor, pp 41–52

EU Linking Directive (2004) Directive 2004/101/EC of the European parliament and of the council of 27 October 2004 amending directive 2003/87/EC establishing a scheme for greenhouse gas emission allowance trading within the community, in respect of the Kyoto protocol's project mechanisms. http://eur-lex.europa.eu/LexUriServ/LexUriServ.do?uri=CELEX:32004L0101:EN:NOT

Gluck P, Rayner J, Cashore B (2005) Changes in the governance of forest resources. In: Mary G, Alfaro R, Kanninen M, Lobovikov M (eds) Forests in the global balance – changing paradigms. IUFRO World Series 17. IUFRO, Vienna, (Chapter 4), pp 39–74

Griscom B, Shoch D, Stanley B, Cortez R, Virgilio N (2009) Sensitivity of amounts and distribution of tropical forest carbon credits depending on baseline rules. Environ Sci Policy 12:897–911

Hunter D, Salzman J, Zaelke D (2007) International environmental law and policy. Foundation Press, New York

Intergovernmental Panel on Climate Change (2007) In: Pachauri RK, Reisinger A (eds) Contribution of working groups I, II and III to the fourth assessment report of the intergovernmental panel on climate change

Karnowski PJ, McDermott CL, Cashore BW (2011) Implementing REDD+: lessons from analysis of forest governance. Environ Sci Policy 14:111–117

Levin K, McDermott C, Cashore B (2008) The climate regime as global forest governance: can reduced emissions from deforestation and forest degradation REDD pass a 'dual effectiveness' test. Int For Rev 10:538–549

Olander L, Gibbs H, Steininger M, Swenson J, Murray BC (2007) Data and methods to estimate national historical deforestation baselines in support of UNFCCC REDD. Nicholas Institute for Environmental Policy Solutions, Durham

Parker C, Mitchell A, Trivedi M, Marda N (2008) The little REDD book: a guide to governmental and non-governmental proposals for reducing emissions from deforestation and degradation. Global Canopy Programme, Oxford

Paulsson E (2009) A review of the CDM literature: from fine-tuning to critical scrutiny? Int Environ Agreem-Politics Law Econ 9:63–80

Pearson T, Walker S, Brown S (2006) Guidebook for the formulation of afforestation and reforestation projects under the clean development mechanism. ITTO Technical Series 25, Yokohama

Phelps J, Webb EL, Agrawal A (2010) Does REDD+ threaten to recentralize forest governance? Science 328:312–313

Potvin C, Bovarnick A (2008) Reducing emissions from deforestation and forest degradation in developing countries: key actors, negotiations and actions. Carbon Clim Law Rev 2:264–272

Schlamadinger B, Bird N, Johns T, Brown S, Canadell J, Ciccarese L, Dutschke M, Fiedler J, Fischlin A, Fearnside P, Forner C, Freibauer A, Frumhoff P, Hoehne N, Kirschbaum MUF, Labat A, Marland G, Michaelowa A, Montanarella L, Moutinho P, Murdiyarso D, Pena N, Pingoud K, Rakonczay Z, Rametsteiner E, Rock J, Sanz MJ, Schneider UA, Shuidenko A, Skutsch M, Smith P, Somogyi Z, Trines E, Ward M, Yamagata Y (2007) A synopsis of land use, land-use change and forestry (LULUCF) under the Kyoto protocol and Marrakech accords. Environ Sci Policy 10:271–282

Sikor T, Stahl J, Enters T, Ribot JC, Singh N, Sunderlin WD, Wollenberg L (2010) REDD-plus, forest people's rights and nested climate governance. Glob Environ Change 20:423–425

Skutsch M, Bird N, Trines E, Dutschke M, Frumhoff P, de Jong BHJ, van Laake P, Masera O, Murdiyarso D (2007) Clearing the way for reducing emissions from tropical deforestation. Environ Sci Policy 10:322–334

Speth JG, Haas P (2006) Global environmental governance. Island Press, Washington, DC

Stone CD (2004) Common but differentiated responsibilities in international law. Am J Int Law 98:276–301

Strassburg B, Turner K, Fisher B, Schaeffer R, Lovett A (2008) Reducing emissions from deforestation – the "combined incentives" mechanism and empirical simulations. Glob Environ Change 19:265–278

Subsidiary Body for Scientific and Technological Advice (SBSTA) (2008) Views on outstanding methodological issues related to policy approaches and incentives to reduce emissions from deforestation and forest degradation in developing countries. http://unfccc.int/documentation/documents/advanced_search/items/3594.php#beg

The Nature Conservancy (2009) Reference emission levels for REDD: incentive implications for differing country circumstances

Thies C, Czebiniak R (2008) Forests for climate: developing a hybrid approach for REDD. Greenpeace International, Amsterdam

UNDP (2008) United Nations collaborative programme on reducing emissions from deforestation and forest degradation in developing countries: framework document

UNFCCC (1992) The United Nations framework convention on climate change. http://unfccc.int/essential_background/convention/items/2627.php

UNFCCC (1997) Kyoto protocol to the United Nations framework convention on climate change. http://unfccc.int/kyoto_protocol/items/2830.php

UNFCCC (2007) Decision-/CP.13 Bali Action Plan

UNFCCC (2009a) Clean development mechanism statistics. http://cdm.unfccc.int/Statistics/Registration/RegisteredProjByScopePieChart.html

UNFCCC (2009b) Parties and observers. http://unfccc.int/parties_and_observers/items/2704.php

Viana V (2009) Financing REDD: meshing markets with government funds. International Institute for Environment and Development, London

Wainwright R, Ozinga S, Dooley K, Leal I (2008) From green ideals to REDD money: a history of schemes to save forests for their carbon. Forests and the European Union Resource Network

Wise M, Calvin K, Thomson A, Clarke L, Bond-Lamberty B, Sands R, Smith SJ, Janetos A, Edmonds J (2009) Implications of limiting CO_2 concentrations for land use and energy. Science 324:1183–1186

Synthesis and Conclusions

18

Mary L. Tyrrell, Mark S. Ashton, Deborah Spalding, and Bradford Gentry

Executive Summary

If the world wants to meet its climate mitigation goals, forests – as both a sink and source – must be included. According to the 2007 IPCC report, deforestation and land use change currently account for a considerable portion of total anthropogenic global greenhouse gas emissions. Any comprehensive climate change policy must address this issue. At the same time, forests and their by-products chiefly for shelter and energy have a significant potential to sequester carbon. Their inclusion in a climate regime could have an immediate impact.

About half of terrestrial carbon is stored in forests, which can act as a sink or a source of carbon under different conditions and across temporal and spatial gradients. Best current estimates are that the terrestrial biome is acting as a small carbon sink, most likely occurring in forested ecosystems. Boreal and temperate forests are sequestering carbon (net sinks), while tropical forests are a net source of CO_2 emissions due to deforestation (land use change).

Understanding the role of forests in global carbon budgets requires quantifying several components of the carbon cycle, including how much carbon is stored in the world's forests (carbon pools), gains and losses of carbon in forests due to natural and anthropogenic processes (carbon fluxes), exchanges between the terrestrial carbon and other sinks and sources, and the ways in which such processes may be altered by climate change.

This extensive review of the literature on forest carbon science, management, and policy has produced many important conclusions, and elucidated what we currently do and do not know about forests, carbon, and climate change. Below we have taken select findings from each chapter and summarized them as bullets in the order of the topic areas of the book. In addition we provide a concise set of core recommendations for action. Taken together, both act as a synopsis of the contents of this book and its presentation of our current knowledge base of how to preserve the carbon stock in the world's forests and potentially maintain forests as CO_2 sinks into the future.

1 The Science of Carbon Uptake and Cycling in Forests

In order to better understand the ways in which future forests will change and be changed by shifting climates, it is necessary to understand the underlying **drivers of forest development** and the ways these drivers are affected by changes in atmospheric carbon dioxide (CO_2) concentrations, temperature, precipitation, and nutrient levels. Successional forces lead to somewhat predictable **changes in forest stands** throughout the

M.L. Tyrrell (✉) • M.S. Ashton • D. Spalding • B. Gentry
Yale School of Forestry and Environmental Studies, 360 Prospect St, New Haven, CT 06511, USA
e-mail: mary.tyrrell@yale.edu

M.S. Ashton et al. (eds.), *Managing Forest Carbon in a Changing Climate*,
DOI 10.1007/978-94-007-2232-3_18, © Springer Science+Business Media B.V. 2012

world. These changes can cause corresponding shifts in the dynamics of carbon uptake, storage, and release.

- Forest stands accumulate carbon as they progress through successional stages. Most studies show that the greatest rate of carbon uptake occurs during the stem exclusion stage, but mature stands also sequester and store significant quantities of carbon. This is even the case for old growth – particularly when such old forests represent significant portions of large areas such as the Amazon and Congo basins.
- Free Air Carbon dioxide Exchange (FACE) experiments are suggesting that forest net primary productivity, and thus carbon uptake, usually increases with higher levels of atmospheric carbon dioxide, likely due to factors such as increased nitrogen use and water use efficiency and competitive advantages of shade tolerant species.
- Experiments dealing with drought and temperature change are providing evidence that water availability, especially soil moisture, may be the most important factor driving forest carbon dynamics.

Caveats

- Although we understand the stages of stand development, there is considerable unpredictability in the actual nature of species composition, stocking, and rates of development at each stage because of numerous positive and negative feedbacks that make precise understanding of future stand development difficult.
- Forest ecosystem experiments, such as FACE programs, have not been operating long enough to predict long term responses of forest ecosystems to increases in carbon dioxide. The expense and time constraints of field experiments force scientists to rely on multi-factor models (the majority of which account for five or fewer variables) leading to results based on broad assumptions.

Soil organic carbon (SOC) stored and cycled under forests is a significant portion of the global total carbon stock, but remains poorly understood due to complex storage mechanisms and inaccessibility at depth.

- Alterations of soil carbon cycling by land use change or disturbance may persist for decades or centuries, confounding results of short-term field studies. Such differences must be characterized, and sequestration mechanisms elucidated, to inform realistic climate change policy directed at carbon management in existing native forests, plantations, and agroforestry systems, as well as reforestation and afforestation projects.
- Fine roots are the main source of carbon additions to soils, whether through root turnover or via exudates to associated mycorrhizal fungi and the rhizosphere.
- Bacterial and fungal, as well as overall faunal community composition, have significant effects (+/–) on soil carbon dynamics; fossil fuel burning, particulate deposition from forest fires, and wind erosion of agricultural soils are thought to affect microbial breakdown of organic matter and alter forest nutrient cycling.

Caveat

- The global nature of the carbon cycle requires a globally-distributed and coordinated research program, but thus far research has been largely limited to the developed world, the top 30 cm of the soil profile, temperate biomes, and agricultural soils. Forest soils in tropical moist regions are represented by only a handful of studies and even fewer have examined sequestration of mineral soil carbon at depth.

2 Tropical, Temperate and Boreal Forest Science

2.1 Tropical Forest Science

Tropical forests play an important role in affecting world climate. Their diversity in structure and composition is largely because of variations in regional climate and soil. Existing literature on climate and tropical forests suggests that, compared to temperate and boreal forest biomes, tropical forests play a disproportionate role in contributing to emissions that both affect and mitigate climate but that only a few tropical forest biomes have been studied sufficiently to understand some of the nuances.

- The difference between the annual stand level growth (uptake: 2%) and mortality (release: 1.6%) of Amazonia is currently estimated to be 0.4%, which is just about enough carbon sequestered to compensate for the carbon emissions of deforestation in the region. This means that either a small decrease in growth or a small increase in mortality in mature forests could convert Amazonia from a sink to a source of carbon.
- CO_2 emitted from tropical soils is positively correlated with both temperature and soil moisture, suggesting that tropical rain forest oxisols are very sensitive to carbon loss with land use change.
- Old growth ever-wet and semi-evergreen forests are experiencing accelerated stand dynamics and their biomass is increasing, particularly in Amazonian and Central African forests, potentially in response to increased atmospheric CO_2.
- Contrary to past assumptions, a significant portion of stored carbon exists below ground in tropical forests. Current estimates of root soil carbon in tropical forests could be underestimated by as much as 60%.
- If drought becomes more common in tropical ever-wet and semi-evergreen forests, as some climate models predict, the likelihood of human-induced fires escaping and impacting large portions of the landscape increases.

Caveats

- Tropical dry deciduous and montane forests are almost a complete unknown because so little research has been done on these forest types. While the majority of dry deciduous forests in the Americas and Asia have been cleared, there is still a significant amount remaining in Africa.
- Uncertainties in both the estimates of biomass and rates of deforestation contribute to a wide range of estimates of carbon emissions in the tropics. Only three studies have analyzed land surface-atmosphere interactions in tropical forest ecosystems. It is essential to understand how carbon is taken up by plants and the pathways of carbon release, and how increasing temperatures could affect these processes and the balance between them.
- Sufficient controversy remains regarding net growth of ever-wet and semi-evergreen forests of Amazonia and Central Africa in response to increased CO_2 that further study is warranted.

2.2 Temperate Forest Science

Twenty-five percent of the world's forests are in the temperate biome. They include a wide range of forest types, and the exact boundaries with boreal forests to the north and tropical forests to the south are not always clear. There is a great variety of species, soil types, and environmental conditions which lead to a diversity of factors affecting carbon storage and flux. Deforestation is not a major concern at the moment, and the biome is currently estimated to be a carbon sink of about 0.2–0.4 Pg C/year, with most of the sink occurring in North America and Europe.

- The future of the temperate forest biome as a carbon reservoir and atmospheric CO_2 sink rests mainly on its productivity and resilience in the face of disturbance. The small "sink" status (0.2–0.4 Pg C/year) of temperate forests could easily change to a "source" status if the balance between photosynthesis and respiration shifts even slightly.
- There is tremendous variability in carbon stocks between forest types and age classes; carbon stocks could easily be lost if disturbance or land use change shifts temperate forests to younger age classes or if climate change shifts the spatial extent of forest types. On the other hand, if temperate forests are managed for longer rotations, or more area in old growth reserves, then the carbon stock will increase.
- Soil carbon under temperate forests appears to be stable under most disturbances, such as logging, wind storms, and invasive species, but not with land use change. Huge losses can occur when converting forests to agriculture or development.
- Temperate forests are strongly seasonal, with a well-defined growing season that depends primarily on light (day length) and temperature. This is probably the most important determinant, along with late-season moisture,

of temperate forest productivity and hence carbon sequestration.

- Natural disturbances, particularly windstorms, ice storms, insect outbreaks, and fire are significant determinants of temperate forest successional patterns. The frequency of stand-leveling windstorms (hurricanes, tornadoes) is expected to increase under a warmer climate in temperate moist broadleaf and coniferous forest regions, so that fewer stands would reach old-growth stages of development.
- If changing climate alters the frequency and intensity of fires, re-vegetation and patterns of carbon storage will likely be affected, particularly in interior coniferous forests.
- Storage of carbon in forests has played a major role in U.S. emission reduction efforts, particularly in the voluntary carbon markets. Considerable efforts have been underway to reduce emissions of greenhouse gases at the regional (Northeastern U.S.), state (California), municipal, corporate, and individual levels.

Caveats

- Atmospheric pollution, primarily in the form of nitrogen oxides (NOx) emitted from burning fossil fuels, and ozone (O_3) is a chronic stressor in temperate forest regions. Because most temperate forests are considered nitrogen-limited, nitrogen deposition may also act as a growth stimulant (fertilizer effect). Under current ambient levels, nitrogen deposition is most likely enhancing carbon sequestration; however, the evidence regarding long-term chronic nitrogen deposition effects on carbon sequestration is mixed.
- Data on mineral soil carbon stocks in temperate forests can only be considered approximations at this time as there is very little research on deep soil carbon (more than 100 cm).
- Global circulation models predict that increasing concentrations of atmospheric CO_2 will increase the severity and frequency of drought in regions where temperate forests are found. However, there is a great deal of uncertainty about how drought will affect carbon cycles.
- Although afforestation and reforestation projects are being considered under various global and national carbon policies, it is important to consider whether it is ecologically beneficial for the land to support trees. Afforestation or reforestation activities that require soil drainage or conversion of wetlands, as well as those that add stress to water-scarce areas, could create more public detriment than benefits.

2.3 Boreal Forest Science

As one of the largest and most intact biomes, the boreal forest occupies a prominent place in the global carbon budget. While it contains about 13% of global terrestrial biomass, its organic-rich soils hold 43% of the world's soil carbon. At present this forest biome acts as a weak sink for atmospheric carbon. However, the conditions that make this true are tenuous, and evidence of rapid climate change at northern latitudes has raised concern that the boreal forest could change to a net source if the ecophysiological processes facilitating carbon uptake are sufficiently disrupted.

- Increased fire frequency could greatly increase carbon release, especially if it increases the decomposition of "old" carbon from the soil pool by increasing soil temperatures, degrading permafrost, and enhancing the rate of heterotrophic respiration.
- While fire is recognized as the dominant natural disturbance type over much of the boreal forest, insect outbreaks (and "background" insect damage during non-outbreak years) are also critically important. In some forest types, insect outbreaks exert the primary influence on age class distribution.
- It appears that climatic warming is shortening the fire return interval in many boreal forests, and speeding up the life cycles of damaging insects. This could result in a large release of carbon, quickly turning the boreal forests from a sink to a source of carbon. Canadian forests in particular are poised to release massive amounts of carbon as the result of die-off from insect infestations.
- The question of whether moisture availability will decline with climatic warming will probably determine whether warming enhances the boreal carbon sink or turns it into a source.

- Lichens and bryophytes in lowland saturated sites contain upwards of 20% of the above ground carbon. These communities have important effects on how carbon is stored in boreal soils. Thick moss layers limit heat gain from the atmosphere, creating cold and wet conditions that promote the development of permafrost, with limited decomposition, thus are important for carbon storage.
- If all the carbon pools, inputs and outputs are considered together, it appears that clearcut stands in boreal forests are carbon sources for the first decade after harvest (thanks to transient increases in respiration), after which they switch to sinks.

Caveats

- There is a tremendous amount of uncertainty in estimates of boreal carbon pools, because there have been so few studies in relation to the vast extent of the biome, and most have been done only in Canada and Fennoscandia.
- There is little quantifiable information about several important carbon pools, including fine root biomass and mycorrhizae, bryophyte and understory layers, and coarse woody debris and litter in Russia.
- Considering the importance of fire in boreal carbon dynamics, there is much that is not well understood, including extent, frequency, and intensity across the biome; and the interactions among fire intensity, nitrogen, and carbon.

3 Carbon Budgeting and Measurement

Quantifying carbon sources and sinks is a particular challenge in forested ecosystems due to the role played by biogeochemistry, climate, disturbance and land use, as well as the spatial and temporal heterogeneity of carbon sequestration across regions and forest types.

- While forests have the capacity to sequester significant amounts of carbon, the natural and anthropogenic processes driving carbon fluxes in forests are complex and difficult to measure. Nevertheless, accurate measurement of carbon stocks and flux in forests is one of the most important scientific bases for successful climate and carbon policy implementation. A measurement framework for monitoring carbon storage and emissions from forests should provide the core tool to qualify country and project level commitments under the United Nations Framework Convention on Climate Change, and to monitor the implementation of the Kyoto Protocol.
- Current consensus is that carbon emissions from land use change have remained fairly steady over the last few decades; however, there have been significant regional variations within this trend. Specifically, deforestation rates in the tropics, particularly in Asia, have grown substantially. In contrast, forests outside the tropics have been sequestering incremental carbon due to increased productivity (possibly because of CO_2 fertilization, although the evidence is not clear) and forest re-growth on lands that had been cleared for agriculture prior to industrialization.
- There are four categories of methods currently used to measure terrestrial carbon stocks and flows: (i) the inventory method, based on biomass measurement data; (ii) remote sensing techniques using satellite data; (iii) eddy covariance method using CO_2 flux data from small experimental sites; and (iv) the inverse method, using CO_2 concentration data and transport models. Each has advantages and disadvantages and varying degrees of accuracy and precision. No single method can meet the accuracy and resolution requirements of all users. A country, user or site will make a choice of method based on the specifics of the circumstances.
- Climate change is likely to generate both positive and negative feedbacks in forest carbon cycling. Positive feedbacks may include increased fire and tree mortality from drought stress, insect outbreaks, and disease. Negative feedbacks may include increased productivity from CO_2 enrichment. While the net result from positive and negative climate feedbacks is generally thought to be greater net carbon

emissions from forests, the timing and extent of these net emissions are difficult to determine.

Caveats

- If a standardized verification system across projects, countries, and regions is ever to be attained, policymakers should be aware that there are different basic approaches to measuring forest carbon, which have advantages and disadvantages, and varying degrees of accuracy and precision.
- Land use change is widely considered the most difficult component to quantify in the global carbon budget. The underlying data is often incomplete and may not be comparable across countries or regions due to different definitions of forest cover and land uses. Deforestation rates in the tropics are particularly difficult to determine due to these factors as well as differences in the way land degradation, such as selective logging and fuelwood removals, are accounted for in national statistics.
- Knowledge of the amount of carbon stored within each pool and across forest types is limited. Even estimates using broad categories such as carbon in vegetation versus soils vary widely due to a lack of data or assumptions about where carbon is stored within the forest and at what rate carbon is sequestered or released.

4 Managing Forests for Carbon

4.1 Tropical Forest Management and Reforestation

In tropical regions deforestation and degradation are contributing about 15% of total annual global greenhouse gas emissions. As policy makers work to develop solutions that address climate change, there has been considerable focus on incorporating forests into the overall climate solution. Silvicultural practices of natural forests (especially tropical forests) will need to be an integral part of reducing carbon loss and improving carbon storage if we are to solve this global challenge while meeting resource needs. In addition though temperate regions contain most of the existing planted and naturally regenerating forests, establishment of new forests is fastest in the tropics, especially Southeast Asia and Latin America. Policy discussions around climate change mitigation must address such newly planted and young second growth forests that are now becoming extensive.

- Reduced impact logging (RIL) is an important practice to lessen carbon loss, but it is necessary to move beyond RIL to substantially increase carbon storage by developing more sophisticated, planned forest management schemes with silvicultural treatments that ensure regeneration establishment, post establishment release, and extended rotations of new stands.
- Land managers should not manage tropical forests only for timber production, but also to maximize and diversify the services and products they obtain from their forests. This approach will provide an increase in net present value and a possible solution to the problem of exploitation and land conversion.
- The largest potential source of carbon sequestration in the tropics is the development of second growth forests on old agricultural lands and crop plantations that have proven unsustainable. Every incentive should be provided to encourage this process. Many logged over and second growth forests are ideal candidates for rehabilitation through enrichment planting of supplemental long-lived canopy trees for carbon sequestration.

Caveats

- More research is needed on how the application of silvicultural practices affects carbon uptake and storage in tropical forests at all levels. Some work has been done in the rainforest regions (ever-wet and semi-evergreen), but only in very specific places; almost none has been done in montane or seasonal (dry deciduous) forests.

4.2 Temperate and Boreal Forest Management

Increasing forest carbon stocks in temperate and boreal regions is a matter of making adjustments to existing forests vs. undergoing extensive reforestation/afforestation. Most boreal and temperate forests are second growth and land conversion is minimal when compared to other regions of the world. Therefore, providing additional carbon storage is a matter of refining silvicultural practices to take advantage of site nuances and enhancement potential.

- Many forest management activities result in net carbon release and thus cannot demonstrate carbon additionality. Mechanisms should be developed to credit managers who can reduce carbon loss, not simply increase carbon gain.
- Resiliency treatments (such as fuel reduction thinning and prescribed fire) result in lowered vegetative carbon storage, but they help produce forests that are significantly less susceptible to catastrophic disturbance (with accompanying drastic carbon release).
- Regeneration harvests significantly reduce the carbon stocks in vegetation and also cause a transient increase in soil respiration, although the annual rate of carbon uptake will be greater in the regenerating stand. Harvested areas often remain net carbon sources for 10–30 years, after which they return to sinks.
- Drainage of wetlands for increased tree production can result in either net carbon gain or loss, depending on how deep the drainage is.
- Studies have shown that many forest practices have a minimal impact on the soil carbon pool, which is the most difficult pool to measure. Thus, it may be possible that offsets involving certain forestry practices could go forward without strict quantification of this pool. This should be tempered by the fact that little is known about the effects of harvesting on deep soil carbon pools
- Managing stands for maximum sustained yield or financially optimum rotation can result in non-optimal carbon storage. Such rotations are often too short to allow the stand to attain maximum biomass. As such, it is often possible to increase carbon sequestration by extending rotations.

Caveat

- If old forests *already exist,* however, it is almost never better to convert them to younger forests. Old-growth forests, especially in productive zones, often have very large pools of vegetative, bryophyte, and soil carbon in comparison to younger, managed forests.

4.3 Forest Products, Recycling and Substitution

Harvested wood products can be long term reservoirs of carbon; however, solid wood products, paper, and paperboard manufacturing require large energy and heat inputs, and end-of-life pathways can further or hinder carbon sequestration, depending on management.

- Forest products are a minor, but growing component of the global carbon budget; nevertheless, harvested wood products can be long term reservoirs of carbon, particularly through substitution for more fossil carbon-intensive materials.
- Recycling postpones carbon emissions of even short-lived harvested wood products, and is especially effective when products are transformed multiple times within a tight recycling chain and finally converted into bioenergy.

Caveats

- New processed wood products and paper manufacturing require large energy and heat inputs, making wood products and carbon a complex topic.
- Landfilling harvested wood products creates high levels of methane, and if capture systems for energy are not in place, then the potential of landfills to act as carbon sinks becomes very unlikely. Therefore, landfill gas capture systems must be required if this end-of-use pathway is to be promoted as a way to reduce carbon emissions.
- The substitution effects on greenhouse gas emissions of wood for other construction materials (e.g., steel and concrete) may be up to 11 times larger than the total amount of carbon sequestered in forest products annually. However, quantification of substitution effects relies on many assumptions about particular counterfactual scenarios, most importantly linkages between increased/decreased forest

products consumption and total extent of forestland.

5 Forests: Society and Policy

5.1 Tropical Forests

Tropical forests contribute nearly half of the total terrestrial gross primary productivity and contain about 40% of the stored carbon in the terrestrial biosphere, with vegetation accounting for 58% and soil 41%. This ratio of vegetation carbon to soil carbon varies greatly by tropical forest type. About 8% of the total atmospheric carbon dioxide cycles through these forests annually. Vast areas of the world's large intact forests are in the tropics. Nevertheless, because of high rates of deforestation, tropical forests play a disproportionate role in contributing to terrestrial biome CO_2 emissions that both affect and mitigate climate.

- First and foremost, the primary risk to the carbon stored in tropical forests is deforestation, particularly converting forests to agriculture. Expanding crop and pasture lands have a profound effect on the global carbon cycle as tropical forests typically store 20–100 times more carbon per unit area than the agriculture that replaces them.
- In addition to the important role the remaining large intact forests play in the global carbon cycle, their protection from land conversion yields highly significant co-benefits. Evidence suggests that large, intact forests have significant cooling effects on both regional and global climates through the accumulation of clouds from forest evapo-transpiration, which also recycles water and contributes to the region's precipitation.
- Intact forests exist because of the geography of remoteness: low populations, lack of foreign investment, and lack of government presence have resulted in poor infrastructure development and the inability to integrate these regions into larger market structures.
- The significant drivers of deforestation (transportation infrastructure, agricultural commodity prices, national economic policies, agricultural technologies) are frequently context-specific and are affected by local political, socioeconomic, cultural, and biophysical factors. The roles of population growth and poverty in driving deforestation have often been overstated for certain regions.
- The overarching issues to be decided in developing an international policy to reduce emissions from tropical deforestation and forest degradation (REDD+) include: the scope of the forestry activities to be covered; the scale of accounting for forestry activities and the baseline for measuring reference emissions levels; the type of financing to be provided for REDD+ activities; how to address fundamental issues of capacity and governance; and the consideration of co-benefits.

Caveats

- Large intact forests of the tropics are increasingly at risk of deforestation attributable to governmental stimulus plans, road building programs, and subsidies for livestock production.
- A lack of governance, coupled with the presence of infrastructure, is often a precondition for widespread illegal operations that promote deforestation (e.g. logging, illicit drug trade). On the other hand, a lack of governance with no infrastructure inhibits illegal operations that promote deforestation.
- It is clear that REDD+ policies are only part of the solution to reduce deforestation and promote carbon sequestration. What is required is a combination of policies and market mechanisms that simultaneously promote sustainable economic growth and reduce poverty and economic inequalities, while protecting forests from further clearing for agriculture.

5.2 Temperate Forests: The United States

The capacity of temperate and boreal forests to continue to serve as a carbon sink makes them useful as mitigation tools to offset the damaging effects of greenhouse gas emissions. However, policymakers must recognize that urbanization and development in the U.S. and many other developed regions will continually pressure forests, leading to reduced forest cover and fragmented landscapes. From a purely economic standpoint, development is often the short-term highest and best use of land, particularly if financial returns are the primary driver in land use decision-making. Finding the right balance between competing land uses has become an area of focus for economists and policymakers particularly in long-term land-use planning.

- Temperate forests have been severely impacted by human use – throughout history, all but about 1% have been logged-over, converted to agriculture, intensively managed, grazed, or fragmented by sprawling development. Nevertheless, they have proven to be resilient – mostly second growth forests now cover about 40–50% of the original extent of the temperate forest biome.
- Residential and commercial development often represent the "highest and best use" for a parcel of land, resulting in the permanent conversion of forestlands, with negative results on U.S. carbon stocks.
- Subsidies and other government programs alter the balance between forestry, agriculture, and development, including which land use is most profitable at any point in time. Adding forest carbon into the mix of values a landowner can derive from the land may make forests more economically viable.

Caveats

- The economic viability of forest carbon projects is still unproven. While models have been developed to predict landowner behaviors when carbon is introduced at various prices, these models have not been widely tested. Additionally, price and project risks continue to challenge the economic attractiveness of potential carbon projects.
- Information on land use changes across the country is incomplete. While general trends in land use change can be determined from national inventory, satellite and remote sensing data, local data is not consistently available at a scale useful to land use planning. Analysis must include not only site specific data, but also local rules and regulatory structures that impact behavior.

6 Economics and Policy

Under the existing treaty, the Kyoto Protocol and its "flexible mechanisms" – particularly the Clean Development Mechanism (CDM) – have succeeded in building a billion dollar market for emission reduction projects in developing countries. The role of forests and land use in those markets has been a major source of controversy, however. As a result, forests currently play an insignificant role in the markets for CDM credits – even though the greenhouse gas emissions from tropical deforestation are larger than those from the global transportation sector. Current global and US policy on climate change is at best rudimentary. Future policies need to take into account a variety of incentives, rules and mechanisms to make forests and the forestry sector a useful contributor to greenhouse gas mitigation and adaptation.

- If and when the U.S. adopts federal climate policy, forests and other land uses are likely to play a major part in both the market and public funding approaches adopted.
- A wide range of land uses seems likely to be included, such as afforestation/ reforestation and managed forests, as well as soil carbon in farm and range lands. The inclusion of harvested wood products as approved project activities seems less likely at this time.
- Both domestic and international offsets from forest projects seem likely to be included. One open question is whether credits from international projects should be discounted compared to those from domestic projects.
- Substantial requirements will be imposed to help ensure that the offsets are "real." Finding the right balance between lower cost and higher accuracy will be difficult in the areas of monitoring and verification.
- While any policy will refer to the need to address leakage, few concrete measures to do so outside of project or entity boundaries seem likely to be required.
- Some combination of dedicating land to carbon storage for a lengthy period of time (through a conservation easement or contractual arrangement) and requiring that a portion

of the credits be held for use as a buffer against unexpected changes seems likely. While there is some discussion of temporary credits, experience in the CDM market suggests that other ways should be used to address permanence issues.

- The scope of the REDD mechanism is likely to include deforestation, degradation, and "plus" (+) activities, i.e. sustainable management of forests, conservation and enhancement of carbon stocks.
- Sub-national level accounting is likely to be allowed under the REDD+ mechanism as an interim measure while countries build technical capacity; however, there is consensus that a national level baseline must ultimately be reached. Therefore, the approach of scaling up from sub-national to national for countries that need time and investment to develop monitoring is likely.
- A hybrid financing system is likely to take shape that accommodates the varied interests and circumstances of states, as well as the needs of the different types of funders.
- The success of a REDD+ mechanism hinges on the ability of countries to address the drivers of deforestation in their countries; in many cases, addressing these issues will require significant investments not just in technical capacity, but in governance reforms and institutional capacity-building.

Caveats

- Which carbon pools to be included in a REDD+ mechanism have not been discussed in significant technical detail and will likely be worked out post-Copenhagen.
- While the majority of the proposals argue for the use of historic baselines, many include provisions for "national circumstances" or "development adjustment factors" that would be incorporated into the calculation of a baseline in some way, although at this stage most parties have not articulated a methodology for achieving these adjusted baselines.
- It is unclear how social safeguards or biodiversity standards might be incorporated into criteria or eligibility for REDD+ funding.

7 Summary

Forests are critical to the global carbon budget, and every effort should be made to conserve intact forests, whether they are primary tropical and boreal forests, or second growth, temperate forests (see Box 18.1 for our prioritized recommendations). All evidence points to the global forest estate being a weak sink for atmospheric CO_2, as a result of a tenuous balance between the carbon sink from productivity in the temperate and boreal biomes and the net CO_2 emissions from the tropics due to large-scale deforestation. Changes in disturbance regimes (fire, storms, insect outbreaks, harvesting) in any of the major forested regions could easily tilt this balance one way or the other. And as these forests mature, their capacity to take up increasing levels of carbon commensurate with increases in CO_2 emissions will diminish. Future climate change effects on the forest carbon balance are difficult to predict: however, higher temperatures are likely to significantly influence the factors driving disturbance such as moisture, storms, and pest species ranges. Evidence of a "CO_2 fertilization effect" on forests is mixed, therefore it is difficult to predict whether or not continued increases in atmospheric CO_2 will counteract the negative influence of changes in disturbance frequency and intensity. Land use change, however, overwhelms all other factors, since continued deforestation in the tropics will most certainly push the "global forest" to being a net source of carbon emissions to the atmosphere instead of the sink it could be.

Box 18.1. Top 10 Recommendations for Preserving Carbon Stocks and Sinks in the World's Forests[1]

1. **Keeping forests as forests** (i.e. preventing deforestation) is the most straightforward way of maintaining carbon stocks and promoting sequestration.
   (a) It is especially important to conserve intact primary forests.
   (b) Laws and economic policies that facilitate deforestation and forest degradation must be changed (for example, land tenure laws that promote deforestation or concession systems that allow poor harvesting practices and cause forest degradation).
2. **Reforestation on appropriate sites** is a viable means to enhance carbon sequestration.
   (a) Where NOT to plant: naturally treeless areas montane grasslands, steppe, prairie, and tropical peatlands
   (b) In afforestation/reforestation projects, soil carbon must be included in carbon stock accounting.
3. Forests are dynamic systems. In order to maintain resilient forests with lower risk of catastrophic carbon loss, it is sometimes necessary to **undertake management practices that lower carbon stocks** (e.g. fuel reduction thinnings in fire prone forests).
4. **Setting a baseline (of carbon stock) against which to measure future gains** for carbon sequestration projects is an important policy choice, and will influence which "carbon positive" activities are implemented by landowners.
5. Forest carbon projects **must not damage other ecosystem services** (water quality/yield, biodiversity, air quality).
6. When implementing forest carbon sequestration projects, efforts need to be made to **minimize shifting of deforestation to other areas (leakage).**
   (a) Activity leakage: There is a risk that by delaying forest harvest in one place (through carbon sequestration projects), it will simply be shifted to other areas.
   (b) Market leakage: The desire to increase carbon sequestration in forests should not discourage wood use in favor of more fossil-carbon intensive products.
7. **U.S. climate policy should include international forests** (as offsets and/or through a fund).
8. In order for all countries to participate in a forest carbon regime, many will need **capacity building related to monitoring, forest management (e.g. zoning, operations and planning), and governance.**
9. **Equity: Forest dependent communities should be included in REDD policy decision-making and receive some of the benefits from carbon projects.**
10. Market vs. Fund Mechanism: A **hybrid financing system** allows for a variety of forestry and climate change objectives to be met:
   (a) Markets can serve as a direct and consistent means for carbon offset credit values and transactions between suppliers and buyers over the long-term.
   (b) Funds can support activities like capacity building, pilot projects, and conservation that may not be intrinsically valued in a market framework.

[1]Compiled by participants in the Yale School of Forestry & Environmental Studies graduate seminar *Managing Forests for Carbon Sequestration: Science, Business, and Policy*: Benjamin Blom, Jaime Carlson, Matthew Carroll, Ian Cummins, Cecilia Del Cid Liccardi, Lauren Goers, Lisa Henke, Thomas Hodgman, Tim Kramer, Janet Larson, Brian Milakovsky, Jacob Munger, Caitlin O'Brady, and Ramon Olivas.

Glossary

Aboveground biomass Living vegetation above the soil, including stem, stump, branches, bark, seeds, and foliage.

Additionality A criterion often applied to greenhouse gas (GHG) reduction projects, stipulating that project-based GHG reductions should only be quantified if the project activity would not have happened in the absence of the revenue from carbon credits and that only credits from projects that are "additional to" the business-as-usual scenario represent a net environmental benefit.

Afforestation Planting of trees on historically non-forested land, e.g. native grasslands.

Agroforestry system A mixed agricultural system that can combine planting of trees with agricultural commodities such as crops or grasses.

Agrosilvopastoral system An agricultural system combining trees and livestock with agricultural crops and pasture.

Albedo A surface's reflectivity of the sun's radiation. White surfaces, such as snow, cement/pavement or bare soil, have a high albedo, reflecting the sun's radiation; dark surfaces, such as tree foliage or water bodies, have low albedo, absorbing more of the sun's radiation.

Allometry The study of the relationship between size and shape of organisms; in forestry, generally the relationship between tree diameter, height, crown size and biomass.

Anoxic Soil conditions without oxygen.

Annex I countries Parties to the United Nations Framework Convention on Climate Change (UNFCCC) that include the industrialized countries that were members of the OECD (Organisation for Economic Cooperation and Development) in 1992 as well as countries with economies in transition (the EIT Parties), including the Russian Federation, the Baltic States, and several Central and Eastern European States.

Autotroph An organism which synthesizes organic materials from inorganic sources such as light (phototrophic) or chemical processes (chemotrophic); green plants and bacteria are autotrophs.

Belowground biomass The living biomass of roots greater than 2 mm diameter.

Biomass The total mass of living and/or dead organic matter found within a unit area usually measured as dry mass in grams, kilograms or tons per meter squared or per hectare.

Bromeliad A diverse family of plants found chiefly in the tropical Americas that usually use the support of trees for their position in a forest canopy. Such plants are called epiphytic. Other bromeliads grow on the ground. Most have leaves arranged as rosettes. Bromeliads include the pineapple family, Spanish moss and various ornamentals.

Carbon allowance Government- issued authorization to emit a certain amount of carbon into the atmosphere. In carbon markets, an allowance is commonly denominated as one metric ton of carbon dioxide, or carbon dioxide equivalent (CO_2e).

Carbon offset A financial instrument aimed at a reduction in greenhouse gas emissions. Carbon offsets are measured in metric tons of carbon dioxide-equivalent (CO_2e) and are frequently generated by projects in sectors such as renewable energy and forestry. Offsets can be sold either in voluntary or compliance

M.S. Ashton et al. (eds.), *Managing Forest Carbon in a Changing Climate*,
DOI 10.1007/978-94-007-2232-3, © Springer Science+Business Media B.V. 2012

markets to an individual or company in order to compensate for greenhouse gas emissions or to comply with caps placed on emissions in certain sectors.

Carbon sequestration The removal and storage of carbon from the atmosphere in carbon sinks (such as oceans, forests or soils) through physical or biological processes, such as photosynthesis.

Carbon Tracker A system developed by the National Oceanic and Atmospheric Administration that calculates carbon dioxide uptake and release at the Earth's surface over time.

Cation exchange capacity The capacity of a soil for ion exchange of cations (positively charged ion) between the soil and the soil solution and is used as a measure of fertility, nutrient retention capacity, and the capacity to protect groundwater from contamination. Plant nutrients such as calcium and potassium are cations, as are toxic metals such as aluminum.

Clean Development Mechanism (CDM) A project-based mechanism defined in Article 12 of the Kyoto Protocol which allows a country with an emission-reduction or emission-limitation commitment under the Kyoto Protocol to implement emission-reduction projects in developing countries. Such projects can earn saleable certified emission reduction (CER) credits, each equivalent to one ton of CO_2, which can be counted toward meeting Kyoto targets.

Chronosequence A sequence of related soils or vegetation that differ from one another in certain properties primarily as a result of time as a soil-forming factor or succession, respectively.

Coppice A traditional method of woodland management in which young tree stems are repeatedly cut down to near ground level so they will sprout into vigorous re-growth of young stems.

Deadwood Non-living woody biomass either standing, lying on the ground (but not including litter).

Deforestation Cutting down all the trees on a piece of land to convert it to another land use, or the long-term reduction of the tree canopy cover below a minimum 10 percent threshold.

Developed land Urban and built-up areas.

Disturbance Any event such as fire, wind, disease, insects, ice, flood, or landslide that disrupts the vegetation and abiotic environment in an area.

DOC Dissolved organic carbon – see below.

DOM Dissolved organic matter comprises carbon compounds in water solution, generally from decomposition of plant and animal tissues in soils.

Ecological succession The relatively predictable change in the composition and/or structure of an ecological community, which may be initiated either by formation of new, unoccupied habitat (such as a severe landslide) or by some form of disturbance (such as fire, severe windthrow, logging) of an existing community.

Ectomycorrhizal fungi A symbiotic association between a fungus and the roots of a plant that forms an important part of soil life and nutrient uptake in some forests.

Eddy covariance A method of carbon measurement from forests that samples three-dimensional wind speed and CO_2 concentrations over a forest canopy at a high frequency and determines the CO_2 flux by the statistical relationship (covariance) of vertical wind velocity and CO_2 concentration.

Epiphytes A plant that grows upon another plant (such as a tree) non-parasitically or sometimes upon some other object (such as a building or a telegraph wire), derives its moisture and nutrients from the air and rain and sometimes from debris accumulating around it, and is found in the temperate zone (such as mosses, liverworts, lichens and algae) and in the tropics.

Extensive agriculture System of crop cultivation using small amounts of labor and capital in relation to area of land being farmed. The crop yield in extensive agriculture depends primarily on the natural fertility of the soil, terrain, climate, and the availability of water.

Ex-ante accounting A method of accounting for emissions reductions in which money is given up-front for the guarantee that a given activity will be carried out and emissions reductions will occur in the future.

Ex-post accounting A method of accounting for emissions reductions in which money is given for an emissions reductions activity after it has delivered its emission reduction.

Fine root turnover The period of time for the fine roots of plants to form, function and then die.

Floristics A sub domain of botany and biogeography that studies distribution and relationships of plant species over geographic areas.

Forest Defined by the Food and Agriculture Organization as land spanning more than 0.5 ha with trees higher than 5 m and a canopy cover of more than 10%, or trees able to reach these thresholds *in situ*. It does not include land that is predominantly under agricultural or urban land use.

Forest degradation Changes within the forest which negatively affect the structure or function of the stand or site, and thereby lower the capacity to supply products and/or services.

Forest dynamics Describes the underlying physical and biological forces that shape and change a forest over time, or the continuous state of change that alters the composition and structure of a forest. Two basic elements of forest dynamics are forest succession and forest disturbance.

Free Air Carbon Dioxide Enrichment (FACE) A method and infrastructure used to experimentally enrich the atmosphere enveloping portions of a terrestrial ecosystem with controlled amounts of carbon dioxide (and in some cases, other gases), without using chambers or walls.

Fragmentation The transformation of a contiguous patch of forest into several smaller, disjointed patches surrounded by other land uses.

Greenhouse gas Gas that traps heat in the atmosphere. The main greenhouse gases in the Earth's atmosphere are carbon dioxide, methane, nitrous oxide, and ozone.

Gross primary productivity (GPP) The total amount of carbon compounds produced by photosynthesis of plants in an ecosystem in a given period of time.

Heterotroph An organism capable of deriving energy for life processes only from the decomposition of organic compounds, and incapable of using inorganic compounds as sole sources of energy or for organic synthesis. Most animals are heterotrophic and rely on directly or indirectly (carnivores) eating most plants that are "autotrophic."

Highest and Best Use (HBU) An appraisal and zoning concept that evaluates all the possible, permissible, and profitable uses of a property to determine the use that will provide the owner with the highest net return on investment in the property, consistent with existing neighboring land uses.

Infiltration The process by which water on the ground surface enters the soil.

Intensive agriculture An agricultural system with high productivity per unit area. Intensive agricultural systems also frequently have high input requirements per unit area, relying upon the use of mechanization, fertilizers, and agrochemicals.

Kyoto Protocol A protocol to the United Nations Framework Convention on Climate Change (UNFCCC). It is an international environmental treaty negotiated in 1997 with the goal of stabilizing the concentration of greenhouse gases in the atmosphere at a level that would prevent dangerous anthropogenic interference with the climate system.

Land rent An economic term defined as the total net revenue or benefits received from a parcel of land.

Land-use change The shift from one use of a land area to another, such as from forestry to agriculture.

Leakage Term applied when activities that reduce greenhouse gas emissions in one place and time result in increases in emissions elsewhere or at a later date. For example, reduction of deforestation in one area of a country may lead to displacement of that deforestation to another region of the country.

Liana Any of various long-stemmed, usually woody vines that are rooted in the soil at ground level and use trees, as well as other means of vertical support, to climb up to the forest canopy in order to get access to light; they are especially characteristic of tropical moist deciduous forests and rainforests.

Lignin A complex chemical compound most commonly found in wood, and an integral part of the secondary cell walls of plants and some algae.

Litter Forest carbon pool that includes the detritus, leaves, small dead biomass lying on the ground, and humus layers of the soil surface.

Net primary productivity (NPP) The amount of carbon retained in an ecosystem (increase in biomass); it is equal to the difference between the amount of carbon produced through photosynthesis (GPP) and the amount of energy that is used for respiration (R).

Non-Annex I countries Term referring to parties to the United Nations Framework Convention on Climate Change that are considered developing countries and were not required by the Kyoto Protocol to undertake national targets to quantify emissions reductions.

Non-timber forest products Any commodity obtained from the forest that does not involve harvesting trees for wood products or pulp (paper products), such as game animals, nuts and seeds, berries, mushrooms, oils, foliage, medicinal plants, or fuelwood.

Orographic precipitation Rain, snow, or other precipitation produced when moist air is lifted as it moves over a mountain range.

Parcelization The breaking up of a land area under single ownership into multiple smaller parcels, usually for resale.

Pasture Agricultural systems containing forage crops and used for grazing animals.

Peatland (peat swamp forests) Tropical moist forests where waterlogged soils prevent dead leaves and wood from fully decomposing, which over time creates thick layers of acidic peat (organic matter).

Permanence The longevity of a carbon pool and the stability of its carbon stocks within its environment.

Photosynthesis The process by which a plant combines sunlight, water, and carbon dioxide to produce oxygen and sugar (stored energy and growth structure).

Photosynthetically active radiation The spectral range (wave band) of solar radiation from 400 to 700 nanometers that photosynthetic organisms (e.g. plants) are able to use in the process of photosynthesis.

Pioneer species Species which colonize previously bare or disturbed land, usually leading to ecological succession. Since uncolonized land may have thin, poor quality soils with few nutrients, pioneer species are often plants with adaptations such as long roots and root nodes containing nitrogen-fixing bacteria, and tend to grow well in open high-light environments.

Plantation Forests planted as crops for the production of timber fruit, latex, oil or pulpwood. Many large industrial plantations are monocultures.

Primary forest "Old" forests that have not been cleared by humans for a long period of time and have developed under natural ecological processes.

Radiocarbon A radioactive isotope of carbon that is the most common for radiometric dating techniques.

REDD+ A climate mitigation policy being negotiated under the United Nations Framework Convention on Climate Change consisting of policy measures to incentivize reduction of emissions from deforestation and forest degradation, conservation, sustainable management of forests, and enhancement of forest carbon stocks in developing countries.

Reforestation Planting trees on land that was previously forested.

Resiliency The capacity of a system to absorb disturbance and reorganize while undergoing change so as to retain essentially the same function, structure, and ecosystem services.

Respiration The process by which animals and plants use up stored foods (mostly complex carbohydrates) by combustion with oxygen to produce energy for body maintenance.

Rhizosphere The area immediately around plant roots, including the roots itself that comprises intense microbial activity, where plants, microorganisms, other soil organisms, and soil structure and chemistry, interact in complex ways.

Roundwood Harvested trees intended for use in products such as solid wood products, engineered wood products, and paper.

Secondary forests Forests that have regenerated by natural processes following the clearance of primary forests by humans or a change in land use, for example, to agriculture, and then abandoned to revert back to forest.

Silviculture The art and science of controlling the establishment, growth, composition, health, and quality of trees (woody plants) to meet diverse needs and values of the many landowners, societies, and cultures.

Stand A population of trees that is defined together by its age-class distribution, composition and site quality.

Structural adjustment Term used to describe the policy changes implemented by the International Monetary Fund (IMF) and the World Bank in developing countries. These policy changes are conditions for getting new loans from the IMF or World Bank, or for obtaining lower interest rates on existing loans.

Soil organic carbon (SOC) The carbon pool that includes all organic material in soil, but excluding the coarse roots of the belowground biomass pool.

Soil microorganisms There are five major groups of soil microorganisms. Bacteria, fungi, actinomycetes, algae, protozoa. Viruses form a small portion of soil microflora. They can be classified as autotrophs (utilize inorganic minerals) and heterotrophs (utilize organic matter).

Thermokarst A land surface that forms as ice-rich permafrost thaws. It occurs extensively in Arctic areas, and on a smaller scale in mountainous areas such as the Himalayas and the Swiss Alps.

Thinning The common term for the process of judiciously removing certain individual trees to improve the remaining quality and tree vigor in the plantation or forest; thinning can reduce the risk of a reversal of carbon sequestration due to fire, windthrow, insect infestations and disease.

Throughfall The process by which precipitation has fallen through the vegetative (forest) canopy, including rain or fog that collects on leaves and branches.

Editor Bios

Mark S. Ashton is the Morris K. Jessup Professor of Silviculture and Forest Ecology at the School of Forestry and Environmental Studies, Yale University. Professor Ashton conducts research on the biological and physical processes governing the regeneration of natural forests and on the creation of their agroforestry analogs. The results of his research have been applied to the development and testing of silvicultural techniques for restoration of degraded lands and for the management of natural forests for a variety of timber and non-timber products. Field sites include tropical forests in Sri Lanka and Panama, temperate forests in India and New England, and boreal forests in Saskatchewan, Canada. He has authored or edited over ten books and monographs and over 100 peer-review papers relating to forest regeneration and natural forest management.

Mary L. Tyrrell is the Executive Director of the Global Institute of Sustainable Forestry at the Yale School of Forestry & Environmental Studies. Her work focuses on land use change, forest fragmentation, sustainable forest management, and U.S. private lands, with a particular emphasis on review and synthesis of scientific research, and making scientific information available to forest managers, policy makers, and conservationists. She is the project manager of the Sustaining Family Forests Initiative, a national coalition focused on research and education about family forest owners in the United States. Ms. Tyrrell is a member of the Board of Advisors of the New England Forestry Foundation and the Board of Directors of the Hamden Land Conservation Trust..

Deborah Spalding is a founder and Managing Partner at Working Lands Investment Partners, LLC, which specializes in the investment and long-term stewardship of sustainably-managed working lands. She has worked in the financial industry for more than 17 years, serving in senior executive positions in the U.S. and overseas. Ms. Spalding is the Coordinator for Special Projects at the Yale School Forests, and has served on several boards, including the National Wildlife Federation, where she is a member of the Executive Committee, the Connecticut Forest & Park Association, and the Guilford Land Conservation Trust. She is a Trustee of the NWF Endowment and the Robert & Patricia Switzer Foundation, where she chairs the investment committee.

Bradford S. Gentry is the Director of the Center for Business and the Environment, as well as a Senior Lecturer and Research Scholar, at the Yale School of Forestry and Environmental Studies. Trained as a biologist and a lawyer, his work focuses on strengthening the links between private investment and improved environmental performance. He is also an advisor to GE, Baker & McKenzie, Suez Environnement, and the UN Climate Secretariat, as well as a member of Working Lands Investment Partners and Board Chair for the Cary Institute of Ecosystem Studies.

M.S. Ashton et al. (eds.), *Managing Forest Carbon in a Changing Climate*,
DOI 10.1007/978-94-007-2232-3, © Springer Science+Business Media B.V. 2012

Author Bios

Benjamin Blom is fluent in Spanish and Bahasa Indonesia. His concentration at Yale was in forest management and climate mitigation in tropical forests of Asia and Latin America. He now is a program specialist for Bureau of Land Management in Colorado.

Mark Bradford is Assistant Professor of Terrestrial Ecosystem Ecology at the Yale School of Forestry and Environmental Studies. Mark is primarily interested in how global change (e.g. climate warming) will affect plants, animals and microorganisms in grasslands and forests, and what the consequences are for soil carbon storage.

Matthew Carroll has twelve years experience as a Wildland firefighter, both as a hotshot and smokejumper for the USFS. His concentration at Yale was in forest management on public lands of the U.S. inland West in relation to climate change. He is currently a smokejumper based in McCall, Idaho.

Jaime Carlson has been involved for over ten years with conservation and development throughout Central America. She is fluent in Spanish and had a concentration at Yale in forest finance and green investment. She is currently a senior Advisor for Finance and Performance and Recovery Act Fellow at the U.S. Department of Energy.

Ben Cashore is a Professor of Environmental Governance & Political Science; and Director, Program on Forest Policy and Governance at the Yale School of Forestry and Environmental Studies. Ben's major research interests include the emergence of private authority, its intersection with traditional governmental regulatory processes, and the role of firms, non-state actors, and governments in shaping these trends. His book, Governing Through Markets: Forest Certification and the Emergence of Non-state Authority is a well known example of his work.

Jeffrey Chatellier has extensive experience in West Africa and SE Asia. His concentration at Yale was on agroforestry, community development, and biofuels. After a Fulbright Fellowship in Indonesia studying effects of biofuel development on climate he is now Director of Carbon Forestry for Forest Carbon, a forestry services company for climate change mitigation based in Indonesia.

Anton Chiono is a Policy Analyst, Pacific Forest Trust. Prior to joining PFT, Anton's work spanned both the science and policy underlying natural resource management. As a researcher with the U.S. Forest Service and The Nature Conservancy, Anton explored the science behind the restoration of historic fire regimes in western forests. Anton's tenure with The Wilderness Society and U.S. Senator Gordon H. Smith provided him with experience in the policy and economics behind resource management.

Kristofer Covey is currently a doctoral candidate at Yale University in forest stand dynamics. His focus at Yale is in forest management and resource issues of mixed temperate forests.

Dylan Craven has over ten years of field experience in agroforestry and reforestation in Central America. He is currently a doctoral candidate at the Yale School of Forestry and Environmental Studies working on the physiological ecology of tropical forest succession.

Ian Cummins is fluent in Spanish, with Peace Corp experience in Peru. His focus at Yale was on tropical forest management and climate change. He is currently the Forest Project Manager and Remote Sensing Specialist for Forest Carbon, a forestry services company for climate change mitigation based in Indonesia.

Cecilia Del Cid-Liccardi has past work experience in community development in Guatemala. Her focus at Yale was on tropical forest management and restoration. She is currently the teaching coordinator of the Environmental Leadership and Training Initiative, an organization focused on training people in tropical forest conservation.

Alexander Evans is Director of Science, Forest Guild. Zander's current research includes guidelines for successful and responsible biomass removal projects, the carbon impact of using forest biomass for energy and heat, and climate change adaptation and mitigation strategies for the forestry sector.

Mark Evidente is an Assistant Professor at De La Salle University, Philippines, where he teaches on sustainable development and international relations. His focus at Yale was environmental law and policy, environmental design, and social ecology. He is also Director of Two Eco Inc., a tourism development and consulting firm.

Alex Finkral is research forester for the Forestland Group, one North America's largest forestland investment and management companies. He was Assistant Professor of Forest Management, Northern Arizona University. His research focuses on how economic considerations affect and are affected by different forest management strategies for a variety of products and services.

Brent Frey is an Assistant Professor of Forest Ecology and Silviculture at Mississippi State University. He studies the regeneration ecology and dynamics of temperate and boreal forests and has strong research and field experience in the interior spruce-aspen forests of Canada.

Eva Garen is a social ecologist with over a decade of experience in Latin America, primarily in Central America and the Caribbean. Her post-doctoral research teaching included an analysis of the relationship between trees and rural farmers in Panama's Azuero Peninsula and as the Neotropics coordinator for the Environmental Leadership and Training Program (ELTI). She is now the lead social scientist for the carbon and climate programs of Conservation International.

Lauren Goers-Williams is currently at the World Resources Institute where she is a Research Analyst studying the governance of forests in developing countries. At Yale Lauren focused on sustainable development and land-use planning. She has experience in evaluating the impacts of climate change mitigation and adaptation strategies on issues relating to community sustainability and development.

Thomas Graedel is the Clifton R. Musser Professor of Industrial Ecology, Professor of Chemical Engineering, Professor of Geology and Geophysics, and Director of the Center for Industrial Ecology at the Yale School of Forestry and Environmental Studies. Professor Graedel was elected to the U.S. National Academy of Engineering for "outstanding contributions to the theory and practice of industrial ecology, 2002." His research is centered on developing and enhancing industrial ecology, the organizing framework for the study of the interactions of the modern technological society with the environment.

Bronson Griscom serves as the chief Forest Carbon Scientist for the Climate Team at The Nature Conservancy. He designs and implements research for the development of projects and policies to reduce emissions from deforestation and degradation (REDD) in Brazil, Indonesia, and elsewhere. He also coordinates carbon accounting aspects of The Nature Conservancy's REDD pilot initiatives.

Heather Griscom is an Assistant Professor in Restoration Ecology at James Madison University. Heather has long term research experience on dry tropical forest ecology and restoration, particularly in Panama.

Jefferson Hall is a Staff Scientist, Smithsonian Tropical Research Institute. Jeff Hall conducts research on tropical forest dynamics. His focus in Central Africa is on the ecology and management of primary tropical rain forests where he has been working with colleagues for over 20 years. In Latin America his focus is on reforestation with native species with particular emphasis on benefits of carbon sequestration and ecosystem services.

Lisa Henke worked as an intern on forestland investment for Equator Environmental LLC. She is currently with Washington chapter of The Nature Conservancy focused on land conservation.

Thomas Hodgeman has past work experience as a forester in timberland reforestation with Equator Environmental. His interest at Yale was in forest management and finance in ecosystem services. He is currently a project manager for forest ecosystem services with the Nature Conservancy.

Thomas James is a doctoral candidate at Yale University. Tom has on-going research investigating climate change impacts on the sub-boreal larch-spruce forests of Mongolia.

Matthew Kelty is a Professor of Silviculture and Forest Ecology at the University of Massachusetts. Matt conducts research on the dynamics of mixed forests in temperate Eastern North America. His focus has been on understanding the facilitative and competitive interactions of species mixtures particularly in relation to ecosystem productivity.

Elif Kendirli is a consultant based in Washington, D.C. with economics and finance expertise and a focus on international trade and development.

Tim Kramer worked several years as an environmental technician for the Antarctica polar experiment station. His focus at Yale was on soil carbon of forests and how invasive grasses may change carbon ecology of forests. He is now a doctoral student at the University of Washington.

Christopher Larson has past work experience in land conservation and management in northern California where he served as founder and principal at Madrone Conservation Finace. He currently Director, Real Assets and Sustainable Agriculture at New Island Capital Management.

Janet Lawson has past field experience with Peace Corps, in Paraguay. She speaks fluent Spanish. Her focus at Yale was on agroforestry and community development in tropical Latin America and Asia. She is currently an Agricultural and Development Officer with USAID based in Cambodia.

Xuhui Lee is currently a Professor of Meteorology at the Yale School of Forestry and Environmental Studies. Professor Lee's research concerns the states and principles that govern the exchanges of radiation, heat, water, and trace gases between vegetation and the atmosphere. His areas of interest include forest meteorology, boundary-layer meteorology, air quality, micrometeorological instrumentation, and remote sensing. His current research projects focus on surface-air exchange in non ideal conditions, the dynamics of air motion in vegetation, forest-water relations using isotopes, carbon sequestration by terrestrial ecosystems.

Reid Lifset is Associate Director of the Industrial Environmental Management Program, and Editor-in-Chief of the Journal of Industrial Ecology at the Yale School of Forestry and Environmental Studies. Mr. Lifset's research and teaching focus on the emerging field of industrial ecology, the study of the environmental consequences of production and consumption. He focuses on the application of industrial ecology to novel problems and research areas and the evolution of extended producer responsibility (EPR).

Eliot Logan-Hines has widespread experience throughout Latin America in organic agriculture and agrofroestry systems. His focus at Yale was on sustainable development, international trade, and agroforestry. He is currently Executive Director of the Runa Foundation, Ecuador, an organization dedicated to promoting Kichwa farmers to "participate in the global economy using renewable resources" while conserving the rainforest.

Kyle Meister has a focus on tropical forest management with extensive experience throughout Latin America, in particular in Colombia and Mexico with community forestry. He currently is a forester for international programs of Scientific Certification Systems.

Brian Milakovsky has worked as a forester for the Baskeahegan Company, ME, and the Manomet Conservation Sciences Center. At Yale he focused on boreal forest management issues, particularly in Russia and the Ukraine. He is fluent in Russian. After a year as a Fulbright fellow in the Ukraine he is now is a research forester for the World Wildlife Fund Siberia Program.

Jacob Munger has past work experience with Americorp and in community development. His focus at Yale was in forest land conservation issues of New England. He is currently a remote sensing specialist in forest conservation at the University of Wisconsin, Madison.

Ramon Olivas has been a project engineer in green energy. His focus at Yale was on energy and economics He worked at the World Business Council for Sustainable Development conducting research in the field of biofuels for the member companies. Prior to Yale, Ramon worked for the consultancy firm Econergy Int. Corp., where he supported the activities of the company in renewable energy projects in Mexico and Latin America, both in the public and the private sector. He is now a senior associate at Emerging Energy and Environment.

Chad Oliver is the Pinchot Professor of Forestry and Environmental Studies, and Director of Yale's Global Institute of Sustainable Forestry. Professor Oliver's initial research focused on the basic understanding of how forests develop and how silviculture can be applied to ecological systems most effectively. He is currently working on landscape approaches to forest management and is involved in the technical tools, the policies, the management approaches, and the educational needs.

Caitlin O'Brady has interests in watershed management and land use planning. Her work now concerns the development of the science and management of ecosystem services to the grassland interior woodland systems of the interior Western, U.S. She is a staff scientist at EcoTrust, Portland, Oregon.

Joseph Orefice is an Assistant Professor at Paul Smith's College in forest management. He has a focus on forest management and operations in New England and up-state New York – particularly in the Adirondacks.

Samuel Price focused on forest finance and management at Yale. His past work experience was in agricultural engineering research. He worked as a consultant in forest finance and land management in China and is now a forester for Longview, Inc, New Hampshire.

Jeffrey Ross has worked at the Montana Cooperative Wildlife Research Unit and the Rocky Mountain Elk Foundation. His focus at Yale was in traditional ecological knowledge and the ecology of temperate forests. He is currently a research fellow at the Agricultural University, Norway.

Xin Zhang is a doctoral candidate at Yale. Her focus is studying the effects of agricultural crops on climate change.

Yong Zhao is a doctoral candidate at Yale. His focus of study is on the carbon flux of salt marsh estuaries

Index

M.S. Ashton et al. (eds.), *Managing Forest Carbon in a Changing Climate*, DOI 10.1007/978-94-007-2232-3, © Springer Science+Business Media B.V. 2012

M

Printed by Publishers' Graphics LLC
LMO140101.15.15.93 20140101